GEOCHEMICAL AND BIOGEOCHEMI… ELING

An indispensable primer and reference textbook, the third edition of *Geochemical and Biogeochemical Reaction Modeling* carries the reader from the field's origins and theoretical underpinnings through to a collection of fully worked examples. A clear exposition of the underlying equations and calculation techniques is balanced by real-world example calculations. The book depicts geochemical reaction modeling as a vibrant field of study applicable to a wide spectrum of issues of scientific, practical, and societal concern. The new edition offers a thorough description of surface complexation modeling, including two- and three-layer methods; broader treatment of kinetic rate laws; the effect of stagnant zones on transport; and techniques for determining gas partial pressures. This handbook demystifies and makes broadly accessible an elegant technique for portraying chemical processes in the geosphere. It will again prove to be invaluable for geochemists, environmental scientists and engineers, aqueous and surface chemists, microbiologists, university teachers, and government regulators.

CRAIG M. BETHKE is the R.E. Grim Professor emeritus at the University of Illinois and principal author of *The Geochemist's Workbench®* software. He is recipient of the M. King Hubbert, O.E. Meinzer, and Waldemar Lindgren awards. Drawing on a four-decade career in industry, academia, consulting, and entrepreneurship, he is an authority on multicomponent chemical modeling.

GEOCHEMICAL AND BIOGEOCHEMICAL REACTION MODELING

Third Edition

CRAIG M. BETHKE
University of Illinois at Urbana-Champaign

CAMBRIDGE
UNIVERSITY PRESS

University Printing House, Cambridge CB2 8BS, United Kingdom

One Liberty Plaza, 20th Floor, New York, NY 10006, USA

477 Williamstown Road, Port Melbourne, VIC 3207, Australia

314–321, 3rd Floor, Plot 3, Splendor Forum, Jasola District Centre, New Delhi – 110025, India

103 Penang Road, #05–06/07, Visioncrest Commercial, Singapore 238467

Cambridge University Press is part of the University of Cambridge.

It furthers the University's mission by disseminating knowledge in the pursuit of education, learning, and research at the highest international levels of excellence.

www.cambridge.org
Information on this title: www.cambridge.org/9781108790864
DOI: 10.1017/9781108807005

First edition published as Geochemical Reaction Modeling by Oxford University Press, 1996

Second edition published as Geochemical and Biogeochemical Reaction Modeling by Cambridge University Press, 2010

Third edition published 2022

Printed in the United Kingdom by TJ Books Limited, Padstow Cornwall

A catalogue record for this publication is available from the British Library.

Library of Congress Cataloging-in-Publication Data
Names: Bethke, Craig, author.
Title: Geochemical and biogeochemical reaction modeling / Craig M. Bethke, University of Illinois at Urbana-Champaign.
Description: Third edition. | Cambridge, United Kingdom ; New York, NY : Cambridge University Press, 2022. | Includes bibliographical references and index.
Identifiers: LCCN 2021044807 | ISBN 9781108790864 (hardback)
Subjects: LCSH: Geochemical modeling. | Chemical reactions–Simulation methods. | BISAC: SCIENCE / Earth Sciences / General
Classification: LCC QE515.5.G43 B48 2022 | DDC 551.9–dc23
LC record available at https://lccn.loc.gov/2021044807

ISBN 978-1-108-79086-4 Paperback

Additional resources for this publication at www.cambridge.org/bethke3

For Abby, Henry, Gabrielle, and Claire

Contents

Preface

When in 1995 I sent my editor the typescript for the first edition of this book—by parcel post, of course—I remember thinking, "Well, I'll never do that again!" Writing such a detailed, technical monograph had proved all-consuming. Lacking further insight I felt I could contribute, I was anxious to leave reaction modeling behind and try new things.

And I did. The microbiological revolution swept over the geosciences shortly following the first edition's debut. Colleagues and I quickly found a place for ourselves at the interface between aqueous geochemistry and environmental microbiology. Computers were traversing their own wild revolution, doubling and redoubling in capability on intervals measured in months. The work a privileged few of us were doing in the early 1990s using enormous, vastly expensive computers became possible at any scientist's or engineer's desk. Simulations that seemed out of reach became almost routine.

A decade after its publication, as a result, the first edition no longer seemed complete. The modern geochemist needed to understand redox disequilibrium, the foundation of chemolithotrophic life. The kinetics of mineral dissolution and precipitation remained important, but the rates at which redox reactions proceed and at which microbes catalyze chemical transformation, and the dynamics of microbial communities seemed just as central to the field. As I prepared the first edition I suspected only a handful of my hardiest colleagues would care about the ins and outs of constructing reactive transport models of geochemical systems, an endeavor impractical for most. But microprocessors replaced supercomputers and an obscure field of study lurched into the mainstream. Buoyed by my editor's enthusiasm, the second edition I had not expected to write emerged page by page, example by example. This time a mouse click sent the finished typescript, now ten chapters longer, on its way to publication.

History repeated itself, as it is apt to do, and a decade on the text once again seemed inadequate. Surface complexation theory had expanded as a field and its treatment needed to be generalized: polydentate complexes, single and double pK_a paradigms, electrically charged complexing sites, and the triple-layer and CD-MUSIC formalisms all demanded full consideration. Activity coefficient theory had evolved, especially in light of wider adoption of the SIT method. The relationship between partial pressure and gas fugacity could no longer remain implied, but had to be set out in a quantitative sense. The modern modeler wanted to account for the kinetics of the formation and breakdown of aqueous and surface complexes, and the transfer of gas species between aqueous and gaseous phases. The discussion of reactive transport modeling needed to better convey the nature of numerical dispersion, and to introduce the role of stagnant zones within the domain—the dual-porosity effect.

None of this work would have been possible had it not been for the good fortune of being able to benefit from the expert, good-spirited, and untiring assistance of my most valued colleagues. Brian Farrell and Melika Sharifironizi spent countless months sifting through the latest developments in surface complexation theory.

Melika took the lead in deciphering three-layer modeling; I remain pleasantly bemused to see our work on the subject laid out in only a dozen book pages. Brian and Jia Wang similarly sorted through SIT electrolyte theory and helped translate literature pages into lines of computer code. Helge Moog and Frank Bok blazed the way as we implemented and documented three methods for computing fugacity coefficients, work that once again was improbably distilled into a few book pages.

Wolfgang Voight, Pat Brady, Jon Chorover, Blaine McCleskey, Bill Casey, and Nita Sahai all stepped in at just the right moment with thoughts I could build upon, or the perfect literature reference. The work Brian, Melika, Jia, Frank, Helge, Blaine, and Pat proffered as they reviewed and proofread revision after revision contributed more than I can describe to the final product. Neither my editors, Cambridge University Press, nor the University of Illinois ever wavered in their support. Finally, I thank the community of more than 10,000 users of our codes for letting us know the moment anything seemed confusing or off-kilter: You are the ultimate quality control on everything we do.

Preface to Second Edition

In the decade since I published the first edition of this book,[1] the field of geochemical reaction modeling has expanded sharply in its breadth of application, especially in the environmental sciences. The descriptions of microbial activity, surface chemistry, and redox chemistry within reaction models have become more robust and rigorous. Increasingly, modelers are called upon to analyze not just geochemical but biogeochemical reaction processes.

At the same time, reaction modeling is now commonly coupled to the problem of mass transport in groundwater flows, producing a subfield known as reactive transport modeling. Whereas a decade ago such modeling was the domain of specialists, improvements in mathematical formulations and the development of more accessible software codes have thrust it squarely into the mainstream.

I have, therefore, approached preparation of this second edition less as an update to the original text than an expansion of it. I pay special attention to developing quantitative descriptions of the metabolism and growth of microbial species, understanding the energy available in natural waters to chemosynthetic microorganisms, and quantifying the effects microorganisms have on the geochemical environment. In light of the overwhelming importance of redox reactions in environmental biogeochemistry, I consider the details of redox disequilibrium, redox kinetics, and effects of inorganic catalysts and biological enzymes.

I expand treatment of sorption, ion exchange, and surface complexation, in terms of the various descriptions in use today in environmental chemistry. And I integrate all the above with the principles of mass transport, to produce reactive transport models of the geochemistry and biogeochemistry of the Earth's shallow crust. As in the first edition, I try to juxtapose derivation of modeling principles with fully worked examples that illustrate how the principles can be applied in practice.

In preparing this edition, I have drawn on the talents and energy of a number of colleagues. First and foremost, discussion of the kinetics of redox reactions and microbial metabolism is based directly on the work my former graduate student Qusheng Jin undertook in his years at Illinois. My understanding of microbiology stems in large part from the tireless efforts of my colleague Robert Sanford. In modeling the development of zoned microbial communities, I use the work of my students Qusheng Jin, Jungho Park, Meng Li, Man Jae Kwon, and Dong Ding. Tom Holm found in the literature he knows so well sorption data for me to use, and Barbara Bekins shared data from her biotransformation experiments. Finally, I owe a large combined debt to the hundreds of people who have over the years reviewed our papers, commented on our software, sent email, talked to us at meetings, and generally pointed out the errors and omissions in our group's thinking.

I owe special thanks to colleagues who reviewed draft chapters: Patrick Brady at Sandia National Laboratories; Glenn Hammond, Pacific Northwest National Laboratory; Thomas Holm, Illinois Water Survey;

[1] *Geochemical Reaction Modeling*, Oxford University Press, 1996.

Qusheng Jin, University of Oregon; Thomas McCollom, University of Colorado; David Parkhurst, US Geological Survey; Robert Sanford, University of Illinois; Lisa Stillings, US Geological Survey; and Brian Viani, Lawrence Livermore National Laboratories.

Finally, the book would not have been possible without the support of the institutions that underwrote it: the Centre for Water Research at the University of Western Australia and UWA's Gledden Fellowship program, the Department of Geology at the University of Illinois, the US Department of Energy (grant DE-FG02-02ER15317), and a consortium of research sponsors (Chevron, Conoco Phillips, Exxon Mobil, Idaho National Engineering and Environmental Laboratory, Lawrence Livermore Laboratories, Sandia Laboratories, SCK CEN, Texaco, and the US Geological Survey).

Preface to First Edition

Geochemists have long recognized the need for computational models to trace the progress of reaction processes, both natural and artificial. Given a process involving many individual reactions (possibly thousands), some of which yield products that provide reactants for others, how can we know which reactions are important, how far each will progress, what overall reaction path will be followed, and what the path's endpoint will be?

These questions can be answered reliably by hand calculation only in simple cases. Geochemists are increasingly likely to turn to quantitative modeling techniques to make their evaluations, confirm their intuitions, and spark their imaginations.

Computers were first used to solve geochemical models in the 1960s, but the new modeling techniques disseminated rather slowly through the practice of geochemistry. Even today, many geochemists consider modeling to be a "black art," perhaps practiced by digital priests muttering mantras like "Newton–Raphson" and "Runge–Kutta" as they sit before their cathode ray altars. Others show little fear in constructing models but present results in a way that adds little understanding of the problem considered. Someone once told me, "Well, that's what came out of the computer!"

A large body of existing literature describes either the formalism of numerical methods in geochemical modeling or individual modeling applications. Few references, however, provide a perspective of the modeling specialty, and some that do are so terse and technical as to discourage the average geochemist. Hence, there are few resources to which someone wishing to construct a model without investing a career can turn.

I have written this book in an attempt to present in one place both the concepts that underpin modeling studies and the ways in which geochemical models can be applied. Clearly, this is a technical book. I have tried to present enough detail to help the reader understand what the computer does in calculating a model, so that the computer becomes a useful tool rather than an impenetrable black box. At the same time, I have tried to avoid submerging the reader in computational intricacies. Such details I leave to the many excellent articles and monographs on the subject.

I have devoted most of this book to applications of geochemical modeling. I develop specific examples and case studies taken from the literature, my experience, and the experiences over the years of my students and colleagues. In the examples, I have carried through from the initial steps of conceptualizing and constructing a model to interpreting the calculation results. In each case, I present complete input to computer programs so that the reader can follow the calculations and experiment with the models.

The reader will probably recognize that, despite some long forays into hydrologic and basin modeling (a topic for another book, perhaps), I fell in love with geochemical modeling early in my career. I hope that I have communicated the elegance of the underlying theory and numerical methods as well as the value of calculating models of reaction processes, even when considering relatively simple problems.

I first encountered reaction modeling in 1980 when working in Houston at Exxon Production Research Company and Exxon Minerals Company. There, I read papers by Harold Helgeson and Mark Reed and experimented with the programs EQ3/EQ6, written by Thomas Wolery, and PATH, written by Ernest Perkins and Thomas Brown.

Computing time was expensive then (about a dollar per second!). Computers filled entire rooms but were slow and incapacious by today's standards, and graphical tools for examining results almost nonexistent. A modeler sent a batch job to a central CPU and waited for the job to execute and produce a printout. If the model ran correctly, the modeler paged through the printout to plot the results by hand. But even at this pace, geochemical modeling was fun!

I returned to modeling in the mid 1980s when my graduate students sought to identify chemical reactions that drove sediment diagenesis in sedimentary basins. Computing time was cheaper, graphics hardware more accessible, and patience generally in shorter supply, so I set about writing my own modeling program, GT, which I designed to be fast enough to use interactively. A student programmer, Thomas Dirks, wrote the first version of a graphics program GTPLOT. With the help of another programmer, Jeffrey Biesiadecki, we tied the programs together, creating an interactive, graphical method for tracing reaction paths.

The program was clearly as useful as it was fun to use. In 1987, at the request of a number of graduate students, I taught a course on geochemical reaction modeling. The value of reaction modeling in learning geochemistry by experience rather than rote was clear. This first seminar evolved into a popular course, "Groundwater Geochemistry," which our department teaches each year.

The software also evolved as my group caught the interactive modeling bug. I converted the batch program GT to REACT, which was fully interactive. The user entered the chemical constraints for his problem and then typed "go" to trigger the calculation. Ming-Kuo Lee and I added Pitzer's activity model and a method for tracing isotope fractionation. Twice I replaced GTPLOT with new, more powerful programs. I wrote ACT2 and TACT to produce activity–activity and temperature–activity diagrams, and RXN to balance reactions and compute equilibrium constants and equations.

In 1992, we bundled these programs together into a package called "The Geochemist's Workbench®" which is owned by The Board of Trustees of the University of Illinois and can be licensed inexpensively for educational or commercial purposes. Within a few months of its completion the software was in use at dozens of universities and companies around the world.

We find that the programs allow us to try fresh approaches to teaching aqueous geochemistry. Once a student can reliably balance reactions by hand, the task quickly becomes a chore. After calculating a few Eh–pH diagrams, what does one learn by manually producing more plots? For many students, trees quickly come to obscure a beautiful forest. The computer can take over the mechanics of basic tasks, once they have been mastered, freeing the student to absorb the big picture and find the broad perspective. This approach has proved popular with students and professors. Many examples given in this book were developed originally as class assignments and projects.

I should not, however, give the impression that geochemical modeling is of any greater value in education than in scientific and practical application. The development of our modeling software, as evident in the case studies in this book, reflects the practical needs of petroleum geology and environmental geochemistry expressed to us over nearly a decade by a consortium of industrial and governmental affiliates to the Hydrogeology Program. These affiliates, without whom neither the software nor this book would exist, are: Amoco Production Research; ARCO Oil and Gas Company; British Petroleum Research; Chevron Petroleum Technology Company; Conoco, Incorporated; Du Pont Company; Exxon Production Research; Hewlett Packard, Incorporated; Illinois State Geological Survey; Japan National Oil Company; Lawrence Livermore National

Laboratory; Marathon Oil Company; Mobil Research and Development; Oak Ridge National Laboratory; Sandia National Laboratories; SiliconGraphics Computer Systems; Texaco, Incorporated; Union Oil Company of California; and the United States Geological Survey.

I can thank just a few of my colleagues and students who helped develop the case studies in this book. John Yecko and William Roy of the Illinois State Geological Survey first modeled degradation of the injection wells at Marshall, Illinois. Rachida Bouhlila provided analyses of the brines at Sebkhat El Melah, Tunisia. Amy Berger helped me write Chapter 10 (Surface Complexation), and Chapter 35 (Acid Drainage) is derived in part from her work. Edward Warren and Richard Worden of British Petroleum's Sunbury lab contributed data for calculating scaling in North Sea oil fields, Richard Wendlandt first modeled the effects of alkali floods on clastic reservoirs, and Kenneth Sorbie helped write Chapter 34 (Petroleum Reservoirs). I borrowed from Elisabeth Rowan's study of the genesis of fluorite ores at the Albigeois district, Wendy Harrison's study of the Gippsland basin, and a number of other published studies, as referenced in the text.

The book benefited enormously from the efforts of a small army of colleagues who served as technical reviewers: Stephen Altaner, Tom Anderson, and Amy Berger (University of Illinois); Greg Anderson (University of Toronto); Paul Barton, Jim Bischoff, Neil Plummer, Geoff Plumlee, and Elisabeth Rowan (US Geological Survey); Bill Bourcier (Livermore); Patrick Brady and Kathy Nagy (Sandia); Ross Brower and Ed Mehnert (Illinois State Geological Survey); David Dzombak (Carnegie Mellon University); Ming-Kuo Lee (Auburn University); Peter Lichtner (Desert Research Institute); Benoit Madé and Jan van der Lee (Ecole des Mines); Mark Reed (University of Oregon); Kenneth Sorbie (Heriot-Watt University); Carl Steefel (Battelle); Jim Thompson (Harvard University); and John Weare (University of California, San Diego). I learned much from them. I also thank Mary Glockner, who read and corrected the entire manuscript; my editor Joyce Berry, and Lisa Stallings at Oxford for their unwavering support; and Bill Bourcier and Randy Cygan, who have always been willing to lend a hand, and often have.

I thank the two institutions that supported me while I wrote this book: the Department of Geology at the University of Illinois and the Centre d'Informatique Géologique at Ecole Nationale Supérieure des Mines de Paris in Fontainebleau, France. I began writing this book in Fontainebleau while on sabbatical leave in 1990 and completed it there under the sponsorship of the Académie des Sciences and Elf Aquitaine in 1995.

A Note About Software

The geochemical modeler's milieu is software and the computer on which it runs. A number of computer programs have been developed over past decades to facilitate geochemical modeling; Appendix A lists the availability of current versions of a number of popular applications. Each program has its own capabilities, limitations, and indeed, personality. There is no best software, only the software that best meets a modeler's needs.

No discussion of geochemical modeling would be fully useful without specific examples showing how models are configured and run. In setting up the examples in this book, I employ a group of interactive programs that my colleagues and I have written over the past thirty-five or so years. The programs, RXN, ACT2, TACT, SPECE8, REACT, GTPLOT, X1T, X2T, XTPLOT, PHASE2, P2PLOT, TEDIT, and GSS, are known collectively as The Geochemist's Workbench®, or simply the GWB. The GWB is available in a versatile Community Edition as a free download, as well as in several paid packages; the latter options will be needed to reproduce the more advanced examples in latter sections of the book.

I chose to use this software for reasons that extend beyond familiarity and prejudice: the programs are interactive and take simple commands as input. As such, I can include within the text of this book scripts that in a few lines show the precise steps taken to calculate each result[1]. Readers can, of course, reproduce a given calculation by using any of a number of other modeling programs, such as those listed in Appendix A. Following the steps shown in the text, they should be able to construct input in the format recognized by the chosen program.

[1] The input scripts developed in this text are installed along with the GWB 2022 and subsequent software releases, and available for download as an Online Resource from Cambridge University Press.

1
Introduction

As geochemists, we frequently need to describe the chemical states of natural waters, including how dissolved mass is distributed among aqueous species, and to understand how such waters will react with minerals, gases, and fluids of the Earth's crust and hydrosphere. We can readily undertake such tasks when they involve simple chemical systems, in which the relatively few reactions likely to occur can be anticipated through experience and evaluated by hand calculation. As we encounter more complex problems, we must rely increasingly on quantitative models of solution chemistry and irreversible reaction to find solutions.

The field of geochemical modeling has grown rapidly since the early 1960s, when the first attempt was made to predict by hand calculation the concentrations of dissolved species in seawater. Today's challenges might be addressed by using computer programs to trace many thousands of reactions in order, for example, to predict the solubility and mobility of forty or more elements in buried radioactive waste.

Geochemists now use quantitative models to understand sediment diagenesis and hydrothermal alteration, explore for ore deposits, determine which contaminants will migrate from mine tailings and toxic waste sites, predict scaling in geothermal wells and the outcome of steam-flooding oil reservoirs, solve kinetic rate equations, manage injection wells, evaluate laboratory experiments, and study acid rain, among many examples. Teachers let their students use these models to learn about geochemistry by experiment and experience.

Many hundreds of scholarly articles have been written on the modeling of geochemical systems, giving mathematical, geochemical, mineralogical, and practical perspectives on modeling techniques. Dozens of computer programs, each with its own special abilities and prejudices, have been developed (and laboriously debugged) to analyze various classes of geochemical problems. In this book, I attempt to treat geochemical modeling as an integrated subject, progressing from the theoretical foundations and computational concerns to the ways in which models can be applied in practice. In doing so, I hope to convey, by principle and by example, the nature of modeling and the results and uncertainties that can be expected.

1.1 Development of Chemical Modeling

Hollywood may never make a movie about geochemical modeling, but the field has its roots in top-secret efforts to formulate rocket fuels in the 1940s and 1950s. Anyone who reads cheap novels knows that these efforts involved brilliant scientists endangered by spies, counter-spies, hidden microfilm, and beautiful but treacherous women.

The rocket scientists wanted to be able to predict the thrust that could be expected from a fuel of a certain composition (see historical sketches by Zeleznik and Gordon, 1968; van Zeggeren and Storey, 1970; Smith and Missen, 1982). The volume of gases exiting the nozzle of the rocket motor could be used to calculate the expected thrust. The scientists recognized that by knowing the fuel's composition, the temperature at which it burned, and the pressure at the nozzle exit, they had uniquely defined the fuel's equilibrium volume, which they set about calculating.

Aspects of these early calculations carry through to geochemical modeling. Like rocket scientists, we define a system of known composition, temperature, and pressure in order to calculate its equilibrium state. Much of the impetus for carrying out the calculations remains the same, too. Theoretical models allowed rocket scientists to test fuels without the expense of launching rockets, and even to consider fuels that had been formulated only on paper. Similarly, they allow geoscientists to estimate the results of a hydrothermal experiment without spending time and money conducting it, test a chemical stimulant for an oil reservoir without risking damage to the oil field, or help evaluate the effectiveness of a scheme to immobilize contaminants leaking from buried waste before spending and perhaps wasting millions of dollars and years of effort.

Chemical modeling also played a role in the early development of electronic computers. Early computers were based on analog methods in which voltages represented numbers. Because the voltage could be controlled to within only the accuracy of the machine's components, numbers varied in magnitude over just a small range. Chemical modeling presented special problems because the concentrations of species vary over many orders of magnitude. Even species in small concentrations, such as H^+ in aqueous systems, must be known accurately, since concentrations appear not only added together in mass balance equations, but multiplied by each other in the mass action (equilibrium constant) equations. The mathematical nature of the chemical equilibrium problem helped to demonstrate the limitations of analog methods, providing impetus for the development of digital computers.

1.1.1 Controversy Over Free-Energy Minimization

Brinkley (1947) published the first algorithm to solve numerically for the equilibrium state of a multicomponent system. His method, intended for a desk calculator, was soon applied on digital computers. The method was based on evaluating equations for equilibrium constants, which, of course, are the mathematical expression of the minimum point in Gibbs free energy for a reaction.

In 1958, White *et al.* published an algorithm that used optimization theory to solve the equilibrium problem by "minimizing the free energy directly." Free-energy minimization became a field of study of its own, and the technique was implemented in a number of computer programs. The method had the apparent advantage of not requiring balanced chemical reactions. Soon, the chemical community was divided into two camps, each of which made extravagant claims about guarantees of convergence and the simplicity or elegance of differing algorithms (Zeleznik and Gordon, 1968).

According to Zeleznik and Gordon, tempers became so heated that a panel that was convened in 1959 to discuss equilibrium computation had to be split in two. Both sides seemed to have lost sight of the fact that the equilibrium constant is a mathematical expression of minimized free energy. As noted by Smith and Missen (1982), the working equations of Brinkley (1947) and White *et al.* (1958) are suspiciously similar. As well, the complexity of either type of formulation depends largely on the choice of components and independent variables, as described in Chapter 3.

Not surprisingly, Zeleznik and Gordon (1960, 1968) and Brinkley (1960) proved that the two methods were computationally and conceptually equivalent. The balanced reactions of the equilibrium constant method are counterparts to the species compositions required by the minimization technique; in fact, given the same choice of components, the reactions and expressions of species compositions take the same form.

Nonetheless, controversy continues even today among geochemical modelers. Colleagues sometimes take sides on the issue, and claims of simplified formulations and guaranteed convergence by minimization are still heard. In this book, I formalize the discussion in terms of equilibrium constants, which are familiar to geochemists and widely reported in the literature. Quite properly, I treat minimization methods as being computationally equivalent to the equilibrium constant approach, and do not discuss them as a separate group.

1.1.2 Application in Geochemistry

When they calculated the species distribution in seawater, Garrels and Thompson (1962) were probably the first to apply chemical modeling in the field of geochemistry. Modern chemical analyses give the composition of seawater in terms of dissociated ions (Na^+, Ca^{++}, Mg^{++}, HCO_3^-, and so on), even though the solutes are distributed among complexes such as $MgSO_4$(aq) and $CaCl^+$ as well as the free ions. Before advent of the theory of electrolyte dissociation, seawater analyses were reported, with equal validity, in terms of the constituent salts NaCl, $MgCl_2$, and so on. Analyses can, in fact, be reported in many ways, depending on the analyst's choice of chemical components.

Garrels and Thompson's calculation, computed by hand, is the basis for a class of geochemical models that predict species distributions, mineral saturation states, and gas fugacities from chemical analyses. This class of models stems from the distinction between a chemical analysis, which reflects a solution's bulk composition, and the actual distribution of species in a solution. Such *equilibrium models* have become widely applied, thanks in part to the dissemination of reliable computer programs such as SOLMNEQ (Kharaka and Barnes, 1973) and WATEQ (Truesdell and Jones, 1974).

Garrels and Mackenzie (1967) pioneered a second class of models when they simulated the reactions that occur as a spring water evaporates. They began by calculating the distribution of species in the spring water, and then repeatedly removed an aliquot of water and recomputed the species distribution. From concepts of equilibrium and mass transfer, the *reaction path model* was born. This class of calculation is significant in that it extends geochemical modeling from considering state to simulating process.

Helgeson (1968) introduced computerized modeling to geochemistry. Inspired by Garrels and Mackenzie's work, he realized that species distributions and the effects of mass transfer could be represented by general equations that can be coded into computer programs. Helgeson and colleagues (Helgeson *et al.*, 1969, 1970) demonstrated a generalized method for tracing reaction paths, which they automated with their program PATHI ("path-one") and used to study weathering, sediment diagenesis, evaporation, hydrothermal alteration, and ore deposition.

Two conceptual improvements have been made since this early work. First, Helgeson *et al.* (1970) posed the reaction path problem as the solution to a system of ordinary differential equations. Karpov and Kaz'min (1972) and Karpov *et al.* (1973) recast the problem algebraically so that a reaction path could be traced by repeatedly solving for a system's equilibrium state as the system varied in composition or temperature. Wolery's (1979) EQ3/EQ6, the first software package for geochemical modeling to be documented and distributed, and Reed's (1977, 1982) SOLVEQ and CHILLER programs used algebraic formulations. This refinement simplified the formulations and codes, separated consideration of mass and heat transfer from the chemical equilibrium calculations, and eliminated the error implicit in integrating differential equations numerically.

Second, modelers took a broader view of the choice of chemical components. Aqueous chemists traditionally think in terms of elements (and electrons) as components, and this choice carried through to the formulations of PATHI and EQ3/EQ6. Morel and Morgan (1972), in calculating species distributions, described composition by using aqueous species for components (much like the seawater analysis described at the beginning of this section; see also Morel, 1983). Reed (1982) formulated the reaction path problem similarly, and Perkins and Brown's (1982) PATH program also used species and minerals as components. Chemical reactions now served double duty by giving the compositions of species and minerals in the system in terms of the chosen component set. This choice, which allowed models to be set up without even acknowledging the existence of elements, simplified the governing equations and provided for easier numerical solutions.

1.2 Scope of This Book

In setting out to write this book, I undertook to describe reaction modeling both in its conceptual underpinnings and its applications. Anything less would not be acceptable. Lacking a thorough introduction to underlying theory, the result would resemble a cookbook, showing the how but not the why of modeling. A book without detailed examples spanning a range of applications, on the other hand, would be sterile and little used.

Of necessity, I limited the scope of the text to discussing reaction modeling itself. I introduce the thermodynamic basis for the equations I derive, but do not attempt a complete development of the field of thermodynamics. A number of texts already present this beautiful body of theory better than I could aspire to in these pages. Among my favorites: Prigogine and Defay (1954), Pitzer and Brewer (1961), Denbigh (1981), Anderson and Crerar (1993), Nordstrom and Munoz (1994), Anderson (2017), and Ganguly (2020). I present (in Chapter 8) but do not derive models for estimating activity coefficients in electrolyte solutions. The reader interested in more detail may refer to Robinson and Stokes (1968), Helgeson *et al.* (1981), Pitzer (1987), and Grenthe *et al.* (1997); Anderson and Crerar (1993, their chapter 17) present a concise but thorough overview of the topic.

Finally, I do not discuss questions of the measurement, estimation, evaluation, and compilation of the thermodynamic data upon which reaction modeling depends. Nordstrom and Munoz (1994, their chapters 11 and 12) provide a summary and overview of this topic, truly a specialty in its own right. Haas and Fisher (1976), Helgeson *et al.* (1978), and Johnson *et al.* (1991) treat aspects of the subject in detail.

2

Modeling Overview

A *model* is a simplified version of reality that is useful as a tool. A successful model strikes a balance between realism and practicality. Properly constructed, a model is neither so simplified that it is unrealistic nor so detailed that it cannot be readily evaluated and applied to the problem of interest.

Geologic maps constitute a familiar class of models. To map a sedimentary section, a geologist collects data at certain outcrops. He or she casts these observations in terms of the local stratigraphy, which is itself a model that simplifies reality by allowing groups of sediments to be lumped together into formations. Our geologist then interpolates among the data points (and projects beneath them) to infer positions for formation contacts, faults, and so on across the field area.

The final map is detailed enough to show the general arrangement of formations and major structures, but simplified enough, when drawn to scale, that small details do not obscure the overall picture. The map, despite its simplicity, is without argument a useful tool for understanding the area's geology. To be successful, a geochemical model should also portray the important features of the problem of interest without necessarily attempting to reproduce each chemical or mineralogical detail.

2.1 Conceptual Models

The first and most critical step in developing a geochemical model is conceptualizing the system or process of interest in a useful manner. By *system*, we simply mean the portion of the Universe that we decide is relevant. The composition of a *closed system* is fixed, but mass can enter and leave an *open system*. A system has an *extent*, which the modeler defines when he sets the amounts of fluid and mineral considered in the calculation. A system's extent might be a droplet of rainfall, the groundwater and sediments contained in a unit volume of an aquifer, or the world's oceans.

The "art" of geochemical modeling is conceptualizing the model in a useful way. Figure 2.1 shows schematically the basis for constructing a geochemical model. The heart of the model is the *equilibrium system*, which remains in some form of chemical equilibrium, as described below, throughout the calculation. The equilibrium system contains an aqueous fluid and optionally one or more minerals. The temperature and composition of the equilibrium system are known at the beginning of the model, which allows the system's equilibrium state to be calculated. Pressure also affects the equilibrium state, but usually in a minor way under the near-surface conditions considered in this book (e.g., Helgeson, 1969; but also see Hemley *et al.*, 1986), unless a gas phase is present.

In the simplest class of geochemical models, the equilibrium system exists as a closed system at a known temperature. Such equilibrium models predict the distribution of mass among species and minerals, as well as the species' activities, the fluid's saturation state with respect to various minerals, and the fugacities of

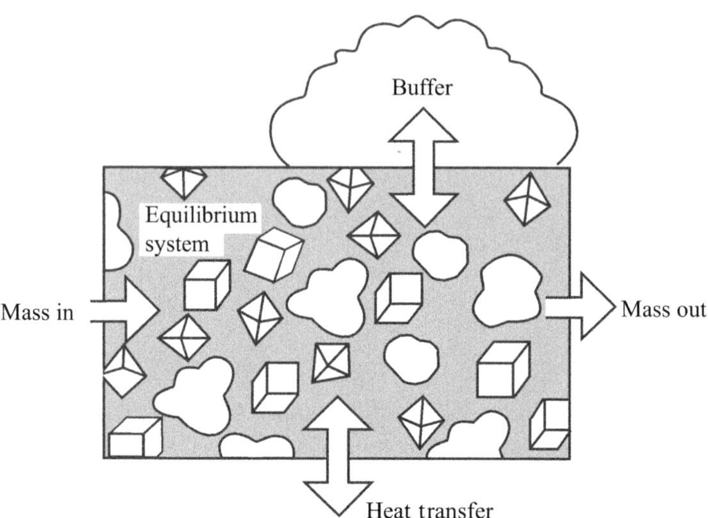

Figure 2.1 Schematic diagram of a reaction model. The heart of the model is the equilibrium system, which contains an aqueous fluid and, optionally, one or more minerals. The system's constituents remain in chemical equilibrium throughout the calculation. Transfer of mass into or out of the system and variation in temperature drive the system to a series of new equilibria over the course of the reaction path. The system's composition may be buffered by equilibrium with an external gas reservoir, such as the atmosphere.

different gases that can exist in the chemical system. In this case, the initial equilibrium system constitutes the entire geochemical model.

More complicated models account for the transport of mass or heat into or out of the system, so that its composition or temperature, or both, vary over the course of the calculation. The system's initial equilibrium state provides the starting point for this class of reaction path models. From this point, the model traces how mass entering and leaving the system, or changes in temperature, affect the system's equilibrium state.

Conceptualizing a geochemical model is a matter of defining (1) the nature of equilibrium to be maintained, (2) the initial composition and temperature of the equilibrium system, and (3) the mass transfer or temperature variation to occur over the course of the reaction process envisioned.

2.1.1 Types of Equilibrium

It is useful at this point to differentiate among the ways in which we can define equilibrium. In a classic sense (e.g., Pitzer and Brewer, 1961; Denbigh, 1981), a system is in equilibrium when it occupies a specific region of space within which there is no spontaneous tendency for change to occur. In this case, which we will call *complete equilibrium*, all possible chemical reactions are in equilibrium. Assuming complete equilibrium, for example, we can predict the distribution of dissolved species in a sample of river water, if the water is not supersaturated with respect to any mineral.

Geochemical models can be conceptualized in terms of certain false equilibrium states (Barton *et al.*, 1963; Helgeson, 1968). A system is in *metastable equilibrium* when one or more reactions proceed toward equilibrium at rates that are vanishingly small on the time scale of interest. Metastable equilibria commonly figure in geochemical models. In calculating the equilibrium state of a natural water from a reliable chemical analysis, for example, we may find that the water is supersaturated with respect to one or more minerals. The calculation predicts that the water exists in a metastable state because the reactions to precipitate these minerals have not progressed to equilibrium.

In tracing a reaction path, likewise, we may find a mineral in the calculation results that is unlikely to form in a real system. Quartz, for example, would be likely to precipitate too slowly to be observed in a laboratory experiment conducted at room temperature. A model can be instructed to seek metastable solutions by not considering (*suppressing*, in modeling parlance) certain minerals in the calculation, as would be necessary to model such an experiment.

A system in complete equilibrium is spatially continuous, but this requirement can be relaxed as well. A system can be in internal equilibrium but, like Swiss cheese, have holes. In this case, the system is in *partial equilibrium*. The fluid in a sandstone, for example, might be in equilibrium itself, but may not be in equilibrium with the mineral grains in the sandstone or with just some of the grains. This concept has provided the basis for many published reaction paths, beginning with the work of Helgeson *et al.* (1969), in which a rock gradually reacts with its pore fluid.

The species dissolved in a fluid may be in partial equilibrium, as well. Many redox reactions equilibrate slowly in natural waters (e.g., Lindberg and Runnells, 1984). The oxidation of methane,

$$CH_4(aq) + 2\,O_2(aq) \rightarrow HCO_3^- + H^+ + H_2O\,, \tag{2.1}$$

is notorious in this regard. Shock (1988), for example, found that although carbonate species and organic acids in oil-field brines appear to be in equilibrium with each other, these species are clearly out of equilibrium with methane. To model such a system, the modeler can *decouple* redox pairs such as $HCO_3^- - CH_4$ (e.g., Wolery, 1983), denying the possibility that oxidized species react with reduced species.

A third variant is the concept of *local equilibrium*, sometimes called mosaic equilibrium (Thompson, 1959, 1970; Valocchi, 1985; Knapp, 1989). This idea is useful when temperature, mineralogy, or fluid chemistry vary across a system of interest. By choosing a small enough portion of a system, according to this assumption, we can consider that portion to be in equilibrium. The concept of local equilibrium can also be applied to model reactions occurring in systems open to groundwater flow, using the "flow-through" and "flush" models described in the next section. The various types of equilibrium can sometimes be combined in a single model. A modeler, for example, might conceptualize a system in terms of partial and local equilibrium.

2.1.2 The Initial System

Calculating a model begins by computing the initial equilibrium state of the system at the temperature of interest. By convention but not requirement, the initial system contains a kilogram of water and so, accounting for dissolved species, a somewhat greater mass of fluid. The modeler can alter the system's extent by prescribing a greater or lesser water mass. Minerals may be included as desired, up to the limit imposed by the phase rule, as described in the next chapter. Each mineral included will be in equilibrium with the fluid, thus providing a constraint on the fluid's chemistry.

The modeler can constrain the initial equilibrium state in many ways, depending on the nature of the problem, but the number of pieces of information required is firmly set by the laws of thermodynamics. In general, the modeler sets the temperature and provides one compositional constraint for each chemical component in the system. Useful constraints include

- The mass of solvent water (1 kg by default),
- The amounts of any minerals in the equilibrium system,
- Any known gas fugacities or partial pressures,
- The amount of any component dissolved in the fluid, such as Na^+ or HCO_3^-, as determined by chemical analysis, and
- The activities of a species such as H^+, as would be determined by pH measurement, or the oxidation state given by an Eh determination.

Unfortunately, the required number of constraints is not negotiable. Regardless of the difficulty of determining these values in sufficient number or the apparent desirability of including more than the allowable number, the system is mathematically underdetermined if the modeler uses fewer constraints than components, or overdetermined if he sets more.

Sometimes the calculation predicts that the fluid as initially constrained is supersaturated with respect to one or more minerals, and hence, is in a metastable equilibrium. If the supersaturated minerals are not suppressed, the model proceeds to calculate the equilibrium state, which it needs to find if it is to follow a reaction path. By allowing supersaturated minerals to precipitate, accounting for any minerals that dissolve as others precipitate, the model determines the stable mineral assemblage and corresponding fluid composition. The model output contains the calculated results for the supersaturated system as well as those for the system at equilibrium.

2.1.3 Mass and Heat Transfer: The Reaction Path

Once the initial equilibrium state of the system is known, the model can trace a reaction path. The reaction path is the course followed by the equilibrium system as it responds to changes in composition and temperature (Fig. 2.1). The measure of reaction progress is the variable ξ, which varies from zero to one from the beginning to end of the path. The simplest way to specify mass transfer in a reaction model (Chapter 14) is to set the mass of a reactant to be added or removed over the course of the path. In other words, the reaction rate is expressed in reactant mass per unit ξ. To model the dissolution of feldspar into a stream water, for example, the modeler would specify a mass of feldspar sufficient to saturate the water. At the point of saturation, the water is in equilibrium with the feldspar, and no further reaction will occur. The results of the calculation are the fluid chemistry and masses of precipitated minerals at each point from zero to one, as indexed by ξ.

Any number of reactants may be considered, each of which can be transferred at a positive or negative rate. Positive rates cause mass to be added to the system; at negative rates it is removed. Reactants may be minerals, aqueous species (in charge-balanced combinations), oxide components, or gases. Since the role of a reactant is to change the system composition, it is the reactant's composition, not its identity, that matters. In other words, quartz, cristobalite, and SiO_2(aq) behave alike as reactants.

Mass transfer can be described in more sophisticated ways. By taking ξ in the previous example to represent time, the rate at which feldspar dissolves and product minerals precipitate can be set using kinetic rate laws, as discussed in Chapter 17. The model calculates the actual rates of mass transfer at each step of the reaction progress from the rate constants, as measured in laboratory experiments, and the fluid's degree of undersaturation or supersaturation.

The fugacity or partial pressure of one or more gases such as CO_2 and O_2 can be buffered (Fig. 2.1; see Chapter 15) so that they are held constant over the reaction path. In this case, mass transfer between the equilibrium system and the gas buffer occurs as needed to maintain the buffer. Adding acid to a CO_2-buffered

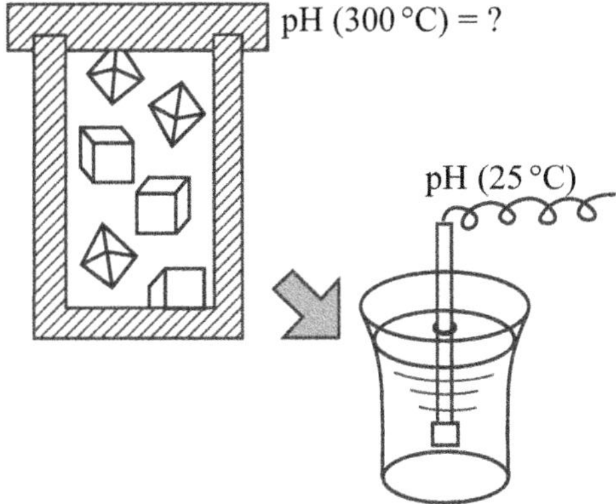

Figure 2.2 Example of a polythermal path. Fluid from a hydrothermal experiment is sampled at 300 °C and analyzed at room temperature. To reconstruct the fluid's pH at high temperature, the calculation equilibrates the fluid at 25 °C and then carries it as a closed system to the temperature of the experiment.

system, for example, would be likely to dissolve calcite,

$$\underset{calcite}{CaCO_3} + 2\,H^+ \rightarrow Ca^{++} + H_2O + CO_2(g)\,. \tag{2.2}$$

Carbon dioxide will pass out of the system into the buffer to maintain the buffered fugacity or partial pressure.

Reaction paths can be traced at steady or varying temperature; the latter case is known as a *polythermal* path. Strictly speaking, heat transfer occurs even at constant temperature, albeit commonly in small amounts, to offset reaction enthalpies. For convenience, modelers generally define polythermal paths in terms of changes in temperature rather than heat fluxes.

2.2 Configurations of Reaction Models

Reaction models, despite their simple conceptual basis (Fig. 2.1), can be configured in a number of ways to represent a variety of geochemical processes. Each type of model imposes on the system some variant of equilibrium, as described in the previous section, but differs from others in the manner in which mass and heat transfer are specified. This section summarizes the configurations that are commonly applied in geochemical modeling.

2.2.1 Closed-System Models

Closed-system models are those in which no mass transfer occurs. Equilibrium models, the simplest of this class, describe the equilibrium state of a system composed of a fluid, any coexisting minerals, and, optionally, a gas buffer. Such models are not true reaction models, however, because they describe state instead of process.

Polythermal reaction models (Section 15.1), however, are commonly applied to closed systems, as in studies of groundwater geothermometry (Chapter 27), and interpretations of laboratory experiments. In hydrothermal experiments, for example, researchers sample and analyze fluids from runs conducted at high temperature, but can determine pH only at room temperature (Fig. 2.2). To reconstruct the original pH

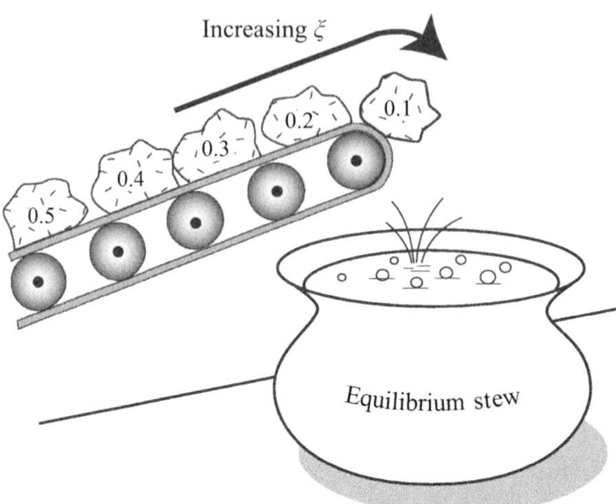

Figure 2.3 Configuration of a reaction path as a titration model. One or more reactants are gradually added to the equilibrium system, as might occur as the grains in a rock gradually react with a pore fluid.

(e.g., Reed and Spycher, 1984), assuming that gas did not escape from the fluid before it was analyzed, an experimentalist can calculate the equilibrium state at room temperature and follow a polythermal path to estimate the fluid chemistry at high temperature.

There is no restriction against applying polythermal models in open systems. In this case, the modeler defines mass transfer as well as the heating or cooling rate in terms of ξ. Realistic models of this type can be hard to construct (e.g., Bowers and Taylor, 1985), however, because the heating or cooling rates need to be balanced somehow with the rates of mass transfer.

2.2.2 Titration Models

The simplest open-system model involves a reactant which, if it is a mineral, is undersaturated in an initial fluid. The reactant is gradually added into the equilibrium system over the course of the reaction path (Fig. 2.3). The reactant dissolves irreversibly. The process may cause minerals to become saturated and precipitate, or drive minerals that already exist in the system to dissolve. The equilibrium system continues to evolve until the fluid reaches saturation with the reactant or the reactant is exhausted. A model of this nature can be constructed with several reactants, in which case the process proceeds until each reactant is saturated or exhausted.

This type of calculation is known as a *titration model* because the calculation steps forward through reaction progress ξ, adding an aliqout of the reactant at each step $\Delta\xi$. To predict, for example, how a rock will react with its pore fluid, we can titrate the minerals that make up the rock into the fluid. The solubility of most minerals in water is rather small, so the fluid in such a calculation is likely to become saturated after only a small amount of the minerals has reacted. Reacting on the order of 10^{-3} moles of a silicate mineral, for example, is commonly sufficient to saturate a fluid with respect to the mineral.

In light of the small solubilities of many minerals, the extent of reaction predicted by this type of calculation may be smaller than expected. Considerable amounts of diagenetic cements are commonly observed, for example, in sedimentary rocks, and crystalline rocks can be highly altered by weathering or hydrothermal fluids. A titration model may predict that the proper cements or alteration products form, but explaining the quantities of these minerals observed in nature will probably require that the rock react repeatedly as its pore fluid is replaced. Local equilibrium models of this nature are described later in this section.

2.2.3 Fixed-Fugacity and Sliding-Fugacity Models

Many geochemical processes occur in which a fluid remains in contact with a gaseous phase. The gas, which could be the Earth's atmosphere or a subsurface gas reservoir, acts to buffer the system's chemistry. By dissolving gas species from the buffer or exsolving gas into it, the fluid will, if reaction proceeds slowly enough, maintain equilibrium with the buffer.

A reaction path in which the fugacity or partial pressure of one or more gases is buffered by an external reservoir (Fig. 2.1) is known as a *fixed-fugacity path* (Section 15.2). Because Garrels and Mackenzie (1967) assumed a fixed CO_2 fugacity when they calculated their pioneering reaction path by hand (see Chapter 28), they also calculated a fixed-fugacity path. Numerical modelers (e.g., Delany and Wolery, 1984) have more recently programmed buffered gas fugacities as options in their software.

The results of fixed-fugacity paths can differ considerably from those of simple titration models. Consider, for example, the oxidation of pyrite to goethite,

$$\underset{\textit{pyrite}}{FeS_2} + {}^5\!/_2\, H_2O + {}^{15}\!/_4\, O_2(aq) \rightarrow \underset{\textit{goethite}}{FeOOH} + 4\, H^+ + 2\, SO_4^{--}\,, \tag{2.3}$$

in a surface water. In a simple titration model, pyrite dissolves until the water's dissolved oxygen is consumed. Water equilibrated with the atmosphere contains about 10 mg kg^{-1} $O_2(aq)$, so the amount of pyrite consumed is small. In a fixed-fugacity model, however, the concentration of $O_2(aq)$ remains in equilibrium with the atmosphere, allowing the reaction to proceed almost indefinitely.

When one or more gas fugacities or partial pressures vary along ξ instead of remaining fixed, the path is called a *sliding-fugacity path* (Section 15.3). This type of path is useful when changes in total pressure allow gases to exsolve from the fluid (e.g., Leach *et al.*, 1991). For example, layers of dominantly carbonate cement are observed along the tops of geopressured zones in the US Gulf of Mexico basin (e.g., Hunt, 1990). The cement is apparently precipitated (Fig. 2.4) as fluids slowly migrate from overpressured sediments in overlying strata at hydrostatic pressure. The pressure drop allows CO_2 to exsolve,

$$Ca^{++} + 2\, HCO_3^- \rightarrow \underset{\textit{calcite}}{CaCO_3} + H_2O + CO_2(g)\,, \tag{2.4}$$

causing calcite to precipitate.

Fixed-activity and *sliding-activity paths* (Sections 15.2–15.3) are analogous to their counterparts in fugacity, except that they apply to aqueous species instead of gases. Fixed-activity paths are useful for simulating, for example, a laboratory experiment controlled by a pH-stat, a device that holds pH constant. Sliding-activity paths make easy work of calculating speciation diagrams, as described in Chapter 15.

2.2.4 Kinetic Reaction Models

In *kinetic reaction paths* (discussed in Chapters 17–21), the rates at which one or more reactions proceed are set by kinetic rate laws, rather than the constraints of equilibrium. Such laws might regulate how quickly

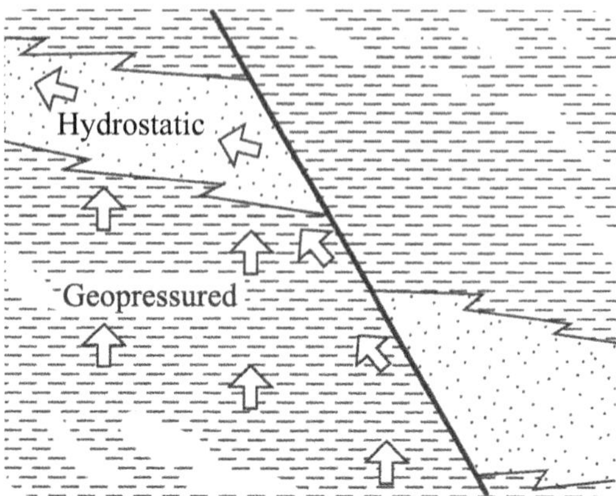

Figure 2.4 Example of a sliding-fugacity path. Deep groundwaters of a geopressured zone in a sedimentary basin migrate upward to lower pressures. During migration, CO_2 exsolves from the water so that its partial pressure follows the variation in total pressure. The loss of CO_2 causes carbonate cements to form.

minerals dissolve and precipitate, redox reactions and microbial transformations progress, complexes form and break apart, or gases pass into and out of solution. In this class of models, reaction progress is measured in time instead of by the nondimensional variable ξ.

The rates at which a mineral dissolves into or precipitates from the equilibrium system, for example, might be set by a kinetic law. According to the rate law, as would be expected, the mineral dissolves into a fluid in which it is undersaturated and precipitates where supersaturated. The rate of dissolution or precipitation in the calculation depends on the variables in the rate law: the reaction's rate constant, the mineral's surface area, the degree to which the mineral is undersaturated or supersaturated in the fluid, and the activities of any catalyzing and inhibiting species.

Kinetic and equilibrium-controlled reactions can be readily combined into a single model. The two descriptions might seem incompatible, but kinetic theory (Chapter 17) provides a conceptual link: the equilibrium point of a reaction is the point at which dissolution and precipitation rates balance.

For practical purposes, geochemical reactions fall into three groups: those that proceed slowly enough relative to the time period of interest that they can be ignored altogether, those whose rates are rapid enough to maintain equilibrium, and the remaining reactions. Only reactions in the latter group require a kinetic description. Brezonik (1994), Lasaga (1998), Zhang (2008), Brantley *et al.* (2008), and Rimstidt (2014) provide useful and insightful reviews of kinetic theory and its application in geoscience.

2.2.5 Special Model Configurations

A reaction model may be conceptualized in terms of several special configurations, as developed further in Section 14.3. The configurations are designed to isolate a portion of the system's mass over the course of a simulation, as if that part had been physically separated from the reacting system.

In a *flow-through* reaction path, the model isolates from the system minerals that form over the course of the calculation, preventing them from reacting further. Garrels and Mackenzie (1967) suggested this configuration, and Wolery (1979) implemented it in the EQ3/EQ6 code. In terms of the conceptual model (Fig. 2.1), the process of isolating product minerals is a special case of transferring mass out of the system.

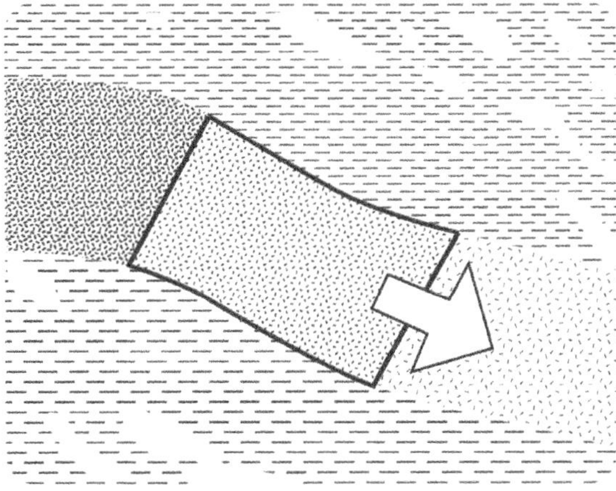

Figure 2.5 "Flow-through" configuration of a reaction path. A packet of fluid reacts with an aquifer as it migrates. Any minerals that form as reaction products are left behind and, hence, isolated from further reaction.

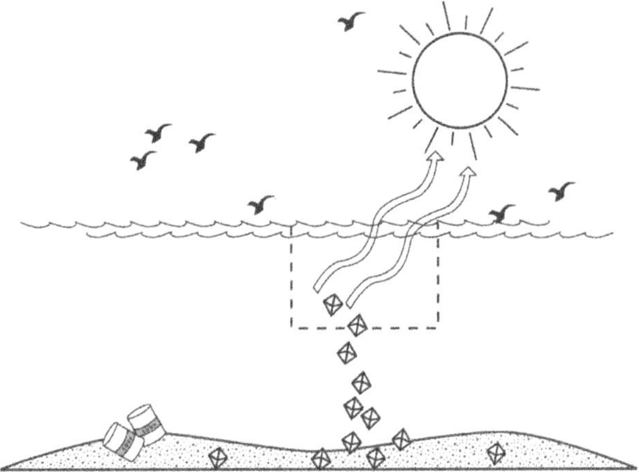

Figure 2.6 Example of a "flow-through" path. Titrating water from a unit volume of seawater increases the seawater's salinity until evaporite minerals form. The product minerals sink to the sea floor, where they are isolated from further reaction.

Rather than completely discarding the removed mass, however, the software tracks the cumulative amount of each mineral isolated from the system over the reaction path.

Using a flow-through model, for example, we can follow the evolution of a packet of fluid as it traverses an aquifer (Fig. 2.5). Fresh minerals in the aquifer react to equilibrium with the fluid at each step in reaction progress. The minerals formed by this reaction are kept isolated from the fluid packet, having been left behind as the packet moves along the aquifer.

In a second example of a flow-through path, we model the evaporation of seawater (Fig. 2.6). The equilibrium system in this case is a unit mass of seawater. Water is titrated out of the system over the course of the path, concentrating the seawater and causing minerals to precipitate. The minerals sink to the sea floor as they form, and so are isolated from further reaction. We carry out such a calculation in Chapter 28.

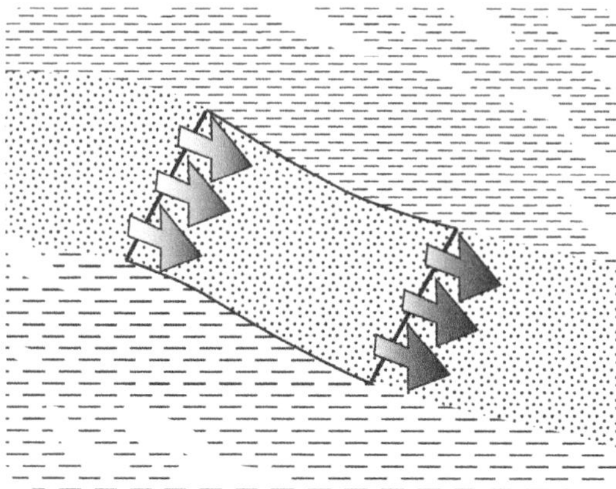

Figure 2.7 "Flush" configuration of a reaction model. Unreacted fluid enters the equilibrium system, which contains a unit volume of an aquifer and its pore fluid, displacing the reacted fluid.

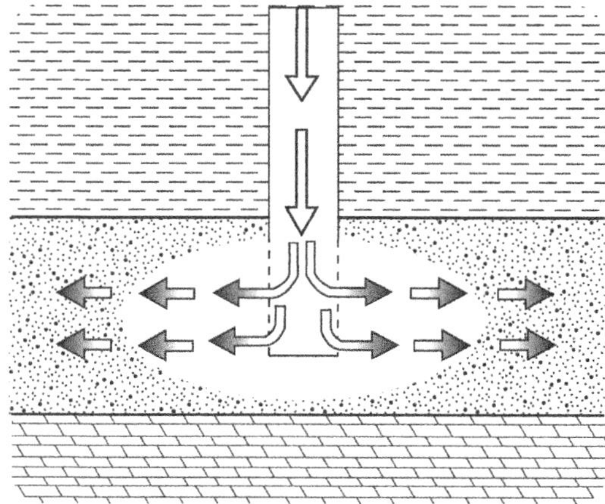

Figure 2.8 Example of a "flush" model. Fluid is pumped into a petroleum reservoir as a stimulant, or industrial waste is pumped into a disposal well. Unreacted fluid enters the formation, displacing the fluid already there.

In a *flush* model, on the other hand, the model tracks the evolution of a system through which fluid migrates (Fig. 2.7). The equilibrium system in this case might be a specified volume of an aquifer, including rock grains and pore fluid. At each step in reaction progress, an increment of unreacted fluid is added to the system, displacing the existing pore fluid. The model is analogous to a "mixed-flow reactor" as applied in chemical engineering (Levenspeil, 1972; Hill, 1977).

Flush models are useful for applications such as studying the diagenetic reactions resulting from groundwater flow in sedimentary basins (see Chapter 29) and predicting formation damage in petroleum reservoirs and injection wells (Fig. 2.8; see Chapters 33 and 34). Stimulants intended to increase production from oil wells (including acids, alkalis, and hot water) as well as the industrial wastes pumped into injection wells commonly react strongly with geologic formations (e.g., Hutcheon, 1984). Reaction models are likely to find increased application as well operators seek to minimize damage from caving and the loss of permeability due to the formation of oxides, clay minerals, and zeolites in the formation's pore space.

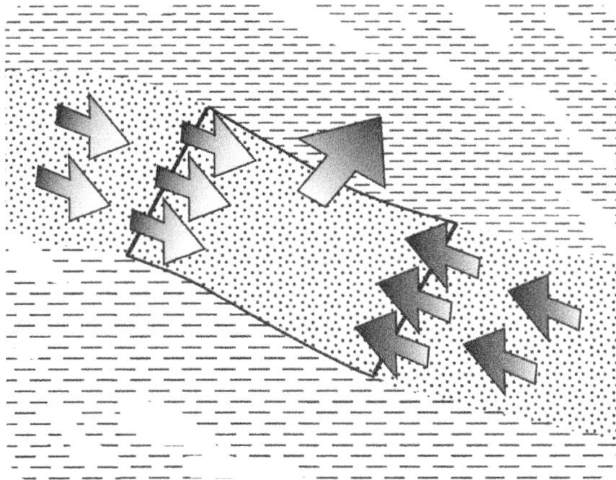

Figure 2.9 Use of a "flush" model to simulate dispersive mixing. Two fluids enter a unit volume of an aquifer where they react with each other and with minerals in the aquifer, displacing the mixed and reacted fluid.

Flush models can also be configured to simulate the effects of dispersive mixing. Dispersion is the physical process by which groundwaters mix in the subsurface (Freeze and Cherry, 1979). With mixing, the groundwaters react with each other and the aquifer through which they flow (e.g., Runnells, 1969). In a flush model, two fluids can flow into the equilibrium system, displacing the mixed and reacted fluid (Fig. 2.9).

A final variant of local equilibrium models is the *dump* option (Wolery, 1979). Here, once the equilibrium state of the initial system is determined, the minerals in the system are jettisoned. The minerals present in the initial system, then, are not available over the course of the reaction path. The dump option differs from the flow-through model in that while the minerals present in the initial system are prevented from back-reacting, those that precipitate over the reaction path are not.

As an example of how the dump option might be used, consider the problem of predicting whether scale will form in the wellbore as groundwater is produced from a well (Fig. 2.10). The fluid is in equilibrium with the minerals in the formation, so the initial system contains both fluid and minerals. The dump option simulates movement of a packet of fluid from the formation into the wellbore, since the minerals in the formation are no longer available to the packet. As the packet ascends the wellbore, it cools, perhaps exsolving gas as it moves toward lower pressure, and leaves behind any scale produced. The reaction model, then, is a polythermal, sliding-fugacity, and flow-through path combined with the dump option.

2.2.6 Reactive Transport Models

Beginning in the late 1980s, a number of groups have worked to develop *reactive transport models* of geochemical reaction in systems open to groundwater flow. As models of this class have become more sophisticated, reliable, and accessible, they have assumed increased importance in the geosciences (e.g., Steefel *et al.*, 2005). The models are a natural marriage (Rubin, 1983; Bahr and Rubin, 1987) of the local equilibrium and kinetic models already discussed with the mass transport models traditionally applied in hydrology and various fields of engineering (e.g., Bear, 1972; Bird *et al.*, 2006).

By design, this class of models predicts the distribution in space and time of the chemical reactions that occur along a groundwater flow path. Among the many papers discussing details of how the reactive transport

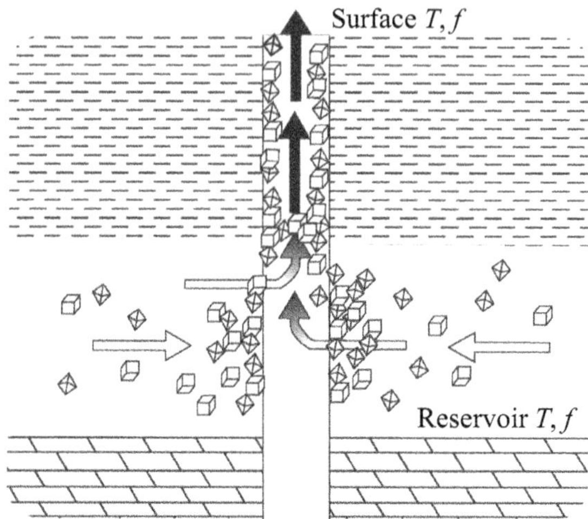

Figure 2.10 Use of the "dump" option to simulate scaling. The pore fluid is initially in equilibrium with minerals in the formation. As the fluid enters the wellbore, the minerals are isolated (dumped) from the system. The fluid then follows a polythermal, sliding-fugacity path as it ascends the wellbore toward lower temperatures and pressures, depositing scale.

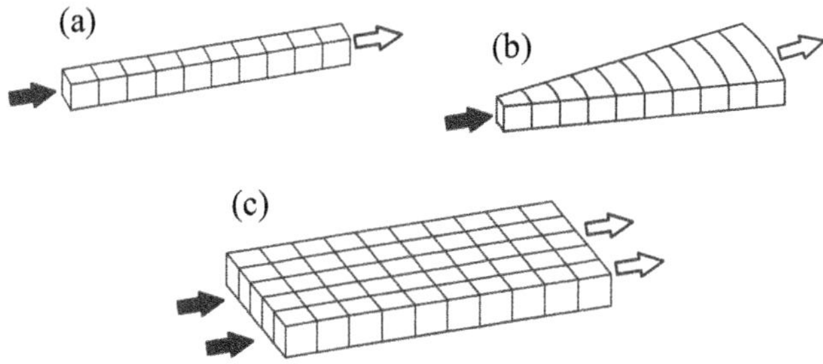

Figure 2.11 Configurations of reactive transport models of water–rock interaction in a system open to groundwater flow: (a) linear domain in one dimension, (b) radial domain in one dimension, and (c) linear domain in two dimensions. Domains are divided into nodal blocks, within each of which the model solves for the distribution of chemical mass as it changes over time, in response to transport by the flowing groundwater. In each case, unreacted fluid enters the domain and reacted fluid leaves it.

problem can be formulated are those of Berner (1980), Lichtner (1985, 1988, 1996), Lichtner *et al.* (1986, 1996), Cederberg *et al.* (1985), Ortoleva *et al.* (1987), Cheng and Yeh (1988), Yeh and Tripathi (1989), Liu and Narasimhan (1989a, 1989b), Steefel and Lasaga (1992, 1994), Yabusaki *et al.* (1998), Malmstrom *et al.* (2004), and Steefel *et al.* (2015); Maher and Mayer (2019) and Druhan and Tournassat (2019) compile studies showing how the technique can be applied to address a spectrum of problems in geoscience.

In a reactive transport model, the domain of interest is divided into nodal blocks, as shown in Figure 2.11. Fluid enters the domain across one boundary, reacts with the medium, and discharges at another boundary.

In many cases, reaction occurs along fronts that migrate through the medium until they either traverse it or assume a steady-state position (Lichtner, 1988). As noted by Lichtner (1988), models of this nature predict that reactions occur in the same sequence in space and time as they do in simple reaction path models. The reactive transport models, however, predict how the positions of reaction fronts migrate through time, provided that reliable input is available about flow rates, the permeability and dispersivity of the medium, and reaction rate constants.

Reactive transport models are, naturally, more challenging to set up and compute than are simple reaction models. The model results reflect the kinetic rate constants taken to describe chemical reaction, as well as the hydrologic properties assumed for the medium. Notably, these data normally comprise the most poorly known parameters in the natural system.

Since a valid reaction model is a prerequisite for a continuum model, the first step in any case is to construct a successful reaction model for the problem of interest. The reaction model provides the modeler with an understanding of the nature of the chemical process in the system. Armed with this information, he or she is prepared to undertake more complex calculations. Chapters 23–25 of this book treat in detail the construction of reactive transport models.

2.3 Uncertainty in Geochemical Modeling

Calculating a geochemical model provides not only results, but uncertainty about the accuracy of the results. Uncertainty, in fact, is an integral part of modeling that deserves as much attention as any other aspect of a study. To evaluate the sources of error in a study, a modeler should consider a number of questions:

- *Is the chemical analysis used sufficiently accurate to support the modeling study?* The chemistry of the initial system in most models is constrained by a chemical analysis, including perhaps a pH determination and some description of the system's oxidation state. The accuracy and completeness of available chemical analyses, however, vary widely. Routine tests made of drinking water supplies and formation fluids from oil wells are commonly too rough and incomplete to be of much use to a modeler. Sets of analyses retrieved unselectively from water quality databases such as WATSTORE at the US Geological Survey are generally not suitable for modeling applications (Hem, 1985). Careful analyses such as those of Iceland's geothermal waters made for scientific purposes (Arnórsson *et al.*, 1983; see Chapter 27), on the other hand, are invaluable.

  As Hem (1985) notes, a chemical analysis with concentrations reported to two or three, and sometimes four or five, significant figures can be misleadingly authoritative. Analytical accuracy and precision are generally in the range of ± 2 to $\pm 10\%$, but depend on the technique used, the skill of the analyst, and whether or not the constituent was present near the detection limit of the analytical method. The third digit in a reported concentration is seldom meaningful, and confidence should not necessarily be placed on the second.

  Care should be taken in interpreting reported pH values, which may have been determined in the field or in the laboratory after the sample had been stored for an unknown period of time. Only the field measurement of pH is meaningful and, in the case of a groundwater, even the field measurement is reliable only if it is made immediately after sampling, before the water can exchange CO_2 with the atmosphere.

  Significant error is introduced when a sample is acidified to "preserve" it, if the sample is not first carefully filtered to remove sediment and colloids (as illustrated in Section 6.2). Until the 1950s, it was normal procedure to sample unfiltered waters, and this practice continues in some organizations today. Even today's common practice of passing samples through a 0.45-μm filter in the field fails to remove colloidal aluminum and iron (e.g., Kennedy *et al.*, 1974); a 0.10-μm filter is usually required to separate these colloids. By adding acid, the sampler dissolves any colloids and some of any suspended sediments, the constituents of which will appear in the chemical analysis as though they had originally been in solution.

Samples of formation water taken from drill-stem tests during oil exploration are generally contaminated by drilling fluids. The expense of keeping a drilling rig idle generally precludes pumping the formation fluid long enough to produce uncontaminated fluid, a procedure that might require weeks. As only rough knowledge of groundwater composition is needed in exploration, there is little impetus to improve procedures. Samples obtained at the well head after a field has been in production (e.g., Lico *et al.*, 1982) may be preferable to analyses made during drill-stem tests, but care must be taken: samples obtained in this way may have already exsolved CO_2 or other gases before sampling.

- *Does the thermodynamic dataset contain the species and minerals likely to be important in the study?* A set of thermodynamic data, especially one intended to span a range of temperatures, is by necessity a balance between completeness and accuracy. The modeler is responsible for assuring that the database includes the predominant species and important minerals in the problem of interest.

The following example shows why this is important. The calculations in this book make use of the dataset compiled by Thomas Wolery, Ken Jackson, and numerous co-workers at Lawrence Livermore National Laboratory (the LLNL dataset; Delany and Lundeen, 1989), which is based in part on a dataset developed by Helgeson *et al.* (1978). The dataset includes a number of Cu-bearing species and minerals, including the cupric species Cu^{++} and $Cu(OH)^{+}$ that are dominant at room temperature under oxidized conditions in acidic and neutral solutions.

At pH values greater than about 9.5, the species $Cu(OH)_2$, $Cu(OH)_3^-$, and $Cu(OH)_4^{--}$ dominate the solubility of cupric copper by some orders of magnitude (Baes and Mesmer, 1976); these species, however, are not included in the database version used in this book. To construct a valid model of copper chemistry in an oxidizing, alkaline solution, the modeler would need to extend the database to include these species.

The same requirement extends to the minerals considered in the calculation. Minerals in nature occur as solid solutions in which elements substitute for one another in the mineral's crystal structure, but thermodynamic datasets generally contain data for pure minerals of fixed composition. A special danger arises in considering the chemistry of trace metals. In nature, these would be likely to occur as ions substituted into the framework of common minerals or sorbed onto mineral or organic surfaces, but the chemical model would consider only the possibility that the species occur as dissolved species or as the minerals of these elements that are seldom observed in nature.

- *Are the equilibrium constants for the important reactions in the thermodynamic dataset sufficiently accurate?* The collection of thermodynamic data is subject to error in the experiment, chemical analysis, and interpretation of the experimental results. Error margins, however, are seldom reported and never seem to appear in data compilations. Compiled data, furthermore, have generally been extrapolated from the temperature of measurement to that of interest (e.g., Helgeson, 1969). The stabilities of many aqueous species have been determined only at room temperature, for example, and mineral solubilities are commonly measured at high temperatures where reactions approach equilibrium most rapidly. Evaluating the stabilities and sometimes even the stoichiometries of complex species is especially difficult and prone to inaccuracy.

For these reasons, the thermodynamic data on which a model is based vary considerably in quality. At the minimum, data error limits the resolution of a geochemical model. The energetic differences among groups of silicates, such as the clay minerals, are commonly smaller than the error implicit in estimating mineral stability. A clay mineralogist, therefore, might find less useful information in the results of a model than expected.

- *Can the species' activity coefficients be calculated accurately?* An activity coefficient relates each dissolved species' concentration to its activity. Most commonly, a modeler uses an extended form of the Debye–Hückel equation to estimate values for the coefficients. Helgeson (1969) correlated the activity coefficients to this equation for dominantly NaCl solutions having concentrations up to 3 molal. The resulting equations are probably reliable for electrolyte solutions of general composition (i.e., those dominated by salts other than NaCl) where ionic strength is less than about 1 molal (Wolery, 1983; see Chapter 8). Calculated activity coefficients are less reliable in more concentrated solutions. As an alternative to the Debye–Hückel method, the modeler can use virial equations (the "Pitzer equations") designed to predict activity coefficients for electrolyte brines. These equations have their own limitations, however, as discussed in Chapter 8.
- *Do the kinetic rate constants and rate laws apply well to the system being studied?* Using kinetic rate laws to describe the dissolution and precipitation rates of minerals adds an element of realism to a geochemical model but can be a source of substantial error. Much of the difficulty arises because a measured rate constant reflects the dominant reaction mechanism in the experiment from which the constant was derived, even though an entirely different mechanism may dominate the reaction in nature (see Chapter 17).

  Rate constants for the dissolution and precipitation of quartz, for example, have been measured in deionized water (Rimstidt and Barnes, 1980). Dove and Crerar (1990), however, found that reaction rates increased by as much as one and a half orders of magnitude when the reaction proceeded in dilute electrolyte solutions. In addition, reaction rates determined in the laboratory from hydrothermal experiments on "clean" systems differ substantially from those that occur in nature, where clay minerals, oxides, and other materials may coat mineral surfaces and hinder reaction.
- *Is the assumed nature of equilibrium appropriate?* The modeler defines an equilibrium system that forms the core of a geochemical model, using one of the equilibrium concepts already described. The modeler needs to ask whether the reactions considered in an equilibrium system actually approach equilibrium. If not, it may be necessary to decouple redox reactions, suppress minerals from the system, or describe mineral dissolution and precipitation using a kinetic rate law in order to calculate a realistic chemical model.
- *Most importantly, has the modeler conceptualized the reaction process correctly?* The modeler defines a reaction process on the basis of a concept of how the process occurs in nature. Many times the apparent failure of a calculation indicates a flawed concept of how the reaction occurs rather than error in a chemical analysis or the thermodynamic data. The "failed" calculation, in this case, is more useful than a successful one because it points out a basic error in the modeler's understanding.

  Errors in conceptualizing a problem are easy to make but can be hard to discover. A modeler, distracted by the intricacies of his tools or the complexities of her results, can too easily lose sight of the nature of the conceptual model used or the assumptions implicit in deriving it. A mistake in the study of sediment diagenesis, for example, is to try to explain the origin of cements in a marine orthoquartzite by heating the original quartz grains and seawater along a polythermal path, to simulate burial.

  The rock in question might contain a large amount of calcite cement, but the reaction path predicts that only a trace of calcite forms during burial. Considering this contradiction, the modeler realizes that this model could not have been successful in the first place: there is not enough calcium or carbonate in seawater to have formed that amount of cement. The model in this case was improperly conceptualized as a closed rather than open system.

Given this array of error sources, how can a geochemical modeler cope with the uncertainties implicit in his or her calculations? The best answer is probably that the modeler should begin work by integrating experimental results and field observations into the study. Having successfully explained the experimental or field data, the modeler can extrapolate to make predictions with greater confidence.

Modelers should also take heart that their work provides an impetus to determine more accurate thermodynamic data, derive better activity models for electrolyte solutions, and measure reaction rates under more realistic conditions.

PART I

EQUILIBRIUM IN NATURAL WATERS

3

The Equilibrium State

Aqueous geochemists work daily with equations that describe the equilibrium points of chemical reactions among dissolved species, minerals, and gases. To study an individual reaction, a geochemist writes the familiar expression, known as the *mass action equation*, relating species' activities to the reaction's equilibrium constant. In this chapter we carry this type of analysis a step farther by developing expressions that describe the conditions under which not just one but all of the possible reactions in a geochemical system are at equilibrium.

We consider a geochemical system comprising at least an aqueous solution in which the species of many elements are dissolved. We generally have some information about the fluid's bulk composition, perhaps directly because we have analyzed it in the laboratory. The system may include one or more minerals, up to the limit imposed by the phase rule (see Section 3.4), that coexist with, and are in equilibrium with the aqueous fluid. The fluid's composition might also be buffered by equilibrium with a gas reservoir (perhaps the atmosphere) that contains one or more gases. The gas buffer is large enough that its composition remains essentially unchanged if gas exsolves from or dissolves into the fluid.

How can we express the equilibrium state of such a system? A direct approach would be to write each reaction that could occur among the system's species, minerals, and gases. To solve for the equilibrium state, we would determine a set of concentrations that simultaneously satisfy the mass action equation corresponding to each possible reaction. The concentrations would also have to add up, together with the mole numbers of any minerals in the system, to give the system's bulk composition. In other words, the concentrations would also need to satisfy a set of *mass balance equations*.

Such an approach, however, is unnecessarily difficult to carry out. Dissolving even a few elements in water produces many tens of species that need to be considered, and complex solutions contain many hundreds of species. Each species represents an independent variable, namely its concentration, in our scheme. For any but the simplest of chemical systems, the problem would contain too many unknown values to be solved conveniently.

Fortunately, few of these variables are truly independent. Geochemists have developed a variety of numerical schemes to solve for equilibrium in multicomponent systems, each of which features a reduction in the number of independent variables carried through the calculation. The schemes are alike in that each solves sets of mass action and mass balance equations. They vary, however, in their choices of thermodynamic components and independent variables, and how effectively the number of independent variables has been reduced.

In this chapter we develop a description of the equilibrium state of a geochemical system in terms of the fewest possible variables and show how the resulting equations can be applied to calculate the equilibrium states of natural waters. We reserve for the next two chapters discussion of how these equations can be solved by using numerical techniques.

3.1 Thermodynamic Description of Equilibrium

To this point we have used a number of terms familiar to geochemists without giving the terms rigorous definitions. We have, for example, discussed thermodynamic components without considering their meaning in a strict sense. Now, as we begin to develop an equilibrium model, we will be more careful in our use of terminology. We will not, however, develop the basic equations of chemical thermodynamics, which are broadly known and clearly derived in a number of texts (as mentioned in Chapter 2).

3.1.1 Phases and Species

A geochemical system can be thought of as an assemblage of one or more phases of given bulk composition. A *phase* is a region of space that is physically distinct, mechanically separable, and homogeneous in its composition and properties. Phases are separated from one another by very thin regions known as *surfaces* over which properties and commonly composition change abruptly (e.g., Pitzer and Brewer, 1961; Nordstrom and Munoz, 1994).

The ancient categories of water, earth, and air persist in classifying the phases that make up geochemical systems. For purposes of constructing a geochemical model, we assume that our system will always contain a fluid phase composed of water and its dissolved constituents, and that it may include the phases of one or more minerals and be in contact with a gas phase. If the fluid phase occurs alone, the system is *homogeneous*; the system when composed of more than one phase is *heterogeneous*.

Species are the molecular entities, such as the gases CO_2 and O_2 in a gas, or the electrolytes Na^+ and SO_4^{--} in an aqueous solution, that exist within a phase. Species, unlike phases, do not have clearly identifiable boundaries. In addition, species may exist only for the most fleeting of moments. Arriving at a precise definition of what a species is, therefore, can be less than straightforward.

In aqueous solutions, geochemists generally recognize dissociated electrolytes and their complexes as species. For example, we can take Ca^{++} as a species in itself, rather than combined with its sphere of hydration as $Ca^{++} \cdot n\ H_2O$. Similarly, we can represent the neutral species of dissolved silica $SiO_2(aq)$ as H_4SiO_4, or silicic acid, just by adjusting our concept of the species' boundaries. In electrolyte brines, cations and anions occur so close together that the degree of complexing among ions, and hence the extent of a species in solution, is difficult to determine. Keeping in mind the unlikeliness of arriving at a completely unambiguous definition, we will define (following, e.g., Smith and Missen, 1982) a *species* as a chemical entity distinguishable from other entities by molecular formula and structure, and by the phase within which it occurs.

3.1.2 Components and the Basis

The overall composition of any system can be described in terms of a set of one or more chemical *components*. We can think of components as the ingredients in a recipe. A certain number of moles of each component go into making up a system, just as the amount of each ingredient is specified in a recipe. By combining the components, each in its specified mass, we reproduce the system's bulk composition.

Whereas species and phases exist as real entities that can be observed in nature, components are simply mathematical tools for describing composition. Expressed another way, a component's stoichiometry but not identity matters: water, ice, and steam serve equally well as component H_2O. Since a component needs no identity, it may be either fictive or a species or phase that actually exists in the system. When we express the composition of a fluid in terms of elements or the composition of a rock in terms of oxides, we do not imply that elemental sodium occurs in the fluid, or that calcium oxide is found in the rock. These are fictive components. If we want, we can invent components that exist nowhere in nature.

On the other hand, when we express the chemical analysis of a fluid in terms of the ions Na^+, Ca^{++}, HCO_3^-, and so on, we use a set of components with the stoichiometries of species that really appear in the fluid. In this case, the distinction between species and component is critical. The bicarbonate component, for example, is distributed among a number of species: HCO_3^-, CO_2(aq), CO_3^{--}, $CaHCO_3^+$, $NaCO_3^-$, etc. Hence, the number of moles of component HCO_3^- in the system would differ, perhaps by orders of magnitude, from that of the HCO_3^- species. Similarly, the other components would be distributed among the various species Na^+, NaCl, $CaSO_4$, and so on.

As a second example, we choose quartz (or any silica polymorph) as a component for a system containing an aqueous fluid and quartz. Now the mole number for the quartz component includes not only the silica in the quartz mineral, the real quartz, but the silica in solution in species such as SiO_2(aq) and $H_3SiO_4^-$. Again, the mole numbers of component quartz and real quartz are not the same. A common mistake in geochemical modeling is confusing the components used to describe the composition of a system with the species and phases that are actually present.

The set of components used in a geochemical model is the calculation's *basis*. The basis is the coordinate system chosen to describe composition of the overall system of interest, as well as the individual species and phases that make up the system (e.g., Greenwood, 1975). There is no single basis that describes a given system. Rather, the basis is chosen for convenience from among an infinite number of possibilities (e.g., Morel, 1983). Any useful basis can be selected, and the basis may be changed at any point in a calculation to a more convenient one. We discuss the choice of basis species in the next section.

3.1.3 Chemical Potentials, Activities, and Fugacities

The tools for calculating the equilibrium point of a chemical reaction arise from the definition of the chemical potential. If temperature and pressure are fixed, the equilibrium point of a reaction is the point at which the Gibbs free-energy function G is at its minimum (Fig. 3.1). As with any convex-upward function, finding the minimum G is a matter of determining the point at which its derivative vanishes.

To facilitate this analysis, we define the *chemical potential* μ of each species that makes up a phase. The chemical potential of a species B,

$$\mu_B \equiv \frac{\partial G_B}{\partial n_B}, \tag{3.1}$$

is the derivative of the species' free energy G_B with respect to its mole number n_B. The value of μ depends on temperature, pressure, and the mole numbers of each species in the phase. Since μ is defined as a partial derivative, we take its value holding constant each of these variables except n_B.

Knowing the chemical potential function for each species in a reaction defines the reaction's equilibrium point. Consider a hypothetical reaction,

$$bB + cC \rightleftarrows dD + eE, \tag{3.2}$$

among species B, C, etc., where b, c, and so on are the reaction coefficients. The free energy is at a minimum

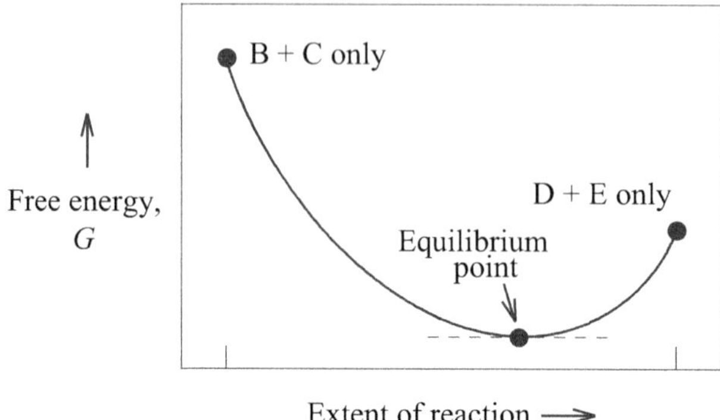

Figure 3.1 Variation in free energy G with reaction progress for the reaction $b\mathrm{B} + c\mathrm{C} \rightleftarrows d\mathrm{D} + e\mathrm{E}$. The reaction's equilibrium point is the minimum along the free-energy curve.

at the point where driving the reaction by a small amount forward or backward has no effect on G. From the definition of chemical potential (Eq. 3.1), the point of minimum G satisfies

$$d\mu_{\mathrm{D}} + e\mu_{\mathrm{E}} - b\mu_{\mathrm{B}} - c\mu_{\mathrm{C}} = 0\,, \tag{3.3}$$

since d moles of D and e moles of E are produced in the reaction for each b moles of B and c moles of C consumed.

We can find the reaction's equilibrium point from Equation 3.3 as soon as we know the form of the function representing chemical potential. The theory of ideal solutions (e.g., Pitzer and Brewer, 1961; Denbigh, 1981) holds that the chemical potential of a species can be calculated from the potential μ°_{B} of the species in its pure form at the temperature and pressure of interest. According to this result, a species' chemical potential is related to its standard potential by

$$\mu_{\mathrm{B}} = \mu^{\circ}_{\mathrm{B}} + RT_{\mathrm{K}}\,\ln X_{\mathrm{B}}\,. \tag{3.4}$$

Here, R is the gas constant, T_{K} is absolute temperature, and X_{B} is the mole fraction of B in the solution phase. Using this equation, we can calculate the equilibrium point of reactions in ideal systems directly from tabulated values of standard potentials μ°.

Unfortunately, phases of geochemical interest are not ideal. Also, aqueous species do not occur in a pure form, since their solubilities in water are limited, so a new choice for the standard state is required. For this reason, the chemical potentials of species in solution are expressed less directly (Stumm and Morgan, 1996, and Nordstrom and Munoz, 1994, for example, give complete discussions), although the form of the ideal solution equation (Eq. 3.4) is retained.

Aqueous Species

The chemical potential of an aqueous species A_i is given by

$$\mu_i = \mu^{\circ}_i + RT_{\mathrm{K}}\,\ln a_i\,. \tag{3.5}$$

The mole fraction X in the previous equation is replaced with a new unitless variable a_i, the species' *activity*. The standard potentials μ°_i are defined at a new standard state: a hypothetical one-molal solution of the species in which activity and molality are equal, and in which the species properties have been extrapolated to infinite dilution.

This choice of a standard state seems like impossible mental gymnastics, but it allows activity to follow a molal scale, so that in dilute solutions activity and molality—despite the fact that activity is unitless—are equivalent numerically. A species' molality m_i, the number of moles of the species per kilogram of solvent, is related to its activity by

$$a_i = \gamma_i m_i \,. \tag{3.6}$$

The constant of proportionality γ_i is the species' *activity coefficient*, which accounts for the nonideality of the aqueous solution. The species' activity coefficients approach unity in very dilute solutions,

$$\gamma_i \rightarrow 1 \quad \text{and} \quad a_i \rightarrow m_i \,, \tag{3.7}$$

so that the species' activities and molalities assume nearly equal values.

Minerals

Chemical potentials for the constituents of minerals are defined in a similar manner. All minerals contain substitutional impurities that affect their chemical properties. Impurities range from trace substitutions, as might be found in quartz, to widely varying fractions of the end-members of solid solutions series. Solid solutions of geologic significance include clay minerals, zeolites, and plagioclase feldspars, which are important components in most geochemical models.

The chemical potential of each end-member component of a mineral,

$$\mu_k = \mu_k^{\circ} + RT_{\mathrm{K}} \, \ln a_k \,, \tag{3.8}$$

is given in terms of a standard potential μ_k°, representing the end-member in pure form at the temperature and pressure of interest, and an activity a_k. A geochemical model constructed in the most general manner would account for the activities of all of the constituents in each stable solid solution.

Models can be constructed in this manner (e.g., Bourcier, 1985), but most modelers choose for practical reasons to consider only minerals of fixed composition. The data needed to calculate activities in even binary solid solutions are, for the most part, lacking at temperatures of interest. The solid solutions can sometimes be assumed to be ideal, so that activities equate to mole fractions, but in many cases this assumption leads to errors more severe than those produced by ignoring the solutions altogether. As well as this, there are several conceptual and theoretical problems (e.g., Glynn *et al.*, 1990) that increase the difficulty of incorporating solid solution theory into reaction modeling in a meaningful way.

In our models, we will consider only minerals of fixed composition. Each mineral, then, exists in its standard state, so that its chemical potential and standard potential are the same,

$$\mu_k = \mu_k^{\circ} \,, \tag{3.9}$$

and its activity is unity,

$$a_k = 1 \,. \tag{3.10}$$

This equality will allow us to eliminate the a_k terms from the governing equations. We will carry these variables through the mathematical development, however, so that the results can be readily extended to account for solid solutions, even though we will not apply them in this manner.

Gases

The chemical potential of a gas species,

$$\mu_m = \mu_m^o + RT_K \ln f_m \,, \tag{3.11}$$

is given in terms of a standard potential of the pure gas at 1 atm and the temperature of interest, and the gas's fugacity f_m. Fugacity is related to partial pressure,

$$f_m = \varphi_m P_m \,, \tag{3.12}$$

by a fugacity coefficient φ_m, as described in Section 3.3.8. At low pressures,

$$\varphi_m \rightarrow 1 \quad \text{and} \quad f_m \rightarrow P_m \,, \tag{3.13}$$

so that fugacity and partial pressure become numerically equivalent.

3.1.4 The Equilibrium Constant

The equilibrium constant expresses the point of minimum free energy for a chemical reaction, as set forth in Equation 3.3, in terms of the chemical potential functions above. The criterion for equilibrium becomes

$$d\mu_D^o + e\mu_E^o - b\mu_B^o - c\mu_C^o = -RT_K\,(d \ln a_D + e \ln a_E - b \ln a_B - c \ln a_C) \,, \tag{3.14}$$

when we substitute the chemical potential functions into Equation 3.3. (If the reaction involves a gas species, we would replace the appropriate activity with the gas fugacity.)

The left side of this equation is the reaction's standard free energy,

$$\Delta G^o = d\mu_D^o + e\mu_E^o - b\mu_B^o - c\mu_C^o \,. \tag{3.15}$$

The equilibrium constant is defined in terms of the standard free energy as

$$\ln K = -\frac{\Delta G^o}{RT_K} \,. \tag{3.16}$$

Equation 3.14 can be written

$$\ln K = d \ln a_D + e \ln a_E - b \ln a_B - c \ln a_C \,, \tag{3.17}$$

or, equivalently,

$$K = \frac{a_D^d\, a_E^e}{a_B^b\, a_C^c} \,, \tag{3.18}$$

which is the familiar mass action equation.

3.2 Choice of Basis

The first decision to be made in constructing a geochemical model is how to choose the basis, the set of thermodynamic components used to describe composition. Thermodynamics provides little guidance in our choice. Given this freedom, we choose a basis for convenience, subject to three rules:

- We must be able to form each species and phase considered in our model from some combination of the components in the basis.
- The number of components in the basis is the minimum necessary to satisfy the first rule.
- The components must be linearly independent of one another. In other words, we should not be able to write a balanced reaction to form one component in terms of the others.

The third rule is, in fact, a logical consequence of the first and second, but we write it out separately because it provides a useful test of a basis choice.

The way we select components to make up the basis is similar to the way a restaurant chef might decide what foodstuffs to buy. The chef needs to be able to prepare each item on the menu from a pantry of ingredients. For various reasons (to simplify ordering, account for limited storage, minimize costs, allow the menu to be changed from day to day, and keep the ingredients fresh), the chef keeps only the minimum number of ingredients on hand. Therefore, the pantry contains no ingredient that can be prepared from the other ingredients on hand. There is no need to store cake mix, since a mixture of flour, sugar, eggs, and so on, serves the same purpose. The chef chooses foodstuffs the same way we choose chemical components.

A straightforward way to choose a basis is to select elements as components. Accounting for redox reactions, the basis also includes the electron or some measure of oxidation state. Clearly, this choice satisfies the three rules mentioned, since any species or phase is composed of elements, and reactions converting one element to another are the stuff of alchemy or nuclear physics, both of which are beyond the scope of this book.

Such a straightforward choice, although commonly used, is seldom the most convenient way to formulate a geochemical model. The chef in our restaurant, if talented enough, could prepare any dish from such ingredients as elemental carbon, hydrogen, and oxygen, instead of flour, eggs, sugar, and so on. (But our analogy is not perfect, since there are more basic ingredients in a kitchen than chemical elements in the foodstuffs; the reader should not take it too literally.) Like the chef's work, our job gets easier if we pick as components certain species or phases that actually go into making up the system of interest.

3.2.1 Convention for Choosing the Basis

Throughout this book, we will choose the following species and phases as components:

- Water, the solvent species,
- Each mineral in equilibrium with the system of interest,
- Any gas species set at known fugacity or partial pressure in the external buffer, and
- Enough aqueous species, preferably those abundantly present in solution, to complete the basis set.

The aqueous species included in the basis are known as *basis species*, while the remaining species in solution comprise the set of *secondary species*.

This choice of basis follows naturally from the steps normally taken to study a geochemical reaction by hand. An aqueous geochemist balances a reaction between two species or minerals in terms of water, the minerals that would be formed or consumed during the reaction, any gases such as O_2 or CO_2 that remain at known fugacity as the reaction proceeds, and, as necessary, the predominant aqueous species in solution. We will show later that formalizing our basis choice in this way provides for a simple mathematical description of equilibrium in multicomponent systems and yields equations that can be evaluated rapidly.

Choosing the basis in this manner sometimes leads to some initial confusion, because we select species present in the system to serve as components. There is a risk of confusing the amount of a component, which describes bulk composition but not the actual state of the system, with the amount of a species or mineral that exists in reality.

3.2.2 Components With Negative Masses

Perhaps the most clear-cut distinction between components and species occurs when a component is present at negative mass. To see how this can occur, we return to our restaurant analogy. The dessert menu includes cakes, which contain whole eggs, and meringue pies made from egg whites. The chef could stock both egg yolks and whites in his pantry, but this would hardly be convenient. He would prefer to stock whole eggs and discard the yolks when necessary. His cake and meringue, then, contain a positive number of eggs, but the meringue contains negative egg yolks.

The same principle applies to a chemical system. Let us consider an alkaline water and assume a component set that includes H_2O and H^+. Each hydroxyl ion,

$$H_2O - H^+ \rightarrow OH^- , \tag{3.19}$$

is made up of a water molecule less a hydrogen ion. Since the solution contains more hydroxyl than hydrogen ions, the overall solution composition is described in terms of a positive amount of water component and a negative amount of the H^+ component. The molality of the H^+ species itself, of course, is positive.

As a second example, consider a solution rich in dissolved H_2S. If our basis includes SO_4^{--}, H^+, and O_2(aq), then H_2S(aq) is formed,

$$SO_4^{--} + 2\,H^+ - 2\,O_2(aq) \rightarrow H_2S(aq) , \tag{3.20}$$

from a negative amount of O_2(aq). The overall solution composition might well include a negative amount of O_2(aq) component, although the O_2(aq) species would be present at a small but positive concentration.

3.3 Governing Equations

At this point we can derive a set of governing equations that fully describes the equilibrium state of the geochemical system. To do this we will write the set of independent reactions that can occur among species, minerals, and gases in the system and set forth the mass action equation corresponding to each reaction. Then we will derive a mass balance equation for each chemical component in the system. Substituting the mass action equations into the mass balance equations gives a set of governing equations, one for each component, that can be solved directly for the system's equilibrium state.

3.3.1 Independent Reactions

To derive the governing equations we need to identify each independent chemical reaction that can occur in the system. It is possible to write many more reactions than are independent in a geochemical system. The

Table 3.1 *Constituents of a geochemical model: the cast of characters*

A_w	Water, the solvent
A_i	Aqueous species in the basis, the basis species
A_j	Other aqueous species, the secondary species
A_k	Minerals in the system
A_l	All minerals, even those that do not exist in the system
A_m	Gases of known fugacity
A_n	All gases

remaining or dependent reactions, however, are linear combinations of the independent reactions and need not be considered.

Since a geochemical model needs to be cast in general form, the species occurring in reactions are represented symbolically (Table 3.1). Depending on the nature of the problem, we have chosen a basis,

$$\mathbf{B} = (A_w,\ A_i,\ A_k,\ A_m)\ , \tag{3.21}$$

according to the convention in the previous section. Here, A_w is water, A_i are the aqueous species, A_k the minerals, and A_m the gases in the basis. These variables are labels rather than numerical values. For example, A_w is "H_2O" and A_i might be "Na^+".

The independent reactions are those between the secondary species and the basis. In general form, the reactions are

$$A_j \rightleftarrows \nu_{wj} A_w + \sum_i \nu_{ij} A_i + \sum_k \nu_{kj} A_k + \sum_m \nu_{mj} A_m\ . \tag{3.22}$$

Here, ν represents the reaction coefficients: ν_{wj} is the number of moles of water in the reaction to form A_j, ν_{ij} is the number of moles of the basis species A_i, and so on for the minerals and gases.

Need we consider any other reactions? From the previous section, according to the criteria for a valid basis set, we know that no reactions can be written among the basis species themselves. We could write reactions among the secondary species, but such reactions would not be independent. In other words, since we have already written Equation 3.22 specifying equilibrium between each secondary species and the basis, any reaction written among the secondary species is redundant (Fig. 3.2).

Taking a basis that contains H_2O, H^+, and HCO_3^-, for example, the reactions for secondary species $CO_2(aq)$ and CO_3^{--} are

$$CO_2(aq) \rightleftarrows HCO_3^- + H^+ - H_2O \tag{3.23}$$

and

$$CO_3^{--} \rightleftarrows HCO_3^- - H^+\ . \tag{3.24}$$

The relation between these two secondary species,

$$CO_2(aq) + H_2O \rightleftarrows CO_3^{--} + 2H^+\ , \tag{3.25}$$

is simply the first reaction above less the second, and so need not be considered independently.

It is fortunate that we do not have to consider the dependent reactions. Given N_j secondary species, there

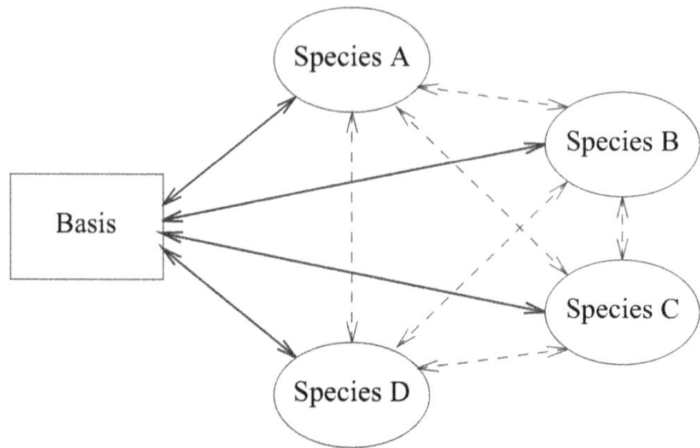

Figure 3.2 Independent (solid lines) and dependent (dashed lines) reactions in a chemical system composed of a basis and four secondary species A through D. Only the independent reactions need be considered.

are just N_j reactions with the basis, but $(N_j^2 - N_j)/2$ reactions could be written among the secondary species. The formula for the latter number, for example, is the number of handshakes if everyone in a group shook everyone else's hand. This is practical at a small party, but impossible at a convention. In chemical systems with many hundreds of species, taking the dependent reactions into account might tax even the most powerful computers.

It is worth noting that the Reaction 3.22 serves two purposes. First, it defines the compositions of all the species A_j in terms of the current component set, the basis **B**. Second, it represents the chemical reactions, each of which has its own equilibrium constant, between the secondary species A_j and the basis species A_w, A_i, A_k, and A_m.

If we had chosen to describe composition in terms of elements, we would need to carry the elemental compositions of all species, minerals, and gases, as well as the coefficients of the independent chemical reactions. Our choice of components, however, allows us to store only one array of reaction coefficients, thereby reducing memory use on the computer and simplifying the forms of the governing equations and their solution. In fact, it is possible to build a complete chemical model (excluding isotope fractionation) without acknowledging the existence of elements in the first place!

3.3.2 Mass Action Equations

Each independent Reaction 3.22 in the system has an associated equilibrium constant K_j at the temperature of interest and, hence, a mass action equation of the form,

$$K_j = \frac{a_w^{\nu_{wj}} \prod^{i} (\gamma_i m_i)^{\nu_{ij}} \prod^{k} a_k^{\nu_{kj}} \prod^{m} f_m^{\nu_{mj}}}{\gamma_j m_j}. \tag{3.26}$$

Here, we have represented the activities of aqueous species with the product γm of the species' activity coefficients and molalities, according to Equation 3.6. The symbol Π in this equation is the product function, the analog in multiplication to the summation Σ. Table 3.2 lists the meaning of each variable in this and following equations.

Table 3.2 *Variables in the governing equations*

Bulk composition, moles	
M_w	Water component
M_i	Species components
M_k	Mineral components
M_m	Gas components
Solvent mass, molalities, mole numbers	
n_w	Solvent mass, kg
m_i	Molalities of basis species
m_j	Molalities of secondary species
n_k	Mole numbers of minerals
Activities and fugacities	
a_w	Water activity
a_i	Activities of basis species
a_j	Activities of secondary species
a_k	Mineral activities
f_m	Gas fugacities
Activity coefficients	
γ_i	Basis species
γ_j	Secondary species
Reaction coefficients	
ν_{wj}, ν_{ij}, ν_{kj}	Secondary species
ν_{wl}, ν_{il}, ν_{kl}	Minerals
ν_{wn}, ν_{in}, ν_{kn}	Gases
Equilibrium constants	
K_j	Secondary species
K_l	Minerals
K_n	Gases

A goal in deriving the governing equations is to reduce the number of independent variables by eliminating the molalities m_j of the secondary species. To this end, we can rearrange the equation above to give the value of m_j,

$$m_j = \frac{1}{K_j \gamma_j} \left[a_w^{\nu_{wj}} \prod^{i} (\gamma_i m_i)^{\nu_{ij}} \prod^{k} a_k^{\nu_{kj}} \prod^{m} f_m^{\nu_{mj}} \right] , \tag{3.27}$$

in terms of the molality and activity coefficient of each aqueous species in the basis and the activity or fugacity of each of the other basis entries. This expression is the mass action equation in its final form.

3.3.3 Mass Balance Equations

The mass balance equations express conservation of mass in terms of the components in the basis. The mass of each chemical component is distributed among the species and minerals that make up the system. The water component, for example, is present in free water molecules of the solvent and as the water required to make up the secondary species. According to Equation 3.22, each mole of species A_j is composed of ν_{wj} moles of the water component. The mole number M_w of water component is given by

$$M_w = n_w \left(55.5 + \sum_j \nu_{wj} m_j \right) , \tag{3.28}$$

where 55.5 (more precisely, 55.5087) is the number of moles of H_2O in a kilogram of water. Multiplying molality units by the mass n_w of solvent water gives result in moles, as desired.

Similar logic gives the mass balance equations for the species components. The mass of the ith component is distributed among the single basis species A_i and the secondary species in the system. By Equation 3.22, there are ν_{ij} moles of component i in each mole of secondary species A_j. There is one mole of Na^+ component, for example, per mole of the basis species Na^+, one per mole of the ion pair NaCl, two per mole of the aqueous complex Na_2SO_4, and so on. Mass balance for species component i, then, is expressed

$$M_i = n_w \left(m_i + \sum_j \nu_{ij} m_j \right) , \tag{3.29}$$

in terms of the solvent mass n_w and the molalities m_i and m_j.

Mineral components are distributed among the mass of actual mineral in the system and the amount required to make up the dissolved species. In a system containing a mole of quartz, for example, there is (in the absence of other silica-bearing components) somewhat more than a mole of component quartz. The additional component mass is required to make up species such as SiO_2(aq) and $H_3SiO_4^-$. Since ν_{kj} moles of mineral component k go into making up each mole of secondary species j, mass balance is expressed as

$$M_k = n_k + n_w \sum_j \nu_{kj} m_j , \tag{3.30}$$

where n_k is the mole number of the mineral corresponding to the component.

Mass balance on gas components is somewhat less complicated because the gas buffer is external to the system. In this case, we need only consider the gas components that make up secondary species:

$$M_m = n_w \sum_j \nu_{mj} m_j , \tag{3.31}$$

where ν_{mj} is the reaction coefficient from Equation 3.22.

3.3.4 Substituted Equations

The final form of the governing equations is given by substituting the mass action equation (Eq. 3.27) for each occurrence of m_j in the mass balance equations (Eqs. 3.28–3.31). The substituted equations are

$$M_w = n_w \left\{ 55.5 + \sum_j \frac{\nu_{wj}}{K_j \gamma_j} \left[a_w^{\nu_{wj}} \prod^i (\gamma_i m_i)^{\nu_{ij}} \prod^k a_k^{\nu_{kj}} \prod^m f_m^{\nu_{mj}} \right] \right\} \quad (3.32)$$

$$M_i = n_w \left\{ m_i + \sum_j \frac{\nu_{ij}}{K_j \gamma_j} \left[a_w^{\nu_{wj}} \prod^i (\gamma_i m_i)^{\nu_{ij}} \prod^k a_k^{\nu_{kj}} \prod^m f_m^{\nu_{mj}} \right] \right\} \quad (3.33)$$

$$M_k = n_k + n_w \sum_j \frac{\nu_{kj}}{K_j \gamma_j} \left[a_w^{\nu_{wj}} \prod^i (\gamma_i m_i)^{\nu_{ij}} \prod^k a_k^{\nu_{kj}} \prod^m f_m^{\nu_{mj}} \right] \quad (3.34)$$

$$M_m = n_w \sum_j \frac{\nu_{mj}}{K_j \gamma_j} \left[a_w^{\nu_{wj}} \prod^i (\gamma_i m_i)^{\nu_{ij}} \prod^k a_k^{\nu_{kj}} \prod^m f_m^{\nu_{mj}} \right] . \quad (3.35)$$

Writing the appropriate governing equation for each chemical component produces a set of equations that completely describes the equilibrium state of the chemical system. As such, the set will include Equation 3.32 written once, Equation 3.33 written individually for each species component, and Equations 3.34 and 3.35 for each mineral and gas component.

Like all formulations of the multicomponent equilibrium problem, these equations are nonlinear by nature because the unknown variables appear in product functions raised to the values of the reaction coefficients. (Nonlinearity also enters the problem because of variation in the activity coefficients.) Such nonlinearity, which is an unfortunate fact of life in equilibrium analysis, arises from the differing forms of the mass action equations, which are product functions, and the mass balance equations, which appear as summations. The equations, however, occur in a straightforward form that can be evaluated numerically, as discussed in Chapter 4.

We have considered a large number of values (including the molality of each aqueous species, the mole number of each mineral, and the mass of solvent water) to describe the equilibrium state of a geochemical system. In Equations 3.32–3.35, however, this long list has given way to a much smaller number of values that constitute the set of independent variables. Since there is only one independent variable per chemical component, and hence per equation, we have succeeded in reducing the number of unknowns in the equation set to the minimum possible. In addition, Equation 3.34 is linear with respect to n_k and, as discussed below, Equation 3.35 need only be evaluated for M_m and hence is linear in its unknown. As we will discuss in the next chapter, the partial linearity of the governing equations leaves them especially easy to evaluate.

To see how the governing equations might be solved, we consider a system that contains an aqueous fluid and several minerals but has no gas buffer. If we know the system's bulk composition in terms of M_w, M_i, and M_k, we can evaluate Equations 3.32–3.34 to give values for the unknown variables: the solvent mass n_w, the basis species' molalities m_i, and the mineral mole numbers n_k.

The other variables in Equations 3.32–3.34 are either known values, such as the equilibrium constants K and reaction coefficients ν, or, in the case of the activity coefficients γ_i, γ_j and activities a_w, a_k, values that can be considered to be known. In practice, the model updates the activity coefficients and activities during the numerical solution so that their values have been accurately determined by the time the iterative procedure is complete.

In solving the equations, we can consider the set of bulk compositions (M_w, M_i, M_k) to be the "boundary conditions" from which we determine the system's equilibrium state. The result is given in terms of the values of (n_w, m_i, n_k). Once these values are known, the dependent variables m_j can be set immediately using Equation 3.27. Note that we have demonstrated the conjecture of the first chapter: that the equilibrium state of any system at known temperature and pressure can be calculated once the system's bulk composition is known.

It is commonly convenient, however, to apply some of the governing equations in the reverse manner. A modeler may specify the value for one or more of the variables n_w, m_i, or n_k that we considered independent in the previous paragraph. Such situations are quite common. The modeler may know the mass of solvent water or of the minerals in the system, or the molalities or activities of certain species. She may wish to constrain a_{H^+}, for example, on the basis of a pH measurement.

In these cases, the equation in question is evaluated to give the mole number M_w, and so on, of the corresponding component. In the presence of a gas buffer, the values of one or more fugacities f_m are fixed. Now, the mole number M_m of the gas component remains to be determined. In general, the value of either M_w or n_w needs to be set to evaluate Equation 3.32, and either M_i or m_i is required for each Equation 3.33. Each Equation 3.34 can be solved knowing either M_k or n_k, whereas Equation 3.35 is generally evaluated directly for M_m.

3.3.5 Charge Balance

The principle of electroneutrality requires that the ionic species in an electrolyte solution remain charge balanced on a macroscopic scale. The requirement of electroneutrality arises from the large amount of energy required to separate oppositely charged particles by any significant distance against coulombic forces (e.g., Denbigh, 1981). Because of this requirement, we cannot obtain a flask of sodium ions at the chemistry supply room, nor can we measure the activity coefficients of individual ions directly.

The electroneutrality condition can be expressed by the condition of charge balance among the species in solution, according to

$$\sum_i z_i m_i + \sum_j z_j m_j = 0\,. \qquad (3.36)$$

Here, z_i and z_j are the ionic charges on basis and secondary species. It is useful to note, however, that electroneutrality is assured when the components in the basis are charge balanced.

To see this, we can use Equation 3.22 to write the ionic charge on a secondary species,

$$z_j = \sum_i \nu_{ij} z_i\,, \qquad (3.37)$$

in terms of the charges on the basis species. Substituting, the electroneutrality condition becomes

$$\sum_i z_i \left(m_i + \sum_j \nu_{ij} m_j \right) = 0\,. \qquad (3.38)$$

According to the mass balance Equation 3.29, the expression in parentheses is M_i. Further, the charge Z_i on a species component is the same as the charge z_i on the corresponding basis species, since components and species share the same stoichiometry. Substituting, the electroneutrality condition becomes

$$\sum_i Z_i M_i = 0\,, \tag{3.39}$$

which requires that components be charge balanced.

This relation is useful because it effectively removes the requirement that M_i be known for one of the basis species. Instead of setting this value directly, it can be determined by balance from the mole numbers of the other basis species. When charged species appear in the basis, in fact, it is customary for equilibrium models to force charge balance by adjusting M_i for a component chosen either by the modeler or the computer program.

The electroneutrality condition is almost always used to set the bulk concentration of the species in abundant concentration for which the greatest analytic uncertainty exists. In practice, this component is generally Cl^- because most commercial labs, unless instructed otherwise, report a chloride concentration calculated by a rough charge balance (i.e., one that excludes the H^+ component and perhaps others) rather than a value resulting from direct analysis. If we were to use the reported chloride content to constrain by rigorous charge balance the concentration of another component, we would at best be propagating the error in the laboratory's rough calculation.

A special danger of the automatic implementation of the electroneutrality condition within computer codes is that the feature can be used to give calculated values for species in small concentration, pH, and even oxidation state. Such values are, however, almost always meaningless because they merely reflect analytical error or even the rough charge balance calculation made by the analytical lab. To see why such calculations fail, consider an attempt to back-calculate the pH of a neutral groundwater. Hydrogen ions are present at a concentration of about 10^{-7} molal. In order to resolve such a small concentration by charge balance, the analyses of all of the other components would have to be accurate to at least 10^{-8} molal, which is of course impossible. Even worse, if the lab reported a chloride concentration calculated to give the appearance of charge balance, then the computed pH would merely reflect the rounding error in the lab's calculation!

3.3.6 Mineral Saturation States

Once we have calculated the distribution of species in the fluid, we can determine the degree to which it is undersaturated or supersaturated with respect to the many minerals in the thermodynamic database. Only a few of the minerals can exist in equilibrium with the fluid, which is therefore undersaturated or supersaturated with respect to each of the rest. For any mineral A_l, we can write a reaction,

$$A_l \rightleftarrows \nu_{wl} A_w + \sum_i \nu_{il} A_i + \sum_k \nu_{kl} A_k + \sum_m \nu_{ml} A_m\,, \tag{3.40}$$

in terms of the basis. Here, A_l is a mineral that can be formed by combining components in the basis. We could not, for example, write a reaction to form muscovite in a system devoid of potassium.

Reaction 3.40 has an activity product Q_l in the form,

$$Q_l = \frac{a_w^{\nu_{wl}} \prod^i (\gamma_i m_i)^{\nu_{il}} \prod^k a_k^{\nu_{kl}} \prod^m f_m^{\nu_{ml}}}{a_l}\,, \tag{3.41}$$

where a_k and a_l have values of unity, since we are considering only minerals of fixed composition, and are carried as a formality. Since this equation has the same form as the mass action equation, the reaction is in equilibrium if Q_l equals the reaction's equilibrium constant K_l. In this case, the fluid is saturated with respect to A_l. As a test of our calculations, for example, we would expect the fluid to be saturated with respect to any mineral in the equilibrium system.

The fluid is undersaturated if Q_l is less than K_l. This condition indicates that Reaction 3.40 has not proceeded to the right far enough to reach the saturation point, either because the water has not been in contact with sufficient amounts of the mineral or because it has not reacted with the mineral long enough. Values of Q_l greater than K_l, on the other hand, indicate that the reaction needs to proceed to the left to reach equilibrium. In this case, the fluid is supersaturated with respect to the mineral.

A fluid's saturation with respect to a mineral A_l is commonly expressed in terms of the *saturation index*,

$$\mathrm{SI}_l = \log Q_l - \log K_l = \log(Q_l/K_l)\,, \tag{3.42}$$

which is the ratio of activity product to equilibrium constant, expressed as a logarithm. From this equation, an undersaturated mineral has a negative saturation index, a supersaturated mineral has a positive index, and a mineral at the point of saturation has an index of zero. A positive saturation index indicates that the calculated state of the system is a metastable equilibrium because of the thermodynamic drive for Reaction 3.40 to precipitate the supersaturated mineral.

3.3.7 Gas Fugacities

Having determined the distribution of species in solution, we can also calculate the fugacity of the various gases with respect to the fluid. For any gas A_n in the database that can be composed from the component set, we can write a reaction,

$$A_n \rightleftarrows \nu_{wn} A_w + \sum_i \nu_{in} A_i + \sum_k \nu_{kn} A_k + \sum_m \nu_{mn} A_m\,, \tag{3.43}$$

in terms of the basis. By mass action, the fugacity f_n of this gas is

$$f_n = \frac{a_w^{\nu_{wn}} \prod^{i} (\gamma_i m_i)^{\nu_{in}} \prod^{k} a_k^{\nu_{kn}} \prod^{m} f_m^{\nu_{mn}}}{K_n}\,, \tag{3.44}$$

where K_n is the equilibrium constant for Reaction 3.43.

The fugacities calculated in this way are those that would be found in a gas phase that is in equilibrium with the system, if such a gas phase were to exist. Whether a gas phase exists or is strictly hypothetical depends on how the modeler has defined the system, but not on the gas fugacities given by Equation 3.44.

3.3.8 Partial Pressure

The partial pressure P_n of a gas A_n is the ratio f_n/φ_n of its fugacity to fugacity coefficient, according to Equation 3.12. The fluid's vapor pressure P_v is the sum of the partial pressures of the dissolved gases,

$$P_\mathrm{v} = \sum_n P_n = \sum_n \frac{f_n}{\varphi_n}\,. \tag{3.45}$$

The fugacity coefficients, then, allow us to determine the partial pressures of gases in the system, as well as the vapor pressure.

Geochemical modeling programs in common use employ one of several methods described below to evaluate fugacity coefficients as functions of temperature and pressure; the GWB programs (Bethke *et al.*, 2021) allow the user to choose from among the three.

Tsonopoulos

The Tsonopoulos method (Tsonopoulos, 1974; Tsonopoulos and Heidman, 1990) is formulated in terms of a gas's critical pressure P_c (bar), critical temperature T_c (K), and acentric factor ω. Two additional factors can be carried for polar gases: a to account for the role of intermolecular attraction in decreasing partial pressure and b to represent irreducible molecular volume, when the gas is taken to be nonideal.

In the method, the ratio of a gas's second virial coefficient B (liter mol^{-1}) to the gas constant R (0.08314 liter bar $\text{K}^{-1}\text{mol}^{-1}$) is given by

$$\frac{B}{R} = \left(f^{(0)} + \omega f^{(1)} + f^{(2)}\right)\frac{T_c}{P_c}, \tag{3.46}$$

where

$$f^{(0)} = 0.1445 - \frac{0.330}{T_r} - \frac{0.1385}{T_r^2} - \frac{0.0121}{T_r^3} - \frac{0.000607}{T_r^8} \tag{3.47}$$

$$f^{(1)} = 0.0637 + \frac{0.331}{T_r^2} - \frac{0.423}{T_r^3} - \frac{0.008}{T_r^8}, \tag{3.48}$$

and, for a polar gas,

$$f^{(2)} = \frac{a}{T_r^6} - \frac{b}{T_r^8}. \tag{3.49}$$

Here, T_r is reduced temperature, the ratio T_K/T_c of absolute temperature (K) to the gas's critical temperature. The gas's fugacity coefficient is then given,

$$\ln \varphi_n = \frac{BP}{RT_K}, \tag{3.50}$$

at the pressure P (bar) and temperature of interest. Table 3.3 shows values of P_c, T_c, ω, a, and b for different gases.

Peng–Robinson

In the Peng–Robinson method (Peng and Robinson, 1976), as with Tsonopoulos' method, a gas's fugacity coefficient follows from its critical pressure P_c, critical temperature T_c, and acentric factor ω, listed in Table 3.3. The method is based upon the equation of state,

$$P = \frac{RT_K}{V_m - b} - \frac{a}{V_m^2 + 2\,V_m b - b^2}, \tag{3.51}$$

for a pure gas, where R is the gas constant (83.14 bar cm^3 K^{-1} mol^{-1}), V_m is real molar volume (cm^3 mol^{-1}), and a (bar cm^6 mol^{-2}) and b (cm^3 mol^{-1}) again represent intermolecular attraction and irreducible size, respectively. The value of a is given by

$$a = \frac{0.45724\,(RT_c)^2}{P_c}\left(1 + \kappa\left(1 - \sqrt{T_r}\right)\right)^2, \tag{3.52}$$

Table 3.3 *Parameters for calculating fugacity coefficients using the Tsonopoulos and Peng–Robinson methods, from Poling* et al. *(2001) and Tsonopoulos and Heidman (1990)*

	P_c (bar)	T_c (K)	ω	a[1]	b[1]
Ar(g)	48.98	150.86	−0.002	—	—
CH_4(g)	45.99	190.56	0.011	—	—
CO_2(g)	73.74	304.12	0.225	—	—
H_2(g)	12.93	32.98	−0.217	—	—
H_2O(g)	220.64	647.14	0.344	−0.0109	0.0
H_2S(g)	89.63	373.40	0.090	—	—
He(g)	2.27	5.19	−0.390	—	—
N_2(g)	33.98	126.20	0.037	—	—
NH_3(g)	113.53	405.40	0.257	—	—
O_2(g)	50.43	154.58	0.025	—	—
SO_2(g)	78.84	430.80	0.256	—	—

[1]Tsonopoulos method only.

where T_r is again reduced temperature and κ for all gases follows the correlation,

$$\kappa = 0.37464 + 1.54226\,\omega - 0.26992\,\omega^2\,. \tag{3.53}$$

Knowing a and noting that $b = 0.07780\,RT_c/P_c$, V_m can be calculated by using Cardano's method (O'Connell and Haile, 2005) to find the cubic root of Equation 3.51 at the pressure and temperature of interest.

To apply the method to a gas mixture, we need the sums

$$a_{\text{mix},i} = \sum_j x_j a_{ij} \quad \text{and} \quad a_{\text{mix}} = \sum_i \sum_j x_i x_j a_{ij} \tag{3.54}$$

over the binary interaction coefficients a_{ij} for each gas pairing. The mole fraction x_i of each gas is the ratio P_i/P_v of its partial pressure to vapor pressure, and the binary terms are found,

$$a_{ij} = \sqrt{a_i a_j}\left(1 - \delta_{ij}\right), \tag{3.55}$$

from the intermolecular attraction term a_i for each gas. Parameters $\delta_{H_2O-CO_2} = 0.1896$, $\delta_{H_2O-H_2S} = 0.1903 - 0.05965\,T_{r,H_2S}$, $\delta_{H_2O-CH_4} = 0.4850$, and $\delta_{H_2O-N_2} = 0.4778$ describe interaction between the more abundant gases and water molecules (Søreide and Whitson, 1992); for other pairings, δ_{ij} is zero. A further sum,

$$b_{\text{mix}} = \sum_i x_i b_i\,, \tag{3.56}$$

spans the irreducible sizes b_i of the component gases.

The fugacity coefficient φ_n for a gas A_n is then given by

$$\ln \varphi_n = \frac{b_n}{b_{\text{mix}}}(Z-1) - \ln(Z-B) - \frac{A}{2\sqrt{2}B}\left(\frac{2a_{\text{mix},n}}{a_{\text{mix}}} - \frac{b_n}{b_{\text{mix}}}\right)\ln\left(\frac{Z+2.414B}{Z-0.414B}\right), \tag{3.57}$$

where the gas's compressibility factor $Z = PV_m/RT_K$ follows from its real molar volume, $A = a_{mix}P/(RT_K)^2$, and $B = b_{mix}P/RT_K$. Appelo *et al.* (2014) implemented the Peng–Robinson method within the PHREEQC software.

Spycher–Reed

Spycher and Reed (1988) represented the fugacity coefficient of a pure gas as

$$\ln \varphi_n = \left(\frac{a}{T_K^2} + \frac{b}{T_K} + c\right) P + \left(\frac{d}{T_K^2} + \frac{e}{T_K} + f\right) \frac{P^2}{2}, \tag{3.58}$$

in terms of a set of coefficients a–f,

	a	b	$c \times 10^5$	$d \times 10^2$	$e \times 10^5$	$f \times 10^8$
$CH_4(g)$	−537.779	1.54946	−92.7827	120.861	−370.814	333.804
$CO_2(g)$	−1 430.87	3.59800	−227.376	347.644	−1 042.47	846.271
$H_2(g)$	−12.5908	0.259789	−7.2473	0.471947	−2.69962	2.15622
$H_2O(g)$	−6 191.41	14.8528	−914.267	−6 633.26	18 277	−13 274

specific to the gas in question. The coefficients are valid in the range 50 °C–340 °C, up to the vapor pressure of water. Additional terms are required to represent nonideal mixing, but Wolery (1995) implemented Equation 3.58 directly in EQ3/EQ6, noting the mixing correction is significant only at very high pressure.

3.3.9 Electrical Conductivity

The electrical conductivity of an electrolyte solution reflects which cations and anions appear in the fluid, in what abundances, and how mobile each is in its aqueous milieu. Conductivity is easily measured, commonly reported as part of a chemical analysis, and quite sensitive to the presence of complex species. Comparing measured conductivity to a calculated estimate, then, we can readily appraise the validity of the species distribution predicted for a natural water (e.g., Rice *et al.*, 2017).

A number of methods for calculating conductivity have been proposed, but few encompass the ranges in electrolyte composition, concentration, pH, and temperature important to the geochemist (e.g., Pawlowicz, 2008). As well, most cannot account explicitly for the effect of ion pairs and other complex species in solution, and hence are poorly suited for integrating into a geochemical modeling code.

Addressing this issue, McCleskey *et al.* (2012) developed a technique for estimating conductivity directly from a calculated species distribution. In their method, electrical conductivity k_{25} (μS cm^{-1}) referenced to 25 °C is given as a function of the measurement temperature T (°C) and species concentration m_i (molal) as

$$k_{25} = \frac{1000}{1 + \alpha(T - 25\,°\mathrm{C})} \sum_i \left(\lambda_i^{\circ} - \frac{a_i\sqrt{I}}{1 + b_i\sqrt{I}}\right) m_i\,. \tag{3.59}$$

In the equation, α is the temperature compensation factor,

$$\alpha = 3.0 \times 10^{-7} T^2 + 5.76 \times 10^{-5} T + 0.0193\,, \tag{3.60}$$

(ISO, 1985) and I (molal) is ionic strength, where $I = 1/2 \sum_i m_i z_i^2$ and z_i is ion charge. Several terms are specific to individual species in solution: λ_i° is molal ionic conductivity,

$$\lambda_i^{\circ} = \lambda_i^{(0)} T^2 + \lambda_i^{(1)} T + \lambda_i^{(2)}\,, \tag{3.61}$$

(μS kg cm^{-1}mol^{-1}) at infinite dilution, a_i varies with temperature according to,

$$a_i = a_i^{(0)} T^2 + a_i^{(1)} T + a_i^{(2)} , \tag{3.62}$$

and b_i is an empirical constant; McCleskey *et al.* give the data needed to evaluate these terms for species of interest in their Table 1. The method is parameterized from 0 °C to 95 °C, at ionic strengths as high as 1 molal.

3.3.10 The pe and Eh

The pe and Eh are equivalent electrochemical descriptions of oxidation state for a system in equilibrium. For an aqueous solution, any half-cell reaction

$$n\,\mathrm{e}^- \rightleftarrows \nu_{wn} A_w + \sum_i \nu_{in} A_i + \sum_k \nu_{kn} A_k + \sum_m \nu_{mn} A_m , \tag{3.63}$$

where e$^-$ is the electron and n its reaction coefficient, sets the pe and Eh. The electron, of course, does not exist as a free species in solution (e.g., Thorstenson, 1984) and so has no concentration. Species in the fluid can donate and accept electrons from a metallic electrode, however, so we can define and measure the electron's free energy and equilibrium activity (Hostettler, 1984). For example, the reaction

$$\mathrm{e}^- + 1/4\,\mathrm{O_2(aq)} + \mathrm{H}^+ \rightleftarrows 1/2\,\mathrm{H_2O} , \tag{3.64}$$

which has a log equilibrium constant K_{e^-} of about 21.5 at 25 °C, fixes the equilibrium electron activity when pH and the oxygen and water activities are known.

The pe, by analogy to pH, is defined as

$$\begin{aligned} \mathrm{pe} &= -\log a_{\mathrm{e}^-} \\ &= -\frac{1}{n} \log \frac{Q_{\mathrm{e}^-}}{K_{\mathrm{e}^-}} , \end{aligned} \tag{3.65}$$

where n is the number of electrons consumed in the half-cell reaction and Q_{e^-} is the activity product for the half-cell reaction, calculated accounting for each species except the electron. The analogy to pH is imperfect because whereas H$^+$ is a species that exists in solution, e$^-$ does not. The Nernst equation,

$$\begin{aligned} \mathrm{Eh} &= -\frac{2.303\, R T_{\mathrm{K}}}{nF} \log \frac{Q_{\mathrm{e}^-}}{K_{\mathrm{e}^-}} \\ &= \frac{2.303\, R T_{\mathrm{K}}}{F}\, \mathrm{pe} , \end{aligned} \tag{3.66}$$

gives the Eh value corresponding to any half-cell reaction (e.g., Cicconi *et al.*, 2020). Here, R is the gas constant, T_{K} absolute temperature, and F the Faraday constant.

Many natural waters, including most waters at low temperature, do not achieve redox equilibrium (e.g., Lindberg and Runnells, 1984; see Chapter 7). In this case, no single value of pe or Eh can be used to represent the redox state. Instead, there is a distinct value for each redox couple in the system. Applying the Nernst equation to Reaction 3.64 gives a pe or Eh representing the electrolysis of water. Under disequilibrium conditions, this value differs from those calculated from reactions such as

$$8\,\mathrm{e}^- + \mathrm{SO_4^{--}} + 9\,\mathrm{H}^+ \rightleftarrows \mathrm{HS}^- + 4\,\mathrm{H_2O} \tag{3.67}$$

and

$$\mathrm{e^-} + \underset{goethite}{\mathrm{FeOOH}} + 3\,\mathrm{H^+} \rightleftarrows \mathrm{Fe^{++}} + 2\,\mathrm{H_2O}\,, \tag{3.68}$$

which represent redox couples for sulfur and iron. The variation among the resulting values of pe or Eh provides a measure of the extent of disequilibrium in a system. Techniques for modeling waters in redox disequilibrium are discussed in Chapter 7.

3.4 Number of Variables and the Phase Rule

The most broadly recognized theorem of chemical thermodynamics is probably the phase rule derived by Gibbs in 1875 (see Guggenheim, 1967; Denbigh, 1981). Gibbs' phase rule defines the number of pieces of information needed to determine the state, but not the extent, of a chemical system at equilibrium. The result is the number of degrees of freedom N_F possessed by the system.

The phase rule says that for each phase beyond the first that occurs at equilibrium in a system, N_F decreases by one. Expressed in general form, the phase rule is

$$N_\mathrm{F} = N_\mathrm{C} - N_\phi + 2\,, \tag{3.69}$$

where N_C is the number of chemical components in the system, and N_ϕ is the number of phases. If temperature and pressure in the system are fixed (i.e., they have equilibrated with some external medium), as we have assumed here, the rule takes the simplified form

$$N_\mathrm{F} = N_\mathrm{C} - N_\phi\,. \tag{3.70}$$

The proof of the phase rule is actually implicit in the derivation of the governing equations (Eqs. 3.32–3.35), and is not repeated here. It is interesting, nonetheless, to compare this well-known result with the governing equations, if only to demonstrate that we have reduced the problem to the minimum number of independent variables.

The number of components in our geochemical system is given by

$$N_\mathrm{C} = 1 + N_i + N_k + N_m\,, \tag{3.71}$$

where 1 accounts for water, N_i is the number of aqueous species serving as components (the basis species), and N_k and N_m are the numbers of mineral and gas components. Phases in the system include the fluid, each mineral, and each gas at known fugacity or partial pressure, so

$$N_\phi = 1 + N_k + N_m\,. \tag{3.72}$$

Since the gases are buffered independently, each counts as a separate phase.

The phase rule (Eq. 3.70), then, predicts that our system has $N_\mathrm{F} = N_i$ degrees of freedom. In other words, given a constraint on the concentration or activity of each basis species, we could determine the system's equilibrium state. To constrain the governing equations, however, we need N_C pieces of information, somewhat more than the degrees of freedom predicted by the phase rule.

The extra pieces of information describe the extent of the system – the amounts of fluid and minerals that are present. It is not necessary to know the system's extent to determine its equilibrium state, but in reaction modeling (see Chapter 14) we generally want to track the masses of solution and minerals in the system; we also must know these masses to search for the system's stable phase assemblage (as described in Section 4.4).

Providing an additional piece of information about the size of each phase predicts that a total of $N_i + N_\phi$, or N_C, values is needed to constrain the system's state and extent. This total matches the number of variables we must supply in order to solve the governing equations. Hence, although we can make no claim that we have cast the governing equations in simplest form, we can say that we have reduced the number of independent variables to the minimum allowed by thermodynamics.

4

Solving for the Equilibrium State

In Chapter 3, we developed equations that govern the equilibrium state of an aqueous fluid and coexisting minerals. The principal unknowns in these equations are the mass of water n_w, the concentrations m_i of the basis species, and the mole numbers n_k of the minerals.

If the governing equations were linear in these unknowns, we could solve them directly using linear algebra. However, some of the unknowns in these equations appear raised to exponents and multiplied by each other, so the equations are nonlinear. Chemists have devised a number of numerical methods to solve such equations (e.g., van Zeggeren and Storey, 1970; Smith and Missen, 1982). All the techniques are iterative and, except for the simplest chemical systems, require a computer. The methods include optimization by steepest descent (White *et al.*, 1958; Boynton, 1960) and gradient descent (White, 1967), back substitution (Kharaka and Barnes, 1973; Truesdell and Jones, 1974), and progressive narrowing of the range of the values allowed for each variable (the monotone sequence method; Wolery and Walters, 1975).

Geochemists, however, seem to have reached a consensus (e.g., Karpov and Kaz'min, 1972; Morel and Morgan, 1972; Crerar, 1975; Reed, 1982; Wolery, 1983) that Newton–Raphson iteration is the most powerful and reliable approach, especially in systems where mass is distributed over minerals as well as dissolved species. In this chapter, we consider the special difficulties posed by the nonlinear forms of the governing equations and discuss how the Newton–Raphson method can be used in geochemical modeling to solve the equations rapidly and reliably.

4.1 Governing Equations

The governing equations are composed of two parts: mass balance equations that require mass to be conserved, and mass action equations that prescribe chemical equilibrium among species and minerals. Water A_w, a set of species A_i, the minerals in the system A_k, and any gases A_m of known fugacity or partial pressure make up the basis $\mathbf{B}$,

$$\mathbf{B} = (A_w, A_i, A_k, A_m) \; . \tag{4.1}$$

The remaining aqueous species are related to the basis entries by the reaction

$$A_j = \nu_{wj} A_w + \sum_i \nu_{ij} A_i + \sum_k \nu_{kj} A_k + \sum_m \nu_{mj} A_m \, , \tag{4.2}$$

which has an equilibrium constant K_j.

The mass balance equations corresponding to the basis entries are

$$M_w = n_w\left(55.5 + \sum_j \nu_{wj} m_j\right) \tag{4.3}$$

$$M_i = n_w\left(m_i + \sum_j \nu_{ij} m_j\right) \tag{4.4}$$

$$M_k = n_k + \sum_j \nu_{kj} m_j \tag{4.5}$$

$$M_m = n_w \sum_j \nu_{mj} m_j \,. \tag{4.6}$$

Here, M_w, M_i, M_k, and M_m give the system's composition in terms of the basis **B**, and m_j is the concentration of each secondary species A_j.

At equilibrium, m_j is given by the mass action equation,

$$m_j = \frac{a_w^{\nu_{wj}} \prod^{i}(\gamma_i m_i)^{\nu_{ij}} \prod^{k} a_k^{\nu_{kj}} \prod^{m} f_m^{\nu_{mj}}}{K_j \gamma_j} \,. \tag{4.7}$$

Here the a_w and a_k are the activities of water and minerals, and the f_m are gas fugacities. We assume that each a_k equals one, and that a_w and the species' activity coefficients γ can be evaluated over the course of the iteration and thus can be treated as known values in posing the problem.

By substituting this equation for each occurrence of m_j in the mass balance equations, we find a set of equations, one for each basis entry, that describes the equilibrium state in terms of the principal variables. The form of the substituted equations appears in Chapter 3 (Eqs. 3.32–3.35), but in this chapter we will carry the variable m_j with the understanding that it represents the result of evaluating Equation 4.7.

In some cases, we set one or more of the values of the principal variables as constraints on the system. For example, specifying pH sets m_{H^+}, and setting the mass of solvent water or quartz in the system fixes n_w or n_{qtz}. To set the bulk compositions M_w and so on, in these cases, we need only evaluate the corresponding equations after the values of the other variables have been determined. Gases appear in the basis as a constraint on fugacity f_m, so Equation 4.6 is always evaluated in this manner.

We pose the problem for the remaining equations by specifying the total mole numbers M_w, M_i, and M_k of the basis entries. Our task in this case is to solve the equations for the values of n_w, m_i, and n_k. The solution is more difficult now because the unknown values appear raised to their reaction coefficients and multiplied by each other in the mass action Equation 4.7. In the next two sections we discuss how such nonlinear equations can be solved numerically.

4.2 Solving Nonlinear Equations

There is no general method for finding the solution of nonlinear equations directly. Instead, such problems need to be solved indirectly by iteration. The set of values that satisfies a group of equations is called the group's *root*. An iterative solution begins with a guess of the root's value, which the solution procedure tries to improve incrementally until the guess satisfies the governing equations to the desired accuracy.

Of such schemes, two of the most robust and powerful are Newton's method for solving an equation with

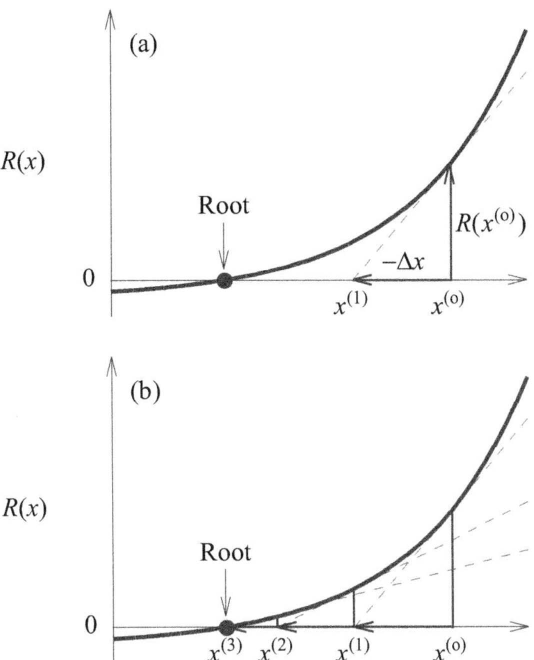

Figure 4.1 Newton's method for solving a nonlinear equation with one unknown variable. The solution, or root, is the value of x at which the residual function $R(x)$ crosses zero. In (a), given an initial guess $x^{(o)}$, projecting the tangent to the residual curve to zero gives an improved guess $x^{(1)}$. By repeating this operation (b), the iteration approaches the root.

one unknown variable, and Newton–Raphson iteration, which treats systems of equations in more than one unknown. I will briefly describe these methods here before I approach the solution of chemical problems. Further details can be found in a number of texts on numerical analysis, such as Carnahan *et al.* (1969).

4.2.1 Newton's Method

In Newton's method, we seek a value of x that satisfies

$$f(x) = a\,, \tag{4.8}$$

where f is an arbitrary function that we can differentiate, and a is a constant. To start the iteration, we provide a guess $x^{(o)}$ at the value of the root. Unless we are incredibly lucky (or have picked an easy problem that we have already solved), our guess will not satisfy the equation. The inequality between the two sides of the equation is the *residual*,

$$R(x) = f(x) - a\,. \tag{4.9}$$

We can think of the residual as a measure of the "badness" of our guess.

The method's goal is to make the residual vanish by successively improving our guess. To find an improved value $x^{(1)}$, we take the tangent line to the residual function at point $x^{(o)}$ and project it to the zero line (Fig. 4.1). We repeat the projection from $x^{(1)}$ to give $x^{(2)}$, and so on. The process continues until we reach a value $x^{(q)}$ on the qth iteration that satisfies our equation to within a small tolerance.

The method can be expressed mathematically by noting that the slope of the residual function plotted against x is $\mathrm{d}f/\mathrm{d}x$. Geometrically, the slope is rise over run, so

$$\frac{\mathrm{d}f}{\mathrm{d}x} = -\frac{R(x^{(q)})}{x^{(q+1)} - x^{(q)}} = -\frac{R(x^{(q)})}{\Delta x} \,. \tag{4.10}$$

At any iteration (q), then, the correction Δx is

$$\Delta x = -\frac{R(x^{(q)})}{\mathrm{d}f/\mathrm{d}x} \,. \tag{4.11}$$

As an example, we seek a root to the function

$$2x^3 - x^2 = 1 \,, \tag{4.12}$$

which we can see is satisfied by $x = 1$. We iterate from an initial guess of $x = 10$ following the Fortran program

```
implicit real*8 (a-h,o-z)
x = 10.

do iter = 1, 99
   resid = 2.*x**3 - x**2 - 1.
   write (6,*) iter-1, x, resid
   if (abs(resid).lt.1.d-10) stop
   dfdx = 6.*x**2 - 2.*x
   x = x - resid/dfdx
end do

end
```

with the results

Iteration, (q)	$x^{(q)}$	$R(x^{(q)})$
0	10	1900
1	6.72	560
2	4.55	170
3	3.10	49
4	2.15	14
5	1.54	3.9
6	1.19	0.94
7	1.033	0.14
8	1.001 3	5.2×10^{-3}
9	1.000 002 1	8.5×10^{-6}
10	1.000 000 000 005 7	2.3×10^{-11}

Some words of caution are in order. Many nonlinear functions have more than one root. The choice of an initial guess controls which root will be identified by the iteration. Also, the method is likely to work only when the function is somewhat regular between initial guess and root. A rippled function, for example, would

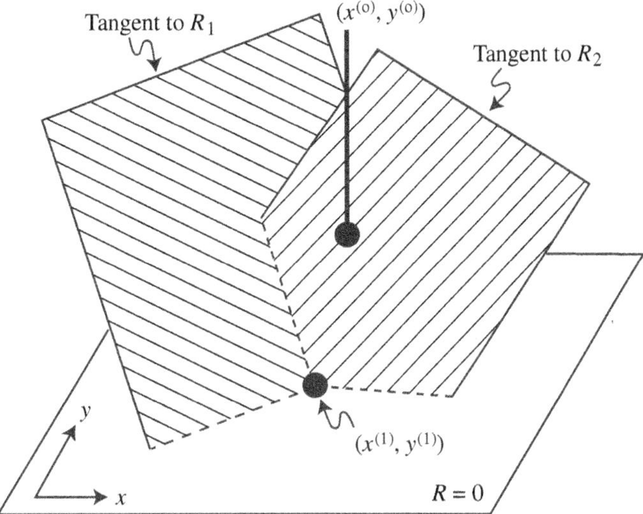

Figure 4.2 Newton–Raphson iteration for solving two nonlinear equations containing the unknown variables x and y. Planes are drawn tangent to the residual functions R_1 and R_2 at an initial estimate $(x^{(o)}, y^{(o)})$ of the value of the root. The improved guess $(x^{(1)}, y^{(1)})$ is the point at which the tangent planes intersect each other and the plane $R = 0$.

produce tangent lines that project along various positive and negative slopes. Iteration in this case might never locate a root. In fact, Newton's method can diverge or cycle indefinitely, given a poorly chosen function or initial guess.

4.2.2 Newton–Raphson Iteration

The multidimensional counterpart to Newton's method is Newton–Raphson iteration. A mathematics professor once complained to me, with apparent sincerity, that he could visualize surfaces in no more than twelve dimensions. My perspective on hyperspace is less incisive, as perhaps is the reader's, so we will consider first a system of two nonlinear equations $f = a$ and $g = b$ with unknowns x and y.

To solve the equations, we want to find x and y such that the residual functions

$$\begin{aligned} R_1(x, y) &= f(x, y) - a \\ R_2(x, y) &= g(x, y) - b \end{aligned} \tag{4.13}$$

nearly vanish. Imagine that x and y lie along a table top, and z is normal to it with the table surface representing z of zero. Plotting the values of R_1 and R_2 along z produces two surfaces that might extend above and below the table surface. The surfaces intersect along a curved line that, if the problem has a solution, passes through the table top at one or more points. Each such point is a root (x, y) to our problem.

To improve an initial guess $(x^{(o)}, y^{(o)})$, we reach above this point and project tangent planes (Fig. 4.2) from the surfaces of R_1 and R_2. The improved guess is the point $(x^{(1)}, y^{(1)})$ where these tangent planes intersect each other and the table top. We repeat this process until each residual function is less than a negligible value.

The corrections Δx and Δy to x and y are those that will project to $R = 0$ along the tangent planes, according to

$$\begin{aligned} \frac{\partial R_1}{\partial x}\Delta x + \frac{\partial R_1}{\partial y}\Delta y &= -R_1 \\ \frac{\partial R_2}{\partial x}\Delta x + \frac{\partial R_2}{\partial y}\Delta y &= -R_2 \,. \end{aligned} \tag{4.14}$$

Written in matrix form,

$$\begin{pmatrix} \dfrac{\partial R_1}{\partial x} & \dfrac{\partial R_1}{\partial y} \\ \dfrac{\partial R_2}{\partial x} & \dfrac{\partial R_2}{\partial y} \end{pmatrix} \begin{pmatrix} \Delta x \\ \Delta y \end{pmatrix} = \begin{pmatrix} -R_1 \\ -R_2 \end{pmatrix} , \tag{4.15}$$

the corrections Δx and Δy are given by the solution to two linear equations. This equation can be seen to be the counterpart in two dimensions to the equation (Eq. 4.11) giving Δx in Newton's method. The equations differ in that there are now vectors of values for the residuals and corrections, and a matrix of partial derivatives replacing the ordinary derivative $\mathrm{d}f/\mathrm{d}x$.

In general, the matrix, known as the *Jacobian*, contains entries for the partial derivative of each residual function R_i with respect to each unknown variable x_i. For a system of n equations in n unknowns, the Jacobian is an $n \times n$ matrix with n^2 entries,

$$(\mathbf{J}) \equiv \begin{pmatrix} \dfrac{\partial R_1}{\partial x_1} & \dfrac{\partial R_1}{\partial x_2} & \cdots & \dfrac{\partial R_1}{\partial x_n} \\ \dfrac{\partial R_2}{\partial x_1} & \dfrac{\partial R_2}{\partial x_2} & \cdots & \dfrac{\partial R_2}{\partial x_n} \\ \vdots & \vdots & \ddots & \vdots \\ \dfrac{\partial R_n}{\partial x_1} & \dfrac{\partial R_n}{\partial x_2} & \cdots & \dfrac{\partial R_n}{\partial x_n} \end{pmatrix} . \tag{4.16}$$

Writing the residual functions and corrections as vectors,

$$(\mathbf{R}) = (R_1, R_2, \ldots, R_n) \tag{4.17}$$

$$(\Delta\mathbf{x}) = (\Delta x_1, \Delta x_2, \ldots, \Delta x_n) \ , \tag{4.18}$$

gives the general equation for determining the correction,

$$(\mathbf{J})\,(\Delta\mathbf{x}) = -\,(\mathbf{R}) \ , \tag{4.19}$$

in a Newton–Raphson iteration.

Like Newton's method, the Newton–Raphson procedure has just a few steps. Given an estimate of the root to a system of equations, we calculate the residual for each equation. We check to see if each residual is negligibly small. If not, we calculate the Jacobian matrix and solve the linear Equation 4.19 for the correction vector. We update the estimated root with the correction vector,

$$(\mathbf{x})^{(q+1)} = (\mathbf{x})^{(q)} + (\Delta\mathbf{x}) \ , \tag{4.20}$$

and return to calculate new values for residuals. The solution procedure, then, reduces to the repetitive solution of a system of linear equations, a task well suited to modern computers.

4.3 Solving the Governing Equations

In this section we consider how Newton–Raphson iteration can be applied to solve the governing equations listed in Section 4.1. There are three steps to setting up the iteration: (1) reducing the complexity of the problem by setting aside for later evaluation the equations that can be solved linearly, (2) computing the residuals, and (3) calculating the Jacobian matrix. Because reserving the equations with linear solutions reduces the number of basis entries carried in the iteration, the solution technique described here is known as the "reduced basis method."

4.3.1 The Reduced Problem

The computing time required to evaluate Equation 4.19 in a Newton–Raphson iteration increases with the cube of the number of equations considered (Dongarra *et al.*, 1979). The numerical solution to Equations 4.3–4.6, therefore, can be found most rapidly by reserving from the iteration any of these equations that can be solved linearly. There are four cases in which equations can be reserved:

- If the mass of water n_w is a constraint on the system, Equation 4.3 can be evaluated directly for M_w.
- When the system chemistry is constrained by the concentration m_i (or activity a_i) of a basis species, Equation 4.4 gives M_i directly.
- Equation 4.5 can *always* be reserved, because the mineral mass n_k is linear in the equation.
- Equation 4.6 can also be reserved because the gas fugacities f_m are known.

The nonlinear portion of the problem, then, consists of just two parts:

- Equation 4.3, when n_w is unknown, and
- Equation 4.4, for each basis species at unknown concentration m_i.

The basis entries corresponding to these two cases are given by the "reduced basis,"

$$\mathbf{B}_\mathrm{r} = (A_w, A_i)_\mathrm{r} \,. \tag{4.21}$$

We will carry the subscript r (for "reduced") to indicate that a vector or matrix includes only entries conforming to one of the two nonlinear cases.

4.3.2 Residual Functions

The residual functions measure how well a guess $(n_w, m_i)_\mathrm{r}$ satisfies the governing Equations 4.3–4.4. The form of the residuals can be written

$$R_w = n_w \left(55.5 + \sum_j \nu_{wj} m_j \right) - M_w \tag{4.22}$$

$$R_i = n_w \left(m_i + \sum_j \nu_{ij} m_j \right) - M_i \,, \tag{4.23}$$

to represent the inequalities involved in evaluating these equations. The vector of residuals corresponding to the reduced basis $\mathbf{B}_\mathrm{r}$ is

$$\mathbf{R}_\mathrm{r} = (R_w, R_i)_\mathrm{r} \,. \tag{4.24}$$

Concurrent loops

```
      do 3   i1  =1, nbasis_spec
      do 2   i2  =1, nbasis_spec
```

Vector loop

```
      sum =  0.0
      do 1   j = 1, n_species
1     sum  =  sum  + n_{i1,j} * n_{i2,j} * m_j

2     J_{i1,i2} = n_w * sum/m_{i2}
3     J_{i1,i1} = J_{i1,i1} + n_w
```

Figure 4.3 Calculation of the entries $J_{ii'}$ in the Jacobian matrix on a vector-parallel computer, using a concurrent-outer, vector-inner (COVI) scheme. Each summation in the Jacobian can be calculated as a vector pipeline as separate processors calculate the entries in parallel.

4.3.3 Jacobian Matrix

The Jacobian matrix contains the partial derivatives of the residuals with respect to each of the unknown values $(n_w, m_i)_r$. To derive the Jacobian, it is helpful to note that

$$\frac{\partial m_j}{\partial n_w} = 0 \quad \text{and} \quad \frac{\partial m_j}{\partial m_i} = \nu_{ij} \frac{m_j}{m_i}, \tag{4.25}$$

as can be seen by differentiating Equation 4.7. The Jacobian entries, given by differentiating Equations 4.22–4.23, are

$$J_{ww} = \frac{\partial R_w}{\partial n_w} = 55.5 + \sum_j \nu_{wj} m_j \tag{4.26}$$

$$J_{wi} = \frac{\partial R_w}{\partial m_i} = \frac{n_w}{m_i} \sum_j \nu_{wj} \nu_{ij} m_j \tag{4.27}$$

$$J_{iw} = \frac{\partial R_i}{\partial n_w} = m_i + \sum_j \nu_{ij} m_j \tag{4.28}$$

and

$$J_{ii'} = \frac{\partial R_i}{\partial m_{i'}} = n_w \left(\delta_{ii'} + \sum_j \nu_{ij} \nu_{i'j} m_j / m_{i'} \right), \tag{4.29}$$

where $\delta_{ii'}$ is the Kronecker delta function,

$$\delta_{ii'} = \begin{cases} 1 & \text{if } i = i' \\ 0 & \text{otherwise} \end{cases}. \tag{4.30}$$

From a computational point of view, the forms of the Jacobian entries above are welcome because they conform to the architectural requirements of vector, parallel, and vector-parallel computers (Fig. 4.3)

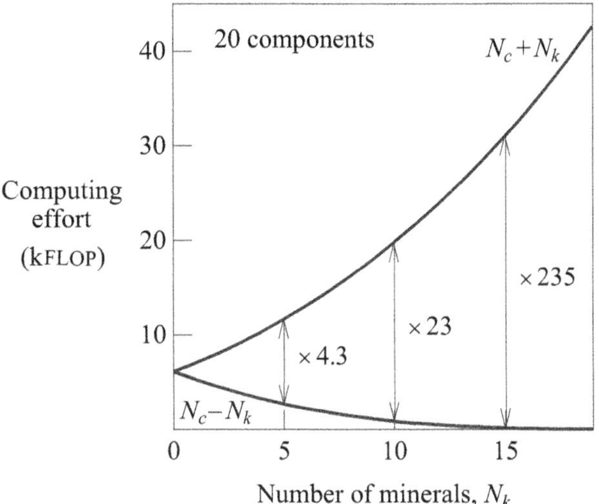

Figure 4.4 Comparison of the computing effort, expressed in thousands of floating point operations (kFLOP), required to factor the Jacobian matrix for a 20-component system ($N_c = 20$) during a Newton–Raphson iteration. For a technique that carries a nonlinear variable for each chemical component and each mineral in the system (top line), the computing effort increases as the number of minerals increases. For the reduced basis method (bottom line), however, less computing effort is required as the number of minerals increases.

4.3.4 Newton–Raphson Iteration

The Newton–Raphson iteration works by incrementally improving an estimate to values $(n_w, m_i)_r$ of the unknown variables in the reduced basis. The procedure begins with a guess at the variables' values. The first guess might be supplied by the modeler, but more commonly the model sets the guess using an ad hoc procedure such as assigning 90% of the mole numbers M_w and M_i to the basis species. If the procedure is invoked while tracing a reaction path, the result for the previous step along the path probably provides the best first guess. The guess can be optimized to better assure convergence in difficult cases, as described later in this section.

From Equation 4.19, the correction to an estimated solution is given by solving the system

$$\begin{pmatrix} J_{ww} & J_{wi} \\ J_{iw} & J_{ii'} \end{pmatrix}_r \begin{pmatrix} \Delta n_w \\ \Delta m_i \end{pmatrix}_r = \begin{pmatrix} -R_w \\ -R_i \end{pmatrix}_r \tag{4.31}$$

for $(\Delta n_w, \Delta m_i)_r$. The equation system can be solved by a variety of methods using widely available software, such as the Linpack library (Dongarra *et al.*, 1979). The correction is added to the current values of the unknown variables, and the iteration continues until the magnitudes of the residual functions fall below a prescribed tolerance.

At this point we can see the advantage of working with the reduced problem. Most published algorithms carry a nonlinear variable for each chemical component plus one for each mineral in the system. The number of nonlinear variables in the method presented here, on the other hand, is the number of components minus the number of minerals. Depending on the size of the problem, the savings in computing effort in evaluating Equation 4.31 can be dramatic (Fig. 4.4).

4.3.5 Non-negativity

There are, in fact, a number of solutions to the governing equations, but usually (see Chapter 13) only one with positive mole numbers and concentrations. Fortunately, the latter answer is of interest to all but the most abstract-thinking geochemist. The requirement that the iteration produce positive masses is known in chemical modeling as the non-negativity constraint.

One method of assuring positive values is to carry the logarithms of the unknown variables through the solution, since this function is not defined for negative numbers (van Zeggeren and Storey, 1970; Wolery, 1979). An alternative and perhaps more straightforward method is to begin the iteration with positive values for the unknown variables and then to scale back corrections that would drive any value negative. This technique in numerical analysis is called *under-relaxation*. The method is the computational equivalent of the childhood riddle: If Marco Polo traveled each day half way to China, how long would it take to get there? Marco's journey would shorten by half each day so that he would never quite reach his destination. Similarly, values in the under-relaxed iteration can approach zero but never become negative.

We force non-negativity upon a Newton–Raphson iteration by defining an under-relaxation factor,

$$\frac{1}{\delta_{\mathrm{UR}}} = \max\left(1, -\frac{\Delta n_w}{1/2\, n_w^{(q)}}, -\frac{\Delta m_i}{1/2\, m_i^{(q)}}\right)_{\mathrm{r}} . \tag{4.32}$$

The updated values are calculated according to

$$\begin{pmatrix} n_w \\ m_i \end{pmatrix}_{\mathrm{r}(q+1)} = \begin{pmatrix} n_w \\ m_i \end{pmatrix}_{\mathrm{r}(q)} + \delta_{\mathrm{UR}} \begin{pmatrix} \Delta n_w \\ \Delta m_i \end{pmatrix}_{\mathrm{r}} . \tag{4.33}$$

By these equations, the correction is allowed to reduce any variable by no more than half its value.

4.3.6 Examples of Convergence

The resulting iteration scheme converges strongly, with each iteration likely to reduce the residuals by an order of magnitude or more. Figure 4.5 shows how the iteration converges when the initial residuals take on large negative and positive values. The figure also shows the convergence when the activity coefficients and water activity are set to one. The extra nonlinearity introduced by the activity coefficients slows convergence by two or three iterations in these tests. From experience, the iteration can be expected to converge within about ten iterations; solutions requiring more than a hundred iterations are rather uncommon.

4.3.7 Optimizing the Starting Guess

The Newton–Raphson iteration usually converges to a solution rapidly enough that the choice of a starting guess is of little practical importance. Sometimes, however, an aqueous species that will end up at extremely low concentration appears in the basis. For example, the basis used to solve for the species distribution in a reduced fluid might contain O_2(aq). The oxidized species begins at high concentration, so by the mass action equations the reduced species such as H_2(aq) and H_2S(aq) start with impossibly large molalities. Hence, the residual for this basis entry becomes extremely large, sometimes in excess of 10^{100}.

Such a situation is dangerous because even though the iteration in many cases converges nicely, seemingly against impossible odds, the algorithm sometimes diverges to the pseudoroot $n_w = 0$ to Equation 4.23. An effective strategy is to repeatedly halve the starting guess for the basis species corresponding to very large positive residuals until the residuals reach a manageable size, perhaps less than 10^3.

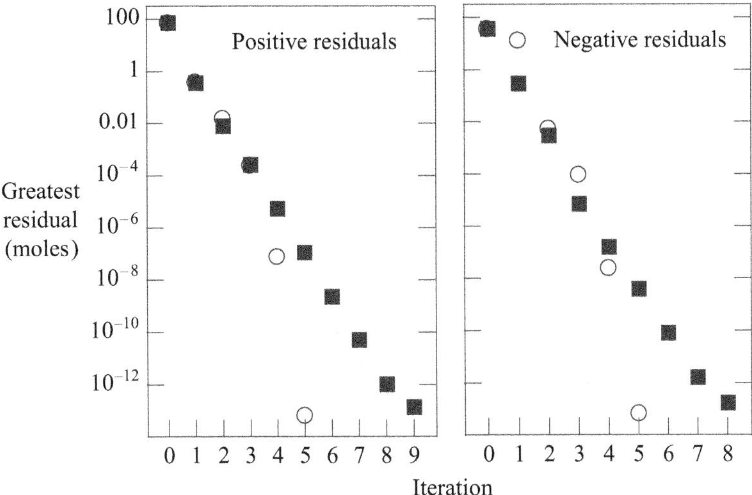

Figure 4.5 Convergence of the iteration to very small residuals in the reduced basis method, for systems contrived to have large positive and negative residuals at the start of the iteration. One test assumes Debye–Hückel activity coefficients (■); the other treats an ideal solution in which the activity coefficients are unity (○).

4.3.8 Activity Coefficients

To this point, we have assumed that the activity coefficients γ_i and γ_j as well as the activity of water a_w are known values. In fact, these values vary with m_i. Our strategy is to ignore this variation while calculating the Jacobian matrix and then update the activity coefficients and water activity after each step in the iteration.

Such a scheme is sometimes called a "soft" Newton–Raphson formulation because the partial derivatives in the Jacobian matrix are incomplete. We could, in principle, use a "hard" formulation in which the Jacobian accounts for the derivatives $\partial\gamma/\partial m_i$ and $\partial a_w/\partial m_i$. The hard formulation sometimes converges in fewer iterations, but in tests, the advantage was more than offset by the extra effort in computing the Jacobian. The soft method also allows us to keep the method for calculating activity coefficients (see Chapter 8) separate from the Newton–Raphson formulation, which simplifies programming.

4.3.9 Charge Balance

Our numerical solution must also honor charge balance among the dissolved species. As shown in Chapter 3, species are charge balanced when the components balance. In calculations constrained by known values of M_i, charge balance can be checked before beginning the solution.

A modeler, however, sometimes constrains one or more components in terms of a basis species' free concentration m_i (i.e., by specifying pH). Charge balance cannot be assured a priori because the system's bulk composition is not known until the iteration has converged. To force electrical neutrality, the model adjusts the mole number M_i of a charged component such as Cl^- after each iteration. This adjustment may be of little practical importance, because laboratories commonly report chloride concentrations computed from charge balance rather than from direct analysis of the element.

4.3.10 Mineral Masses

As we have noted, the mole numbers n_k of minerals in the system appear as linear terms in Equation 4.5. For this reason, these equations are omitted from the reduced basis. After the iteration is complete, the values of n_k, when unknown, are calculated according to

$$n_k = M_k - n_w \sum_j \nu_{kj} m_j \,, \tag{4.34}$$

which is obtained by reversing Equation 4.5. Note that there is nothing in the solution procedure that prevents negative values of n_k, a useful feature in determining the stable mineral assemblage (see Section 4.4).

4.3.11 Bulk Composition

The remaining step is to compute the system's bulk composition, if it is not fully known, according to the mass balance equations. The mole numbers M_w, M_i, and M_k are not known when the modeler has constrained the corresponding variable n_w, m_i, or n_k. In these cases, the mole numbers are determined directly from Equations 4.3–4.5. Where gases appear in the basis, the mole numbers M_m of gas components are similarly calculated from Equation 4.6.

4.4 Finding the Stable Phase Assemblage

The calculation described to this point does not predict the assemblage of minerals that is stable in the current system. Instead, the assemblage is assumed implicitly by setting the basis **B** before the calculation begins. A solution to the governing equations constitutes the equilibrium state of the system if two conditions are met: (1) no mineral in the system has dissolved away and become undersaturated, and (2) the fluid is not supersaturated with respect to any mineral.

A calculation procedure could, in theory, predict at once the distribution of mass within a system and the equilibrium mineral assemblage. Brown and Skinner (1974) undertook such a calculation for petrologic systems. For an n-component system, they calculated the shape of the free-energy surface for each possible solid solution in a rock. They then raised an n-dimensional hyperplane upward, allowing it to rotate against the free-energy surfaces. The hyperplane's resting position identified the stable minerals and their equilibrium compositions. Inevitably, the technique became known as the "crane plane" method.

Such a method has seldom been used with systems containing an aqueous fluid, probably because the complexity of the solution's free-energy surface and the wide range in aqueous solubilities of the elements complicate the numerics of the calculation (e.g., Harvie *et al.*, 1987). Instead, most models employ a procedure of elimination. If the calculation described fails to predict a system at equilibrium, the mineral assemblage is changed to swap undersaturated minerals out of the basis or supersaturated minerals into it, following the steps in the previous chapter; the calculation is then repeated.

4.4.1 Undersaturated Minerals

Minerals that have become undersaturated are revealed in the iteration results by negative mole numbers n_k. A negative mass, of course, is not meaningful physically beyond demonstrating that the mineral was completely consumed, perhaps to form another mineral, in the approach to equilibrium.

Minerals that develop negative masses are removed from the basis one at a time, and the solution is then recalculated. When a mineral is removed, an aqueous species must be selected from among the secondary species A_j to replace it in the basis. The species selected should be in high concentration to assure numerical stability in the iterative scheme described here and must, in combination with the other basis entries, form a valid component set (see Section 3.2). The species best fitting these criteria satisfies

$$\max_j \left(m_j \left| \nu_{kj} \right| \right) , \tag{4.35}$$

where ν_{kj} is the reaction coefficient for the undersaturated mineral in the reaction to form the secondary species.

4.4.2 Supersaturated Minerals

The saturation state of each mineral that can form in a model must be checked once the iteration is complete to identify supersaturated minerals. A mineral A_l, which is not in the basis, forms by the reaction

$$A_l = \nu_{wl} A_w + \sum_i \nu_{il} A_i + \sum_k \nu_{kl} A_k + \sum_m \nu_{ml} A_m , \tag{4.36}$$

which has an associated equilibrium constant K_l. The saturation state is given by the ratio Q_l / K_l, where Q_l is the mineral's activity product,

$$Q_l = \frac{a_w^{\nu_{wl}} \prod^i (\gamma_i m_i)^{\nu_{il}} \prod^k a_k^{\nu_{kl}} \prod^m f_m^{\nu_{ml}}}{a_l} . \tag{4.37}$$

Here a_k and a_l assume values of unity for minerals A_k and A_l of fixed composition. Ratios greater than one identify supersaturated minerals that need to be swapped into the basis and allowed to precipitate.

Choosing the location in the basis for the new mineral is a matter of identifying a basis species A_i that is similar in composition to the mineral to be removed and preferably in small concentration. The best species to be displaced from the basis satisfies

$$\max_i \left(\frac{|\nu_{il}|}{m_i} \right) , \tag{4.38}$$

where ν_{il} are the coefficients for the basis species in the reaction to form the supersaturated mineral.

4.4.3 Swap Procedure

A calculated solution may have just one supersaturated or undersaturated mineral, in which case calculating the solution for the new mineral assemblage will give the equilibrium state. Not uncommonly, however, more than one mineral appears under- or supersaturated. Such a situation is best handled one swap at a time, according to a set algorithm coded into the model.

No algorithm is guaranteed to arrive at the equilibrium state in this manner (Wolery, 1979), but the procedure outlined in Figure 4.6 gives good results in solving a wide range of problems. The procedure first checks for undersaturated minerals. If any exist, the one with the most negative n_k is removed from the basis and the governing equations solved once again.

Once there are no undersaturated minerals, the procedure checks for supersaturated minerals. If any exist, the most supersaturated mineral, identified by the largest Q_l / K_l, is swapped into the basis and the governing

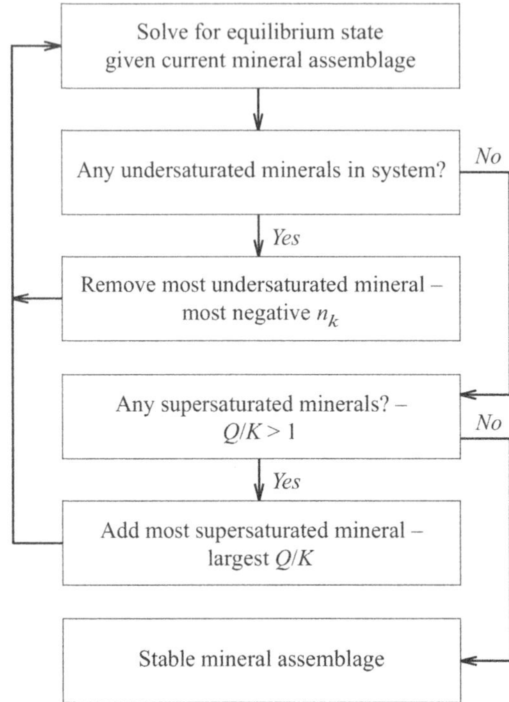

Figure 4.6 Procedure for finding the stable mineral assemblage in a system of known composition.

equations are solved. Precipitating a new mineral, however, may dissolve another away, so now the process begins anew by checking for undersaturated minerals. Once a solution has been found that includes neither undersaturated nor supersaturated minerals, the true equilibrium state has been located.

4.4.4 Apparent Violation of the Phase Rule

Sometimes when a mineral becomes supersaturated, there is no logical aqueous species in the basis with which to swap the mineral. Such a situation occurs when no species appear in the reaction to form the mineral. Wolery (1979) and Reed (1982) refer to such a situation as an "apparent violation of the phase rule," because adding the mineral to the basis would produce more phases in the system than there are components.

To include the supersaturated mineral in the basis in this case, another mineral must be removed. Although different schemes have been suggested for identifying the mineral to be removed, the most straightforward is to recognize that the reaction to form the supersaturated mineral,

$$\nu_{wl} A_w + \sum_k \nu_{kl} A_k + \sum_m \nu_{ml} A_m \rightarrow A_l \, , \tag{4.39}$$

will proceed until a mineral A_k in the basis is exhausted. The mineral satisfying

$$\min_k \left(\frac{n_k}{\nu_{kl}} \right) \tag{4.40}$$

is the first mineral exhausted and hence the entry to be swapped out of the basis.

5

Changing the Basis

To this point, we have assumed the existence of a basis of chemical components that corresponds to the system to be modeled. The basis, as discussed in the previous chapter, includes water, each mineral in the equilibrium system, each gas at known fugacity, and certain aqueous species. The basis serves two purposes: each chemical reaction considered in the model is written in terms of the members of the basis set, and the system's bulk composition is expressed in terms of the components in the basis.

Since we could not possibly store each possible variation on the basis, it is important for us to be able at any point in the calculation to adapt the basis to match the current system. It may be necessary to change the basis (make a *basis swap*, in modeling vernacular) for several reasons. This chapter describes how basis swaps can be accomplished in a computer model, and Chapter 12 shows how this technique can be applied to automatically balance chemical reactions and calculate equilibrium constants.

The modeler first encounters basis swapping in setting up a model, when it may be necessary to swap the basis to constrain the calculation. The thermodynamic dataset contains reactions written in terms of a preset basis that includes water and certain aqueous species (Na^+, Ca^{++}, K^+, Cl^-, HCO_3^-, SO_4^{--}, H^+, and so on) normally encountered in a chemical analysis. Some of the members of the original basis are likely to be appropriate for a calculation. When a mineral appears at equilibrium or a gas at known fugacity appears as a constraint, however, the modeler needs to swap the mineral or gas in question into the basis in place of one of these species.

Over the course of a reaction model, a mineral may dissolve away completely or become supersaturated and precipitate. In either case, the modeling software must alter the basis to match the new mineral assemblage before continuing the calculation. Finally, the basis sometimes must be changed in response to numerical considerations (e.g., Coudrain-Ribstein and Jamet, 1989). Depending on the numerical technique employed, the model may have trouble converging to a solution for the governing equations when one of the basis species occurs at small concentration. Including such a species in the basis can lead to numerical instability because making small corrections to its molality leads to large deviations in the molalities of the secondary species, when they are calculated using the mass action equations. In such a case, the modeling software may swap a more abundant species into the basis.

Fortunately, the process of changing the basis is straightforward and quickly performed on a computer using linear algebra, as we will see in this chapter. Modeling software, furthermore, performs basis changes automatically, so that the details need be of little concern in practice. Nonetheless, the nature of the process should be clear to a modeler. There are four steps in changing the basis: finding the transformation matrix, rewriting reactions to reflect the new basis, altering the equilibrium constants for the reactions, and re-expressing bulk composition in terms of the new basis. The process is familiar to mathematicians, who will recognize it as a linear, nonorthogonal transformation of coordinates.

5.1 Determining the Transformation Matrix

Initially, a basis vector

$$\mathbf{B} = (A_w, A_i, A_k, A_m) \tag{5.1}$$

describes a system. We wish to transform $\mathbf{B}$ to a new basis $\mathbf{B}'$. The new basis, which also contains water, is

$$\mathbf{B}' = (A_w, A'_i, A'_k, A'_m)\,. \tag{5.2}$$

Each entry A'_i, A'_k, or A'_m is a species or mineral that can be formed according to a *swap reaction* as a combination of the entries in the original basis. If A_j, a secondary species under the original basis, is to be swapped into basis position A'_i, the corresponding swap reaction is

$$A'_i = A_j = \nu_{wj} A_w + \sum_i \nu_{ij} A_i + \sum_k \nu_{kj} A_k + \sum_m \nu_{mj} A_m\,. \tag{5.3}$$

Alternatively, the reaction for swapping a mineral A_l or gas A_n into position A'_k or A'_m is

$$A'_k = A_l = \nu_{wl} A_w + \sum_i \nu_{il} A_i + \sum_k \nu_{kl} A_k + \sum_m \nu_{ml} A_m \tag{5.4}$$

$$A'_m = A_n = \nu_{wn} A_w + \sum_i \nu_{in} A_i + \sum_k \nu_{kn} A_k + \sum_m \nu_{mn} A_m\,. \tag{5.5}$$

An equilibrium constant K^{sw} is associated with each swap reaction. The swap reactions, written ensemble, form a matrix equation,

$$\begin{pmatrix} A_w \\ A'_i \\ A'_k \\ A'_m \end{pmatrix} = (\boldsymbol{\beta}) \begin{pmatrix} A_w \\ A_i \\ A_k \\ A_m \end{pmatrix}, \tag{5.6}$$

that gives the new basis in terms of the old. Here, $(\boldsymbol{\beta})$ is a matrix of the stoichiometric coefficients,

$$(\boldsymbol{\beta}) = \begin{pmatrix} 1 & 0 & 0 & 0 \\ \nu_{wj} & \nu_{ij} & \nu_{kj} & \nu_{mj} \\ \nu_{wl} & \nu_{il} & \nu_{kl} & \nu_{ml} \\ \nu_{wn} & \nu_{in} & \nu_{kn} & \nu_{mn} \end{pmatrix}, \tag{5.7}$$

from the swap reactions. Writing the old basis in terms of the new, then, is simply a matter of reversing the equation to give

$$\begin{pmatrix} A_w \\ A_i \\ A_k \\ A_m \end{pmatrix} = (\boldsymbol{\beta})^{-1} \begin{pmatrix} A_w \\ A'_i \\ A'_k \\ A'_m \end{pmatrix}. \tag{5.8}$$

Here $(\boldsymbol{\beta})^{-1}$, the inverse of $(\boldsymbol{\beta})$, is the *transformation matrix*, which is applied frequently in petrology (e.g., Brady, 1975; Greenwood, 1975; Thompson, 1982), but somewhat less commonly in aqueous geochemistry.

5.1.1 Example: Calculating the Transformation Matrix

These equations are abstract, but an example makes their meaning clear. In calculating a model, we might wish to convert a basis

$$\mathbf{B} = \left(\mathrm{H_2O}, \mathrm{Ca^{++}}, \mathrm{HCO_3^-}, \mathrm{H^+}\right) \tag{5.9}$$

to a new basis,

$$\mathbf{B}' = \left(\mathrm{H_2O}, \mathrm{Ca^{++}}, \mathrm{Calcite}, \mathrm{CO_2(g)}\right) , \tag{5.10}$$

in order to take advantage of a known CO_2 fugacity and the assumption of equilibrium with calcite. The swap reactions (including "null" swaps for H_2O and Ca^{++}) are

$$\begin{aligned} \mathrm{H_2O} &= \mathrm{H_2O} \\ \mathrm{Ca^{++}} &= \mathrm{Ca^{++}} \\ \mathrm{Calcite} &= \mathrm{Ca^{++}} - \mathrm{H^+} + \mathrm{HCO_3^-} \\ \mathrm{CO_2(g)} &= -\mathrm{H_2O} + \mathrm{H^+} + \mathrm{HCO_3^-} , \end{aligned} \tag{5.11}$$

or, in matrix form,

$$\begin{pmatrix} \mathrm{H_2O} \\ \mathrm{Ca^{++}} \\ \mathrm{Calcite} \\ \mathrm{CO_2(g)} \end{pmatrix} = \begin{pmatrix} 1 & 0 & 0 & 0 \\ 0 & 1 & 0 & 0 \\ 0 & 1 & 1 & -1 \\ -1 & 0 & 1 & 1 \end{pmatrix} \begin{pmatrix} \mathrm{H_2O} \\ \mathrm{Ca^{++}} \\ \mathrm{HCO_3^-} \\ \mathrm{H^+} \end{pmatrix} . \tag{5.12}$$

Inverting the matrix of coefficients gives reactions to form the old basis in terms of the new,

$$\begin{pmatrix} \mathrm{H_2O} \\ \mathrm{Ca^{++}} \\ \mathrm{HCO_3^-} \\ \mathrm{H^+} \end{pmatrix} = \begin{pmatrix} 1 & 0 & 0 & 0 \\ 0 & 1 & 0 & 0 \\ 1/2 & -1/2 & 1/2 & 1/2 \\ 1/2 & 1/2 & -1/2 & 1/2 \end{pmatrix} \begin{pmatrix} \mathrm{H_2O} \\ \mathrm{Ca^{++}} \\ \mathrm{Calcite} \\ \mathrm{CO_2(g)} \end{pmatrix} . \tag{5.13}$$

Written in standard form, the four reactions represented in the matrix equation may appear unusual because of the choice of components, but can be verified to balance. The transformation matrix for this change of basis is

$$(\boldsymbol{\beta})^{-1} = \begin{pmatrix} 1 & 0 & 0 & 0 \\ 0 & 1 & 0 & 0 \\ 1/2 & -1/2 & 1/2 & 1/2 \\ 1/2 & 1/2 & -1/2 & 1/2 \end{pmatrix} . \tag{5.14}$$

5.1.2 Test for a Valid Basis

The process of determining the transformation matrix provides a chance to check that the current basis is thermodynamically valid. In the previous chapter we noted that if a basis is valid, it is impossible to write a balanced reaction to form one entry in terms of the other entries in the basis.

An equivalent statement is that no row of the coefficient matrix $(\boldsymbol{\beta})$ can be formed as a linear combination of the other rows. Since the matrix's determinant is nonzero when and only when this statement is true, we need only evaluate the determinant of $(\boldsymbol{\beta})$ to demonstrate that a new basis $\mathbf{B}'$ is valid. In practice, this test can be accomplished using a linear algebra package, or implicitly by testing for error conditions produced while inverting the matrix, since a square matrix has an inverse if and only if its determinant is not zero.

5.2 Rewriting Reactions

Each time we change the basis, we must rewrite each chemical reaction in the system in terms of the new basis. This task, which might seem daunting, is quickly accomplished on a computer, using the transformation matrix. Consider the reaction to form an aqueous species,

$$A_j = A_w \nu_{wj} + \sum_i \nu_{ij} A_i + \sum_k \nu_{kj} A_k + \sum_m \nu_{mj} A_m \,. \tag{5.15}$$

The reaction can be written in vector form,

$$A_j = \left(\nu_{wj}, \nu_{ij}, \nu_{kj}, \nu_{mj}\right) \begin{pmatrix} A_w \\ A_i \\ A_k \\ A_m \end{pmatrix}, \tag{5.16}$$

in terms of the old basis. Substituting the transformation to the new basis (Eq. 5.6) gives

$$A_j = \left(\nu_{wj}, \nu_{ij}, \nu_{kj}, \nu_{mj}\right) (\boldsymbol{\beta})^{-1} \begin{pmatrix} A'_w \\ A'_i \\ A'_k \\ A'_m \end{pmatrix}. \tag{5.17}$$

The new reaction coefficients for the species A_j, then, are simply the matrix products of the old coefficients and the transformation matrix,

$$\left(\nu'_{wj}, \nu'_{ij}, \nu'_{kj}, \nu'_{mj}\right) = \left(\nu_{wj}, \nu_{ij}, \nu_{kj}, \nu_{mj}\right) (\boldsymbol{\beta})^{-1} \,. \tag{5.18}$$

Similarly, the products

$$\left(\nu'_{wl}, \nu'_{il}, \nu'_{kl}, \nu'_{mk}\right) = \left(\nu_{wl}, \nu_{il}, \nu_{kl}, \nu_{ml}\right) (\boldsymbol{\beta})^{-1} \tag{5.19}$$

$$\left(\nu'_{wn}, \nu'_{in}, \nu'_{kn}, \nu'_{mn}\right) = \left(\nu_{wn}, \nu_{in}, \nu_{kn}, \nu_{mn}\right) (\boldsymbol{\beta})^{-1} \tag{5.20}$$

give the revised reaction coefficients for minerals A_l and gases A_m.

Following our example from above, we write the reaction to form, as an example, carbonate ion,

$$\begin{aligned} CO_3^{--} &= (0, 0, 1, -1) \begin{pmatrix} H_2O \\ Ca^{++} \\ HCO_3^- \\ H^+ \end{pmatrix} \\ &= (0, 0, 1, -1) \begin{pmatrix} 1 & 0 & 0 & 0 \\ 0 & 1 & 0 & 0 \\ 1/2 & -1/2 & 1/2 & 1/2 \\ 1/2 & 1/2 & -1/2 & 1/2 \end{pmatrix} \begin{pmatrix} H_2O \\ Ca^{++} \\ \text{Calcite} \\ CO_2(g) \end{pmatrix}, \end{aligned} \tag{5.21}$$

using the transformation matrix that we already calculated (Eq. 5.14). Multiplying the numeric terms gives the reaction to form this species in terms of the new basis,

$$CO_3^{--} = (0, -1, 1, 0) \begin{pmatrix} H_2O \\ Ca^{++} \\ \text{Calcite} \\ CO_2(g) \end{pmatrix}, \tag{5.22}$$

or simply

$$CO_3^{--} = \text{Calcite} - Ca^{++} . \tag{5.23}$$

5.3 Altering Equilibrium Constants

The third step in changing the basis is to set the equilibrium constants for the revised reactions. The new equilibrium constant K'_j for a species reaction can be found from its value K_j before the basis swap according to

$$\log K'_j = \log K_j - \sum_i \nu'_{ij} \log K_i^{\text{sw}} - \sum_k \nu'_{kj} \log K_k^{\text{sw}} - \sum_m \nu'_{mj} \log K_m^{\text{sw}} . \tag{5.24}$$

Here, K_i^{sw}, K_k^{sw}, and K_m^{sw} are the equilibrium constants for the reactions by which we swap species, minerals, and gases into the basis.

In matrix form, the equation is

$$\log K'_j = \log K_j - \left(\nu'_{wj}, \nu'_{ij}, \nu'_{kj}, \nu'_{mj}\right) \begin{pmatrix} 0 \\ \log K_i^{\text{sw}} \\ \log K_k^{\text{sw}} \\ \log K_m^{\text{sw}} \end{pmatrix} . \tag{5.25}$$

The equilibrium constants for mineral and gas reactions are calculated from their revised reaction coefficients in similar fashion as

$$\log K'_l = \log K_l - \left(\nu'_{wl}, \nu'_{il}, \nu'_{kl}, \nu'_{ml}\right) \begin{pmatrix} 0 \\ \log K_i^{\text{sw}} \\ \log K_k^{\text{sw}} \\ \log K_m^{\text{sw}} \end{pmatrix} \tag{5.26}$$

and

$$\log K'_n = \log K_n - \left(\nu'_{wn}, \nu'_{in}, \nu'_{kn}, \nu'_{mn}\right) \begin{pmatrix} 0 \\ \log K_i^{\text{sw}} \\ \log K_k^{\text{sw}} \\ \log K_m^{\text{sw}} \end{pmatrix} . \tag{5.27}$$

Basis entries that do not change over the swap have no effect in these equations, since they are represented by null swap reactions (e.g., $H_2O = H_2O$) with equilibrium constants of unity.

What is the equilibrium constant for CO_3^{--} in the example from the previous section? The value at 25 °C for the reaction written in terms of the original basis is $10^{10.34}$, and the equilibrium constants of the swap reactions for calcite and CO_2(g) are $10^{1.71}$ and $10^{-7.82}$. The new value according to the equation above is

$$\log K'_{CO_3^{--}} = (10.34) - \left(0, -1, 1, 0\right) \begin{pmatrix} 0 \\ 0 \\ 1.71 \\ -7.82 \end{pmatrix} , \tag{5.28}$$

or $10^{8.63}$. To verify this result, we can calculate the equilibrium constant directly by elimination,

$$\begin{array}{rcl@{\qquad}l} CO_3^{--} &=& HCO_3^- - H^+ & \log K = 10.34 \\ HCO_3^- + Ca^{++} - H^+ &=& \text{Calcite} & \log K = -1.71 \\ \hline CO_3^{--} &=& \text{Calcite} - Ca^{++} & \log K = 8.63 \end{array} \tag{5.29}$$

and arrive at the same value.

5.4 Re-expressing Bulk Composition

The final step in changing the basis is to recalculate the system's bulk composition in terms of the new component set. Composition in terms of the old and new bases is related by the stoichiometric coefficients,

$$\begin{pmatrix} M_w \\ M_i \\ M_k \\ M_m \end{pmatrix} = (\boldsymbol{\beta})^{\mathrm{T}} \begin{pmatrix} M'_w \\ M'_i \\ M'_k \\ M'_m \end{pmatrix}, \tag{5.30}$$

where $(\boldsymbol{\beta})^{\mathrm{T}}$ is the transpose of $(\boldsymbol{\beta})$,

$$(\boldsymbol{\beta})^{\mathrm{T}} = \begin{pmatrix} 1 & \nu_{wj} & \nu_{wl} & \nu_{wn} \\ 0 & \nu_{ij} & \nu_{il} & \nu_{in} \\ 0 & \nu_{kj} & \nu_{kl} & \nu_{kn} \\ 0 & \nu_{mj} & \nu_{ml} & \nu_{mn} \end{pmatrix} \tag{5.31}$$

(i.e., the coefficient matrix flipped on its diagonal). This relationship holds because reaction coefficients give the amounts of the old basis entries that go into making up the new entries, as noted in the previous chapter.

Reversing this equation shows that the composition in terms of the new basis is given immediately by the transpose of the transformation matrix,

$$\begin{pmatrix} M'_w \\ M'_i \\ M'_k \\ M'_m \end{pmatrix} = (\boldsymbol{\beta})^{-1\mathrm{T}} \begin{pmatrix} M_w \\ M_i \\ M_k \\ M_m \end{pmatrix} \tag{5.32}$$

(because the inverse of a transposed matrix is the same as the transposed inverse). In other words, the entries in the transformation matrix can simply be flipped (or, in practice, their subscripts reversed) to revise the bulk composition.

Continuing our example from the previous section, we want to use this formula to revise bulk composition in a system in which calcite and $CO_2(g)$ have been swapped for HCO_3^- and H^+. The old and new compositions are related by

$$\begin{pmatrix} M_{H_2O} \\ M_{Ca^{++}} \\ M_{HCO_3^-} \\ M_{H^+} \end{pmatrix} = \begin{pmatrix} 1 & 0 & 0 & -1 \\ 0 & 1 & 1 & 0 \\ 0 & 0 & 1 & 1 \\ 0 & 0 & -1 & 1 \end{pmatrix} \begin{pmatrix} M_{H_2O} \\ M_{Ca^{++}} \\ M_{\text{Calcite}} \\ M_{CO_2(g)} \end{pmatrix}, \tag{5.33}$$

which we write by transposing the coefficient matrix for the swap reactions (from Eq. 5.12).

To find the new composition, we need only flip the elements in the transformation matrix about its diagonal, giving

$$\begin{pmatrix} M_{H_2O} \\ M_{Ca^{++}} \\ M_{Calcite} \\ M_{CO_2(g)} \end{pmatrix} = \begin{pmatrix} 1 & 0 & 1/2 & 1/2 \\ 0 & 1 & -1/2 & 1/2 \\ 0 & 0 & 1/2 & -1/2 \\ 0 & 0 & 1/2 & 1/2 \end{pmatrix} \begin{pmatrix} M_{H_2O} \\ M_{Ca^{++}} \\ M_{HCO_3^-} \\ M_{H^+} \end{pmatrix} . \tag{5.34}$$

In interpreting this equation, it is important to recall that the total mole numbers M of the components provide a mathematical tool for describing the system's composition, but do not give the amounts of species, minerals, or gases actually present in the system.

According to the equation, we expect more water component when the system is defined in terms of the new basis rather than the old because hydrogen occurs in the new basis only in H_2O, whereas it was also found in the H^+ and HCO_3^- components of the old basis. (Of course, the amount of actual water remains the same, since we are expressing the same bulk composition in different terms.) Similarly, the calcium in the original system will be distributed between the Ca^{++} and calcite components of the new system, and so on.

To complete the example numerically, we take as our system a solution of 10^{-3} moles calcium bicarbonate dissolved in one kg (55.5 moles) of water. For simplicity, we set the H^+ component to zero moles, which corresponds to a pH of about 8. The system's composition expressed in terms of the two bases is

$$\begin{pmatrix} M_{H_2O} \\ M_{Ca^{++}} \\ M_{HCO_3^-} \\ M_{H^+} \end{pmatrix} = \begin{pmatrix} 55.5 \\ 0.001 \\ 0.002 \\ 0 \end{pmatrix} \qquad \begin{pmatrix} M_{H_2O} \\ M_{Ca^{++}} \\ M_{Calcite} \\ M_{CO_2(g)} \end{pmatrix} = \begin{pmatrix} 55.501 \\ 0 \\ 0.001 \\ 0.001 \end{pmatrix} , \tag{5.35}$$

where we have calculated the second set of values from the first according to Equation 5.34. We can quickly verify that each composition is charge balanced, and that the two vectors contain the same mole numbers of the elements hydrogen, oxygen, calcium, and carbon. The compositions expressed in terms of the two component sets are, in fact, equivalent.

6

Equilibrium Models of Natural Waters

Having derived a set of equations describing the equilibrium state of a multicomponent system and devised a scheme for solving them, we can begin to model the chemistries of natural waters. In this chapter we construct four models, each posing special challenges, and look in detail at the meaning of the calculation results.

In each case, we use program SPECE8 or REACT and employ an extended form of the Debye–Hückel equation for calculating species' activity coefficients, as discussed in Chapter 8. In running the programs, you work interactively following the general procedure:

- **Swap** into the basis any needed species, minerals, or gases. Table 6.1 shows the basis in its original configuration (as it exists when you start the program). You might want to change the basis by replacing SiO_2(aq) with quartz so that equilibrium with this mineral can be used to constrain the model. Or to set a fugacity buffer you might swap CO_2(g) for either H^+ or HCO_3^-.
- **Set** a constraint for each basis member that you want to include in the calculation. For instance, the constraint might be the total concentration of sodium in the fluid, the free mass of a mineral, or a gas's fugacity or partial pressure. You may also set temperature (25 °C, by default) or special program options.
- **Run** the program by typing `go`.
- **Revise** the basis or constraints and re-execute the program as often as you wish.

In this book, input scripts for running the various programs are set in a "typewriter" typeface. Unless a script is marked as a continuation of the previous script, you should start the program anew or type `reset` to clear your previous configuration.

6.1 Chemical Model of Seawater

For a first chemical model, we calculate the distribution of species in surface seawater, a problem first undertaken by Garrels and Thompson (1962; see also Thompson, 1992). We base our calculation on the major element composition of seawater (Table 6.2), as determined by chemical analysis. To set pH, we assume equilibrium with CO_2 in the atmosphere (Table 6.3). Since the program will determine the HCO_3^- and water activities, setting the CO_2 partial pressure fixes pH according to the reaction

$$H^+ + HCO_3^- \rightleftarrows CO_2(g) + H_2O\,. \qquad (6.1)$$

Similarly, we define oxidation state according to

$$O_2(aq) \rightleftarrows O_2(g)\,, \qquad (6.2)$$

by specifying the partial pressure of O_2(g) in the atmosphere.

Table 6.1 *Basis species in the* LLNL *database*

H_2O	Cr^{+++}	Li^{+}	Ru^{+++}
Ag^{+}	Cs^{+}	Mg^{++}	SO_4^{--}
Al^{+++}	Cu^{+}	Mn^{++}	SeO_3^{--}
Am^{+++}	Eu^{+++}	NO_3^{-}	$SiO_2(aq)$
$As(OH)_4^{-}$	F^{-}	Na^{+}	Sn^{++++}
Au^{+}	Fe^{++}	Ni^{++}	Sr^{++}
$B(OH)_3$	H^{+}	Np^{++++}	TcO_4^{-}
Ba^{++}	HCO_3^{-}	$O_2(aq)$	Th^{++++}
Br^{-}	HPO_4^{--}	Pb^{++}	U^{++++}
Ca^{++}	Hg^{++}	PuO_2^{++}	V^{+++}
Cl^{-}	I^{-}	Ra^{++}	Zn^{++}
Co^{++}	K^{+}	Rb^{+}	

Table 6.2 *Major element composition of seawater (Drever, 1988)*

	Concentration ($mg\ kg^{-1}$)
Cl^{-}	19 350
Na^{+}	10 760
SO_4^{--}	2 710
Mg^{++}	1 290
Ca^{++}	411
K^{+}	399
HCO_3^{-}	142
$SiO_2(aq)$	0.5–10
$O_2(aq)$	0.1–6

Table 6.3 *Partial pressures of some gases in the atmosphere (Hem, 1985)*

Gas	Pressure (atm)
N_2	0.78
O_2	0.21
H_2O	0.001–0.23
CO_2	0.0003
CH_4	1.5×10^{-6}
CO	$(0.06–1) \times 10^{-6}$
SO_2	1×10^{-6}
N_2O	5×10^{-7}
H_2	$\sim 5 \times 10^{-7}$
NO_2	$(0.05–2) \times 10^{-8}$

The latter two assumptions are simplistic, considering the number of factors that affect pH and oxidation state in the oceans (e.g., Sillén, 1967; Holland, 1978; McDuff and Morel, 1980). Consumption and production of CO_2 and O_2 by plant and animal life, reactions among silicate minerals, dissolution and precipitation of carbonate minerals, solute fluxes from rivers, and reaction between convecting seawater and oceanic crust all affect these variables. Nonetheless, it will be interesting to compare the results of this simple calculation to observation.

To calculate the model with REACT, we swap CO_2(g) and O_2(g) into the basis in place of H^+ and O_2(aq), and constrain each basis member. The procedure is

```
swap CO2(g) for H+
swap O2(g) for O2(aq)
P CO2(g)  = 10^-3.5 atm
P O2(g)   = 0.2     atm

Cl-       = 19350 mg/kg
Ca++      =   411 mg/kg
Mg++      =  1290 mg/kg
Na+       = 10760 mg/kg
K+        =   399 mg/kg
SO4--     =  2710 mg/kg
HCO3-     =   142 mg/kg
SiO2(aq)  =     6 mg/kg

print species=long
go
```

Here, the `print` command causes the program to list in the output all of the aqueous species, not just those in greatest concentration. Typing `go` triggers the model to begin calculations and write its results to the output dataset.

Table 6.4 *Calculated molalities* (m)*, activity coefficients* (γ)*, and log activities* (a) *of the most abundant species in seawater*

Species	m	γ	$\log a$
Cl^-	0.5500	0.6276	−0.4619
Na^+	0.4754	0.6717	−0.4958
Mg^{++}	0.3975×10^{-1}	0.3160	−1.9009
SO_4^{--}	0.1607×10^{-1}	0.1692	−2.5657
K^+	0.1033×10^{-1}	0.6276	−2.1881
$MgCl^+$	0.9126×10^{-2}	0.6717	−2.2125
$NaSO_4^-$	0.6386×10^{-2}	0.6717	−2.3676
Ca^{++}	0.5953×10^{-2}	0.2465	−2.8334
$MgSO_4$	0.5767×10^{-2}	1.0	−2.2391
$CaCl^+$	0.3780×10^{-2}	0.6717	−2.5953
NaCl	0.2773×10^{-2}	1.0	−2.5571
HCO_3^-	0.1499×10^{-2}	0.6906	−2.9849
$CaSO_4$	0.8334×10^{-3}	1.0	−3.0792
$NaHCO_3$	0.4450×10^{-3}	1.0	−3.3516
O_2(aq)	0.2178×10^{-3}	1.1735	−3.5925
$MgHCO_3^+$	0.1982×10^{-3}	0.6717	−3.8757
KSO_4^-	0.1869×10^{-3}	0.6717	−3.9013
$MgCO_3$	0.1061×10^{-3}	1.0	−3.9745
SiO_2(aq)	0.8201×10^{-4}	1.1735	−4.0167
KCl	0.5785×10^{-4}	1.0	−4.2377
CO_3^{--}	0.5401×10^{-4}	0.1891	−4.9909

6.1.1 Species Distribution

In the program's text-format output, we find two blocks of results that represent, respectively, the system's state before and after supersaturated minerals precipitate. Each block shows the concentration, activity coefficient, and activity calculated for each aqueous species (Table 6.4), the saturation state of each mineral that can be formed from the basis, the partial pressure and fugacity of each such gas, and the system's bulk composition. The extent of the system is 1 kg of solvent water and the solutes dissolved in it; the solution mass is 1.0364 kg.

We can quickly identify the input constraints in the first block: the partial pressures of CO_2(g) and O_2(g), recalling that an atm is 1.013 bar, and the bulk composition expressed in terms of components Cl^-, Ca^{++}, and so on. Note that the free species concentrations do not correspond to the input constraints, which are bulk or total values. The free concentration of the species Ca^{++}, in other words, accounts for just part of the solution's calcium content.

The predicted pH is 8.34, a value lying within but toward the alkaline end of the range 7.8 to 8.5 observed in seawater (Fig. 6.1). The dissolved oxygen content predicted by the calculation is 210 μmol kg^{-1}, or 6.7 mg kg^{-1}. This value compares well with values measured near the ocean surface (Fig. 6.2).

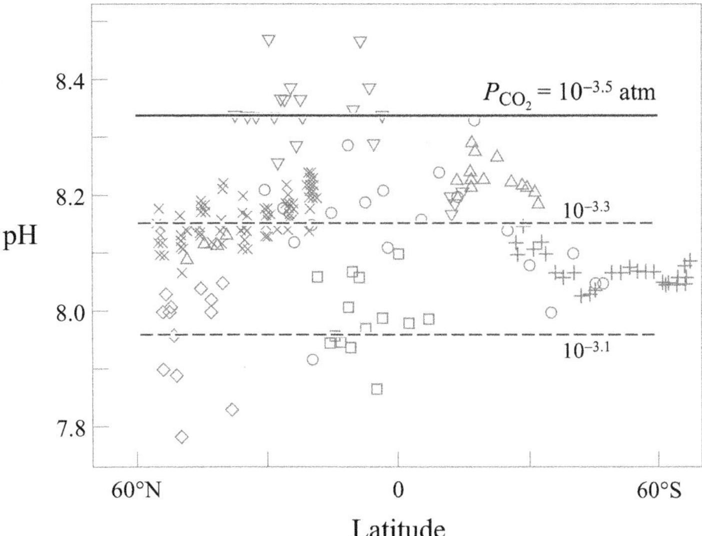

Figure 6.1 pH of surface seawater from the western Pacific Ocean (Skirrow, 1965), as measured in situ during oceanographic cruises (various symbols). Line shows pH predicted by the model for seawater in equilibrium with atmospheric CO_2 at a partial pressure of $10^{-3.5}$ atm. Dashed lines show pH values that result from assuming larger partial pressures of $10^{-3.3}$ atm and $10^{-3.1}$ atm.

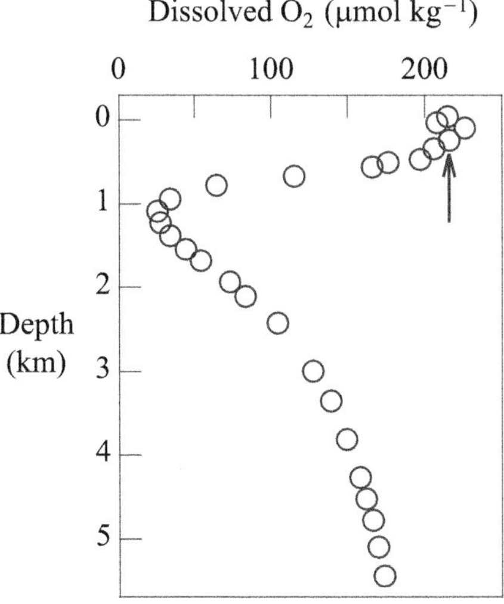

Figure 6.2 Profile of dissolved oxygen versus depth at GEOSECS site 226 (Drever, 1988, p. 267). Arrow marks oxygen content predicted by the model, assuming equilibrium with oxygen in the atmosphere.

Table 6.5 *Extent of complexing of major cations and anions in seawater*

Species	Total concentration mg kg^{-1}	molal	Free concentration molal	% of total	Complexes % of total
Na^+	10 760	0.4850	0.4754	98	2
Mg^{++}	1 290	0.05501	0.03975	72	28
Ca^{++}	411	0.01063	0.005953	56	44
Cl^-	19 350	0.5658	0.5500	97	3
SO_4^{--}	2 710	0.02924	0.01607	55	45
HCO_3^-	142	0.002412	0.001499	62	38

It is clear from the species distribution that the dissolved components in seawater react to varying extents to form complex species (Table 6.5). Components Na^+ and Cl^- are present almost entirely as free ions. Only a few percent of their masses appear in complexes, most notably the ion pairs $MgCl^+$, $NaSO_4^-$, $CaCl^+$, and NaCl (Table 6.4). Components Ca^{++}, SO_4^{--}, and HCO_3^-, on the other hand, complex strongly; complex species account for a third to a half of their total concentrations.

Some of the species concentrations predicted by the mathematical model are too small to be physically meaningful. The predicted concentration of H_2(aq), for example, is 4×10^{-45} molal. Multiplying this value by Avogadro's number (6.02×10^{23} mol^{-1}) and the volume of the Earth's oceans (1400×10^{18} l), the concentration is equivalent to just three molecules in all of the world's seas!

We can calculate a more realistic H_2(aq) concentration from the partial pressure of H_2(g) in the atmosphere (Table 6.3) and the equilibrium constant for the reaction

$$H_2(g) \rightleftarrows H_2(aq)\,, \qquad (6.3)$$

which is $10^{-3.1}$. In this case, the molality of H_2(aq) in equilibrium with the atmosphere is about 5×10^{-10}. This value, while small, is tens of orders of magnitude greater than the value calculated at equilibrium. Clearly, the equilibrium value is a mathematical abstraction.

6.1.2 Mineral Saturation

Thirteen minerals appear supersaturated in the first block of results produced by the chemical model (Table 6.6). These results, therefore, represent an equilibrium achieved internally within the fluid but metastable with respect to mineral precipitation. It is quite common in modeling natural waters, especially when working at low temperature, to find one or more minerals listed as supersaturated. Unfortunately, the error sources in geochemical modeling are large enough that it can be difficult to determine whether a water is in fact supersaturated.

Many natural waters are supersaturated at low temperature, primarily because less stable minerals dissolve more quickly than more stable minerals precipitate. Relatively unstable silica phases such as chalcedony or amorphous silica, for example, may control a fluid's SiO_2 concentration because quartz, the most stable silica mineral, precipitates slowly.

Uncertainty in the calculation, however, affects the reliability of values reported for saturation indices.

Table 6.6 *Calculated saturation indices (SI) of various minerals in seawater and the mass of each precipitate in the stable phase assemblage*

Mineral	Composition	Initial SI (log Q/K)	Amount formed (mg)
Antigorite	$Mg_{24}Si_{17}O_{42.5}(OH)_{31}$	44.02	—
Tremolite	$Ca_2Mg_5Si_8O_{22}(OH)_2$	7.69	—
Talc	$Mg_3Si_4O_{10}(OH)_2$	6.66	—
Chrysotile	$Mg_3Si_2O_5(OH)_4$	4.70	—
Sepiolite	$Mg_4Si_6O_{15}(OH)_2 \cdot 6H_2O$	3.90	—
Dolomite	$CaMg(CO_3)_2$	3.45	50
Dolomite-ord	$CaMg(CO_3)_2$	3.45	—
Anthophyllite	$Mg_7Si_8O_{22}(OH)_2$	3.44	—
Huntite	$CaMg_3(CO_3)_4$	2.12	—
Dolomite-dis	$CaMg(CO_3)_2$	1.91	—
Magnesite	$MgCO_3$	1.02	—
Calcite	$CaCO_3$	0.81	—
Aragonite	$CaCO_3$	0.64	—
Quartz	SiO_2	−0.02	1

Reaction log Ks for many minerals are determined by extrapolating the results of experiments conducted at high temperature to the conditions of interest. The error in this type of extrapolation shows up directly in the denominator of log Q/K. Error in calculating activity coefficients (see Chapter 8), on the other hand, directly affects the computed activity product Q. The effect is pronounced for reactions with large coefficients, such as those for clay minerals.

Error in the input data can also be significant. The saturation state calculated for an aluminosilicate mineral, for example, depends on the analytical concentrations determined for aluminum and silicon. These analyses are difficult to perform accurately. As discussed in the next section, the presence of colloids and suspended particles in solution often affects the analytical results profoundly.

Averaging analytical data from different fluids commonly leads to inflated saturation indices. Scientists generally regard averaging as a method for reducing uncertainty in measured values, but the practice can play havoc in geochemical modeling. Waters with a range of Ca^{++} and SO_4^{--} concentrations, for example, can be in equilibrium with anhydrite ($CaSO_4$), so long as the activity product $a_{Ca^{++}} \times a_{SO_4^{--}}$ matches the equilibrium constant for the dissolution reaction. The averaged composition of two fluids (A and B) with differing Ca^{++} to SO_4^{--} ratios is, in the absence of activity coefficient effects, invariably supersaturated. This point can be shown quickly:

$$Q_A = (a_{Ca^{++}})_A (a_{SO_4^{--}})_A = K \tag{6.4}$$

$$Q_B = (a_{Ca^{++}})_B (a_{SO_4^{--}})_B = K \tag{6.5}$$

and

$$
\begin{aligned}
Q_{\text{ave}} &= \frac{1}{2}\left[(a_{\text{Ca}^{++}})_A + (a_{\text{Ca}^{++}})_B\right] \times \frac{1}{2}\left[(a_{\text{SO}_4^{--}})_A + (a_{\text{SO}_4^{--}})_B\right] \\
&= \frac{1}{2}K + \frac{1}{4}(a_{\text{Ca}^{++}})_A(a_{\text{SO}_4^{--}})_B + \frac{1}{4}(a_{\text{Ca}^{++}})_B(a_{\text{SO}_4^{--}})_A \geq K\,.
\end{aligned}
\tag{6.6}
$$

The activity product Q_{ave} corresponding to the averaged analysis (ignoring variation in activity coefficients) equals the equilibrium constant K only when fluids A and B are identical; otherwise Q_{ave} exceeds K and anhydrite is reported to be supersaturated. To demonstrate this inequality, we can assume arbitrary values for $a_{\text{Ca}^{++}}$ and $a_{\text{SO}_4^{--}}$ that satisfy Equations 6.4–6.5 and substitute them into Equation 6.6.

The physical analogy to the averaging problem occurs when a sample consists of a mixture of fluids, as can occur when a well draws water from two or more producing intervals. In this case, the mixture may be supersaturated even when the individual fluids are not.

In the seawater example (Table 6.6), the saturation indices are inflated somewhat by the choice of a rather alkaline pH, reflecting equilibrium with atmospheric CO_2. If we had chosen a more acidic pH within the range observed in seawater, the indices would be smaller. The choice of large formula units for the phyllosilicate minerals also serves to inflate the saturation indices reported for these minerals; see Section 6.1.5.

Calculating the saturation state of dolomite presents an interesting problem. Geochemists have generally believed that ordered dolomite is supersaturated in seawater but prevented from precipitating by kinetic factors (e.g., Kastner, 1984; Hardie, 1987). The stability of dolomite at 25 °C, however, is poorly known (Carpenter, 1980). The mineral might be more soluble, and hence less supersaturated in seawater, than shown in the LLNL database, which uses an estimate by Helgeson *et al.* (1978) derived by extrapolating data from a metamorphic reaction to room temperature. On the basis of careful experiments at lower temperatures, Lafon *et al.* (1992) suggest that seawater is close to equilibrium with dolomite. The saturation index predicted by our model, therefore, may be unrealistically high.

6.1.3 Mass Balance and Mass Action

Because we formulated the governing equations from the principles of mass balance and mass action, we should now be able to show that our calculation results honor these principles. Demonstrating that the computer performed the calculations correctly is an important step for a geochemical modeler. No programmer is incapable of erring, no storage device is incorruptible, and no computer is infallible; the responsibility of showing correctness ultimately lies with the modeler. In geochemical modeling, fortunately, we can accomplish this relatively easily.

To show mass balance, we add the molalities of each species containing a component (but not species concentrations in mg kg^{-1}, since the mole weight of each species differs) to arrive at the input constraint. Taking component SO_4^{--} as an example, we find the total mole number (M_i) from the molalities (m_i and m_j) of the sulfur-bearing species,

SO_4^{--}	0.01607
$NaSO_4^-$	0.006386
$MgSO_4$	0.005767
$CaSO_4$	0.0008334
KSO_4^-	0.0001869
HSO_4^-	1.8×10^{-9}
H_2SO_4	5.6×10^{-21}
HS^-	1.3×10^{-142}
$H_2S(aq)$	3.5×10^{-144}
S^{--}	1.1×10^{-147}
	0.02924 molal

Converting units,

$$(0.02924 \text{ molal}) \times (1 \text{ kg solvent}) \times (96.058 \text{ g SO}_4 \text{ mol}^{-1}) \times \\ (1000 \text{ mg g}^{-1})/(1.0364 \text{ kg solution}) = 2710 \text{ mg SO}_4 \text{ (kg solution)}^{-1}, \tag{6.7}$$

which is the input value from Table 6.2.

Repeating the procedure for component Ca^{++},

Ca^{++}	0.005953
$CaCl^+$	0.003780
$CaSO_4$	0.0008334
$CaHCO_3^+$	0.00003547
$CaCO_3$	0.00002464
$CaH_3SiO_4^+$	1.28×10^{-7}
$CaOH^+$	9.57×10^{-8}
CaH_2SiO_4	5.74×10^{-9}
$Ca(H_3SiO_4)_2$	2.05×10^{-10}
	0.01063 molal

gives

$$(0.01063 \text{ molal}) \times (1 \text{ kg solvent}) \times (40.080 \text{ g Ca mol}^{-1}) \times \\ (1000 \text{ mg g}^{-1})/(1.0364 \text{ kg solution}) = 411 \text{ mg Ca (kg solution)}^{-1}. \tag{6.8}$$

Again, this is the input value.

To demonstrate mass action, we show that for any possible reaction the activity product Q matches the equilibrium constant K. This step is most easily accomplished by computing $\log Q$ as the sum of the products of the reaction coefficients and log activities of the corresponding species. The reaction for the sodium–sulfate ion pair, for example,

$$NaSO_4^- \rightleftarrows Na^+ + SO_4^{--}, \tag{6.9}$$

has a $\log K$ of -0.694, according to the LLNL database. Calculating $\log Q$,

Na^+	1×-0.4958
SO_4^{--}	1×-2.5657
$NaSO_4^-$	-1×-2.3676
	-0.6939

gives the value of log K, demonstrating that the reaction is indeed in equilibrium.

The procedure for verifying mineral saturation indices is similar. The program reports that log Q/K for the dolomite reaction,

$$\underset{\textit{dolomite}}{CaMg(CO_3)_2} + 2\,H^+ \rightleftarrows Ca^{++} + Mg^{++} + 2\,HCO_3^- , \tag{6.10}$$

is 3.451. The log K for the reaction is 2.5207. Determining log Q as before,

Mg^{++}	1×-1.9009
Ca^{++}	1×-2.8334
HCO_3^-	2×-2.9849
H^+	-2×-8.3379
	5.9717

gives the expected result,

$$\log Q/K = \log Q - \log K = 3.451 . \tag{6.11}$$

6.1.4 Stable Phase Assemblage

After finding the equilibrium distribution of species in the initial fluid, the program sets about determining the system's theoretical state of true equilibrium. To do so, it searches for the stable mineral assemblage, following the procedure described in Section 4.4. A second block of results in the program output shows the equilibrium state corresponding to the predicted stable assemblage of fluid, dolomite, and quartz. These results are largely of academic interest: you could leave a bottle of seawater on a shelf for a very long time without fear that dolomite or quartz would form.

The search procedure, which we can trace through the flow chart shown in Fig. 4.6, is of interest. The steps in the procedure are:

1. Calculate initial species distribution; thirteen minerals are supersaturated.
2. Remove fugacity buffers.
   Swap $MgCO_3$ for $CO_2(g)$, $O_2(aq)$ for $O_2(g)$.
3. Swap antigorite for $MgCO_3$.
   Iteration converges; six minerals are supersaturated.
4. Swap dolomite for $SiO_2(aq)$.
   Iteration converges; antigorite has dissolved away.
5. Swap $SiO_2(aq)$ for antigorite.
   Iteration converges; quartz is supersaturated.
6. Swap quartz for $SiO_2(aq)$.
   Iteration converges; there are no supersaturated minerals.

The program begins (step 1) with the distribution of species, as already described. Before beginning to search for the stable phase assemblage, it sets a closed system (step 2) by eliminating the fugacity buffers for $CO_2(g)$ and $O_2(g)$. It does so by swapping the aqueous species $MgCO_3$ and $O_2(aq)$ into the basis. If these gases had been set as fixed buffers using the `fix` command, however, the program would skip this step.

Next (step 3), the program swaps antigorite, the most supersaturated mineral in the initial fluid, into the basis in place of species $MgCO_3$. With antigorite in the system, six minerals are supersaturated. In the next step (step 4), the program chooses to swap dolomite into the basis in place of $SiO_2(aq)$. This swap seems strange until we write the reaction for dolomite in terms of the current basis,

$$\underset{\textit{dolomite}}{CaMg(CO_3)_2} + 1.6\,H_2O + 0.7\,SiO_2(aq) \rightleftarrows 0.04\,\underset{\textit{antigorite}}{Mg_{24}Si_{17}O_{42.5}(OH)_{31}} + Ca^{++} + 2\,HCO_3^- . \tag{6.12}$$

Since antigorite holds the place of magnesium in the basis, dolomite contains a negative amount of silica.

Dolomite precipitation consumes magnesium and produces H^+, causing the antigorite to dissolve completely. The resulting pH shift also causes quartz, which is most soluble under alkaline conditions, to become supersaturated. The program (step 5) replaces antigorite with $SiO_2(aq)$, leading to a solution supersaturated with respect to only quartz. Including quartz in the mineral assemblage (step 6), the program converges to a saturated solution representing the system's theoretical equilibrium state.

6.1.5 Interpreting Saturation Indices

It is tempting to place significance on the relative magnitudes of the saturation indices calculated for various minerals and then to relate these values to the amounts of minerals likely to precipitate from solution. The data in Table 6.6, however, suggest no such relationship. Thirteen minerals are supersaturated in the initial fluid, but the phase rule limits to ten the number of minerals that can form; only two (dolomite and quartz) appear in the final phase assemblage.

For a number of reasons, using saturation indices as measures of the mineral masses to be formed as a fluid approaches equilibrium is a futile (if commonly undertaken) exercise. First, a mineral's saturation index depends on the choice of its formula unit. If we were to write the formula for quartz as Si_2O_4 instead of SiO_2, we would double its saturation index. Large formula units have been chosen for many of the clay and zeolite minerals listed in the LLNL database, and this explains why these minerals appear frequently at the top of the supersaturation list.

Second, at a given saturation index, supersaturated minerals with high solubilities have the potential to precipitate in greater mass than do less soluble ones. Consider a solution equally supersaturated with respect to halite (NaCl) and gypsum ($CaSO_4{\cdot}2H_2O$). Of the two minerals, halite is the more soluble, and hence more of it must precipitate for the fluid to approach equilibrium.

Third, for minerals with binary or higher-order reactions, there is no assurance that the reactants are available in stoichiometric proportions. We could prepare solutions equally supersaturated with respect to gypsum by using differing Ca^{++} to SO_4^{--} ratios. A solution containing these components in equal amounts would precipitate the most gypsum. Solutions rich in Ca^{++} but depleted in SO_4^{--}, or rich in SO_4^{--} but depleted in Ca^{++}, would produce lesser amounts of gypsum.

Finally, common ion effects link many mineral precipitation reactions, so the reactions do not operate independently. In the seawater example, dolomite precipitation consumed magnesium and produced hydrogen ions, significantly altering the saturation states of the other supersaturated minerals.

6.2 Amazon River Water

We turn our attention to developing a chemical model of water from the Amazon River, using a chemical analysis reported by Hem (1985, p. 9). The procedure in SPECE8 is

```
pH = 6.5
SiO2(aq)  =  7.   mg/kg
Al+++     =   .07 mg/kg
Fe++      =   .06 mg/kg
Ca++      =  4.3  mg/kg
Mg++      =  1.1  mg/kg
Na+       =  1.8  mg/kg
HCO3-     = 19.   mg/kg
SO4--     =  3.   mg/kg
Cl-       =  1.9  mg/kg
O2(aq)    =  5.8  free mg/kg

balance on Cl-
go
```

Here, we use SPECE8, which does not account for precipitation of supersaturated minerals, rather than REACT, since we are not especially interested in the fluid's true equilibrium state.

The resulting species distribution (Table 6.7), as would be expected, differs sharply from that in seawater (Table 6.4). Species approach millimolal instead of molal concentrations, and activity coefficients differ less from unity. In the Amazon River water, the most abundant cation and anion are Ca^{++} and HCO_3^-; in seawater, in contrast, Na^+ and Cl^- predominate. Seawater, clearly, is not simply concentrated river water.

In the river water, as opposed to seawater, the neutral species O_2(aq), CO_2(aq), and SiO_2(aq) are among the species present in greatest concentration. Complexing among species is of little consequence in the river water, so the major cations and anions are present almost entirely as free ions.

The calculation predicts that a number of aluminum-bearing and iron-bearing minerals are supersaturated in the river water (Fig. 6.3). As discussed in the previous section, interpreting saturation indices calculated for natural waters can be problematic, since data errors can affect the predicted values so strongly.

Table 6.7 *Calculated molalities* (m)*, activity coefficients* (γ)*, and log activities* (a) *of the most abundant species in Amazon River water*

Species	m	γ	$\log a$
HCO_3^-	0.181×10^{-3}	0.974	−3.75
$O_2(aq)$	0.181×10^{-3}	1.000	−3.74
Cl^-	0.138×10^{-3}	0.973	−3.87
$CO_2(aq)$	0.130×10^{-3}	1.000	−3.89
$SiO_2(aq)$	0.116×10^{-3}	1.000	−3.93
Ca^{++}	0.106×10^{-3}	0.899	−4.02
Na^+	0.783×10^{-4}	0.973	−4.12
Mg^{++}	0.450×10^{-4}	0.901	−4.39
SO_4^{--}	0.305×10^{-4}	0.898	−4.56

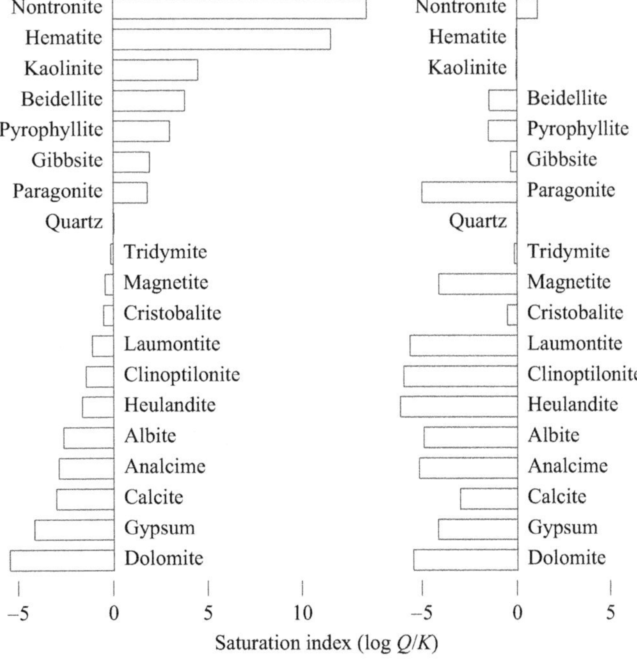

Figure 6.3 Saturation indices of Amazon River water with respect to various minerals (left) calculated directly from a chemical analysis, and (right) computed assuming that equilibrium with kaolinite and hematite controls the fluid's aluminum and iron content.

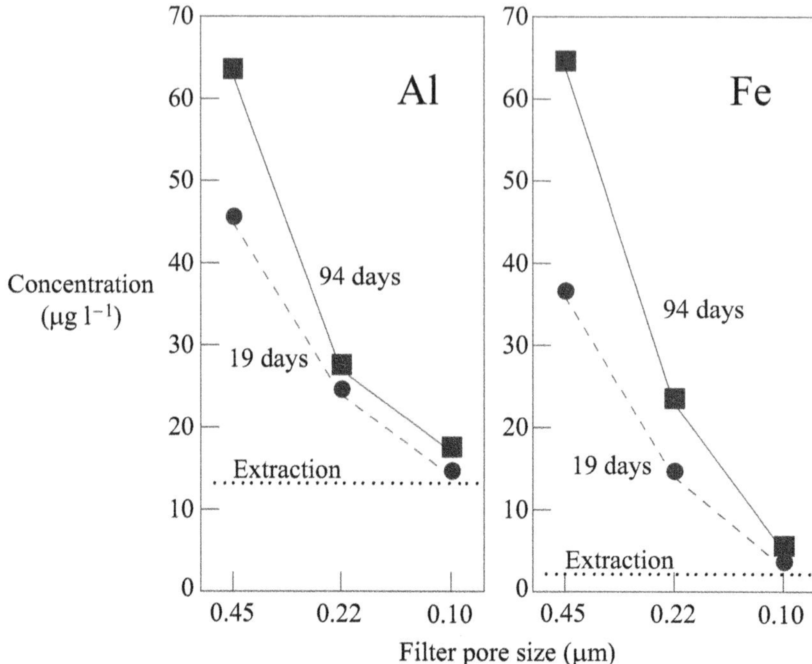

Figure 6.4 Effects of the pore size of filter paper used during fluid sampling on the analytical concentrations reported for aluminum and iron (Kennedy *et al.*, 1974). Samples were acidified, stored for 19 (●) or 94 (■) days, and analyzed by standard wet chemical methods. Dotted lines show dissolved concentrations determined by a solvent extraction technique.

In this case, a likely explanation for the apparent supersaturation is that the chemical analysis included not only dissolved aluminum and iron, but also a certain amount of aluminum and iron suspended in the water as colloids and fine sediments. Analytical error of this type occurs because the standard sampling procedure calls for passing the sample through a rather coarse filter of 0.45 μm pore size and then adding acid to "preserve" it during transport and storage.

Acidifying the sample causes colloids and fine sediments that passed through the filter to gradually dissolve, yielding abnormally high concentrations of elements such as aluminum, iron, silicon, and titanium when the fluid is analyzed. Figure 6.4, from a study of this problem by Kennedy *et al.* (1974), shows how the pore size of the filter paper used during sample collection affects the concentrations determined for aluminum and iron.

To construct an alternative model of Amazon River water, we assume that equilibrium with kaolinite (a clay mineral, $Al_2Si_2O_5(OH)_4$) and hematite (ferric oxide, Fe_2O_3) controls the aluminum and iron concentrations:

```
(cont'd)
swap Kaolinite for Al+++
swap Hematite for Fe++
1 free cm3 Kaolinite
1 free cm3 Hematite
go
```

Here, we arbitrarily specify the amounts of the minerals that coexist with the fluid. The amount chosen (in this

Table 6.8 *Chemical composition of the lower brine, Atlantis II deep, Red Sea (Shanks and Bischoff, 1977)*

Na^+ (mg kg^{-1})	92 600
K^+	1 870
Mg^{++}	764
Ca^{++}	5 150
Cl^-	156 030
SO_4^{--}	840
HCO_3^-	140
Ba^{++}	0.9
Cu^+	0.26
F^-	5
Fe^{++}	81
Pb^{++}	0.63
Zn^{++}	5.4
T (°C)	60
pH (25 °C, measured)	5.5
pH (60 °C, estimated)	5.6

case, 1 cm^3), of course, has no effect on the predicted fluid composition. We set "free" constraints because we wish to specify amounts of actual minerals (n_k), rather than mineral components (M_k).

According to the second model (Fig. 6.3), only the nontronite clay minerals (for instance, $Ca_{.165}Fe_2Al_{.33}Si_{3.67}O_{10}(OH)_2$) are supersaturated by any significant amount. The nontronite minerals appear supersaturated so frequently in modeling natural waters that they are almost certainly erroneously stable in the database. In fact, the stabilities of iron-bearing aluminosilicate minerals are problematic in general becauses of the special difficulties entailed in controlling redox state during solubility experiments.

The second model is perhaps more attractive than the first because the predicted saturation states seem more reasonable. The assumption of equilibrium with kaolinite and hematite can be defended on the basis of known difficulties in analyzing for dissolved aluminum and iron. Nonetheless, on the basis of information available to us, neither model is correct or incorrect; they are simply founded on differing assumptions. The most that we can say is that one model may prove more useful for our purposes than the other.

6.3 Red Sea Brine

We turn our attention now to the hydrothermal brines of the Red Sea. An oceanic survey in 1963 discovered pools of hot, saline, and metal-rich brines along the axial rift of the Red Sea (Degens and Ross, 1969; Hoffmann, 1991). The dense brines pond in the rift's depressions, or deeps. The Atlantis II deep contains the largest pool, which measures 5 × 14 km and holds about 5 km^3 of supersaline brine. The deep holds two layers of brine. The lower brine contains about 25 wt.% dissolved salts and exists at temperatures up to 60 °C. Table 6.8 shows the brine's average composition. A somewhat cooler, less saline water overlies the lower brine, separating it from normal seawater.

According to various lines of evidence (Shanks and Bischoff, 1977), subsea hot springs feed the brine pool by discharging each year about ten billion liters of brine, at temperatures of 150 to 250 °C, upward across the sea floor. The springs have yet to be located.

The Red Sea deeps attracted the attention of geologists because of the metalliferous sediments that have accumulated as muds along the sea floor. The muds contain sulfide minerals including pyrite, chalcopyrite, and sphalerite, as well as hematite, magnetite, barite, and clay minerals. The minerals apparently precipitated from the brine pool over the past 25 000 years, accumulating into a metal-rich layer about 20–30 m thick. Geologists study the deeps to look at the process of ore deposition in a modern environment and thus gain insight to how ancient mineral deposits formed.

Modeling the chemistry of highly saline waters is perilous (as discussed in Chapter 8) because of the difficulty in computing activity coefficients at high ionic strengths. Here, we employ an extension of the Debye–Hückel method (the B-dot equation; see Section 8.1), using an ionic strength of 3 molal, the limiting value for the correlations, instead of the actual ionic strength, which is greater than 5 molal. In Chapter 8, we examine alternative methods for estimating activity coefficients in brines.

We attempt the calculation in the hope that error in estimating activity coefficients will not be so large as to render the results meaningless. In fact, the situation may be slightly better than might be feared; because the activity coefficients appear in the numerator and denominator of the mass action equations, the error tends to cancel itself.

To model the brine's chemistry, we need to estimate its oxidation state. We could use the ratio of sulfate to sulfide species to fix a_{O_2}, but chemical analysis has not detected reduced sulfur in the brine, which is dominated by sulfate species. A less direct approach is to assume equilibrium with a mineral containing reduced iron or sulfur, or with a pair of minerals that form a redox couple. Equilibrium with hematite and magnetite, for example,

$$\underset{\textit{hematite}}{3\ \mathrm{Fe_2O_3}} \rightleftarrows \underset{\textit{magnetite}}{2\ \mathrm{Fe_3O_4}} + {}^{1}\!/_{2}\,\mathrm{O_2(aq)}\,, \tag{6.13}$$

fixes a_{O_2}, as does the coexistence of sphalerite,

$$\mathrm{Zn^{++}} + \mathrm{SO_4^{--}} \rightleftarrows \underset{\textit{sphalerite}}{\mathrm{ZnS}} + 2\,\mathrm{O_2(aq)}\,, \tag{6.14}$$

given constraints on the Zn^{++} and SO_4^{--} concentrations. We will assume the latter. We further constrain the Ba^{++} concentration by assuming equilibrium with barite,

$$\mathrm{Ba^{++}} + \mathrm{SO_4^{--}} \rightleftarrows \underset{\textit{barite}}{\mathrm{BaSO_4}}\ . \tag{6.15}$$

This assumption provides a convenient check on the calculation's accuracy, since we already know the fluid's barium content.

The calculation procedure in REACT is to swap sphalerite and barite into the basis in place of $O_2(aq)$ and Ba^{++}

```
swap Sphalerite for O2(aq)
swap Barite for Ba++
```

and then to set temperature and constrain each basis entry

(cont'd)
```
T = 60
pH = 5.6
TDS = 257000

Na+     =      92600    mg/kg
K+      =       1870    mg/kg
Mg++    =        764    mg/kg
Ca++    =       5150    mg/kg
Cl-     =     156030    mg/kg
SO4--   =        840    mg/kg
HCO3-   =        140    mg/kg
Cu+     =          .26  mg/kg
F-      =          5    mg/kg
Fe++    =         81    mg/kg
Pb++    =          .63  mg/kg
Zn++    =          5.4  mg/kg

1.e-9 free grams Sphalerite
1.e-9 free grams Barite
print species=long
go
```

Here, we set a vanishingly small free mass for each mineral so that only negligible amounts can dissolve during the calculation's second step, when supersaturated minerals precipitate from the fluid.

The species distribution (Table 6.9) calculated for the brine differs from that of seawater and Amazon River water in the large molalities predicted and the predominance of ion pairs such as NaCl, $CaCl^+$, and $MgCl^+$. The complex species make up a considerable portion of the brine's dissolved load.

It is interesting to compare the effects of complexing in the three waters we have studied so far. As shown in Table 6.10, the complexed fraction of each of the major dissolved components increases with salinity. Whereas complex species are of minor importance in the Amazon River water, they are abundant in seawater and account for about three-quarters of the calcium and sulfate and more than half of the magnesium in the Red Sea brine.

The principle of mass action explains the relationship between concentration and complexation. The abundance of ion pairs in aqueous solution is controlled by reactions such as

$$Na^+ + Cl^- \rightleftarrows NaCl \tag{6.16}$$

and

$$Ca^{++} + Cl^- \rightleftarrows CaCl^+ \, . \tag{6.17}$$

Starting with a dilute solution, for each doubling of the activities of the free ions on the left of these reactions, the activities of the ion pairs on the right sides must quadruple. As concentration increases, the ion pairs become progressively more important and eventually can come to overwhelm the free ions in solution. The higher temperature of the Red Sea brine also favors complexing because ion pairs gain stability relative to free ions as temperature increases.

We can check our results for the Red Sea brine against two independent pieces of information. In our results, sulfate species such as $NaSO_4^-$ dominate reduced sulfur species such as H_2S(aq) and HS^-, in seeming

Table 6.9 *Calculated molalities (m), activity coefficients (γ), and log activities (a) of the most abundant species in the Red Sea brine*

Species	m	γ	$\log a$
Cl^-	5.183	0.6125	0.5017
Na^+	4.861	0.7036	0.5341
$NaCl$	0.5512	1.0	−0.2587
$CaCl^+$	0.1276	0.7036	−1.047
K^+	0.5951×10^{-1}	0.6125	−1.438
Ca^{++}	0.4445×10^{-1}	0.1941	−2.064
$MgCl^+$	0.2496×10^{-1}	0.7036	−1.756
Mg^{++}	0.1692×10^{-1}	0.2895	−2.310
$NaSO_4^-$	0.7906×10^{-2}	0.7036	−2.255
KCl	0.4736×10^{-2}	1.0	−2.325
SO_4^{--}	0.2675×10^{-2}	0.0985	−3.579
$CO_2(aq)$	0.1809×10^{-2}	1.0	−2.743

Table 6.10 *Extent of complexing in three natural waters*

	Amazon River	Seawater	Red Sea deep
I (molal)	0.0006	0.65	5.3
TDS (mg kg^{-1})	45	35 000	260 000
Na^+ (% complexed)	0.03	2	10
K^+	—	3	8
Mg^{++}	0.5	28	60
Ca^{++}	0.8	44	74
Cl^-	~0	3	12
SO_4^{--}	2.4	45	77
$HCO_3^- + CO_2(aq)$	0.1	37	28

accord with the failure of analysis to detect reduced sulfur in the brine. The predominance of sulfate over sulfide species in our calculation reflects the oxidation state resulting from our assumption of equilibrium with sphalerite.

To check that this oxidation state might be reasonable, we determine the total concentration of sulfide species to be

Species	Molality
$H_2S(aq)$	5.2×10^{-8}
HS^-	6.6×10^{-9}
S^{--}	1.4×10^{-15}
S_2^{--}	8.6×10^{-19}
S_3^{--}	7.4×10^{-23}
S_4^{--}	2.7×10^{-24}
S_5^{--}	1.4×10^{-28}
S_6^{--}	3.7×10^{-33}
	5.9×10^{-8}

or about 0.002 mg kg^{-1} as H_2S. The detection limit for reduced sulfur in a typical chemical analysis is about 0.01 mg kg^{-1}. The sulfide concentration and oxidation state, therefore, do not appear unreasonable.

Second, the total barium concentration in solution, which is constrained by equilibrium with barite, is 6.6 μmolal or 0.68 mg kg^{-1}. The concentration reported by chemical analysis (Table 6.8) is 0.9 mg kg^{-1}, in close agreement with the calculation. Considering the uncertainties in the calculation, these values are probably fortuitously similar.

The calculation predicts that the brine is supersaturated with respect to seven minerals: bornite (Cu_5FeS_4), chalcopyrite ($CuFeS_2$), chalcocite (Cu_2S), pyrite (FeS_2), fluorite (CaF_2), galena (PbS), and covellite (CuS). The saturation index for bornite, the most supersaturated mineral, is greater than 7, indicating significant supersaturation. The predicted saturation indices reflect in large part the calculated values for activity coefficients and, in the case of the sulfide minerals, the oxidation state. Unfortunately, the activity coefficient model and oxidation state represent two of the principal uncertainties in the calculation.

Uncertainties aside, it is interesting to pursue the question of whether the brine has the potential to precipitate sulfide minerals. In the second step of the calculation, where the model allows precipitation reactions to progress to the system's equilibrium state, three minerals form in small quantities:

Mineral	Mass (g)
Fluorite, CaF_2	7.3×10^{-3}
Chalcocite, Cu_2S	9.3×10^{-6}
Barite, $BaSO_4$	5.8×10^{-8}

Less than 10 μg of sulfide mineral formed as the fluid (which has a total mass of 1.25 kg) equilibrated. Bornite, the most supersaturated mineral, failed to form at all.

From the perspective of ore genesis, these results seem disappointing in light of the fluid's high degree of supersaturation. Shanks and Bischoff (1977), for example, estimate that about 60 mg of sphalerite alone precipitate from each kg of ore fluid feeding the Atlantis II deep. The reaction to form chalcocite,

$$2\,Cu^+ + H_2S(aq) \rightleftarrows \underset{chalcocite}{Cu_2S} + 2\,H^+\,, \tag{6.18}$$

explains why so little of that mineral formed. Since the fluid is nearly depleted in reduced sulfur species, the reaction can proceed to the right by only a minute increment. As noted in Section 6.1, a large saturation index does not guarantee that the reactants needed to form a mineral are available in suitable proportions for the reaction to proceed to any significant extent.

Because little mass can precipitate from it, the brine, if related to deposition of the metalliferous muds, is likely to be a residuum of the original ore fluid. As it discharged into the deep, the ore fluid was richer in metals than in reduced sulfur. Mineral precipitation depleted the fluid of nearly all of its reduced sulfur without exhausting the metals, leaving the metal-rich brine observed in the deep.

7
Redox Disequilibrium

The equilibrium model, despite its limitations, in many ways provides a useful if occasionally abstract description of the chemical states of natural waters. However, if used to predict the state of redox reactions, especially at low temperature, the model is likely to fail. This shortcoming does not result from any error in formulating the thermodynamic model. Instead, it arises from the fact that redox reactions in natural waters proceed at such slow rates that they commonly remain far from equilibrium.

Complicating matters further is the fact that the platinum electrode, the standard tool for measuring Eh directly, does not respond to some of the most important redox couples in geochemical systems. The electrode, for example, responds incorrectly or not at all to the couples SO_4^{--}–HS^-, O_2–H_2O, CO_2–CH_4, NO_3^-–N_2, and N_2–NH_4 (Stumm and Morgan, 1996; Hostettler, 1984). In a laboratory experiment, Runnells and Lindberg (1990) prepared solutions with differing ratios of selenium in the Se^{4+} and Se^{6+} oxidation states. They found that even under controlled conditions the platinum electrode was completely insensitive to the selenium composition. The meaning of an Eh measurement from a natural water, therefore, may be difficult or impossible to determine (e.g., Westall, 2002).

7.1 Redox Potentials in Natural Waters

Geochemists (e.g., Thorstenson *et al.*, 1979; Thorstenson, 1984) have long recognized that at low temperature many redox reactions are unlikely to achieve equilibrium, and that the meaning of Eh measurements is problematic. Lindberg and Runnells (1984) demonstrated the generality of the problem. They compiled from the WATSTORE database more than 600 water analyses that provided at least two measures of oxidation state. The measures included Eh, dissolved oxygen content, concentrations of dissolved sulfate and sulfide, ferric and ferrous iron, nitrate and ammonia, and so on.

They calculated species distributions for each sample and then computed redox potentials for the various redox couples in the analysis, using the Nernst equation,

$$\mathrm{Eh} = -\frac{2.303\, R T_K}{n F} \log \frac{Q}{K} \tag{7.1}$$

(Eq. 3.66). Variable R in this equation is the gas constant (8.3143 J $\mathrm{mol^{-1} K^{-1}}$ or V C $\mathrm{mol^{-1} K^{-1}}$), T_K is absolute temperature (K), n is the number of electrons consumed in the half-reaction, F is the Faraday constant (96 485 C $\mathrm{mol^{-1}}$), Q is the half-reaction's ion activity product, and K is its equilibrium constant.

For example, when they found an analysis reporting concentrations of both sulfate and sulfide, they calculated the Nernst Eh for the reaction,

$$8\,e^- + SO_4^{--} + 9\,H^+ \rightleftarrows HS^- + 4\,H_2O\,, \tag{7.2}$$

according to

$$\text{Eh} = -\frac{2.303\,RT_K}{8F}\left(\log\frac{a_{\text{HS}^-}\,a^4_{\text{H}_2\text{O}}}{a_{\text{SO}_4^{--}}\,a^9_{\text{H}^+}} - \log K\right). \tag{7.3}$$

Given a measurement of dissolved oxygen, they similarly computed the Eh corresponding to the reaction

$$4\,\text{e}^- + \text{O}_2(\text{aq}) + 4\,\text{H}^+ \rightleftarrows 2\,\text{H}_2\text{O}\,, \tag{7.4}$$

according to

$$\text{Eh} = -\frac{2.303\,RT_K}{4F}\left(\log\frac{a^2_{\text{H}_2\text{O}}}{a_{\text{O}_2(\text{aq})}\,a^4_{\text{H}^+}} - \log K\right). \tag{7.5}$$

In this way, they could calculate a redox potential for each redox couple reported for a sample.

Their results show that the redox couples in a sample generally failed to achieve equilibrium with each other. For a given sample, the Nernst Eh values calculated for different redox couples varied over a broad range, by as much as 1000 mV. If the couples had been close to redox equilibrium, they would have yielded Nernst Eh values similar to each other. In addition, the authors could find little relationship between the Nernst values and Eh measured by platinum electrode. Criaud *et al.* (1989) computed similarly discordant Nernst Eh values for low-temperature geothermal fluids from the Paris basin.

There are, fortunately, some instances in which measured Eh values can be interpreted in a quantitative sense. Nordstrom *et al.* (1979), for example, showed that Eh measurements in acid mine drainage accurately reflect the $a_{\text{Fe}^{+++}}/a_{\text{Fe}^{++}}$ ratio. They further noted a number of other studies establishing agreement between measured and Nernst Eh values for various couples. Nonetheless, it is clearly dangerous for a geochemical modeler to assume a priori that a sample is in internal redox equilibrium or that an Eh measurement reflects a sample's redox state.

7.2 Redox Coupling

A flexible method for modeling redox disequilibrium is to divide the reaction database into two parts. The first part contains reactions between the basis species (e.g., Table 6.1) and a number of redox species, which represent the basis species in alternative oxidation states. For example, redox species Fe^{+++} forms a redox pair with basis species Fe^{++}, and HS^- forms a redox pair with SO_4^{--}. These coupling reactions are balanced in terms of an electron donor or acceptor, such as O_2(aq). Table 7.1 shows coupling reactions from the LLNL database.

The second part of the database contains reactions for the various secondary species, minerals, and gases. These reactions are balanced in terms of the basis and redox species, avoiding (to the extent practical) electron transfer. Species and minerals containing ferric iron, for example, are balanced in terms of the redox species Fe^{+++},

$$\underset{hematite}{\text{Fe}_2\text{O}_3} + 6\,\text{H}^+ \rightleftarrows 3\,\text{H}_2\text{O} + 2\,\text{Fe}^{+++}\,, \tag{7.6}$$

whereas those containing ferrous iron are balanced with basis species Fe^{++},

$$\text{FeSO}_4 \rightleftarrows \text{Fe}^{++} + \text{SO}_4^{--}\,. \tag{7.7}$$

The mineral magnetite (Fe_3O_4) contains oxidized and reduced iron, so its reaction,

Table 7.1 *Some of the redox couples in the* LLNL *database*

Redox pair	Coupling reaction
$AsH_3(aq)$–$As(OH)_4^-$	$AsH_3(aq) + H_2O + 3/2\,O_2(aq) \rightleftarrows As(OH)_4^- + H^+$
AsO_4^{---}–$As(OH)_4^-$	$As(OH)_4^- + 1/2\,O_2(aq) \rightleftarrows AsO_4^{---} + H_2O + 2\,H^+$
Au^{+++}–Au^+	$2\,H^+ + Au^+ + 1/2\,O_2(aq) \rightleftarrows Au^{+++} + H_2O$
CH_3COO^-–HCO_3^-	$CH_3COO^- + 2\,O_2(aq) \rightleftarrows 2\,HCO_3^- + H^+$
$CH_4(aq)$–HCO_3^-	$CH_4(aq) + 2\,O_2(aq) \rightleftarrows H_2O + H^+ + HCO_3^-$
ClO_4^-–Cl^-	$Cl^- + 2\,O_2(aq) \rightleftarrows ClO_4^-$
Co^{+++}–Co^{++}	$Co^{++} + H^+ + 1/4\,O_2(aq) \rightleftarrows Co^{+++} + 1/2\,H_2O$
Cr^{++}–Cr^{+++}	$Cr^{++} + H^+ + 1/4\,O_2(aq) \rightleftarrows 1/2\,H_2O + Cr^{+++}$
CrO_4^{--}–Cr^{+++}	$Cr^{+++} + 5/2\,H_2O + 3/4\,O_2(aq) \rightleftarrows CrO_4^{--} + 5\,H^+$
CrO_4^{---}–Cr^{+++}	$Cr^{+++} + 3\,H_2O + 1/2\,O_2(aq) \rightleftarrows CrO_4^{---} + 6\,H^+$
Cu^{++}–Cu^+	$Cu^+ + H^+ + 1/4\,O_2(aq) \rightleftarrows Cu^{++} + 1/2\,H_2O$
Eu^{++}–Eu^{+++}	$Eu^{++} + H^+ + 1/4\,O_2(aq) \rightleftarrows 1/2\,H_2O + Eu^{+++}$
Fe^{+++}–Fe^{++}	$Fe^{++} + H^+ + 1/4\,O_2(aq) \rightleftarrows Fe^{+++} + 1/2\,H_2O$
$H_2(aq)$–$O_2(aq)$	$H_2(aq) + 1/2\,O_2(aq) \rightleftarrows H_2O$
HS^-–SO_4^{--}	$HS^- + 2\,O_2(aq) \rightleftarrows SO_4^{--} + H^+$
Hg_2^{++}–Hg^{++}	$Hg_2^{++} + 2\,H^+ + 1/2\,O_2(aq) \rightleftarrows 2\,Hg^{++} + H_2O$
MnO_4^-–Mn^{++}	$Mn^{++} + 3/2\,H_2O + 5/4\,O_2(aq) \rightleftarrows MnO_4^- + 3\,H^+$
MnO_4^{--}–Mn^{++}	$Mn^{++} + 2\,H_2O + O_2(aq) \rightleftarrows MnO_4^{--} + 4\,H^+$
$N_2(aq)$–NO_3^-	$N_2(aq) + H_2O + 5/2\,O_2(aq) \rightleftarrows 2\,H^+ + 2\,NO_3^-$
NH_4^+–NO_3^-	$NH_4^+ + 2\,O_2(aq) \rightleftarrows NO_3^- + 2\,H^+ + H_2O$
NO_2^-–NO_3^-	$NO_2^- + 1/2\,O_2(aq) \rightleftarrows NO_3^-$
Se^{--}–SeO_3^{--}	$Se^{--} + 3/2\,O_2(aq) \rightleftarrows SeO_3^{--}$
SeO_4^{--}–SeO_3^{--}	$SeO_3^{--} + 1/2\,O_2(aq) \rightleftarrows SeO_4^{--}$
Sn^{++}–Sn^{++++}	$Sn^{++} + 2\,H^+ + 1/2\,O_2(aq) \rightleftarrows Sn^{++++} + H_2O$
U^{+++}–U^{++++}	$U^{+++} + H^+ + 1/4\,O_2(aq) \rightleftarrows U^{++++} + 1/2\,H_2O$
UO_2^+–U^{++++}	$U^{++++} + 3/2\,H_2O + 1/4\,O_2(aq) \rightleftarrows UO_2^+ + 3\,H^+$
UO_2^{++}–U^{++++}	$U^{++++} + H_2O + 1/2\,O_2(aq) \rightleftarrows UO_2^{++} + 2\,H^+$
VO^{++}–V^{+++}	$V^{+++} + 1/2\,H_2O + 1/4\,O_2(aq) \rightleftarrows VO^{++} + H^+$
VO_4^{---}–V^{+++}	$V^{+++} + 3\,H_2O + 1/2\,O_2(aq) \rightleftarrows VO_4^{---} + 6\,H^+$

$$\underset{magnetite}{Fe_3O_4} + 8\,H^+ \rightleftarrows 2\,Fe^{+++} + Fe^{++} + 4\,H_2O\,, \tag{7.8}$$

contains both the basis and redox species.

The modeler controls which redox reactions should be in equilibrium by interactively coupling or decoupling the redox pairs. For each coupled pair, the model uses the corresponding coupling reaction to eliminate redox species from the reactions in the database. For example, if the pair Fe^{+++}–Fe^{++} is coupled, the model adds the coupling reaction to the reaction for hematite,

$$\begin{array}{rcl} Fe_2O_3 + 6\,H^+ & \rightleftarrows & 3\,H_2O + 2\,Fe^{+++} \\ 2 \times [\,Fe^{+++} + 1/2\,H_2O & \rightleftarrows & Fe^{++} + H^+ + 1/4\,O_2(aq)\,] \\ \hline \underset{hematite}{Fe_2O_3} + 4\,H^+ & \rightleftarrows & 2\,H_2O + 2\,Fe^{++} + 1/2\,O_2(aq)\,, \end{array} \tag{7.9}$$

to eliminate the redox species Fe^{+++}. The same procedure is applied to the reactions for the other species and minerals that contain ferric iron.

A coupling reaction commonly links a redox species to a basis species, as in the examples above, but it is also possible to define couples among the redox species themselves. If HCO_3^- appears in the basis, for example, methane might be linked to it by the coupling reaction

$$CH_4(aq) + 2\,O_2(aq) \rightleftarrows HCO_3^- + H^+ + H_2O\,. \tag{7.10}$$

Then, the couple for acetate ion can be written as a dismutation reaction,

$$CH_3COO^- + H_2O \rightleftarrows HCO_3^- + CH_4(aq)\,, \tag{7.11}$$

in terms of the redox species methane, rather than as a simple oxidation to the basis species bicarbonate. Disengaging the first but not the second coupling reaction, a model could be constructed in which acetate is in redox equilibrium with bicarbonate and methane, even though those species are not in equilibrium with each other.

Models of natural waters calculated assuming redox disequilibrium generally require more input data than equilibrium models, in which a single variable constrains the system's oxidation state. The modeler can decouple as many redox pairs as can be independently constrained. A completely decoupled model, therefore, would require analytical data for each element in each of its redox states. Unfortunately, analytical data of this completeness are seldom collected.

7.3 Morro do Ferro Groundwater

As an example of modeling a fluid in redox disequilibrium, we use an analysis, slightly simplified from Nordstrom *et al.* (1992), of a groundwater sampled near the Morro do Ferro ore district in Brazil (Table 7.2). There are three measures of oxidation state in the analysis: the Eh value determined by platinum electrode, the dissolved oxygen content, and the distribution of iron between ferrous and ferric species.

To calculate an equilibrium model in SPECE8, the procedure is

```
T = 22
pH = 6.05
```

Table 7.2 *Chemical analysis of a groundwater from near the Morro do Ferro deposits, Brazil (Nordstrom* et al., *1992)*

HCO_3^- (mg l^{-1})	1.8
Ca^{++}	0.238
Mg^{++}	0.352
Na^+	0.043
K^+	0.20
Fe (II)	0.73
Fe (total)	0.76
Mn^{++}	0.277
Zn^{++}	0.124
SO_4^{--}	1.5
Cl^-	<2.0
Dissolved O_2	4.3
T (°C)	22
pH	6.05
Eh (mV)	504

```
O2(aq)  = 4.3   free mg/l
HCO3-   = 1.8   mg/l
Ca++    = 0.238 mg/l
Mg++    = 0.352 mg/l
Na+     = 0.043 mg/l
K+      = 0.20  mg/l
Fe++    = 0.73  mg/l
Mn++    = 0.277 mg/l
Zn++    = 0.124 mg/l
SO4--   = 1.5   mg/l

balance on Cl-
print species = long
go
```

Here, we set oxidation state in the model using the dissolved oxygen content and calculate the distribution of species assuming redox equilibrium.

To account for the possibility of redox disequilibrium among iron species, we use the analysis for ferrous as well as total iron:

```
(cont'd)
decouple Fe+++
Fe+++   = 0.03  mg/l
Fe++    = 0.73  mg/l
go
```

Table 7.3 *Concentrations (molal) of predominant iron species in Morro do Ferro groundwater, calculated assuming equilibrium and redox disequilibrium*

Species	Equilibrium	Disequilibrium
Ferrous		
Fe^{++}	0.11×10^{-12}	0.13×10^{-4}
$FeSO_4$	0.24×10^{-15}	0.29×10^{-7}
$FeHCO_3^+$	0.20×10^{-16}	0.25×10^{-8}
$FeCl^+$	0.68×10^{-17}	0.13×10^{-8}
$FeOH^+$	0.65×10^{-17}	0.78×10^{-9}
Ferric		
$Fe(OH)_2^+$	0.92×10^{-5}	0.38×10^{-6}
$Fe(OH)_3$	0.39×10^{-5}	0.16×10^{-6}
$FeOH^{++}$	0.29×10^{-7}	0.12×10^{-8}
$Fe(OH)_4^-$	0.93×10^{-9}	0.38×10^{-10}

By decoupling the ferric–ferrous reaction with the `decouple` command, we add Fe^{+++} as a new basis entry in the calculation, setting up a model in which O_2 and iron are held in redox disequilibrium. We constrain the new entry using the difference between the total and ferrous iron contents.

As shown in Table 7.3, the two calculations predict broadly differing species distributions for iron. In the first calculation, the fluid is almost devoid of ferrous iron species, reflecting the high concentration of dissolved oxygen. This result contradicts the dominance of ferrous over ferric species reported in the chemical analysis. The disequilibrium calculation, in which we separately constrain the fluid's ferrous and ferric iron contents, provides a species distribution in which ferrous iron species predominate, in accord with the analytical data.

We can compare the Eh measured for the Morro do Ferro groundwater (Table 7.2) with the Nernst Eh values (Eq. 7.1) given by the reactions for dissolved oxygen and iron oxidation, as reported in the program output,

	Eh (mV)
Measured by electrode	504
$1/4\ O_2(aq) + H^+ + e^- \rightleftarrows 1/2\ H_2O$	861
$Fe^{+++} + e^- \rightleftarrows Fe^{++}$	306

The ratio of ferrous to ferric species represents a redox state considerably less oxidizing than suggested by the dissolved oxygen content. The measured Eh falls between these values. Because the values vary over a range of more than 500 mV, this water clearly is not in redox equilibrium; assuming that it is gives an incorrect distribution of iron species.

7.4 Energy Available for Microbial Respiration

In recent years, geochemists have come to appreciate that chemotrophic microorganisms pervade the upper kilometers of the Earth's crust and play a profound role in controlling the chemistry of water in the hydrosphere (e.g., Banfield and Nealson, 1997; Chapelle, 2001). Although microbes of this class are primary producers of biomass, they take the energy they need to live and grow not from the Sun by photosynthesis, but from chemical energy found in the natural environment. Chemolithotrophs derive energy from inorganic species, whereas chemoheterotrophs obtain energy from organic molecules. Because the ability of the microbes to grow depends on geochemical conditions in the environment, and the microbes in turn can affect the environment so significantly, the study of microbial processes in the geosphere has become an active and fertile area of inquiry.

The chemotrophs take advantage of the redox disequilibrium found in the natural environment to derive the energy they need to synthesize new cells and maintain those already formed. Many such microbes work by respiration, transferring electrons from a reduced donor species to an oxidized acceptor species. To respire, a microbe uses a special enzyme in its cell membrane to strip one or more electrons from the electron donor, leaving behind a more oxidized species. The electron cascades through a series of enzymes and coenzymes in the cell membrane, known as the electron transport chain, before reaching a terminal enzyme. The terminal enzyme transfers one or more electrons onto the accepting species, leaving that species reduced.

As an electron passes through the electron transport chain, it releases energy that the microbe conserves in its cytoplasm in the form of adenosine triphosphate (ATP), which it synthesizes from adenosine diphosphate (ADP) and free phosphate ions. Adenosine triphosphate is a relatively unstable molecule that serves as the cell's primary energy store. By coupling the energetically favorable breakdown of ATP (to form ADP) to energetically unfavorable reactions, a microbe accomplishes critical tasks such as creating the molecules it needs to grow and reproduce. In this way, respiring microbes work by using their enzymes to catalyze electron transfer reactions that would otherwise proceed too slowly to approach equilibrium. From the geochemist's perspective, a microbe might be described as a self-replicating bundle of enzymes, and chemotrophic growth as an autocatalytic reaction.

By analogy to higher animals, we might expect that respiring microbes would use reduced carbon compounds as electron donors and O_2 as the electron acceptor; indeed, aerobic bacteria that do just this are common in oxic environments. Microbes, however, are notably versatile, employing H_2, H_2S, NH_4^+, CH_4, Fe^{++}, and many other species as electron donors. They can similarly use SO_4^{--}, NO_3^-, NO_2^-, HCO_3^-, and so on as electron acceptors. The microbes can even use ferric and oxidized Mn minerals as electron acceptors, in effect "breathing" the rocks and sediments in which they live.

Respiration, as we have described, drives two half-reactions, one to donate electrons and one to accept them. Iron-reducing bacteria, for example, can live on acetate, which is produced during the breakdown of organic matter. Oxidizing acetate provides electrons,

$$\mathrm{CH_3COO^- + 4\,H_2O \rightarrow 2\,HCO_3^- + 9\,H^+ + 8\,e^-}\,, \tag{7.12}$$

to the cell. The electrons are eventually consumed by reducing ferric iron,

$$\underset{\textit{ferric hydroxide}}{\mathrm{Fe(OH)_3}} \quad \mathrm{+ 3\,H^+ + e^- \rightarrow Fe^{++} + 3\,H_2O}\,, \tag{7.13}$$

which is insoluble at neutral pH and hence commonly available in the environment as hydrous ferric oxide. The overall reaction,

$$\begin{aligned} \mathrm{CH_3COO^-} + 8 \underset{\textit{ferric hydroxide}}{\mathrm{Fe(OH)_3}} + 15\,\mathrm{H^+} \rightarrow \\ 2\,\mathrm{HCO_3^-} + 8\,\mathrm{Fe^{++}} + 20\,\mathrm{H_2O}\,, \end{aligned} \tag{7.14}$$

is the sum of the half-cell reactions, weighted to conserve electrons.

A second class of chemotrophs works by fermentation instead of respiration. In fermentation, an energy source such as acetate reacts to form two species, one more reduced than the energy source, and one more oxidized. Acetotrophic methanogens, for example, cleave the acetate ion to form methane and bicarbonate. The energy source for a fermenting microbe serves as both electron donor and acceptor, and the energy available to it can be analyzed in the same way as for a respiring organism by writing the fermentation reaction as the sum of an oxidizing half-reaction and a reducing half-reaction. For an acetotrophic methanogen, for example, we would write a half-reaction by which acetate is oxidized to bicarbonate, and one in which it is reduced to methane.

Understanding energy availability in natural waters requires that we consider many possible combinations of donor and acceptor species. Respiration can proceed only for combinations in which the electron transfer is energetically favorable; i.e., the overall reaction must release free energy (Thauer *et al.*, 1977). It is, furthermore, in the microbe's interest to respire only under conditions where the electron transfer liberates enough energy for a cell to synthesize ATP (e.g., Jin and Bethke, 2002; 2009; Bethke *et al.*, 2011). The possibility of a given microbial metabolism proceeding in the natural environment, therefore, depends directly on the extent of redox disequilibrium there.

The difference $\Delta\mathrm{Eh}$ between the redox potentials for the electron accepting and donating reactions ($\mathrm{Eh_{acc}}$ and $\mathrm{Eh_{don}}$, respectively) provides a convenient measure of the energy available in an environment (e.g., Madigan and Martinko, 2017). This value is a measure of the energy change of reaction ΔG_r (J mol^{-1}, or V C mol^{-1}) for the overall reaction catalyzed by a microbe's metabolism. The relationship between the potential difference and ΔG_r is

$$\Delta\mathrm{Eh} = \mathrm{Eh_{acc}} - \mathrm{Eh_{don}} = -\frac{\Delta G_r}{nF}\,, \tag{7.15}$$

where n is the number of electrons accepted or donated and F is the Faraday constant (96 485 C mol^{-1}).

A microbe can derive energy only from a reaction with a negative ΔG_r; hence the Eh for the electron donor must be smaller than that of the electron acceptor for the microbe's metabolism to proceed. As noted, the reaction must liberate enough energy for the microbe to synthesize ATP. The potential difference required to release this amount of energy varies depending on the microbe's metabolism (Jin and Bethke, 2005), ranging from perhaps 50 mV for hydrogenotrophic methanogens to as much as 500 mV for aerobic respirers.

The SPECE8 input script below describes the analysis of a hypothetical groundwater, assuming equilibrium with ferric hydroxide and a soil gas in which $P_{\mathrm{CO_2}} = 10^{-2}$ bar. In the script, we decouple a number of redox pairs so that we can constrain the amounts of several elements in two or more redox states.

```
decouple CH3COO-
decouple CH4(aq)
decouple Fe+++
decouple H2(aq)
decouple HS-
decouple NH4+
decouple NO2-
```

```
swap CO2(g) for HCO3-
swap Fe(OH)3(ppd) for Fe+++

pH = 6
P CO2(g)  = .01 bar
1 free cm3 Fe(OH)3(ppd)

Cl-      = 15      mg/kg
Na+      = 10      mg/kg
Ca++     = 15      mg/kg
Mg++     =  2      mg/kg
SO4--    = 35      mg/kg
Fe++     =    .2   mg/kg
H2(aq)   =    .004 free umolal
HS-      =    .05  mg/kg
O2(aq)   =    .1   free mg/kg
CH4(aq)  =    .4   mg/kg
NO3-     =  2      mg/kg
NO2-     =    .4   mg/kg
NH4+     =    .1   mg/kg
CH3COO-  =    .3   mg/kg

go
```

The program reports in its output the resulting redox potential for each redox couple, as calculated from the Nernst equation. The Nernst potentials, arranged in decreasing order, are

	Eh (mV)
$e^- + 1/4\ O_2(aq) + H^+ \rightleftarrows 1/2\ H_2O$	836
$2\ e^- + 2\ H^+ + NO_3^- \rightleftarrows H_2O + NO_2^-$	481
$8\ e^- + 10\ H^+ + NO_3^- \rightleftarrows 3\ H_2O + NH_4^+$	443
$6\ e^- + 8\ H^+ + NO_2^- \rightleftarrows 2\ H_2O + NH_4^+$	430
$e^- + Fe^{+++} \rightleftarrows Fe^{++}$	322
$8\ e^- + 9\ H^+ + SO_4^{--} \rightleftarrows 4\ H_2O + HS^-$	−126
$4\ e^- + 9/2\ H^+ + 1/2\ CH_3COO^- \rightleftarrows H_2O + CH_4(aq)$	−145
$8\ e^- + 9\ H^+ + HCO_3^- \rightleftarrows 3\ H_2O + CH_4(aq)$	−187
$2\ e^- + 2\ H^+ \rightleftarrows H_2(aq)$	−199
$8\ e^- + 9\ H^+ + 2\ HCO_3^- \rightleftarrows 4\ H_2O + CH_3COO^-$	−230

Reactions at the top of this list are most favored in this environment to accept electrons, and those at the bottom most prone to donate them.

In considering energy availability, it is helpful to remember that "electrons fall uphill." In other words, if the overall reaction is to liberate energy, the donating half-reaction in a list ordered as shown must appear below the accepting half-reaction, to give a positive ΔEh. The oxidation of acetate ($Eh_{don} = -230$ mV) by ferric iron ($Eh_{don} = +322$ mV) is, then, strongly favored, since ΔEh for this pairing is $+552$ mV. The oxidation of methane by O_2 is strongly favored ($\Delta Eh = +1023$ mV), as we might expect. Hydrogenotrophic methanogenesis in this water yields only $+10$ mV, not enough to synthesize ATP, and reduction of sulfate by ferrous iron is not favored to proceed, since ΔEh is -448 mV.

8
Activity Coefficients

Among the most vexing tasks for geochemical modelers, especially when they work with concentrated solutions, is estimating values for the activity coefficients of electrolyte species. To understand in a qualitative sense why activity coefficients in electrolyte solutions vary, we can imagine how solution concentration affects species activities. In the solution, electrical attraction draws anions around cations and cations around anions. We might think of a dilute solution as an imperfect crystal of loosely packed, hydrated ions that, within a matrix of solvent water, is constantly rearranging itself by Brownian motion. A solution of uncharged, nonpolar species, by contrast, would be nearly random in structure.

The electrolyte solution is lower in free energy G than it would be if the species did not interact electrically because of the energy liberated by moving ions of opposite charge together while separating those of like charge. The chemical potentials μ_i of the species, for the same reason, are smaller than they would be in the absence of electrostatic forces. By the equation

$$\mu_i \equiv \frac{\partial G_i}{\partial n_i} = \mu_i^{\mathrm{o}} + RT_{\mathrm{K}} \ln \gamma_i m_i \,, \tag{8.1}$$

(Eq. 3.5, taking $a_i = \gamma_i m_i$), the reduction in a species' chemical potential is reflected by a decreased value (relative to one in an ideal solution) for its activity coefficient. By Coulomb's law, electrostatic forces vary inversely with the square of the distance of ion separation. For this reason, activity coefficients in dilute fluids decrease as concentration increases because the coulombic forces become stronger as ions pack together more closely.

When electrical attraction and repulsion operate over distances considerably larger than the hydrated sizes of the ions, we can compute species' activities quite well from electrostatic theory, as demonstrated in the 1920s by the celebrated physical chemists Debye and Hückel. At moderate concentrations, however, the ions pack together rather tightly. In a one molal solution, for example, just a few molecules of solvent separate ions. Because electrical interactions at short range are complicated in nature, it is no longer sufficient to treat ions as points of electrical charge, as Debye and Hückel did in their analysis. Rather, we need to account for the distribution of electrical charge throughout the hydration sphere of each ion. Since this distribution is complex, simple analysis cannot account for the energetics of attraction and repulsion among closely packed ions.

In concentrated solutions, the energetic effect of ion repulsion seems to dominate. This effect is sometimes termed "hard core repulsion." With increasing concentration, furthermore, much and then most of the water in the solution is taken up in the species' hydration spheres, reducing the amount of water acting as solvent. These two effects cause the activity coefficients to increase. Increasing concentration, on the other hand, leads to greater degrees of ion association, which serves to decrease a species' free molality and, hence, activity relative to that expected if the ion were fully dissociated. Depending on the activity model employed, the effects of ion association may be accounted for directly by mass action (thereby decreasing activity by lowering species concentration) or indirectly, by adjusting activity coefficients while assuming complete dissociation.

The modeler should bear in mind a further complication: the requirement of electroneutrality (as discussed in Chapter 3) precludes the possibility of observing either the chemical potential of a charged species or its activity coefficient. We can measure only the mean activity coefficient of the ions comprising a salt. This value can be separated into the coefficients for single ions by following any of a number of conventions. The MacInnes convention, which is commonly but not universally employed in geochemical calculations, holds that the coefficients for K^+ and Cl^- are equal in a solution of a given ionic strength. The modeler must guard against mixing activity coefficients determined using differing conventions.

Geochemical modelers currently employ two types of methods to estimate activity coefficients (Plummer, 1992; Wolery, 1992b). The first type consists of applying variants of the Debye–Hückel equation, a simple relationship that treats a species' activity coefficient as a function of the species' size and the solution's ionic strength. Methods of this type take into account the distribution of species in solution and are easy to use, but can be applied with accuracy to modeling only relatively dilute fluids.

Virial methods, the second type, employ coefficients that account for interactions among the individual components (rather than species) in solution. The virial methods are less general, rather complicated to apply, require considerable amounts of data, and allow little insight into the distribution of species in solution. They can, however, reliably predict mineral solubilities even in concentrated brines.

The following sections describe the two estimation techniques. The discussion here leans toward the practical aspects of estimating activity coefficients. For an understanding of the theoretical basis of the activity models, the reader may refer to a thermodynamics text (such as Robinson and Stokes, 1968, or Anderson, 2017) and the papers referenced herein.

8.1 Debye–Hückel Methods

In 1923, Debye and Hückel published their famous papers describing a method for calculating activity coefficients in electrolyte solutions. They assumed that ions behave as spheres with charges located at their center points. The ions interact with each other by coulombic forces. Robinson and Stokes (1968) present their derivation, and the papers are available (Interscience Publishers, 1954) in English translation.

The result of their analysis, known as the Debye–Hückel equation,

$$\log \gamma_i = -\frac{A z_i^2 \sqrt{I}}{1 + \mathring{a}_i B \sqrt{I}}, \tag{8.2}$$

gives the activity coefficient γ_i of an ion with electrical charge z_i. In this chapter we need not differentiate between the basis and secondary species, A_i and A_j. Hence, we will let γ_i, z_i, and so on represent the properties of the aqueous species in general.

Variable $\mathring{a}_i$ in Equation 8.2 is the ion size parameter. In practice, this value is determined by fitting the Debye–Hückel equation to experimental data. Variables A and B are functions of temperature, and I is the solution ionic strength. At 25 °C, given I in molal units and taking $\mathring{a}_i$ in Å, the value of A is 0.5092, and B is 0.3283.

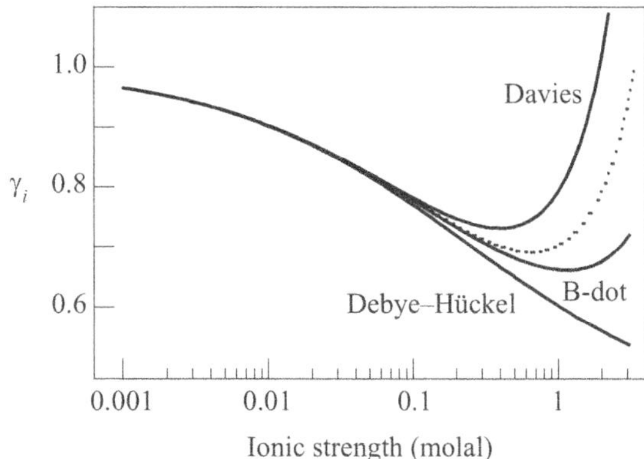

Figure 8.1 Activity coefficients γ_i predicted at 25 °C for a singly charged ion with size $\mathring{a}$ of 4 Å, according to the Debye–Hückel (Eq. 8.2), Davies (Eq. 8.4), and B-dot (Eq. 8.5) equations. Dotted line shows the Davies equation evaluated with a coefficient of 0.2 instead of 0.3.

The ionic strength,

$$I = \frac{1}{2} \sum_i m_i z_i^2 , \tag{8.3}$$

is half the sum of the product of each species molality m_i and the square of its charge. In the literature, ionic strength may be reported in molar or molal units, and may be calculated either by accounting for the effect of ion pairing or assuming that the electrolyte dissociates completely. We use molal units here and follow Helgeson's (1969) terminology regarding the question of ion pairing. Ionic strength I refers to the "true ionic strength" calculated by Equation 8.3, accounting for the role of complexing in reducing the number of free ions in solution. We refer to the value calculated assuming complete dissociation (i.e., neglecting ion pairs) as the "stoichiometric ionic strength," which we label I_S.

Equation 8.2 is notable in that it predicts a species' activity coefficient using only two numbers (z_i and $\mathring{a}_i$) to account for the species' properties and a single value I to represent the solution. As such, it can be applied readily to study a variety of geochemical systems, simple and complex.

Unfortunately, the equation becomes inaccurate at moderate ionic strength, above about 0.1 molal (e.g., Stumm and Morgan, 1996). As can be seen from Equation 8.2, Debye–Hückel activity coefficients approach unity when ionic strength nears zero. With increasing ionic strength, the coefficient decreases monotonically (Fig. 8.1). This decrease reflects the increasing strength of the long-range coulombic forces in solution, when short-range forces and hydration effects are ignored.

8.1.1 Davies Equation

The Davies (1962) equation is a variant of the Debye–Hückel equation (Eq. 8.2) that can be carried to somewhat higher ionic strengths. The equation follows from Equation 8.2 by noting that at 25 °C the product $\mathring{a}_i B$ is about one. Including an empirical term $0.3I$ in the correlation gives

$$\log \gamma_i = -A z_i^2 \left(\frac{\sqrt{I}}{1 + \sqrt{I}} - 0.3I \right) \tag{8.4}$$

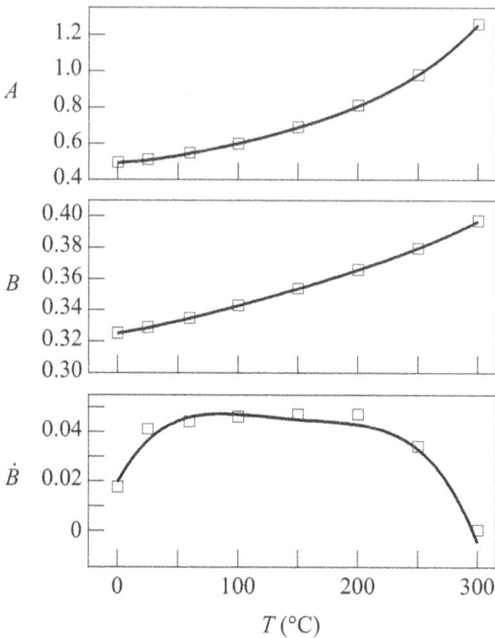

Figure 8.2 Values of A, B, and $\dot{B}$ for the B-dot (modified Debye–Hückel) equation at 0 °C, 25 °C, 60 °C, 100 °C, 150 °C, 200 °C, 250 °C, and 300 °C (squares) and interpolation functions (lines). Values correspond to I taken in molal and $\mathring{a}$ in Å. Data from the LLNL database, after Helgeson (1969) and Helgeson and Kirkham (1974).

(Davies originally used a factor of 0.2, but later corrected the coefficient to 0.3.) The only variable specific to the species in question is the charge z_i, which of course is known. For this reason, the Davies equation is especially easy to apply within geochemical models designed for work at 25 °C, such as WATEQ (Ball *et al.*, 1979) and its successors, and PHREEQE (Parkhurst *et al.*, 1980).

As can be seen in Figure 8.1, the Davies equation does not decrease monotonically with ionic strength, as the Debye–Hückel equation does. Beginning at ionic strengths of about 0.1 molal, it deviates above the Debye–Hückel function and at about 0.5 molal starts to increase in value. The Davies equation is reasonably accurate to an ionic strength of about 0.3 or 0.5 molal.

8.1.2 B-Dot Model

Helgeson (1969; see also Helgeson and Kirkham, 1974) presented an activity model based on an equation similar in form to the Davies equation. The model, adapted from earlier work (see Pitzer and Brewer, 1961, p. 326, p. 578, and Appendix D, and references therein), is parameterized from 0 °C to 300 °C for solutions of up to 3 molal ionic strength in which NaCl is the dominant solute. The model takes its name from the "B-dot" equation,

$$\log \gamma_i = -\frac{A z_i^2 \sqrt{I}}{1 + \mathring{a}_i B \sqrt{I}} + \dot{B} I\,, \tag{8.5}$$

which is an extension of the Debye–Hückel equation (Eq. 8.2). Coefficients A, B, and $\dot{B}$ vary with temperature, as shown in Figure 8.2, whereas the ion size parameter $\mathring{a}_i$ for each species remains constant.

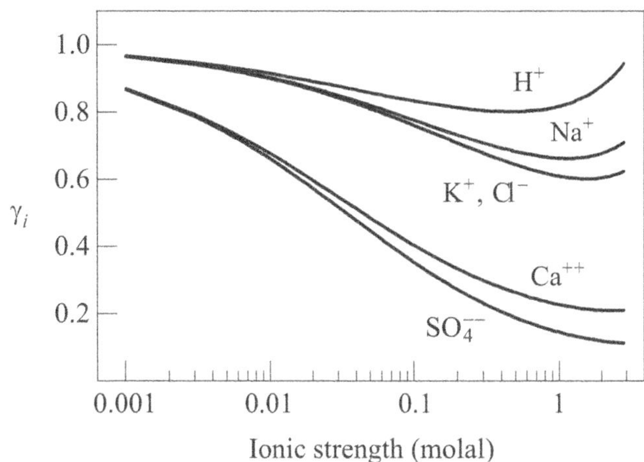

Figure 8.3 Activity coefficients γ_i predicted at 25 °C by the B-dot equation (Eq. 8.5) for some singly and doubly charged ions, as a function of ionic strength. Corresponding $\mathring{a}$ values are 3 Å (K^+ and Cl^-), 4 Å (Na^+ and SO_4^{--}), 6 Å (Ca^{++}), and 9 Å (H^+).

The B-dot equation is widely applied in geochemical models designed to operate over a range of temperatures, such as EQ3/EQ6 (Wolery, 1979; 2003), CHILLER (Reed, 1982), SOLMINEQ (Kharaka *et al.*, 1988), and the GWB programs (Bethke *et al.*, 2021). The equation is considered reasonably accurate in predicting the activities of Na^+ and Cl^- ions to concentrations as large as several molal, and of other species to ionic strengths up to about 0.3 to 1 molal. Figure 8.3 shows the activity coefficients predicted at 25 °C for species of differing charge and ion size.

In the B-dot model, as currently applied (Wolery, 2003), the activity coefficients of electrically neutral, nonpolar species [$B(OH)_3$, O_2(aq), SiO_2(aq), CH_4(aq), and H_2(aq)] are calculated from ionic strength using an empirical relationship,

$$\log \gamma_o = aI + bI^2 + cI^3 . \tag{8.6}$$

Here, a, b, and c are polynomial coefficients that vary with temperature,

	a	b	c
25 °C	0.1127	−0.01049	1.545×10^{-3}
100 °C	0.08018	−0.001503	0.5009×10^{-3}
200 °C	0.09892	−0.01040	1.386×10^{-3}
300 °C	0.1967	−0.01809	-2.497×10^{-3}

Figure 8.4 shows the function plotted against ionic strength at 25 °C, 100 °C, and 300 °C. The rapid increase in value at high ionic strength represents the "salting out" effect by which gas solubility decreases with increasing salinity. In the model, polar neutral species are simply assigned activity coefficients of unity.

The ideality of the solvent in aqueous electrolyte solutions is commonly tabulated in terms of the osmotic coefficient ϕ (e.g., Pitzer and Brewer, 1961, p. 321; Denbigh, 1981, p. 288), which assumes a value of unity in an ideal dilute solution under standard conditions. By analogy to a solution of a single salt, the water activity

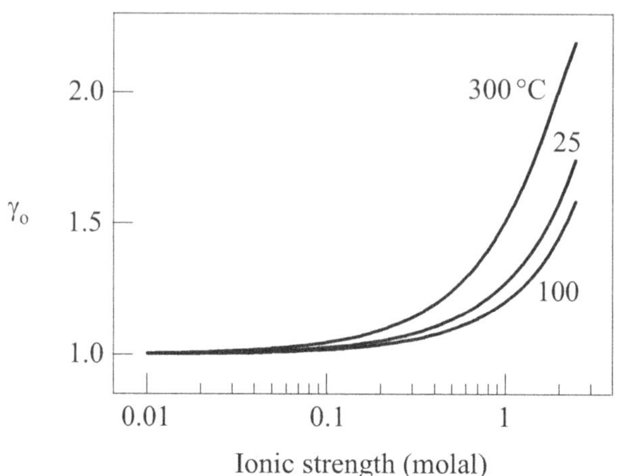

Figure 8.4 Activity coefficients γ_o for neutral, nonpolar species as a function of ionic strength (molal) at 25 °C, 100 °C, and 300 °C, according to the activity model of Helgeson (1969).

can be determined from the osmotic coefficient and the stoichiometric ionic strength I_S according to

$$\ln a_w = -\frac{2I_S\phi}{55.5}\,. \tag{8.7}$$

In the B-dot model, the osmotic coefficient is taken to be described by a power series,

$$\phi = 1 - \frac{2.303A}{a^3 I_S}\left[\hat{b} - 2\ln\hat{b} - \frac{1}{\hat{b}}\right] + \frac{bI_S}{2} + \frac{2cI_S^2}{3} + \frac{3dI_S^3}{4}\,, \tag{8.8}$$

in terms of regression coefficients a, b, c, and d, which differ from those in Equation 8.6. Term $\hat{b}$ in this equation is given as

$$\hat{b} = 1 + a\sqrt{I_S}\,, \tag{8.9}$$

and representative values of the regression coefficients, which vary with temperature, are

	a	b	c	d
25 °C	1.454	0.02236	9.380×10^{-3}	-5.362×10^{-4}
100 °C	1.555	0.03648	6.437×10^{-3}	-7.132×10^{-4}
150 °C	1.623	0.04589	4.522×10^{-3}	-8.312×10^{-4}

Figure 8.5 shows the predicted water activity at 25 °C and 300 °C, plotted against I_S.

The GWB programs differ somewhat from other models in that they set limits to the values for I and I_S used to evaluate the B-dot model (Eqs. 8.5–8.7). Reflecting the fact that there is no basis for extrapolating the B-dot model to high ionic strength, the programs calculate activity coefficients using the lesser value of the actual ionic strength (I or I_S) and the limiting value. The limiting values are carried internally as variables `timax` and `simax`, which, by default, are set to 3 molal, but may be reset by the user.

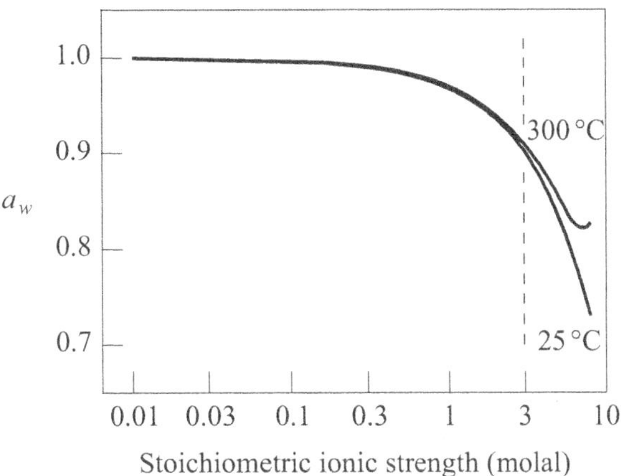

Figure 8.5 Water activity a_w versus stoichiometric ionic strength I_S of NaCl solutions at 25 °C and 300 °C, according to the activity model of Helgeson (1969). Dashed line shows 3 molal limit to the model parameterization; values to right of this line are extrapolations of the original data.

8.2 Virial Methods

Virial equations offer a conceptual alternative to the Debye–Hückel methods for calculating electrolyte activities. The semi-empirical equations are sometimes called specific interaction equations, phenomenological equations, or simply the Pitzer equations, after Kenneth Pitzer, who has been largely responsible for their development since the 1970s (among his and his co-workers' many papers on the subject, see reviews by Pitzer, 1979, 1987). Virial equations are frequently employed in geochemical modeling because they can be applied with accuracy at high ionic strength.

The virial methods differ conceptually from other techniques in that they may take little or no explicit account of the distribution of species in solution. In their purest form, the equations recognize only free ions, as though each salt has fully dissociated in solution. The molality m_i of the Na^+ ion, then, is taken to be the analytical concentration of sodium. All of the calcium in solution is represented by Ca^{++}, the chlorine by Cl^-, the sulfate by SO_4^{--}, and so on.

In many chemical systems, however, it is desirable to include some complex species in the virial formulation. Species that protonate and deprotonate with pH, such as those in the series CO_3^{--}–HCO_3^-–CO_2(aq) and Al^{+++}–$AlOH^{++}$–$Al(OH)_2^+$, typically need to be included to account for amphoteric behavior, and incorporating strong ion pairs such as $CaSO_4$(aq) may improve the model's accuracy at high temperatures. Weare (1987, pp. 148–153) discusses the criteria for selecting complex species to include in a virial formulation.

In the virial methods, therefore, the activity coefficients account implicitly for the reduction in the free ion's activity due to the formation of ion pairs and complex species, except for those specifically included in the formulation. As such, they describe not only the factors traditionally accounted for by activity coefficient models, such as the effects of electrostatic interaction and ion hydration, but also the distribution of species in solution. There is no provision in the method for separating the traditional part of the coefficients from the portion attributable to speciation. For this reason, the coefficients differ (even in the absence of error) in meaning and value from activity coefficients given by other methods. It might be more accurate and less confusing to refer to the virial methods as activity models rather than as activity coefficient models.

8.2.1 Pitzer Equations

The family of implementations known collectively as the Pitzer equations is perhaps the most developed of the virial methods. Programs SOLMINEQ.88 (Kharaka *et al.*, 1988), PHRQPITZ (Plummer *et al.*, 1988), PHREEQC (Parkhurst, 1995), EQ3/EQ6 (Wolery, 1992a), and the GWB programs (Bethke *et al.*, 2021), for example, include provision for using the Pitzer equations.

The methods work by assuming that the solution's excess free energy G^{ex} (i.e., the free energy in excess of that in an ideal solution) can be described by a function of the form

$$\frac{G^{ex}}{n_w RT_K} = f^{dh}(I) + \sum_i \sum_j \lambda_{ij}(I) m_i m_j + \sum_i \sum_j \sum_k \mu_{ijk} m_i m_j m_k . \tag{8.10}$$

Here, i, j, and k are subscripts representing the various species in solution and f^{dh} is a function of ionic strength similar in form to the Debye–Hückel equation. The terms λ_{ij} and μ_{ijk} are second and third virial coefficients, which are intended to account for short-range interactions among ions; the second virial coefficients vary with ionic strength, whereas the third virial coefficients do not.

Equation 8.10 is notable in that it ascribes specific energetic effects to the interactions of the aqueous species taken in pairs (the first summation) and triplets (second summation). The equation's general form is not ad hoc but suggested by statistical mechanics (Anderson and Crerar, 1993, pp. 446–451). The values of the virial coefficients, however, are largely empirical, being deduced from chemical potentials determined from solutions of just one or two salts.

An expression for the ion activity coefficients γ_i follows from differentiating Equation 8.10 with respect to m_i. The result in general form is

$$\ln \gamma_i = \ln \gamma_i^{dh} + \sum_j D_{ij}(I) m_j + \sum_j \sum_k E_{ijk} m_j m_k . \tag{8.11}$$

Here, γ_i^{dh} is a Debye–Hückel term, and D_{ij} and E_{ijk} are second and third virial coefficients, defined for each pair and triplet of ions in solution. As before, the values of D_{ij} vary with ionic strength, whereas the terms E_{ijk} are constant at a given temperature.

One of the most useful implementations for geochemical modeling, published by Harvie *et al.* (1984), is known as the Harvie–Møller–Weare (or HMW) method. The method treats solutions in the Ca–H^+–K–Mg–Na–Cl–CO_2–SO_4–H_2O system at 25 °C. Notably absent from the method are the components SiO_2 and Al^{+++}, which are certainly important in many geochemical studies.

Tables 8.1–8.2 show the calculation procedure for the HMW method (Pitzer, 1991, offers the equations in STEP 8 in an expanded form, for reference), and Tables 8.3–8.6 list the required coefficients. In these tables, subscripts c, c', and M refer to cations, a, a', and X to anions, and n and N to species of neutral charge; subscripts i, j, and k refer to species in general. In Appendix B, we carry through an example calculation by hand to provide a clearer idea of how the method is implemented. The reader should work through the example calculation before attempting to program the method.

Considering the rather large amount of data required to implement virial methods even at 25 °C (e.g., Tables 8.3–8.6), it is tempting to dismiss the methods as no more than statistical fits to experimental data. In reality, however, virial methods take chemical potentials measured from simple solutions containing just one or two salts to provide an activity model capable of accurately predicting species activities in complex fluids.

Table 8.1 *Procedure (Part 1) for evaluating the* HMW *activity model*

GIVEN DATA. The following data are known at the onset of the calculation:

For each cation–anion pair: model parameters $\beta^{(0)}_{MX}$, $\beta^{(1)}_{MX}$, $\beta^{(2)}_{MX}$, C^{ϕ}_{MX}, $\alpha^{(1)}_{MX}$, and $\alpha^{(2)}_{MX}$.
For each cation–cation and anion–anion pair: model parameters θ_{ij}.
For each pairing of an ion with a neutral species: model parameters λ_{ni}.
For each cation–cation–anion and anion–anion–cation triplet: model parameters ψ_{ijk}.
For each species considered: the molality m_i and charge z_i.

STEP 1. Calculate solution ionic strength I and the total molal charge Z:

$$I = \frac{1}{2}\sum_i m_i z_i^2 \quad \text{and} \quad Z = \sum_i m_i \mid z_i \mid .$$

STEP 2. Determine for each possible pairing of like-signed charges the values of the functions ${}^E\theta_{ij}(I)$ and ${}^E\theta'_{ij}(I)$ of I by numerical integration or approximation (for details, see Pitzer, 1987, pp. 130–132; Harvie and Weare, 1980). A computer program for this purpose is listed in Appendix B. The functions are zero for like charges and symmetrical about zero, so only the positive unlike pairings (e.g., 2–1, 3–1, 3–2) need be evaluated.

STEP 3. For each cation–anion pair MX, evaluate functions $g(x)$ and $g'(x)$ for $x = \alpha^{(1)}_{MX}\sqrt{I}$ and $x = \alpha^{(2)}_{MX}\sqrt{I}$,

$$g(x) = \frac{2\left[1-(1+x)\,\mathrm{e}^{-x}\right]}{x^2} \quad \text{and} \quad g'(x) = -\frac{2\left[1-(1+x+x^2/2)\,\mathrm{e}^{-x}\right]}{x^2} \,.$$

STEP 4. Compute for each cation–anion pair the second virial coefficients B_{MX}, B'_{MX}, B^{ϕ}_{MX},

$$B_{MX} = \beta^{(0)}_{MX} + \beta^{(1)}_{MX}\, g(\alpha^{(1)}_{MX}\sqrt{I}) + \beta^{(2)}_{MX}\, g(\alpha^{(2)}_{MX}\sqrt{I})$$

$$B'_{MX} = \beta^{(1)}_{MX}\, g'(\alpha^{(1)}_{MX}\sqrt{I})/I + \beta^{(2)}_{MX}\, g'(\alpha^{(2)}_{MX}\sqrt{I})/I$$

$$B^{\phi}_{MX} = B_{MX} + I B'_{MX} \,.$$

STEP 5. Calculate for each cation–anion pair the third virial coefficient C_{MX},

$$C_{MX} = \frac{C^{\phi}_{MX}}{2\sqrt{\mid z_M\, z_X \mid}} \,.$$

STEP 6. Compute for cation–cation and anion–anion pairs the second virial coefficients Φ_{ij}, Φ'_{ij}, Φ^{ϕ}_{ij},

$$\Phi_{ij} = \theta_{ij} + {}^E\theta_{ij}(I) \;,\quad \Phi'_{ij} = {}^E\theta'_{ij}(I) \;,\text{and}\quad \Phi^{\phi}_{ij} = \Phi_{ij} + I\Phi'_{ij} \,.$$

STEP 7. Determine the intermediate value F,

$$F = -A^{\phi}\left(\frac{\sqrt{I}}{1+1.2\sqrt{I}} + \frac{2}{1.2}\ln\left(1+1.2\sqrt{I}\right)\right)$$
$$+\sum_{c=1}^{N_c}\sum_{a=1}^{N_a} m_c m_a B'_{ca} + \sum_{c=1}^{N_c-1}\sum_{c'=c+1}^{N_c} m_c m_{c'} \Phi'_{cc'} + \sum_{a=1}^{N_a-1}\sum_{a'=a+1}^{N_a} m_a m_{a'} \Phi'_{aa'} \,.$$

Here $A^{\phi} = 2.303A/3$, where A is the Debye–Hückel parameter.

Table 8.2 *Procedure (Part 2) for evaluating the* HMW *activity model*

STEP 8. Calculate activity coefficients γ_M, γ_X, and γ_N for cations, anions, and neutral species,

$$\ln \gamma_M = z_M^2 F + \sum_{a=1}^{N_a} m_a \left(2B_{Ma} + ZC_{Ma}\right) + \sum_{c=1}^{N_c} m_c \left(2\Phi_{Mc} + \sum_{a=1}^{N_a} m_a \psi_{Mca}\right)$$
$$+ \sum_{a=1}^{N_a-1} \sum_{a'=a+1}^{N_a} m_a m_{a'} \psi_{aa'M} + \mid z_M \mid \sum_{c=1}^{N_c} \sum_{a=1}^{N_a} m_c m_a C_{ca} + \sum_{n=1}^{N_n} m_n \left(2\lambda_{nM}\right)$$

$$\ln \gamma_X = z_X^2 F + \sum_{c=1}^{N_c} m_c \left(2B_{cX} + ZC_{cX}\right) + \sum_{a=1}^{N_a} m_a \left(2\Phi_{Xa} + \sum_{c=1}^{N_c} m_c \psi_{Xac}\right)$$
$$+ \sum_{c=1}^{N_c-1} \sum_{c'=c+1}^{N_c} m_c m_{c'} \psi_{cc'X} + \mid z_X \mid \sum_{c=1}^{N_c} \sum_{a=1}^{N_a} m_c m_a C_{ca} + \sum_{n=1}^{N_n} m_n \left(2\lambda_{nX}\right)$$

$$\ln \gamma_N = \sum_{c=1}^{N_c} m_c \left(2\lambda_{Nc}\right) + \sum_{a=1}^{N_a} m_a \left(2\lambda_{Na}\right) .$$

STEP 9. Calculate the osmotic coefficient ϕ according to

$$\sum_i m_i (\phi - 1) = 2\Bigg[\frac{-A^\phi I^{3/2}}{1 + 1.2\sqrt{I}} + \sum_{c=1}^{N_c} \sum_{a=1}^{N_a} m_c m_a \left(B_{ca}^\phi + ZC_{ca}\right)$$
$$+ \sum_{c=1}^{N_c-1} \sum_{c'=c+1}^{N_c} m_c m_{c'} \left(\Phi_{cc'}^\phi + \sum_{a=1}^{N_a} m_a \psi_{cc'a}\right)$$
$$+ \sum_{a=1}^{N_a-1} \sum_{a'=a+1}^{N_a} m_a m_{a'} \left(\Phi_{aa'}^\phi + \sum_{c=1}^{N_c} m_c \psi_{aa'c}\right)$$
$$+ \sum_{n=1}^{N_n} \sum_{a=1}^{N_a} m_n m_a \lambda_{na} + \sum_{n=1}^{N_n} \sum_{c=1}^{N_c} m_n m_c \lambda_{nc}\Bigg] .$$

STEP 10. Calculate the activity a_w of water from the relation

$$\ln a_w = -\frac{W}{1000} \left(\sum_i m_i\right) \phi ,$$

where W is the mole weight of water (18.016 g mol^{-1}).

Table 8.3 HMW *model parameters for cation–anion pairs*

c	a	$\beta_{ca}^{(0)}$	$\beta_{ca}^{(1)}$	$\beta_{ca}^{(2)}$	C_{ca}^{ϕ}	$\alpha_{ca}^{(1)}$	$\alpha_{ca}^{(2)}$
Na^+	Cl^-	0.0765	0.2664*	—	0.00127	2	—
Na^+	SO_4^{--}	0.01958	1.113	—	0.00497	2	—
Na^+	HSO_4^-	0.0454	0.398	—	—	2	—
Na^+	OH^-	0.0864	0.253	—	0.0044	2	—
Na^+	HCO_3^-	0.0277	0.0411	—	—	2	—
Na^+	CO_3^{--}	0.0399	1.389	—	0.0044	2	—
K^+	Cl^-	0.04835	0.2122	—	−0.00084	2	—
K^+	SO_4^{--}	0.04995	0.7793	—	—	2	—
K^+	HSO_4^-	−0.0003	0.1735	—	—	2	—
K^+	OH^-	0.1298	0.320	—	0.0041	2	—
K^+	HCO_3^-	0.0296	−0.013	—	−0.008	2	—
K^+	CO_3^{--}	0.1488	1.43	—	−0.0015	2	—
Ca^{++}	Cl^-	0.3159	1.614	—	−0.00034	2	—
Ca^{++}	SO_4^{--}	0.20	3.1973	−54.24	—	1.4	12
Ca^{++}	HSO_4^-	0.2145	2.53	—	—	2	—
Ca^{++}	OH^-	−0.1747	−0.2303	−5.72	—	2	12
Ca^{++}	HCO_3^-	0.4	2.977	—	—	2	—
Ca^{++}	CO_3^{--}	—	—	—	—	—	—
Mg^{++}	Cl^-	0.35235	1.6815	—	0.00519	2	—
Mg^{++}	SO_4^{--}	0.2210	3.343	−37.23	0.025	1.4	12
Mg^{++}	HSO_4^-	0.4746	1.729	—	—	2	—
Mg^{++}	OH^-	—	—	—	—	—	—
Mg^{++}	HCO_3^-	0.329	0.6072	—	—	2	—
Mg^{++}	CO_3^{--}	—	—	—	—	—	—
MgOH	Cl^-	−0.10	1.658	—	—	2	—
MgOH	SO_4^{--}	—	—	—	—	—	—
MgOH	HSO_4^-	—	—	—	—	—	—
MgOH	OH^-	—	—	—	—	—	—
MgOH	HCO_3^-	—	—	—	—	—	—
MgOH	CO_3^{--}	—	—	—	—	—	—
H^+	Cl^-	0.1775	0.2945	—	0.0008	2	—
H^+	SO_4^{--}	0.0298	—	—	0.0438	—	—
H^+	HSO_4^-	0.2065	0.5556	—	—	2	—
H^+	OH^-	—	—	—	—	—	—
H^+	HCO_3^-	—	—	—	—	—	—
H^+	CO_3^{--}	—	—	—	—	—	—

*Corrected value.

Table 8.4 HMW *model parameters for cation–cation pairs and triplets*

c	c'	$\theta_{cc'}$	$\psi_{cc'\mathrm{Cl}}$	$\psi_{cc'\mathrm{SO_4}}$	$\psi_{cc'\mathrm{HSO_4}}$	$\psi_{cc'\mathrm{OH}}$	$\psi_{cc'\mathrm{HCO_3}}$	$\psi_{cc'\mathrm{CO_3}}$
Na^+	K^+	−0.012	−0.0018	0.010	—	—	−0.003	0.003
Na^+	Ca^{++}	0.07	−0.007	−0.055	—	—	—	—
Na^+	Mg^{++}	0.07	−0.012	−0.015	—	—	—	—
Na^+	MgOH	—	—	—	—	—	—	—
Na^+	H^+	0.036	−0.004	—	−0.0129	—	—	—
K^+	Ca^{++}	0.032	−0.025	—	—	—	—	—
K^+	Mg^{++}	0.0	−0.022	−0.048	—	—	—	—
K^+	MgOH	—	—	—	—	—	—	—
K^+	H^+	0.005	−0.011	0.197	−0.0265	—	—	—
Ca^{++}	Mg^{++}	0.007	−0.012	0.024	—	—	—	—
Ca^{++}	MgOH	—	—	—	—	—	—	—
Ca^{++}	H^+	0.092	−0.015	—	—	—	—	—
Mg^{++}	MgOH	—	0.028	—	—	—	—	—
Mg^{++}	H^+	0.10	−0.011	—	−0.0178	—	—	—
MgOH	H^+	—	—	—	—	—	—	—

Table 8.5 HMW *model parameters for anion–anion pairs and triplets*

a	a'	$\theta_{aa'}$	$\psi_{aa'\mathrm{Na}}$	$\psi_{aa'\mathrm{K}}$	$\psi_{aa'\mathrm{Ca}}$	$\psi_{aa'\mathrm{Mg}}$	$\psi_{aa'\mathrm{MgOH}}$	$\psi_{aa'\mathrm{H}}$
Cl^-	SO_4^{--}	0.02	0.0014	—	−0.018	−0.004	—	—
Cl^-	HSO_4^-	−0.006	−0.006	—	—	—	—	0.013
Cl^-	OH^-	−0.050	−0.006	−0.006	−0.025	—	—	—
Cl^-	HCO_3^-	0.03	−0.015	—	—	−0.096	—	—
Cl^-	CO_3^{--}	−0.02	0.0085	0.004	—	—	—	—
SO_4^{--}	HSO_4^-	—	−0.0094	−0.0677	—	−0.0425	—	—
SO_4^{--}	OH^-	−0.013	−0.009	−0.050	—	—	—	—
SO_4^{--}	HCO_3^-	0.01	−0.005	—	—	−0.161	—	—
HSO_4^-	OH^-	—	—	—	—	—	—	—
HSO_4^-	HCO_3^-	—	—	—	—	—	—	—
HSO_4^-	CO_3^{--}	—	—	—	—	—	—	—
OH^-	HCO_3^-	—	—	—	—	—	—	—
OH^-	CO_3^{--}	0.10	−0.017	−0.01	—	—	—	—
HCO_3^-	CO_3^{--}	−0.04	0.002	0.012	—	—	—	—

Table 8.6 HMW *model parameters for neutral species–ion pairs*

i	$\lambda_{CO_2\,i}$	$\lambda_{CaCO_3\,i}$	$\lambda_{MgCO_3\,i}$
H^+	0.0	—	—
Na^+	0.100	—	—
K^+	0.051	—	—
Ca^{++}	0.183	—	—
Mg^{++}	0.183	—	—
MgOH	—	—	—
Cl^-	−0.005	—	—
SO_4^{--}	0.097	—	—
HSO_4^-	−0.003	—	—
OH^-	—	—	—
HCO_3^-	—	—	—
CO_3^{--}	—	—	—

Eugster *et al.* (1980), for example, used the virial method of Harvie and Weare (1980) to accurately trace the evaporation of seawater almost to the point of desiccation. Using any other activity model, such a calculation could not even be contemplated. Other accomplishments in geochemistry (Weare, 1987) include prediction of mineral precipitation in alkaline lakes and in fluid inclusions within evaporite minerals.

8.2.2 SIT *Method*

The SIT method for calculating activity coefficients is a virial technique significantly simpler in form than the Pitzer equations (Grenthe *et al.*, 1997). The method is notable in the way it neatly bridges the gap between the simplicity of the Debye–Hückel equation and the breadth in applicability of Pitzer methods. Short for Specific ion Interaction Theory, the method is sometimes called the Brønsted–Guggenheim–Scatchard or BGS model; it is currently implemented in PHREEQC, Visual MINTEQ, GEMS, and the GWB programs.

The method employs a single virial coefficient ε_{ca} to describe interaction between a given cation c and anion a; in some cases, coefficients ε_{cn}, ε_{an}, $\varepsilon_{n'n}$ and ε_{nn} are carried to represent interaction between ions and a neutral species n, two neutral species n' and n, or a neutral species n with itself, respectively, but links among like-charged ions are not considered. In this way, the method differs from the Pitzer equations, which carry up to five parameters representing species pairings in all charge combinations, as well as coefficients describing species taken three at a time (Tables 8.3–8.6). (At the same time, common implementations of the SIT method carry more ion pairs and complex species than might be found in a comparable Pitzer compilation.) As you might expect from the extent of parameterization, the method is accurate to higher ionic strengths than the Debye–Hückel equations, but not to the salinities attainable through application of the Pitzer equations.

Plyasunov and Popova (2013) lay out the governing equations succinctly. A cation's activity coefficient,

$$\log \gamma_c = -\frac{z_c^2 A\sqrt{I}}{1 + 1.5\sqrt{I}} + \sum_a \varepsilon_{ca} m_a + \sum_n \varepsilon_{cn} m_n \,, \tag{8.12}$$

is given as a Debye–Hückel term plus summations over the products of the cation's interaction coefficients

Table 8.7 SIT *coefficients* ε_{ca} *for the Ca–H$^+$–K–Mg–Na–Cl–CO$_2$–SO$_4$–H$_2$O system*

	Cl^-	CO_3^{--}	HCO_3^-	HSO_4^-	OH^-	SO_4^{--}
Ca^{++}	0.14	—	—	—	—	—
H^+	0.12	—	—	—	—	—
K^+	0.0	—	—	—	—	—
Mg^{++}	0.19	—	—	—	—	—
Na^+	0.03	−0.08	0.0	−0.01	0.04	−0.12

with anions and neutral species, and those species' molal concentrations. Activity coefficients for anions, similarly,

$$\log \gamma_a = -\frac{z_a^2 A \sqrt{I}}{1 + 1.5\sqrt{I}} + \sum_c \varepsilon_{ca} m_c + \sum_n \varepsilon_{an} m_n , \tag{8.13}$$

follow from interactions with cations and uncharged species. For neutral species, the first term vanishes,

$$\log \gamma_n = \sum_c \varepsilon_{cn} m_c + \sum_a \varepsilon_{an} m_a + \sum_{n' \neq n} \varepsilon_{n'n} m_{n'} + \varepsilon_{nn} m_n , \tag{8.14}$$

since z_n is zero.

As before (Eq. 8.7), the activity of water follows from the osmotic coefficient ϕ,

$$\ln a_w = -\frac{W \sum_i m_i}{1000} \phi , \tag{8.15}$$

where W is the mole weight of water, 18.016 g mol^{-1}. The osmotic coefficient, in turn, can be determined according to

$$\begin{aligned} \frac{\sum_i m_i}{2.303}(\phi - 1) &= -\frac{2A}{1.5^3}\left(t - 2\ln t - t^{-1}\right) \\ &+ \sum_c \sum_a \varepsilon_{ca} m_c m_a + \sum_c \sum_n \varepsilon_{cn} m_c m_n + \sum_a \sum_n \varepsilon_{an} m_a m_n \\ &+ \sum_n \sum_{n' \neq n} \varepsilon_{nn'} m_n m_{n'} + \frac{1}{2} \sum_n \varepsilon_{nn} m_n^2 , \end{aligned} \tag{8.16}$$

where $t = 1 + 1.5\sqrt{I}$. Comparing Equations 8.12–8.16 with those in Tables 8.1–8.2 establishing the HMW model, we see the SIT method requires considerably less effort to code into a software program and less computing time to evaluate than the Pitzer equations.

Available computer codes generally make provision for allowing the virial parameters ε_{ca}, ε_{cn}, ε_{an}, $\varepsilon_{n'n}$, and ε_{nn} in the above relations to vary with temperature and ionic strength, or can be readily made to do so. Electrolyte chemists, nonetheless, are only starting to define how the terms should behave over a range in conditions (e.g., Plyasunov and Popova, 2013), and hence guidance available to the modeler is incomplete currently.

Table 8.7 lays out the interaction coefficients ε_{ca} taken from Grenthe *et al.* (1997) for the Ca–H$^+$–K–Mg–Na–Cl–CO$_2$–SO$_4$–H$_2$O system, which is the system addressed by the HMW activity model discussed in the previous section. Comparing this compilation with the parameters needed to evaluate the HMW model, as shown in Tables 8.3–8.6, we appreciate the simplicity of the SIT method relative to the Pitzer equations.

As an example, we take at 25 °C a 6 molal NaCl solution containing 0.01 molal dissolved $CaSO_4$. The ionic strength is $6\times1^2 + 0.01\times2^2 = 6.04$, so $\sqrt{I}$ is 2.458. We know A from the Debye–Hückel equation (Eq. 8.2) is 0.5092 at this temperature. Taking values for ε_{ca} from Table 8.7, and ignoring protonation and complexation reactions, we can set out a tableau,

		$\varepsilon_{ca}m_c$		$\varepsilon_{ca}m_a$			
	D–H	Na^+	Ca^{++}	Cl^-	SO_4^{--}	$\log\gamma$	γ_c or γ_a
Na^+	−0.2670	—	—	0.18	−0.0012	−0.0882	0.816
Ca^{++}	−1.0681	—	—	0.84	—	−0.2281	0.591
Cl^-	−0.2670	0.18	0.0014	—	—	−0.0856	0.821
SO_4^{--}	−1.0681	−0.72	—	—	—	−1.7882	0.016

showing the calculation process for each of the free ions in solution. Column D–H is the Debye–Hückel factor, the first term in Equations 8.12 and 8.13; the next four columns are products of interaction parameters ε_{ca} and ion concentration, m_c or m_a; the sum of those five columns is carried under $\log\gamma$. The right-hand column exponentiates the latter value to give the activity coefficient, γ_c or γ_a, corresponding to the ion in the left-hand column.

To reproduce the results, we load file `thermo_sit.tdat`, which contains the ThermoChimie database (Giffaut *et al.*, 2014) and the SIT coefficients therein, and enter the commands

```
6   molal Na+
6   molal Cl-
.01 molal Ca++
.01 molal SO4--
go
```

into SPECE8.

8.3 Comparison of the Methods

It is interesting to compare the Debye–Hückel and virial methods, since each has its own advantages and limitations. The Debye–Hückel equations are simple to apply and readily extensible to include new species in solution, since they require few coefficients specific to either species or solution. The method can be applied as well over the range of temperatures most important to an aqueous geochemist. There is an extensive literature on ion association reactions, so there are few limits to the complexity of the solutions that can be modeled.

The Debye–Hückel methods work poorly, however, when carried to moderate or high ionic strength, especially when salts other than NaCl dominate the solute. In the theory, the ionic strength represents all the properties of a solution. For this reason, a Debye–Hückel method applied to any solution of a certain ionic strength (whether dominated by NaCl, KCl, HCl, H_2SO_4, or any salt or salt mixture) gives the same set of activity coefficients, regardless of the solution's composition. This result, except for dilute solutions, is, of course, incorrect. Clearly, we cannot rely on a single value to describe how the properties of a concentrated solution depend on its solute content.

The virial methods, on the other hand, can provide remarkably accurate results over a broad range of solution concentrations and with a variety of dominant solutes. The methods, however, are limited in breadth (e.g., Rowland *et al.*, 2015). Notably incomplete at present are data for redox reactions in saline solutions, as well as for oxides with low solubilities. The vast majority of data used to parameterize virial methods are collected near room temperature (e.g., Møller, 1988; Greenberg and Møller, 1989; Simoes *et al.*, 2017), furthermore, limiting our ability to derive parameters at other temperatures.

Unlike the Debye–Hückel equations, the virial methods may provide little or no information about the distribution of species in solution. Geochemists like to identify the dominant species in solution in order to write the reactions that control a system's behavior. In the virial methods, this information is hidden within the complexities of the virial equations and coefficients. Many geochemists, therefore, find the virial methods to be less satisfying than methods that predict the species distribution. The information given by Debye–Hückel methods about species distributions in concentrated solutions, however, is not necessarily reliable and should be used with caution.

To explore the differences between the methods, we use SPECE8 to calculate at 25 °C the solubility of gypsum ($CaSO_4 \cdot 2H_2O$) as a function of NaCl concentration. We use three datasets: `thermo.tdat` uses the B-dot equation, `thermo_hmw.tdat` is based on the HMW model, and `thermo_sit.tdat` invokes the ThermoChimie database and the SIT method.

To perform the calculation for a 1 molal NaCl solution, for example, we use the `data` command to set the appropriate dataset (`thermo.tdat`, `thermo_hmw.tdat`, or `thermo_sit.tdat`) and then enter

```
swap Gypsum for Ca++
100 free grams Gypsum
balance on SO4--

1 molal Na+
1 molal Cl-
go
```

We then repeat the calculation over a range of NaCl concentrations. (To save effort, we can use REACT to perform the calculation in one step, as we discuss in Chapter 14, by titrating NaCl into an initially dilute solution.)

Figure 8.6 shows the calculation results plotted against measured solubilities from laboratory experiments. Results of applying the B-dot equation coincide with observations only at NaCl concentrations less than about 0.5 molal; at 3 molal salinity, the predicted solubility is about double the measured value. The HMW calculations, in contrast, predict the experimental data closely over the full range in salinity, reflecting the fact that these same data were used to parameterize the model (Harvie and Weare, 1980). Solubility determined using the SIT method tracks the experimental results up to NaCl concentrations of more than 4 molal, a significant improvement over the B-dot results, but is less accurate than the HMW method above this salinity.

The extent to which error in predicting solubility might be considered acceptable depends primarily on the modeler's goals (e.g., Weare, 1987, pp. 160–162; Brantley *et al.*, 1984; Felmy and Weare, 1986). If a model is designed to accurately predict mineral solubility in concentrated solutions, clearly the B-dot equation would be considered incorrect. A model created to explore the behavior of a fluid in equilibrium with gypsum might be useful, however, even though the modeler recognizes that the predicted gypsum solubility is inaccurate by a factor of two or more. Indeed, as discussed in Chapter 2, errors of this magnitude are not uncommon in geochemical modeling.

Figure 8.7 shows how concentrations and activities of the calcium and sulfate species vary with NaCl concentration, depending on the activity model chosen. In the B-dot model, there are three ion pairs ($CaCl^+$,

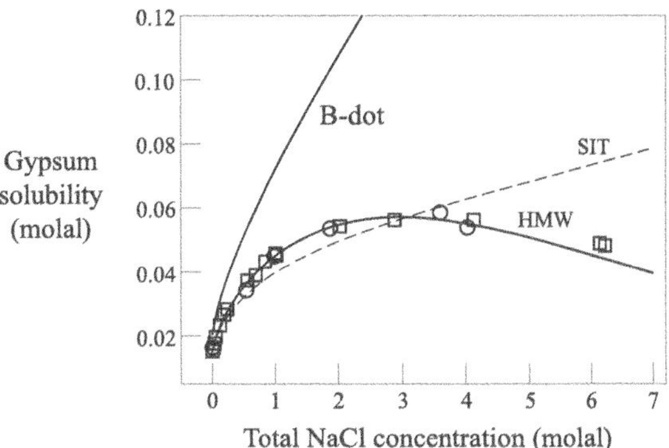

Figure 8.6 Solubility of gypsum ($CaSO_4 \cdot 2H_2O$) at 25 °C as a function of NaCl concentration, calculated according to the B-dot (modified Debye–Hückel) equation; the Harvie–Møller–Weare activity model; and Specific ion Interaction Theory (SIT), according to the ThermoChimie database. Squares and circles, respectively, show experimental determinations by Marshall and Slusher (1966) and Block and Waters (1968).

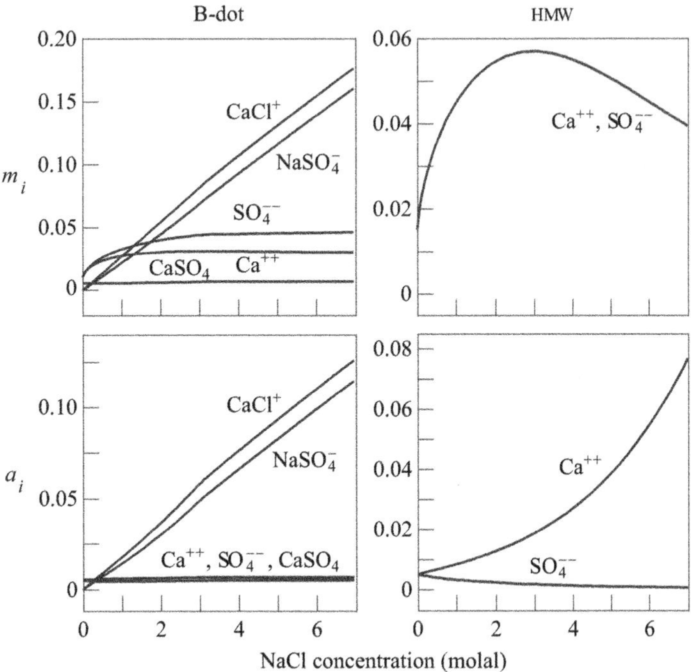

Figure 8.7 Molal concentrations m_i and activities a_i of calcium and sulfate species in equilibrium with gypsum at 25 °C as functions of NaCl concentration, calculated using the B-dot equation (left) and the HMW activity model (right).

Table 8.8 *Chemical compositions (g l^{-1}) of brines from the Sebkhat El Melah brine deposit, Zarzis, Tunisia (Jarraya and El Mansar, 1987)*

Well	K^+	Mg^{++}	Na^+	Ca^{++}	Br^-	Cl^-	SO_4^{--}	HCO_3^-	pH
RZ-2	6.90	52.1	43.1	0.20	2.25	195	27.4	0.14	7.15
RZ-7	7.65	47.2	50.0	0.30	2.11	195	26.9	0.19	7.0
RZ-8	7.46	52.3	37.5	0.20	2.60	192	27.9	0.20	6.8
RZ-9	6.70	47.3	50.8	0.20	2.35	188	30.2	0.17	6.8
RZ-10	6.60	34.4	51.3	0.40	1.36	177	21.7	0.16	6.85
RZ-11	7.85	54.7	32.0	0.20	2.32	206	19.2	0.17	6.0
RZ-16	6.80	46.2	47.0	0.20	3.29	195	18.7	0.20	7.1
RZ-17	8.85	55.9	39.3	0.20	3.20	202	28.2	0.18	7.1
RZ-19	7.23	50.0	44.2	0.50	2.00	200	28.8	0.24	6.9

$NaSO_4^-$, and $CaSO_4$) in addition to the free ions Ca^{++} and SO_4^{--}. The activities of the free ions remain roughly constant with NaCl concentration, and their concentrations increase only moderately, reflecting the decrease in the B-dot activity coefficients with increasing ionic strength (Fig. 8.3). Formation of the complex species $CaCl^+$ and $NaSO_4^-$ drives the general increase in gypsum solubility with NaCl concentration predicted by the B-dot model.

In the HMW model, in contrast, Ca^{++} and SO_4^{--} are the only calcium or sulfate-bearing species considered. The species maintain equal concentration, as required by electroneutrality, and mirror the solubility curve in Figure 8.6. Unlike the B-dot model, the species' activities follow trends dissimilar to their concentrations. The Ca^{++} activity rises sharply while that of SO_4^{--} decreases. In this case, variation in gypsum solubility arises not from the formation of ion pairs, but from changes in the activity coefficients for Ca^{++} and SO_4^{--} as well as in the water activity. The latter value, according to the model, decreases with NaCl concentration from one to about 0.7.

8.4 Brine Deposit at Sebkhat El Melah

As a test of our ability to calculate activity coefficients in natural brines, we consider groundwater from the Sebkhat El Melah brine deposit near Zarzis, Tunisia (Perthuisot, 1980). The deposit occurs in a buried evaporite basin composed of halite (NaCl), anhydrite ($CaSO_4$), and dolomite [$CaMg(CO_3)_2$]. The Tunisian government would like to exploit the brines for their chemical content, especially for the potassium, which is needed to make fertilizer.

Since the deposit contains halite and anhydrite, the brines should be saturated with respect to these minerals and hence provide a good test of the activity models. Table 8.8 shows analyses of brine samples from the deposit. Note that the reported pH values are almost certainly incorrect because pH electrodes do not respond accurately in concentrated solutions. Hence, there is little to be gained by calculating dolomite saturation.

To model the brine, we set the activity method by reading the corresponding dataset (`thermo.tdat`, `thermo_sit.tdat`, or `thermo_hmw.tdat`) and set out the chemical composition. Taking the analysis for well RZ-2 as an example, the procedure in SPECE8 is

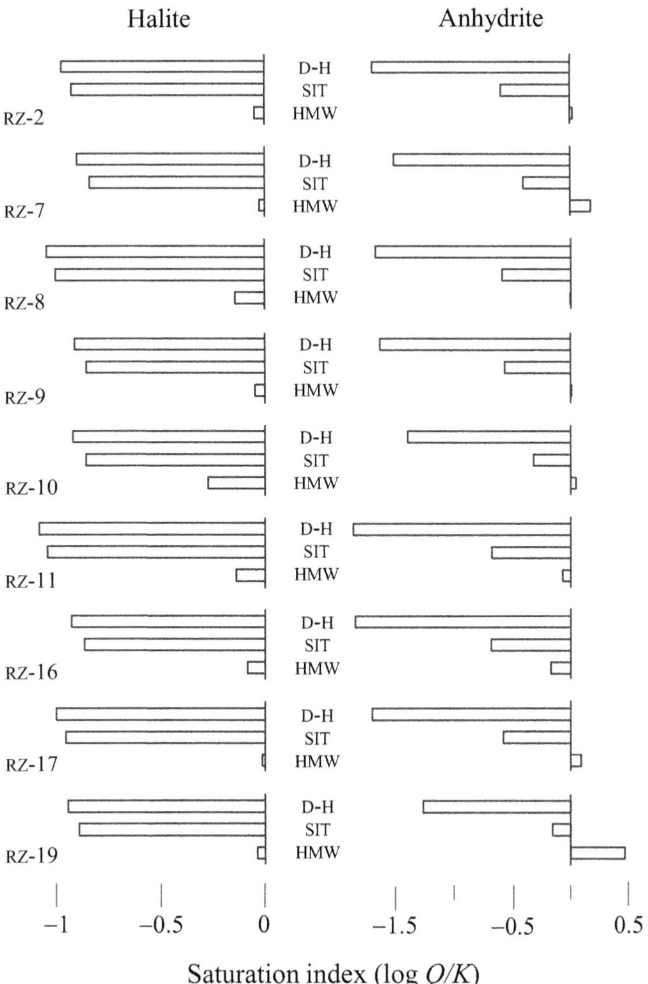

Figure 8.8 Saturation indices of Sebkhat El Melah brine samples with respect to halite (left) and anhydrite (right), calculated using the B-dot (modified Debye–Hückel) equation; Specific ion Interaction Theory (SIT), according to the ThermoChimie database; and the Harvie–Møller–Weare (HMW) model.

```
K+     =   6.9  g/l
Mg++   =  52.1  g/l
Na+    =  43.1  g/l
Ca++   =    .2  g/l
Br-    =   2.25 g/l
Cl-    = 195    g/l
SO4--  =  27.4  g/l
HCO3-  =    .14 g/l
pH     =   7.15

balance on Cl-
go
```

The SIT dataset carries CO_3^{--} rather than HCO_3^- in the basis, so in this case we need to use the command "`CO3-- = .14 g/l as HCO3-`" to represent carbonate. Also, the HMW model does not account for bromide, so we would type `remove Br-` before invoking it. Program GSS can invoke SPECE8 internally, so the most expeditious procedure is to paste the chemical analyses in Table 8.8 into a datasheet and let GSS calculate mineral saturation for the samples all at once.

Figure 8.8 shows the resulting saturation indices for halite and anhydrite, calculated for the Sebkhat El Melah samples. The Debye–Hückel (B-dot) method, which of course is not intended to be used to model saline fluids, predicts that the minerals are significantly undersaturated in the brine samples. Calculations using the SIT method better model anhydrite saturation, but offer little improvement with respect to halite. The Harvie–Møller–Weare model, on the other hand, predicts that halite and anhydrite in most of the samples are near equilibrium with the brine, as we would expect. As usual, we cannot determine whether the residual discrepancies result from analytical imprecision, inaccuracy of the activity model, or error arising from other sources.

9

Sorption and Ion Exchange

In building certain types of geochemical models, especially those applied to environmental problems, we commonly find need to account for the sorption of aqueous species onto the surfaces of sediments, organic matter, and even bacteria (e.g., Zhu and Anderson, 2002). Because of their large surface areas and high reactivities (e.g., Davis and Kent, 1990), many components of the geosphere—especially clay minerals, zeolites, metal oxides and oxyhydroxides, and organics—can sorb considerable masses.

Sorption significantly diminishes the mobility and bioavailability of certain dissolved components in solution, especially those present in minor amounts. The process, for example, may retard the spread of radionuclides near a radioactive waste repository or the migration of contaminants away from a polluting landfill (see Chapters 24 and 36). In acid mine drainages, ferric oxide sorbs heavy metals from surface water, helping limit their downstream movement (see Chapter 35). A geochemical model useful in investigating such cases must provide an accurate assessment of the effects of surface reactions.

In this chapter, we consider several simple models of ion sorption and exchange that can be applied within the context of a geochemical model. These models include distribution coefficients, Freundlich and Langmuir isotherms, and ion exchange theory. In subsequent chapters (Chapters 10 and 11), we consider surface complexation theory, which is more complicated but in many ways more robust than the models presented here.

9.1 Distribution Coefficient (K_d) Approach

The *distribution coefficient approach*—commonly referred to as the K$_d$ *approach*—is the most widely applied method in environmental geochemistry for predicting the sorption of contaminant species onto sediments. The distribution coefficient K_d itself is simply the ratio under specific conditions of the sorbed to the dissolved mass of a contaminant. Sorbed and dissolved mass are expressed in units such as moles per gram of dry sediment and moles per cm^3 fluid, respectively, so K_d has units such as cm^3 g^{-1}.

A K_d is by nature descriptive, but in the K_d approach the value is taken over the range of chemical conditions considered in a model to be constant and predictive. This assumption, of course, cannot hold in the general sense. A K_d value varies sharply with pH, contaminant concentration, ionic composition of the fluid, and so on; its measurement is specific to the fluid and sediment tested. It is imperative that the modeler keep these points in mind.

The limitations of the K_d approach stem in part from the fact that it takes no account of the number of sorbing sites on the sediment, treating them as if they are in excess supply. The approach allows a solute to sorb without limit, without being affected by the sorption of competing species. As well as this, the approach treats sorption as a simple process of attachment. It does not consider the possibility of hydrolysis at the interface between sediment and fluid, so it cannot account for the effects of pH. Nor does the approach consider electrostatic interactions between the surface and charged ions.

For these reasons, the K_d approach works best for describing sorption of trace amounts of nonionized, hydrophobic organic molecules (Stumm and Morgan, 1996). The approach is broadly recognized to describe poorly the behavior of ionic species, especially metal ions, within soils and sediments (e.g., Reardon, 1981; Domenico and Schwartz, 1998). Nonetheless, it is commonly applied for just such purposes, often giving misleading results (Bethke and Brady, 2000).

The K_d approach, as strictly defined, implies but does not specify a chemical reaction. A variant of the approach known as the *reaction* K_d *model* (or *activity* K_d *model*) based on a specific chemical reaction is commonly applied in reaction modeling. For example, the sorption and desorption of Cd^{++} might be taken to occur according to the reaction

$$>\mathrm{Cd}^{++} \rightleftarrows \mathrm{Cd}^{++} , \tag{9.1}$$

where $>Cd^{++}$ is the sorbed form of the cadmium ion. In this model, the distribution coefficient for the reaction is defined to be the ratio of sorbed mass to the activity of the free ion. It can be expressed

$$K'_\mathrm{d} = \frac{m_{>\mathrm{Cd}^{++}}}{a_{\mathrm{Cd}^{++}}} \times \frac{n_w}{n_\mathrm{s}} , \tag{9.2}$$

where $m_{>\mathrm{Cd}^{++}}$ is molal concentration of the sorbed species, $a_{\mathrm{Cd}^{++}}$ is activity of the free ion, n_s is the mass in grams of dry sediment in the system, and n_w, as before, is the mass of solvent water, in kg. The definition of the distribution coefficient in this case differs somewhat from the general definition, so we carry it not as K_d, but as K'_d, which we see has units such as mol g^{-1}. Rearranging gives the mass action equation,

$$m_{>\mathrm{Cd}^{++}} = \frac{K'_\mathrm{d}\, n_\mathrm{s}}{n_w}\, a_{\mathrm{Cd}^{++}} , \tag{9.3}$$

in terms of the distribution coefficient.

The reaction K_d model, as we can see, differs from the general approach in two ways. The activity rather than the concentration of the dissolved species is carried. Distribution coefficients calculated in the traditional manner, therefore, need to be corrected by a factor of the species' activity coefficient. The value of K'_d for Cd^{++} in mol g^{-1} can be determined from a K_d in cm^3 g^{-1} as

$$K'_\mathrm{d} = \frac{\rho_w}{1000\, \gamma_{\mathrm{Cd}^{++}}} K_\mathrm{d} , \tag{9.4}$$

assuming the water is dilute. Here, ρ_w is the density of water and $\gamma_{\mathrm{Cd}^{++}}$ is the species' activity coefficient.

Potentially more significant is the fact that a single ion is used to represent the dissolved form of the contaminant in question, an assumption that can lead to serious error. Cadmium in a model calculated at pH 12, for example, is present primarily as the species $Cd(OH)_2$; almost no free ion Cd^{++} occurs. Employing the reaction K_d model in terms of Cd^{++} in this case would predict a contaminant distribution unlike that suggested by the distribution coefficient, applied in the traditional sense. We see the importance of applying a K_d model to systems similar to that for which it was originally determined.

To generalize this discussion, we start as before with a vector,

$$\mathbf{B} = (A_w, A_i, A_k, A_m) , \tag{9.5}$$

of basis entries. Since the K_d model offers no accounting for the number of sorbing sites available, and uncomplexed sorbing sites do not appear in the sorption reactions, we do not need to include in the basis any entry representing the sorbing surface itself. The reaction for each sorbed species A_q considered,

$$A_q = A_j \, , \tag{9.6}$$

can be written in terms of the basis entries as

$$A_q = \nu_{wj} A_w + \sum_i \nu_{ij} A_i + \sum_k \nu_{kj} A_k + \sum_m \nu_{mj} A_m \, , \tag{9.7}$$

by substituting Equation 3.22. The molal concentration m_q of each sorbed species,

$$m_q = \frac{K'_{\mathrm{d}_q} n_{\mathrm{s}}}{n_w} a_j \, , \tag{9.8}$$

given by Equation 9.3 can be expressed by the mass action equation,

$$m_q = \frac{K'_{\mathrm{d}_q} n_{\mathrm{s}}}{n_{\mathrm{w}}} \left(\frac{1}{K_j} a_w^{\nu_{wj}} \prod^i (\gamma_i m_i)^{\nu_{ij}} \prod^k a_k^{\nu_{kj}} \prod^m f_m^{\nu_{mj}} \right) , \tag{9.9}$$

by substituting Equation 3.27.

The mass balance equations for a system including sorbing species are given as

$$M_w = n_w \left(55.5 + \sum_j \nu_{wj} m_j + \sum_q \nu_{wq} m_q \right) \tag{9.10}$$

$$M_i = n_w \left(m_i + \sum_j \nu_{ij} m_j + \sum_q \nu_{iq} m_q \right) \tag{9.11}$$

$$M_k = n_k + n_w \left(\sum_j \nu_{kj} m_j + \sum_q \nu_{kq} m_q \right) \tag{9.12}$$

$$M_m = n_w \left(\sum_j \nu_{mj} m_j + \sum_q \nu_{mq} m_q \right) . \tag{9.13}$$

These equations are the same as those already considered (3.28–3.31), with the addition of the summations over the masses of the sorbed species.

9.2 Freundlich Isotherms

The *Freundlich isotherm* (or *Freundlich model*) is an empirical description of species sorption similar to the K_{d} approach, but differing in how the ratio of sorbed to dissolved mass is computed. In the model, dissolved mass, the denominator in the ratio, is raised to an exponent less than one. The ratio, represented by the Freundlich coefficient K_{f}, is taken to be constant, as is the exponent, denoted n_{f}, where $0<n_{\mathrm{f}}<1$. As before, the masses of dissolved and sorbed species are entered, respectively, in units such as moles per cm^3 fluid and moles per gram of dry sediment. Since the denominator is raised to an arbitrary exponent n_{f}, the units for K_{f} are not commonly reported, and care must be taken to note the units in which the ratio was determined.

The effect of the exponent n_{f} is to predict progressively less effective sorption as concentration increases.

Where n_f approaches one, the isotherm reverts to the K_d model, in which sorption is equally effective at any concentration. For n_f less than one, a smaller fraction of a component sorbs at high than at low concentration. This effect is taken as reflecting a heterogeneous distribution of sorbing sites in a sediment sample: the more strongly sorbing sites are occupied at low solute concentration, leaving available only weaker sites as concentration increases. Such a pattern can in fact arise not from sorption, but the onset of surface precipitation at high solute concentration.

Geochemical models, as with the K_d approach, are commonly formulated with a variant of the Freundlich isotherm based on a chemical reaction, like Reaction 9.1. In this approach, known as the *reaction Freundlich model* or the *activity Freundlich model*, the extent of sorption by the reaction can be expressed

$$K'_f = \frac{m_{>\mathrm{Cd}++}}{a^{n_f}_{\mathrm{Cd}++}} \times \frac{n_w}{n_s} . \tag{9.14}$$

Here, we represent the Freundlich coefficient as K'_f (mol g^{-1}), as a reminder that we are not working with Freundlich isotherms in the strictly traditional sense. Rearranging gives

$$m_{>\mathrm{Cd}++} = \frac{K'_f\, n_s}{n_w}\, a^{n_f}_{\mathrm{Cd}++} , \tag{9.15}$$

which is the reaction's mass action equation.

To cast the model in general form, we begin with the basis shown in Equation 9.5 and write each sorption reaction in the form of Equation 9.7. The mass action equation corresponding to the reaction for each sorbed species A_q is

$$m_q = \frac{K'_{f_q}\, n_s}{n_w} \left(\frac{1}{K_j}\, a_w^{\nu_{wj}} \prod^{i} (\gamma_i m_i)^{\nu_{ij}} \prod^{k} a_k^{\nu_{kj}} \prod^{m} f_m^{\nu_{mj}} \right)^{n_f} , \tag{9.16}$$

where K'_{f_q} is the reaction's Freundlich coefficient. The mass balance equations, unchanged from the K_d model, are given by Equations 9.10–9.13.

9.3 Langmuir Isotherms

The *Langmuir isotherm* (or *Langmuir model*) provides an improvement over the K_d and Freundlich approaches by maintaining a mass balance on the sorbing sites (Stumm and Morgan, 1996). The model, for this reason, does not predict that species can sorb indefinitely, since the number of sites available is limited. When the calculation carries reactions for the sorption of more than one aqueous species, furthermore, it accounts for competition; such a calculation is known as a *competitive Langmuir model.*

In the Langmuir model, a species sorbs and desorbs according to a reaction such as

$$>\mathrm{L{:}Cd}^{++} \rightleftarrows\ >\mathrm{L{:}} + \mathrm{Cd}^{++} , \tag{9.17}$$

where $>\mathrm{L{:}Cd}^{++}$ is a sorbing site occupied by a cadmium ion and $>\mathrm{L{:}}$ is an unoccupied site. The Langmuir reaction has an equilibrium constant K such that

$$K = \frac{m_{>\mathrm{L:}}\, a_{\mathrm{Cd}++}}{m_{>\mathrm{L:Cd}++}} . \tag{9.18}$$

Here $m_{>\mathrm{L:}}$ and $m_{>\mathrm{L:Cd}++}$ are molal concentrations of the unoccupied and occupied sites, respectively, and $a_{\mathrm{Cd}++}$ is the activity of the free ion. Activity coefficients for the surface sites are not carried in the equation; they are assumed to cancel. Equilibrium constants reported in the literature are in many cases tabulated in terms of the concentrations of free species, rather than their activities, as assumed here, and hence may require adjustment.

To parameterize a Langmuir model, you determine from experimental measurements not only the equilibrium constant K, but the surface's *sorption capacity* (or *exchange capacity*). The latter value is a measure of the number of sorbing sites and is commonly reported in moles or electrical equivalents, per gram of dry sediment. Multiplying the sorption capacity in moles by the mass of sediment in a system gives the mole number of sorbing sites, which is

$$M_{>\mathrm{L:}} = n_{\mathrm{w}} \left(m_{>\mathrm{L:}} + m_{>\mathrm{L:Cd}^{++}} \right) . \tag{9.19}$$

When more than one sorption reaction is considered, of course, the summation includes the other sorbed species.

Combining Equations 9.18 and 9.19 gives

$$m_{>\mathrm{L:Cd}^{++}} = \frac{M_{>\mathrm{L:}}}{n_{\mathrm{w}}} \frac{a_{\mathrm{Cd}^{++}}}{\left(K + a_{\mathrm{Cd}^{++}}\right)} , \tag{9.20}$$

which is the well-known Langmuir equation. Where $a_{\mathrm{Cd}^{++}}$ is much smaller than K, by this relation, the sorbed concentration varies in proportion to the concentration of the dissolved species. As $a_{\mathrm{Cd}^{++}}$ increases to values greater than K, in contrast, the sorbed concentration approaches the concentration of sorbing sites, the limiting value in light of the sediment's sorption capacity.

To cast the model in general terms, we set a vector of basis entries,

$$\mathbf{B} = \left(A_w, A_i, A_k, A_m, A_p \right) , \tag{9.21}$$

that includes the sorbing site >L:, represented as A_p. The reaction for each sorbed species A_q written in terms of the basis is

$$A_q = \nu_{wq} A_w + \sum_i \nu_{iq} A_i + \sum_k \nu_{kq} A_k + \sum_m \nu_{mq} A_m + \nu_{pq} A_p , \tag{9.22}$$

where ν_{pq} is generally one and carried as a formality, for consistency with models to be introduced later. Each such reaction has an associated equilibrium constant K_q and mass action equation,

$$m_q = \frac{1}{K_q} \left(a_w^{\nu_{wq}} \prod^{i} (\gamma_i m_i)^{\nu_{iq}} \prod^{k} a_k^{\nu_{kq}} \prod^{m} f_m^{\nu_{mq}} \, m_p^{\nu_{pq}} \right) . \tag{9.23}$$

The mass balance equations are the same as for the K_{d} model (Eqs. 9.10–9.13), with the addition of

$$M_p = n_{\mathrm{w}} \left(m_p + \sum_q \nu_{pq} m_q \right) , \tag{9.24}$$

where m_p is the molal concentration of unoccupied sites. This equation, the general form of Equation 9.19, enforces mass balance on the sorbing sites A_p.

9.4 Ion Exchange

The *ion exchange model* is most commonly applied in geochemistry to describe the interaction of major cationic species with clay minerals, or the clay mineral fraction of a sediment; it has also been applied to zeolites and other minerals, and to ions besides the major cations (e.g., Viani and Bruton, 1992). As the name suggests, the model treats not the sorption and desorption of a species on the surface and in the interlayers of the clay, but the replacement of one ion there by another.

The exchange of K^+ for Na^+, for example, proceeds according to

$$>\text{X}:\text{Na}^+ + \text{K}^+ \rightleftarrows >\text{X}:\text{K}^+ + \text{Na}^+ , \tag{9.25}$$

where >X: represents the exchanging site. The mass action equation for this reaction is given as

$$K = \frac{\beta_{>\text{X:K}^+}\, a_{\text{Na}^+}}{\beta_{>\text{X:Na}^+}\, a_{\text{K}^+}} , \tag{9.26}$$

where K is the *exchange coefficient* and β represents the activity of a species complexed with an exchange site. In many cases, exchange constants in the literature have been calculated using the molal or molar concentrations of the aqueous species, rather than their activities; such cases require correction by a factor of the ratio of the species' activity coefficients in order to bring the Ks into accord with Equation 9.26.

The activities β may be calculated in different ways (e.g., Appelo and Postma, 2005) and, in collecting exchange constants from the literature, care must be taken to note the method used. By the *Gaines–Thomas convention*, the activity of $>\text{X:Na}^+$, for example, is given in terms of the fraction of the total electrical equivalents of exchange capacity occupied by sodium ions,

$$\beta_{>\text{X:Na}^+} = \frac{\text{meq}_{>\text{X:Na}^+}\ \text{per 100 g sediment}}{\text{CEC}} . \tag{9.27}$$

Here, meq is number of milliequivalents and CEC is the cation exchange capacity, expressed in milliequivalents per 100 g of sediment. According to the *Vanselow convention*, alternatively, the activity is expressed as the fraction of the sites occupied by the ion,

$$\beta_{>\text{X:Na}^+} = \frac{m_{>\text{X:Na}^+}\ \text{per 100 g sediment}}{\text{TEC}} , \tag{9.28}$$

where m is number of moles and TEC is the total exchangeable cations, in moles per 100 g of sediment.

Where only univalent cations are considered, the activities calculated by Equations 9.27 and 9.28 are equivalent in value, but this is not the case if divalent ions are included. The reaction describing exchange of sodium for calcium ions,

$$2 >\text{X}:\text{Na}^+ + \text{Ca}^{++} \rightleftarrows >\text{X}:_2\text{Ca}^{++} + 2\,\text{Na}^+ , \tag{9.29}$$

has an associated mass action equation,

$$K = \frac{\beta_{>\text{X:}_2\text{Ca}^{++}}\, a^2_{\text{Na}^+}}{\beta^2_{>\text{X:Na}^+}\, a_{\text{Ca}^{++}}} . \tag{9.30}$$

Since $>\text{X:}_2\text{Ca}^{++}$ has a double charge, and because the CEC and TEC in this case differ in value, the activities calculated for the species according to the two conventions differ.

It is possible to write exchange reactions for divalent ions in an alternative form,

$$>\text{X}:\text{Na}^+ + {}^1\!/\!_2\,\text{Ca}^{++} \rightleftarrows >\text{X}:\text{Ca}^{++}_{1/2} + \text{Na}^+ , \tag{9.31}$$

in which case the mass balance equation is given as

$$K = \frac{\beta_{>\text{X:Ca}^{++}_{1/2}}\, a_{\text{Na}^+}}{\beta_{>\text{X:Na}^+}\, a^{1/2}_{\text{Ca}^{++}}} . \tag{9.32}$$

The reaction written in this fashion corresponds to the *Gapon convention*, in which case the activities β may be calculated equivalently by either Equation 9.27 or 9.28, since the exchanging site maintains unit charge.

To allow for its numerical solution, we formalize our discussion of the ion exchange model by including in the basis a species A_p (e.g., >X:Na$^+$) representing the exchanging site (Eq. 9.21). This species has a molal concentration m_p, and the total mole number of exchanging sites is given by M_p; the latter value is set using the CEC or TEC provided by the user.

There are furthermore one or more additional species A_q (e.g., >X:K$^+$ and >X:$_2$Ca^{++}) formed by ion exchange with A_p. From the reaction to form each such species,

$$A_q = \nu_{wq} A_w + \sum_i \nu_{iq} A_i + \sum_k \nu_{kq} A_k + \sum_m \nu_{mq} A_m + \nu_{pq} A_p \, , \tag{9.33}$$

we can write a mass action equation,

$$\beta_q = \frac{1}{K_q} \left(a_w^{\nu_{wq}} \prod^i (\gamma_i m_i)^{\nu_{iq}} \prod^k a_k^{\nu_{kq}} \prod^m f_m^{\nu_{mq}} \, \beta_p^{\nu_{pq}} \right) , \tag{9.34}$$

where β_p and β_q are the activities of A_p and A_q, respectively. The activities are calculated according to the Gaines–Thomas convention as

$$\beta_p = \frac{z_p m_p}{z_p m_p + \sum_q z_q m_q} \qquad \beta_q = \frac{z_q m_q}{z_p m_p + \sum_{q'} z_{q'} m_{q'}} \, , \tag{9.35}$$

where we have substituted a mathematical expression of the CEC in the denominators, or by the Vanselow convention as

$$\beta_p = \frac{m_p}{m_p + \sum_q m_q} \qquad \beta_q = \frac{m_q}{m_p + \sum_{q'} m_{q'}} \, , \tag{9.36}$$

in which the denominators represent the TEC. In these expressions, q' represents a counter independent of q.

Substituting, the mass action equation (Eq. 9.34) becomes, for the Gaines–Thomas convention,

$$m_q = \frac{z_p^{\nu_{pq}} / z_q}{K_q} \left(a_w^{\nu_{wq}} \prod^i (\gamma_i m_i)^{\nu_{iq}} \prod^k a_k^{\nu_{kq}} \prod^m f_m^{\nu_{mq}} \, m_p^{\nu_{pq}} \right) \left(z_p m_p + \sum_{q'} z_{q'} m_{q'} \right)^{1-\nu_{pq}} , \tag{9.37}$$

or, for the Vanselow convention,

$$m_q = \frac{1}{K_q} \left(a_w^{\nu_{wq}} \prod^i (\gamma_i m_i)^{\nu_{iq}} \prod^k a_k^{\nu_{kq}} \prod^m f_m^{\nu_{mq}} \, m_p^{\nu_{pq}} \right) \left(m_p + \sum_{q'} m_{q'} \right)^{1-\nu_{pq}} . \tag{9.38}$$

For reactions in which the exchanging ions are of equal charge, the last term in these equations becomes unity because $1 - \nu_{pq} = 0$. In the Gaines–Thomas convention, the last term is the exchange capacity of the surface and remains constant, regardless of the charges on the exchanging ions. For the Vanselow convention, the final term varies with the proportion of monovalent and divalent ions sorbed, and this variation must be accounted for when computing the Jacobian matrix, as described in the next section. The mass balance equations for the ion exchange model are of the form of Equations 9.10–9.13 and 9.24, as already described.

9.5 Numerical Solution

A speciation calculation including one of the sorption models described above, or a combination of two or more sorption models, can be evaluated numerically following a procedure that parallels the technique described in Chapter 4. We begin as before by identifying the nonlinear portion of the problem to form the reduced basis,

$$\mathbf{B}_\mathrm{r} = \left(A_w\,,\ A_i\,,\ A_p\right)_\mathrm{r}\ . \tag{9.39}$$

The basis includes an entry A_p for each sorption model considered, except for the K_d and Freundlich models, which require no special entry.

For each basis entry we cast a residual function, which is the difference between the right and left sides of the mass balance equations (Eqs. 9.10, 9.11, and 9.24),

$$R_w = n_w\left(55.5 + \sum_j \nu_{wj} m_j + \sum_q \nu_{wq} m_q\right) - M_w \tag{9.40}$$

$$R_i = n_w\left(m_i + \sum_j \nu_{ij} m_j + \sum_q \nu_{iq} m_q\right) - M_i \tag{9.41}$$

$$R_p = n_w\left(m_p + \sum_q \nu_{pq} m_q\right) - M_p\ . \tag{9.42}$$

We employ the Newton–Raphson method to iterate toward a set of values for the unknown variables $(n_w,\ m_i,\ m_p)_\mathrm{r}$ for which the residual functions become vanishingly small.

To do so, we calculate the Jacobian matrix, which is composed of the partial derivatives of the residual functions with respect to the unknown variables. Differentiating the mass action equations for aqueous species A_j (Eq. 4.2), we note that

$$\frac{\partial m_j}{\partial n_w} = 0 \qquad \frac{\partial m_j}{\partial m_i} = \nu_{ij}\frac{m_j}{m_i} \qquad \frac{\partial m_j}{\partial m_p} = 0\,, \tag{9.43}$$

which simplifies the derivation. The values of the partial derivatives $\partial m_q/\partial n_w$, $\partial m_q/\partial m_i$, and $\partial m_q/\partial m_p$, as given by differentiating the mass action equations for the sorbed species A_q (Eqs. 9.9, 9.16, 9.23, 9.37, and 9.38), depend on the sorption model chosen.

The entries in the Jacobian matrix are

$$J_{ww} = \frac{\partial R_w}{\partial n_w} = 55.5 + \sum_j \nu_{wj} m_j + \sum_q \nu_{wq} m_q + n_w \sum_q \nu_{wq}\frac{\partial m_q}{\partial n_w} \tag{9.44}$$

$$J_{wi} = \frac{\partial R_w}{\partial m_i} = \frac{n_w}{m_i}\sum_j \nu_{wj}\nu_{ij} m_j + n_w \sum_q \nu_{wq}\frac{\partial m_q}{\partial m_i} \tag{9.45}$$

$$J_{wp} = \frac{\partial R_w}{\partial m_p} = n_w \sum_q \nu_{wq}\frac{\partial m_q}{\partial m_p} \tag{9.46}$$

$$J_{iw} = \frac{\partial R_i}{\partial n_w} = m_i + \sum_j \nu_{ij} m_j + \sum_q \nu_{iq} m_q + n_w \sum_q \nu_{iq}\frac{\partial m_q}{\partial n_w} \tag{9.47}$$

$$J_{ii'} = \frac{\partial R_i}{\partial m_{i'}} = n_w \delta_{ii'} + \frac{n_w}{m_{i'}}\sum_j \nu_{ij}\nu_{i'j} m_j + n_w \sum_q \nu_{iq}\frac{\partial m_q}{\partial m_{i'}} \tag{9.48}$$

$$J_{ip} = \frac{\partial R_i}{\partial m_p} = n_w \sum_q \nu_{iq} \frac{\partial m_q}{\partial m_p} \tag{9.49}$$

$$J_{pw} = \frac{\partial R_p}{\partial n_w} = m_p + \sum_q \nu_{pq} m_q + n_w \sum_q \nu_{pq} \frac{\partial m_q}{\partial n_w} \tag{9.50}$$

$$J_{pi} = \frac{\partial R_p}{\partial m_i} = n_w \sum_q \nu_{pq} \frac{\partial m_q}{\partial m_i} \tag{9.51}$$

$$J_{pp'} = \frac{\partial R_p}{\partial m_{p'}} = n_w \delta_{pp'} + n_w \sum_q \nu_{pq} \frac{\partial m_q}{\partial m_{p'}} \,. \tag{9.52}$$

Here, the Kronecker delta function is defined as

$$\delta_{ii'} = \begin{cases} 1 & \text{if } i = i' \\ 0 & \text{otherwise} \end{cases} \quad \text{and} \quad \delta_{pp'} = \begin{cases} 1 & \text{if } p = p' \\ 0 & \text{otherwise} \end{cases} . \tag{9.53}$$

For the K_d and Freundlich models, as mentioned, there is no basis entry A_p and hence we do not write a residual function of the form Equation 9.42, nor do we carry Jacobian entries for Equations 9.46 or 9.49–9.52.

To evaluate the Jacobian matrix, we need to compute values for $\partial m_q / \partial n_w$, $\partial m_q / \partial m_i$, and $\partial m_q / \partial m_p$. For the K_d and Freundlich models,

$$\frac{\partial m_q}{\partial n_w} = -\frac{m_q}{n_w} \,, \tag{9.54}$$

the effect of which is to negate the first summations over q in Equations 9.44, 9.47, and 9.50. This derivative is of zero value for the Langmuir and ion exchange models. For each model,

$$\frac{\partial m_q}{\partial m_i} = \nu_{iq} \frac{m_q}{m_i} \,, \tag{9.55}$$

except for the Freundlich, in which case,

$$\frac{\partial m_q}{\partial m_i} = n_\mathrm{f} \nu_{iq} \frac{m_q}{m_i} \,. \tag{9.56}$$

No value for $\partial m_q / \partial m_p$ is needed to evaluate the K_d and Freundlich models. For the Langmuir model and the ion exchange model under the Gaines–Thomas and Gapon conventions,

$$\frac{\partial m_q}{\partial m_p} = \nu_{pq} \frac{m_q}{m_p} \,. \tag{9.57}$$

Under the Vanselow convention, as previously mentioned, extra work is needed to calculate this set of derivatives. Differentiating Equation 9.38 gives

$$\frac{\partial m_q}{\partial m_p} = \nu_{pq} \frac{m_q}{m_p} + \left(1 - \nu_{pq}\right) \beta_q \left(1 + \sum_{q'} \frac{\partial m_{q'}}{\partial m_p}\right) , \tag{9.58}$$

or, rearranging,

$$\frac{\partial m_q}{\partial m_p} - \left(1 - \nu_{pq}\right) \beta_q \sum_{q'} \frac{\partial m_{q'}}{\partial m_p} = \nu_{pq} \frac{m_q}{m_p} + \left(1 - \nu_{pq}\right) \beta_q \,. \tag{9.59}$$

Writing this relation for each secondary species q gives the matrix equation

$$\begin{pmatrix} 1-(1-\nu_{p1})\,\beta_1 & -(1-\nu_{p2})\,\beta_2 & \cdots \\ -(1-\nu_{p1})\,\beta_1 & 1-(1-\nu_{p2})\,\beta_2 & \cdots \\ \vdots & \vdots & \ddots \end{pmatrix} \begin{pmatrix} \dfrac{\partial m_1}{\partial m_p} \\ \dfrac{\partial m_2}{\partial m_p} \\ \vdots \end{pmatrix} = \begin{pmatrix} \nu_{p1}\dfrac{m_1}{m_p} + (1-\nu_{p1})\,\beta_1 \\ \nu_{p2}\dfrac{m_2}{m_p} + (1-\nu_{p2})\,\beta_2 \\ \vdots \end{pmatrix}, \tag{9.60}$$

where subscripts 1, 2, ... represent $q = 1$, $q = 2$, ... This equation is solved numerically at each iteration in the solution procedure to give the values required for the derivatives $\partial m_q/\partial m_p$.

At each step in the Newton–Raphson iteration, we evaluate the residual functions and Jacobian matrix. We then calculate a correction vector as the solution to the matrix equation

$$\begin{pmatrix} J_{ww} & J_{wi} & J_{wp} \\ J_{iw} & J_{ii'} & J_{ip} \\ J_{pw} & J_{pi} & J_{pp'} \end{pmatrix}_{\mathrm{r}} \begin{pmatrix} \Delta_w \\ \Delta_i \\ \Delta_p \end{pmatrix}_{\mathrm{r}} = -\begin{pmatrix} R_w \\ R_i \\ R_p \end{pmatrix}_{\mathrm{r}}. \tag{9.61}$$

To assure non-negativity of the unknown variables, we determine an under-relaxation factor δ_{UR} according to

$$\frac{1}{\delta_{\mathrm{UR}}} = \max\left(1,\ -\frac{\Delta_w}{1/2\,n_w^{(q)}},\ -\frac{\Delta_i}{1/2\,m_i^{(q)}},\ -\frac{\Delta_p}{1/2\,m_p^{(q)}}\right)_{\mathrm{r}}, \tag{9.62}$$

and then update values from the current (q) iteration level,

$$\begin{pmatrix} n_w \\ m_i \\ m_p \end{pmatrix}_{\mathrm{r}}^{(q+1)} = \begin{pmatrix} n_w \\ m_i \\ m_p \end{pmatrix}_{\mathrm{r}}^{(q)} + \delta_{\mathrm{UR}} \begin{pmatrix} \Delta_w \\ \Delta_i \\ \Delta_p \end{pmatrix}_{\mathrm{r}}, \tag{9.63}$$

to give those at the new $(q + 1)$ level.

As a final note, a variant of the calculation is useful in many cases. Suppose a chemical analysis of a groundwater is available, giving the amount of a component in solution, and we wish to compute how much of the component is sorbed to the sediment. We can solve this problem by eliminating the summations over the sorbed species (the $\sum$ over q terms) from each of the mass balance equations, except that for surface sites (i.e., from Equations 9.10–9.13, but not from Equation 9.24). Now, the component mole numbers M_w, M_i, etc., refer to the fluid, rather than the fluid plus sorbate. The summation terms no longer appear in the residual functions R_w and R_i (Eqs. 9.40 and 9.41), nor in the derivatives of these functions within the Jacobian matrix (9.44–9.49). The Newton–Raphson iteration proceeds as before, giving the distribution of mass across dissolved and sorbed species, but honoring the fluid composition specified.

9.6 Example Calculations

We consider as a first example sorption of selenate (SeO_4^{--}), as predicted by the reaction K_d, reaction Freundlich, and Langmuir approaches (Sections 9.1–9.3). Alemi *et al.* (1991) observed the partitioning of selenate in batch experiments between 10 g of a loamy soil and 20 ml of a pH 7.5 solution containing small amounts of Na_2SeO_4; their results are shown in Figure 9.1.

In fitting these data, we note that at pH 7.5 selenate is present almost exclusively as the SeO_4^{--} oxyanion,

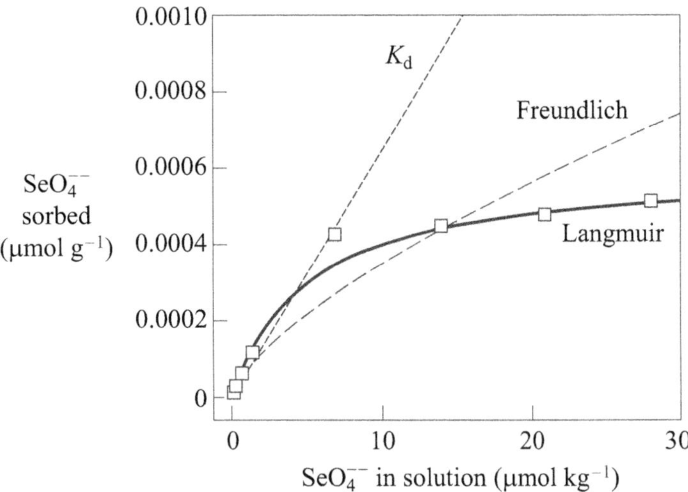

Figure 9.1 Sorption of selenate (SeO_4^{--}) to a loamy soil, showing mass sorbed per gram of dry soil, as a function of concentration in solution. Symbols show results of batch experiments by Alemi *et al.* (1991; their Fig. 1) and lines are fits to the data using the reaction K_d, reaction Freundlich, and Langmuir approaches.

and the species' activity coefficient in the dilute fluid is nearly one. We can, therefore, take the species' activity as equal to its dissolved concentration, in mol kg^{-1}. If this had not been the case, we would need to account for the speciation and activity coefficient in determining the value of $a_{SeO_4^{--}}$ for each experiment.

The parameters of the three sorption models can be determined by linear regression of the experimental observations. The reaction K_d model (Eq. 9.2) holds that

$$K_d' = \frac{m'_{>SeO_4^{--}}}{a_{SeO_4^{--}}}, \tag{9.64}$$

where $m'_{>SeO_4^{--}} = n_w\ m_{>SeO_4^{--}}/n_s$ is the concentration of sorbed selenate, expressed in mol g^{-1} soil. The value of K_d' is the limiting slope of the sorption data, given as the intercept of the best-fit line in a plot of the ratio $a_{SeO_4^{--}}/m'_{>SeO_4^{--}}$ versus $a_{SeO_4^{--}}$.

By the reaction Freundlich model (Eq. 9.16),

$$m'_{>SeO_4^{--}} = K_f'\, a_{SeO_4^{--}}^{n_f}. \tag{9.65}$$

Since

$$\log m'_{>SeO_4^{--}} = \log K_f' + n_f \log a_{SeO_4^{--}}, \tag{9.66}$$

$\log K_f'$ is the intercept of the best-fit line in a plot of $\log m'_{>SeO_4^{--}}$ versus $\log a_{SeO_4^{--}}$, and n_f is the line's slope.

Finally, the Langmuir model (Eq. 9.20) is

$$m_{>L:SeO_4^{--}} = m^T_{>L:} \frac{a_{SeO_4^{--}}}{\left(K + a_{SeO_4^{--}}\right)}, \tag{9.67}$$

where $m^T_{>L:}$ is the total concentration of sorbing sites, complexed and not. Rearranging,

$$\frac{a_{SeO_4^{--}}}{m_{>L:SeO_4^{--}}} = \frac{K}{m^T_{>L:}} + \frac{a_{SeO_4^{--}}}{m^T_{>L:}}. \tag{9.68}$$

We see $m^{\mathrm{T}}_{>\mathrm{L:}}$ is the reciprocal slope of the best-fit line in a plot of $a_{\mathrm{SeO_4^{--}}}/m_{>\mathrm{L:SeO_4^{--}}}$ versus $a_{\mathrm{SeO_4^{--}}}$, and K is the ratio of the line's intercept to its slope. The sorption capacity, in mol g^{-1} soil, then, is given $n_w\ m^{\mathrm{T}}_{>\mathrm{L:}}/n_s$.

The parameters determined in this manner for the three sorption models are

K_d	$K'_d = 0.065 \times 10^{-3}$ mol g^{-1}
Freundlich	$K'_f = 0.88 \times 10^{-6}$ mol g^{-1}
	$n_f = 0.68$
Langmuir	$K = 5.4 \times 10^{-6}$
	Sorption capacity = 0.62×10^{-9} mol g^{-1}

and the resulting isotherms are shown in Figure 9.1. The reaction Freundlich model as stated is equivalent to a standard model with this n_f in which K_f is 0.073×10^{-3}, as reported by Alemi *et al.* (1991), if the sorbed and dissolved concentrations are taken in μmol g^{-1} and μmol kg^{-1}.

To incorporate the isotherms into a geochemical model, we save the parameters into datasets "SeO4_Kd.sdat", "SeO4_Fr.sdat", and "SeO4_La.sdat". To model the Alemi *et al.* (1991) configuration, we note that per kg of water there are 500 g of dry soil, or 189 cm^3, taking the soil density to be 2.65 g cm^{-3}.

If we take the dissolved selenate concentration to be 5 μmolal and use the K_d approach, the procedure in SPECE8 is

```
inert = 189
decouple SeO4--
pH = 7.5
balance on Na+
read SeO4_Kd.sdat

SeO4-- = 5 umolal
go
```

For a concentration of 20 μmolal, we type

```
(cont'd)
SeO4-- = 20 umolal
go
```

The procedure for the Freundlich model is

```
(cont'd)
surface_data remove SeO4_Kd.sdat
read SeO4_Fr.sdat
SeO4-- = 5 umolal
go
```

and

```
(cont'd)
SeO4-- = 20 umolal
go
```

The commands to invoke the Langmuir isotherm are

```
surface_data remove SeO4_Fr.sdat
read SeO4_La.sdat
exchange_capacity = 0.62e-9 mol/g
SeO4-- = 5 umolal
go
```

and

```
SeO4-- = 20 umolal
go
```

The sorbed masses predicted for the soil in equilibrium with 5 μmolal and 20 μmolal selenate solutions are

	SeO_4^{--} sorbed (μmol g^{-1})	
	5 μmolal	20 μmolal
K_d	0.32×10^{-3}	1.30×10^{-3}
Freundlich	0.22×10^{-3}	0.55×10^{-3}
Langmuir	0.31×10^{-3}	0.49×10^{-3}

The values can be seen in Figure 9.1 to correspond to the observed data and fitted isotherms.

As a second example, we consider ion exchange in an aquifer sediment. The aquifer contains fresh water initially, but overpumping causes it to be invaded by seawater. The compositions of the fresh water and seawater are taken to be

	Fresh water	Seawater
pH	7.5	8.3
Ca^{++}(mg kg^{-1})	60	411
Mg^{++}	8	1 290
Na^{+}	20	10 760
HCO_3^{-}	210	142
SO_4^{--}	40	2 710
Cl^{-}	15	19 350

We would like to calculate the amount of Ca^{++}, Mg^{++}, and Na^{+}sorbed at the exchanging sites initially, in contact with the fresh water, and then after the sediment has fully equilibrated with seawater.

We assume Ca^{++}, Mg^{++}, and Na^{+}exchange in the aquifer sediment according to the reactions,

$$2\,{>}X{:}Na^{+} + Ca^{++} \rightleftarrows {>}X{:}_2Ca^{++} + 2\,Na^{+} \tag{9.69}$$

and

$$2\,{>}X{:}Na^{+} + Mg^{++} \rightleftarrows {>}X{:}_2Mg^{++} + 2\,Na^{+}\,, \tag{9.70}$$

for which we take exchange coefficients, reported using the Gaines–Thomas convention, of 0.4 and 0.5, respectively (see Appelo and Postma, 2005, p. 255, for a compilation of exchange coefficients from natural sediments).

To prepare for our calculation, we save these reactions and exchange coefficients in a dataset "CaMgNa_Ix.sdat". We take the sediment's cation exchange capacity to be 4 meq (100 g)$^{-1}$, or 4×10^{-5} eq g^{-1}. We further assume a porosity of 30%, so that per kg of pore fluid, there is about 2300 cm^3 of sediment.

The procedure in SPECE8 to model equilibrium exchange with the fresh water is

```
read CaMgNa_Ix.sdat
exchange_capacity = 4e-5 eq/g
inert = 2300

pH    = 7.5
Ca++  =  60 mg/kg
Mg++  =   8 mg/kg
Na+   =  20 mg/kg
HCO3- = 210 mg/kg
SO4-- =  40 mg/kg
Cl-   =  15 mg/kg
go
```

The corresponding procedure for seawater is

```
(cont'd)
pH    =    8.3
Ca++  =    411 mg/kg
Mg++  =   1290 mg/kg
Na+   =  10760 mg/kg
HCO3- =    142 mg/kg
SO4-- =   2710 mg/kg
Cl-   =  19350 mg/kg
go
```

Since we are honoring the Gaines–Thomas convention, the activity of a sorbed ion in the calculation results is the fraction of the total equivalents of the sediment's exchange capacity the ion occupies. The resulting values are

	Fresh water	Seawater
Ca^{++}	0.833	0.024
Mg^{++}	0.152	0.163
Na^{+}	0.015	0.813

10
Surface Complexation

Sorption from solution, as we noted in the previous chapter (Chapter 9), controls the environmental mobility and bioavailability of certain dissolved species, among them a variety of contaminants and nutrients. A broadly accurate description of sorption, therefore, is prerequisite to the construction of many types of geochemical models.

The empirical models presented in the previous chapter are too simplistic to be incorporated into a geochemical model intended for use under general conditions, such as across a range in pH. The distribution coefficient (K_d) approach and Freundlich and Langmuir isotherms are applied widely in groundwater studies (Domenico and Schwartz, 1998) and have been used with considerable success to describe sorption of uncharged, weakly sorbing organic molecules (Adamson and Gast, 1997). Ion exchange theory can account well for sorption of major cationic species on the clay fraction of a soil or sediment (Stumm and Morgan, 1996; Sposito, 2016).

Those models, nonetheless, take into account neither the control of pH on sorption, nor the importance of the surface's electrical state, which varies sharply with pH, ionic strength, and solution composition. The K_d and Freundlich approaches prescribe no concept of mass balance, so that a surface might be predicted to sorb from solution without limit. The approaches further depend on coefficients that must be measured for each given combination of sediment and fluid, and hence lack generality.

To portray electrolyte sorption in any broad sense, a model needs to describe hydrolysis of the mineral surface, account for electrical charge there, and provide for mass balance on the sorbing sites. In addition, an internally consistent and sufficiently comprehensive database of binding reactions should accompany the theory. Of the approaches available, a class known as *surface complexation models* (e.g., Adamson and Gast, 1997; Stumm, 1992) reflects such an ideal most closely. This class includes the double-layer (also known as the diffuse layer), triple-layer (e.g., Westall and Hohl, 1980; Sverjensky, 1993), and CD-MUSIC (Hiemstra and Van Riemsdijk, 1996) models.

Of these approaches, *double-layer theory* is better developed in the literature (e.g., Dzombak and Morel, 1987) and hence the most commonly used in geochemical modeling. Some researchers use the term two-layer model to describe an application of double-layer theory that accounts for more than one type of sorbing site on a given surface. The *constant capacitance model* refers to an application of double-layer theory in which the sorbing surface's capacitance is specified, rather than taken as a calculation result (e.g., Stumm, 1992). Electrical potential at the surface is prescribed, similarly, in a *constant potential model*. A constant potential model in which potential is set to zero yields a *nonelectrostatic model*.

In this chapter, we discuss double-layer theory and how it can be incorporated into a geochemical model. We consider as an example hydrous ferric oxide ($FeOOH \cdot nH_2O$), which is one of the most important sorbing minerals at low temperature under oxidizing conditions. Sorption by hydrous ferric oxide has been widely studied, and Dzombak and Morel (1990) have compiled an internally consistent database of its complexation reactions. The model we develop, nonetheless, is general and can be applied equally well to surface complexation with other metal oxides for which a reaction database is available. In the next chapter (Chapter 11), we expand our discussion to consider the complexation models derived considering three planes of electrical charge; a later chapter (Chapter 20) treats the reaction kinetics of surface complexation.

10.1 Complexation Reactions

Surface complexation theory is well described in a number of texts on surface chemistry, including Adamson and Gast (1997), Stumm and Morgan (1996), Sposito (2016), and Stumm (1992). According to the theory, the sorbing surface includes metal-hydroxyl functional groups, or sites, >XOH, that can react with species in solution to form surface complexes; the notation ">" represents bonding to the structure of the underlying mineral, and "X" is the metal.

An uncomplexed site >XOH in the theory can protonate or deprotonate to form surface species such as $>XOH_2^+$ and $>XO^-$. The corresponding reactions are

$$>XOH_2^+ \rightleftarrows >XOH + H^+ \tag{10.1}$$

$$>XO^- + H^+ \rightleftarrows >XOH\,. \tag{10.2}$$

Set out in this way, a parameterization is sometimes referred to as a *2-p*K$_a$ *model*, because it accounts for two protonation steps, Reactions 10.1–10.2.

Uncomplexed sites can further react with cations and anions, commonly metals Me^{++} and oxyanions $OxAn^{--}$, from solution to form complexes such as $>XOMe^+$ and $>XOxAn^-$,

$$>XOMe^+ + H^+ \rightleftarrows >XOH + Me^{++} \tag{10.3}$$

$$>XOxAn^- + H_2O \rightleftarrows >XOH + OxAn^{--} + H^+\,. \tag{10.4}$$

Following our convention, we place complexes on the left side of the reactions and the uncomplexed sites to the right.

In a *single-site model*, the simplest case, the reacting surface is populated by a single type >XOH of metal-hydroxyl sites, as shown above. A surface made up of several populations of sites, on the other hand, comprises a *multisite model*. Such a surface, for example, might contain sites >(a)XOH, >(b)XOH, and so on, each of which reacts following its own protonation, deprotonation, and complexation reactions.

*10.1.1 1-p*K$_a$ *Models*

It is also possible to describe a surface's amphoteric behavior using a *1-p*K$_a$ *model*, in which case an electrically charged site such as $>XOH^{-1/2}$ protonates in a single step,

$$>XOH_2^{+1/2} \rightleftarrows >XOH^{-1/2} + H^+\,, \tag{10.5}$$

as suggested by Bolt and Van Riemsdijk (1982) and Gunnarson *et al.* (2002). As before, the site can react,

$$>XOHMe^{+1.5} \rightleftarrows >XOH^{-1/2} + Me^{++}\,, \tag{10.6}$$

for example, to form surface complexes of varying stoichiometry.

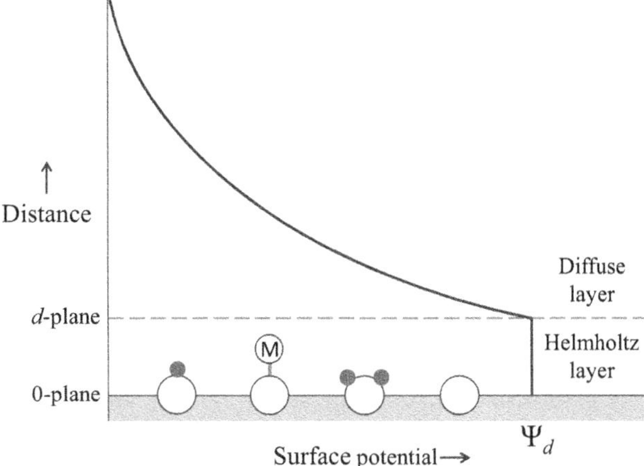

Figure 10.1 Surface complexes in the double-layer model lie within a Helmholtz or Stern layer extending from the 0-plane at the mineral surface to the d-plane, the innermost approach of a diffuse layer of counterions. Surface potential Ψ_d within the diffuse layer decreases with distance from the d-plane as the counterions progressively shield the electrical charge on the mineral surface. Depicted are an uncomplexed hydroxyl site, a metal complexed site, a protonated site, and a deprotonated site.

10.1.2 Dzombak and Morel Compilation

Dzombak and Morel (1990) derived a widely used 2-pK_a model in which hydrous ferric oxide holds two types of surface sites, one weakly and the other strongly binding. In their uncomplexed forms, the sites are labeled >(w)FeOH and >(s)FeOH.

Each site can protonate or deprotonate to form surface species such as >(w)$FeOH_2^+$ and >(w)FeO^-. The corresponding reactions are

$$>(w)FeOH_2^+ \rightleftarrows >(w)FeOH + H^+ \tag{10.7}$$

$$>(w)FeO^- + H^+ \rightleftarrows >(w)FeOH\,. \tag{10.8}$$

As well as this, the sites can react with cations and anions from solution,

$$>(w)FeOCa^+ + H^+ \rightleftarrows >(w)FeOH + Ca^{++} \tag{10.9}$$

$$>(w)FeSO_4^- + H_2O \rightleftarrows >(w)FeOH + SO_4^{--} + H^+\,, \tag{10.10}$$

to form complexes such as >(w)$FeOCa^+$ and >(w)$FeSO_4^-$.

For use with the GWB programs, file "FeOH.sdat" contains the database of surface complexation reactions prepared by Dzombak and Morel (1990), and file "FeOH+.sdat" is the Dzombak and Morel database extended to include reactions for which the binding constants have been only estimated.

10.2 Effect of Surface Charge

The uncomplexed and complexed sites on the mineral surface in double-layer theory lie within what is variously known as the Helmholtz or Stern layer (Fig. 10.1). The Helmholtz layer extends from the 0-plane at the mineral surface to the d-plane, sometimes called the plane of nearest approach. The layer holds a net electrical charge, either positive or negative, that reflects the balance of charge among the surface sites.

A second layer, the diffuse layer, extends outward from the d-plane. This blanket of counterions separates the surface from the bulk fluid. The counterions balance electrical charge on the Helmholtz layer, so that from a distance the surface appears uncharged. Double-layer theory, applied to a mixed ionic solution, does not specify which counterions make up the diffuse layer.

10.2.1 Origin of Surface Charge

As can be seen from Reactions 10.1–10.4, the state of the Helmholtz layer depends on the chemistry of the solution it contacts. As pH decreases, the numbers of protonated sites (e.g., $>XOH_2^+$) and anion-complexed sites (e.g., $>XSO_4^-$) increase. If protonated sites dominate, as is likely under acidic conditions, the surface holds a net positive charge.

Increasing pH, on the other hand, causes sites to deprotonate and complex with bivalent cations, forming species such as $>XO^-$ and $>XOCa^+$. In contact with an alkaline fluid, deprotonated sites are likely to dominate the surface, and the net charge will be negative.

The point of zero charge (PZC) is the pH at which positive and negative complexes balance. The pH at which protonated and deprotonated sites achieve charge balance is the pristine point of zero charge, or PPZC. When there are no sorbing cations or anions, the PZC and PPZC are equivalent.

10.2.2 Surface Charge Density

We can figure the surface charge directly from the molalities m_p and m_q of the uncomplexed and complexed surface sites, respectively, along with the sites' electrical charges z_p and z_q. Here, p indexes the uncomplexed sites A_p, and q enumerates the surface complexes A_q. The surface charge density σ_o (C m^{-2}), the electrical charge per unit area of the mineral surface, then, is given as

$$\sigma_o = \frac{F\,n_w}{A_{sf}}\left(\sum_p z_p m_p + \sum_q z_q m_q\right), \tag{10.11}$$

where F is the Faraday constant (96 485 C mol^{-1}), n_w is the mass of solvent water (kg), and A_{sf} is the sorbing surface area (m^2). By charge balance, the charge density at the d-plane is the negation of that on the 0-plane, so $\sigma_d = -\sigma_o$.

10.2.3 Surface Potential

Since the sorbing surface holds a charge, its electrical potential differs from that of the solution. The potential difference between surface and fluid is known as the surface potential and can be expressed in volts. According to Gouy–Chapman theory, the surface potential Ψ_d (V) at the d-plane is related to the surface charge density by

$$\sigma_o = -\sigma_d = \sqrt{8\,RT_K\,\varepsilon\varepsilon_o\,I \times 10^3}\,\sinh\left(\frac{z_\pm \Psi_d F}{2\,RT_K}\right). \tag{10.12}$$

Here, R is the gas constant (8.3143 J mol^{-1} K^{-1} or V C mol^{-1} K^{-1}), T_K is absolute temperature (K), ε is the dielectric constant (78.5 at 25 °C), ε_o is the permittivity of free space (8.854×10^{-12} C V^{-1} m^{-1}), I is ionic

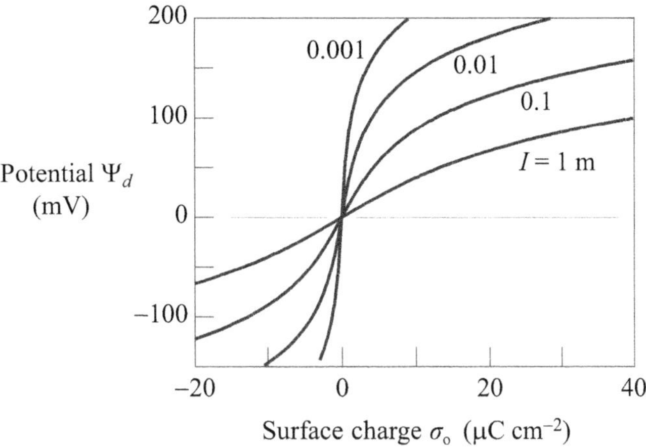

Figure 10.2 Relationship (Eq. 10.12) between surface charge density σ_o and surface potential Ψ_d for a sorbing surface in contact with solutions of differing ionic strengths I (molal).

strength (molal), the factor 10^3 represents fluid density (kg m^{-3}), and $z_\pm$ is the charge on the background solute, commonly taken to be unity. The relationship is known as the Gouy–Chapman equation.

Ionic strength in the equation serves as a proxy for solute concentration, as carried in the original derivation, since the derivation formally considers a solution of a single symmetrical electrolyte, rather than a mixed solution. A correlation,

$$\varepsilon = 87.740 - 0.40008T + 9.398 \times 10^{-4} T^2 - 1.410 \times 10^{-6} T^3 , \tag{10.13}$$

derived by Malmberg and Maryott (1956) can be used to account for variation of the dielectric constant with temperature T (°C), from 0 °C to 100 °C.

Figure 10.2 shows the relationship between Ψ_d and σ_o graphically. The integral capacitance C_1 (farad m^{-2}) of the Helmholtz layer, the charge stored on a unit area per unit potential, is the ratio of charge density σ_o to potential Ψ_d, and hence a calculation result. In a constant capacitance model, in contrast, C_1 is specified in advance. In this case, Ψ_d is given directly as the ratio of σ_o to C_1; there is no need to apply Equation 10.12. This model is appropriately applied where surface charge is small, ionic strength is large, or both, as can be seen in Figure 10.2. In a constant potential model, of course, Ψ_d is a known quantity.

10.2.4 Mass Action Equation

In order for an ion to sorb from, or desorb into solution, it needs to cross the electrical potential field near the mineral surface. In doing so, the ion consumes or liberates energy. Moving a positively charged ion toward a positively charged surface, for example, requires that work be done on the ion, increasing the system's free energy. An ion of the same charge escaping the surface would have the opposite effect. Any reaction that alters surface charge, therefore, drives a change in electrostatic energy.

To write a mass action equation for a surface complexation reaction (such as Reactions 10.1–10.4), we must therefore account for electrostatic as well as chemical contributions to the free-energy change of reaction. In other words, a coulombic term ΔG_{coul},

$$\Delta G_{\text{tot}} = \Delta G^{\circ} + \Delta G_{\text{coul}} \, , \tag{10.14}$$

needs to be added to the chemical term ΔG° to arrive at ΔG_{tot}, the total energy change of reaction under standard conditions.

The work required to bring an elementary charge e from the bulk solution to the sorbing surface is given by the product $e\Psi_d$. Recalling that the Faraday constant F is defined as $N_{\text{A}}e$, where N_{A} is Avogadro's constant (6.022×10^{23} mol^{-1}), the coulombic term is simply

$$\Delta G_{\text{coul}} = \Delta z \, F\Psi_d \, , \tag{10.15}$$

where Δz is the change over the reaction in charge on the surface site.

Following the logic leading to Equation 3.16, the the mass action equation for a surface reaction must be

$$K \exp\left(-\frac{\Delta z \Psi_d F}{RT_{\text{K}}}\right) = Q \, . \tag{10.16}$$

Here the equilibrium constant K represents the chemical effects on standard free energy, whereas the exponential factor in this equation accounts for electrostatic effects; as before, Q is the reaction's ion activity product. The exponential term in this equation is known as the Boltzman factor B.

The product KB gives a complete accounting of a surface reaction's free-energy change under standard conditions and, as such, can be inserted directly into the reaction's mass action equation. Mass action for Reaction 10.1, for example, is written

$$KB = K \exp\left(\frac{\Psi_d \, F}{RT_{\text{K}}}\right) = \frac{m_{>\text{XOH}} \, a_{\text{H}^+}}{m_{>\text{XOH}_2^+}} \, , \tag{10.17}$$

since Δz for the reaction is -1. Note that activity coefficients (i.e., $\gamma_{>\text{XOH}}$ and $\gamma_{>\text{XOH}_2^+}$) are not defined for the surface species.

10.2.5 Multidentate Complexes

When an aqueous species binds to adjacent sites on a mineral surface, it forms a multidentate surface complex; a bidentate complex, the most common case, involves binding to a pair of neighboring sites. Such complexes present special problems to the geochemical modeler, as discussed in detail by Wang and Giammar (2013).

For example, writing a reaction by which a bidentate complex $(>\text{XO})_2\text{Me}$ decomposes to a metal ion and a pair of sites,

$$(>\text{XO})_2\text{Me} + 2\,\text{H}^+ \rightleftarrows 2 >\text{XOH} + \text{Me}^{++} \, , \tag{10.18}$$

in terms of the basis entries, as is our practice, leads to the mass action equation

$$KB = \frac{m^2_{>\text{XOH}} \, a_{\text{Me}^{++}}}{m_{(>\text{XO})_2\text{Me}} \, a^2_{\text{H}^+}} \, , \tag{10.19}$$

assuming a molal standard state. Such a stoichiometric approach is demonstrably incorrect, however, because it holds that the probability of finding two uncomplexed sites together varies as the square of their concentration, as if they were independent molecules in solution, rather than fixed in place on the mineral surface.

To see the inherent inconsistency, consider a surface of given size, in equilibrium with a fluid of a certain composition. If we were to remove half of the fluid's volume, masses of the surface complexes would not change: the surface remains in contact with the same solution. Nonetheless, the molal concentrations of both >XOH and $(\text{>XO})_2\text{Me}$ would double. The former, but not the latter concentration appears squared in Equation 10.19, so Reaction 10.18 would of necessity, and incorrectly, shift to the left.

We might respond by rebalancing Reaction 10.18,

$$(\text{>XO})_2\text{Me} + 2\,\text{H}^+ \rightleftarrows (\text{>XOH})_2 + \text{Me}^{++}\,, \tag{10.20}$$

in terms of a fixed pair $(\text{>XOH})_2$ of complexing sites. While conceptually correct, the reaction is of little utility to us because no general relationship exists between the concentration of the individual sites and those taken in pairs (Benjamin, 2002). The pair-wise concentration depends on geometric factors, such as the number of neighbors that can combine with a given site, as well as the state of the surface, including which sites are already occupied.

Davis and Leckie (1980) proposed a modified mass action equation,

$$KB = \frac{m_{\text{>XOH}}\, a_{\text{Me}^{++}}}{m_{(\text{>XO})_2\text{Me}}\, a^2_{\text{H}^+}}\,, \tag{10.21}$$

in which the concentration of the bidentate complex varies linearly with that of the individual uncomplexed site, rather than as its square. The equation is not subject to the contradiction implicit in Equation 10.19 and as such has been broadly adopted by the modeling community.

Surface chemists (e.g., Sverjensky, 2003) have increasingly questioned the validity of using a molal standard state for surface complexes. Hiemstra and Van Riemsdijk (1996) replaced the molal terms in the mass action equation with mole fractions of sorbing sites. Taking $M_{\text{>XOH}}$ as the overall mole number of sites, whether complexed or not, the mole fraction $X_{\text{>XOH}}$ of uncomplexed sites is $n_w m_{\text{>XOH}}/M_{\text{>XOH}}$, the fraction $X_{(\text{>XO})_2\text{Me}}$ of complexed sites is $n_w m_{(\text{>XO})_2\text{Me}}/M_{\text{>XOH}}$, and mass action becomes

$$KB = \frac{X^2_{\text{>XOH}}\, a_{\text{Me}^{++}}}{X_{(\text{>XO})_2\text{Me}}\, a^2_{\text{H}^+}}\,. \tag{10.22}$$

The CD-MUSIC model, discussed in the next chapter (Chapter 11), makes use of this formalism.

A final alternative, advocated by Appelo and Postma (1999), casts mass action in terms of site coverage, rather than mole fraction. Since a bidentate complex occupies two surface sites, the mass action equation for Reaction 10.18 becomes

$$KB = \frac{X^2_{\text{>XOH}}\, a_{\text{Me}^{++}}}{2X_{(\text{>XO})_2\text{Me}}\, a^2_{\text{H}^+}}\,. \tag{10.23}$$

Recent releases of the PHREEQC software follow this formalism. Like Equation 10.21, Equations 10.22 and 10.23 resolve the inconsistency inherent in the stoichiometric formulation (Equation 10.19). The GWB software can recognize each of the four paradigms: Equations 10.19, 10.21, 10.22, and 10.23.

10.3 Governing Equations

Because it is based on chemical reactions, the double-layer model can be integrated into the equations describing the equilibrium state of a multicomponent system, as developed in Chapter 3. To cast the equations in general terms, recall that a label A_p represents each type of surface site. In the case of the Dzombak and Morel compilation, there are two such entries, >(w)FeOH and >(s)FeOH. The basis appears as before (Table 3.1),

$$\mathbf{B} = \left(A_w,\ A_i,\ A_k,\ A_m,\ A_p\right)\,, \tag{10.24}$$

with the addition of an entry A_p for each type of surface site considered.

There are M_p total moles of each site type in the system, divided between uncomplexed and complexed sites. The value may be figured for each site type as the product of mass (mol) of the sorbing mineral and the site density (moles of sites per mole of mineral). Alternatively, given density in sites per nm^2, the value is known from the sorbing mineral's mass, its specific surface area (nm^2 g^{-1}) and mole weight, and Avogadro's constant N_A.

The possible complexed sites A_q include deprotonated and protonated sites (e.g., >(w)FeO$^-$ and >(w)FeOH$_2^+$) and complexes with cations and anions (>(w)FeOZn$^+$ and >(w)FePO$_4^{--}$, for example). As already noted, the molalities of the uncomplexed and complexed sites, respectively, are m_p and m_q.

10.3.1 Complexation Reactions

Parallel to the reactions to form the secondary species (Eq. 3.22),

$$A_j = \nu_{wj} A_w + \sum_i \nu_{ij} A_i + \sum_k \nu_{kj} A_k + \sum_m \nu_{mj} A_m\,, \tag{10.25}$$

there is a reaction to form each surface complex,

$$A_q = \nu_{wq} A_w + \sum_i \nu_{iq} A_i + \sum_k \nu_{kq} A_k + \sum_m \nu_{mq} A_m + \sum_p \nu_{pq} A_p\,. \tag{10.26}$$

Here, ν_{wq}, ν_{iq}, ν_{kq}, ν_{mq}, and ν_{pq} are coefficients in the reaction, written in terms of the basis $\mathbf{B}$, for surface complex A_q. The coefficients ν_{pq} are zero for each surface site A_p, except the site incorporated into the complex in question. For a surface complex involving a single site, a monodentate complex, that coefficient is one; the value for a bidentate complex is two, and so on.

10.3.2 Mass Action

We have already shown (Eq. 3.27) that the molality of each secondary species is given by a mass action equation,

$$m_j = \frac{1}{K_j \gamma_j}\left(a_w^{\nu_{wj}} \prod^i (\gamma_i m_i)^{\nu_{ij}} \prod^k a_k^{\nu_{kj}} \prod^m f_m^{\nu_{mj}}\right). \tag{10.27}$$

In Reaction 10.26 the change Δz_q in surface charge is

$$\Delta z_q = \sum_p \nu_{pq} z_p - z_q\,. \tag{10.28}$$

With this in mind, we can write a generalized mass action equation cast in the form of Equation 10.17,

$$m_q = \frac{e^{\Delta z_q F \Psi_d / R T_\mathrm{K}}}{K_q}\left(a_w^{\nu_{wq}} \prod^i (\gamma_i m_i)^{\nu_{iq}} \prod^k a_k^{\nu_{kq}} \prod^m f_m^{\nu_{mq}} \prod^p m_p^{\nu_{pq}}\right), \tag{10.29}$$

that sets the molality of each monodentate surface complex. The equation can be recast slightly in the form of one of the Equations 10.21–10.23 to account for multidentate complexation, according to the various formalisms.

10.3.3 Mass Balance

As before, we write mass balance equations for each basis entry. The equations,

$$M_w = n_w \left(55.5 + \sum_j \nu_{wj} m_j + \sum_q \nu_{wq} m_q \right) \tag{10.30}$$

$$M_i = n_w \left(m_i + \sum_j \nu_{ij} m_j + \sum_q \nu_{iq} m_q \right) \tag{10.31}$$

$$M_k = n_k + n_w \left(\sum_j \nu_{kj} m_j + \sum_q \nu_{kq} m_q \right) \tag{10.32}$$

$$M_m = n_w \left(\sum_j \nu_{mj} m_j + \sum_q \nu_{mq} m_q \right) , \tag{10.33}$$

differ from Equations 3.28–3.31 by the inclusion in each of a summation over the surface complexes. The summations account for the sorbed mass of each component. We write an additional mass balance equation,

$$M_p = n_w \left(m_p + \sum_q \nu_{pq} m_q \right) , \tag{10.34}$$

for each entry A_p in the basis. This equation constrains, for each site type, the number of uncomplexed sites and surface complexes to the total number of sites M_p.

Together, the mass action and mass balance relationships form a set of governing equations describing multicomponent equilibrium in the presence of a sorbing mineral surface. The problem can alternatively be posed in terms of fluid, rather than system composition. In this case, Equations 10.30–10.33 are written omitting the second summation, and the values of M_w, M_i, M_k, and M_m reflect the composition of the fluid and coexisting minerals, neglecting the surface complexes.

10.4 Numerical Solution

The procedure for solving the governing equations parallels the technique described in Chapter 4, with the added complication of accounting for electrostatic effects. We begin as before by identifying the nonlinear portion of the problem,

$$\mathbf{B}_\mathrm{r} = \left(A_w, A_i, A_p \right)_\mathrm{r} , \tag{10.35}$$

to form the reduced basis.

10.4.1 Residual Functions

For each entry in the reduced basis we cast a residual function, which is the difference between the right and left sides of the mass balance equations (Eqs. 10.30, 10.31, and 10.34),

$$R_w = n_w \left(55.5 + \sum_j \nu_{wj} m_j + \sum_q \nu_{wq} m_q \right) - M_w \tag{10.36}$$

$$R_i = n_w \left(m_i + \sum_j \nu_{ij} m_j + \sum_q \nu_{iq} m_q \right) - M_i \tag{10.37}$$

$$R_p = n_w \left(m_p + \sum_q \nu_{pq} m_q \right) - M_p \,. \tag{10.38}$$

We employ the Newton–Raphson method to iterate toward a set of values for the unknown variables $(n_w, m_i, m_p)_r$ for which the residual functions become vanishingly small.

10.4.2 Jacobian Matrix

At each iteration, we need to calculate the Jacobian matrix, which is composed of the partial derivatives of the residual functions with respect to the unknown variables. Noting the results of differentiating the mass action equations (Eqs. 10.27 and 10.29),

$$\begin{array}{lll} \dfrac{\partial m_j}{\partial n_w} = 0 & \dfrac{\partial m_j}{\partial m_i} = \nu_{ij} \dfrac{m_j}{m_i} & \dfrac{\partial m_j}{\partial m_p} = 0 \\ \dfrac{\partial m_q}{\partial n_w} = 0 & \dfrac{\partial m_q}{\partial m_i} = \nu_{iq} \dfrac{m_q}{m_i} & \dfrac{\partial m_q}{\partial m_p} = \nu_{pq} \dfrac{m_q}{m_p} \,, \end{array} \tag{10.39}$$

simplifies the derivation. The entries in the Jacobian matrix are

$$J_{ww} = \frac{\partial R_w}{\partial n_w} = 55.5 + \sum_j \nu_{wj} m_j + \sum_q \nu_{wq} m_q \tag{10.40}$$

$$J_{wi} = \frac{\partial R_w}{\partial m_i} = \frac{n_w}{m_i} \left(\sum_j \nu_{wj} \nu_{ij} m_j + \sum_q \nu_{wq} \nu_{iq} m_q \right) \tag{10.41}$$

$$J_{wp} = \frac{\partial R_w}{\partial m_p} = \frac{n_w}{m_p} \sum_q \nu_{wq} \nu_{pq} m_q \tag{10.42}$$

$$J_{iw} = \frac{\partial R_i}{\partial n_w} = m_i + \sum_j \nu_{ij} m_j + \sum_q \nu_{iq} m_q \tag{10.43}$$

$$J_{ii'} = \frac{\partial R_i}{\partial m_{i'}} = n_w \delta_{ii'} + \frac{n_w}{m_{i'}} \left(\sum_j \nu_{ij} \nu_{i'j} m_j + \sum_q \nu_{iq} \nu_{i'q} m_q \right) \tag{10.44}$$

$$J_{ip} = \frac{\partial R_i}{\partial m_p} = \frac{n_w}{m_p} \sum_q \nu_{iq} \nu_{pq} m_q \tag{10.45}$$

$$J_{pw} = \frac{\partial R_p}{\partial n_w} = m_p + \sum_q \nu_{pq} m_q \tag{10.46}$$

$$J_{pi} = \frac{\partial R_p}{\partial m_i} = \frac{n_w}{m_i} \left(\sum_q \nu_{pq} \nu_{iq} m_q \right) \tag{10.47}$$

$$J_{pp'} = \frac{\partial R_p}{\partial m_{p'}} = n_w \delta_{pp'} + \frac{n_w}{m_{p'}} \left(\sum_q \nu_{pq} \nu_{p'q} m_q \right) . \tag{10.48}$$

Here, the Kronecker delta function is defined as

$$\delta_{ii'} = \begin{cases} 1 & \text{if } i = i' \\ 0 & \text{otherwise} \end{cases} \quad \text{and} \quad \delta_{pp'} = \begin{cases} 1 & \text{if } p = p' \\ 0 & \text{otherwise} \end{cases} . \tag{10.49}$$

At each step in the iteration, we evaluate the residual functions and Jacobian matrix. We then calculate a correction vector as the solution to the matrix equation,

$$\begin{pmatrix} J_{ww} & J_{wi} & J_{wp} \\ J_{iw} & J_{ii'} & J_{ip} \\ J_{pw} & J_{pi} & J_{pp'} \end{pmatrix}_{\mathrm{r}} \begin{pmatrix} \Delta_w \\ \Delta_i \\ \Delta_p \end{pmatrix}_{\mathrm{r}} = - \begin{pmatrix} R_w \\ R_i \\ R_p \end{pmatrix}_{\mathrm{r}} . \tag{10.50}$$

To assure non-negativity of the unknown variables, we determine an under-relaxation factor δ_{UR} according to

$$\frac{1}{\delta_{\mathrm{UR}}} = \max \left(1, -\frac{\Delta_w}{1/2\, n_w^{(q)}}, -\frac{\Delta_i}{1/2\, m_i^{(q)}}, -\frac{\Delta_p}{1/2\, m_p^{(q)}} \right)_{\mathrm{r}} , \tag{10.51}$$

and then update values from the current (q) iteration level,

$$\begin{pmatrix} n_w \\ m_i \\ m_p \end{pmatrix}_{\mathrm{r}}^{(q+1)} = \begin{pmatrix} n_w \\ m_i \\ m_p \end{pmatrix}_{\mathrm{r}}^{(q)} + \delta_{\mathrm{UR}} \begin{pmatrix} \Delta_w \\ \Delta_i \\ \Delta_p \end{pmatrix}_{\mathrm{r}} , \tag{10.52}$$

to give those at the new $(q + 1)$ level.

10.4.3 Solving for Electrostatic State

The iteration step is complicated by the need to account for the electrostatic state of the sorbing surface when setting values for m_q. The surface potential Ψ_d affects the binding reactions, according to the mass action equation (Eq. 10.29). In turn, according to Equation 10.11, the concentrations m_q of the sorbed species control the surface charge and hence (by Eq. 10.12) potential. Since the relationships are nonlinear, we must solve numerically (e.g., Westall, 1980) for a consistent set of values for the potential and species concentrations.

The solution, performed at each step in the Newton–Raphson iteration, is accomplished by setting Equation 10.11 equal to Equation 10.12. We seek to minimize a residual function R_Ψ,

$$R_\Psi = \frac{A_{\mathrm{sf}}}{F\, n_w} \sqrt{8\, R T_{\mathrm{K}}\, \varepsilon \varepsilon_{\mathrm{o}}\, I \times 10^3} \sinh \left(\frac{z_\pm \Psi_d F}{2\, R T_{\mathrm{K}}} \right) - \sum_p z_p m_p - \sum_q z_q m_q , \tag{10.53}$$

expressing imbalance in the equality we have written, given an estimate for potential Ψ_d. The function's derivative with respect to Ψ_d is

$$\frac{dR_\Psi}{d\Psi_d} = \frac{z_\pm A_{sf}}{2\,n_w\,RT_K}\sqrt{8\,RT_K\,\varepsilon\varepsilon_o\,I\times 10^3}\cosh\left(\frac{z_\pm\Psi_d F}{2RT_K}\right) - \frac{F}{RT_K}\sum_q \Delta z_q^2 m_q\,. \tag{10.54}$$

Using Newton's method (described in Chapter 4), we can quickly locate the appropriate surface potential by driving down the residual function until it approaches zero. In the constant capacitance and constant potential cases, of course, Ψ_d presents directly, without need for iteration.

10.5 Example Calculation

As an example of an equilibrium calculation accounting for surface complexation, we consider the sorption of mercury, lead, and sulfate onto hydrous ferric oxide at pH 4 and 8. We use ferric hydroxide [$Fe(OH)_3$] precipitate from the LLNL database to represent in the calculation hydrous ferric oxide ($FeOOH\cdot nH_2O$). Following Dzombak and Morel (1990), we assume a sorbing surface area of 600 $m^2\ g^{-1}$ and site densities for the weakly and strongly binding sites, respectively, of 0.2 and 0.005 mol (mol FeOOH)$^{-1}$. We choose a system containing 1 kg of solvent water (the default) in contact with 1 g of ferric hydroxide.

To set up the calculation in SPECE8, we enter the commands

```
read FeOH+.sdat
sorbate include

decouple Fe+++
swap Fe(OH)3(ppd) for Fe+++
1 free gram Fe(OH)3(ppd)
```

First, we read in the dataset of complexation reactions and specify that the initial mass balance calculations should include the sorbed as well as aqueous species. We disable the ferric–ferrous redox couple (since we are not interested in ferrous iron), and specify that the system contains 1 g of sorbing mineral.

We set a dilute NaCl solution containing small concentrations of Hg^{++}, Pb^{++}, and SO_4^{--}. For the first calculation, we set pH to 4

```
(cont'd)
Na+    =  10 mmolal
Cl-    =  10 mmolal
Hg++   = 100 umolal
Pb++   = 100 umolal
SO4--  = 200 umolal
balance on Cl-

pH = 4
go
```

and run the model. We repeat the calculation assuming a pH of 8

```
(cont'd)
pH = 8
go
```

Table 10.1 *Calculated examples of surface complexation*

	pH = 4		pH = 8	
Surface charge ($\mu C\ cm^{-2}$)		16.0		0.4
Surface potential (mV)		168		17.1
SITE OCCUPANCY				
Weak sites	mmolal	% of sites	mmolal	% of sites
$>(w)FeOH_2^+$	1.23	65.9	0.129	6.92
$>(w)FeOH$	0.435	23.2	1.29	69.0
$>(w)FeO^-$	0.351×10^{-2}	0.19	0.295	15.8
$>(w)FeOHg^+$	0.415×10^{-6}	0.000	0.0984	5.26
$>(w)FeOPb^+$	0.386×10^{-3}	0.02	0.0534	2.85
$>(w)FeSO_4^-$	0.117	6.27	0.189×10^{-3}	0.010
$>(w)FeOHSO_4^{--}$	0.0826	4.41	0.377×10^{-2}	0.20
	1.871	100	1.871	100
Strong sites	mmolal	% of sites	mmolal	% of sites
$>(s)FeOH_2^+$	0.00559	12.0	0.505×10^{-5}	0.01
$>(s)FeOH$	0.00197	4.21	0.504×10^{-4}	0.11
$>(s)FeO^-$	0.159×10^{-4}	0.03	0.115×10^{-4}	0.03
$>(s)FeOHg^+$	0.385×10^{-7}	0.000	0.784×10^{-4}	0.17
$>(s)FeOPb^+$	0.0392	83.8	0.0466	99.69
	0.0468	100	0.0468	100
SORBED FRACTIONS				
		% sorbed		% sorbed
Hg^{++}		0.000		98.45
Pb^{++}		39.59		100
SO_4^{--}		99.95		1.98

Table 10.1 summarizes the calculation results.

The predicted state of the sorbing surface in the two calculations differs considerably. At pH 4, the surface carries a positive surface charge and potential. The electrical charge arises largely from the predominance of the protonated surface species $>(w)FeOH_2^+$, which occupies about two-thirds of the weakly binding sites. At pH 8, however, the surface charge and potential nearly vanish because of the predominance of the uncomplexed species >(w)FeOH, which is electrically neutral.

We can quickly verify the calculated charge density using Equation 10.11 and the data in Table 10.1. At pH 4, we add the products of each species' charge and concentration,

		mmolal
$>$(w)$FeOH_2^+$	$+1\ \times$	1.23
$>$(w)FeO^-	$-1\ \times$	0.0035
$>$(w)$FeSO_4^-$	$-1\ \times$	0.117
$>$(w)$FeOHSO_4^{--}$	$-2\ \times$	0.0826
$>$(s)$FeOH_2^+$	$+1\ \times$	0.0056
$>$(s)$FeOPb^+$	$+1\ \times$	0.0392
		0.989

and then calculate the charge density as

$$(96\,485\ \mathrm{C\ mol^{-1}}) \times (1\ \mathrm{kg\ H_2O}) \times (0.989 \times 10^{-3}\ \mathrm{molal})/(600\ \mathrm{m^2}) = 0.16\ \mathrm{C\ m^{-2}}\,, \tag{10.55}$$

or 16 μC cm^{-2}. Taking ionic strength as 0.01 molal, we read the corresponding surface potential from Figure 10.2. We can verify the results at pH 8 in a similar fashion.

According to the calculations, the surface's ability to sorb cations and anions differs markedly between pH 4 and 8, reflecting both electrostatic influences and mass action. Nearly all of the sulfate is sorbed at pH 4, whereas most of the lead and all the mercury remain in solution; the opposite holds at pH 8. At pH 4, the surface carries a positive charge that attracts sulfate ions but repels lead and mercury ions. The electrostatic effect is almost nil at pH 8, however, where the surface charge approaches zero. Also, the complexation reactions,

$$>\mathrm{(w)FeSO_4^-} + \mathrm{H_2O} \rightleftarrows\ >\mathrm{(w)FeOH} + \mathrm{H^+} + \mathrm{SO_4^{--}} \tag{10.56}$$

$$>\mathrm{(s)FeOPb^+} + \mathrm{H^+} \rightleftarrows\ >\mathrm{(s)FeOH} + \mathrm{Pb^{++}} \tag{10.57}$$

$$>\mathrm{(w)FeOHg^+} + \mathrm{H^+} \rightleftarrows\ >\mathrm{(w)FeOH} + \mathrm{Hg^{++}}\,, \tag{10.58}$$

conspire by mass action to favor complexing of sulfate at low pH and lead and mercury at high pH. The reaction to form the surface complex $>$(w)$FeOHSO_4^{--}$,

$$>\mathrm{(w)FeOHSO_4^{--}} \rightleftarrows\ >\mathrm{(w)FeOH} + \mathrm{SO_4^{--}}\,, \tag{10.59}$$

has no dependence on pH. Electrostatic forces and the variation in the amount of SO_4^{--} in solution [depending on the amount sorbed as $>$(w)$FeSO_4^-$] explain the variation in abundance of this species.

10.6 Limitations of the Modeling

Despite the seeming exactitude of the mathematical development, the modeler should bear in mind that surface complexation modeling involves uncertainties and data limitations in addition to those outlined in Chapter 2 (e.g., Lützenkirchen, 2006). Perhaps foremost is the nature of the sorbing material itself. Measurements for a given metal oxide or hydroxide are conducted using a spectrum of materials ranging from fresh laboratory precipitate to minerals aged in the natural environment for millions of years. Since laboratories follow different aging procedures, results of their studies can be difficult to compare.

The Dzombak and Morel (1990) database, for example, was parameterized from experiments performed using synthetic precipitates of hydrous ferric oxide ($FeOOH \cdot nH_2O$). The material ripens with time, however, changing in water content and extent of polymerization, eventually crystallizing to form goethite (FeOOH). It is not clear how closely the synthetic materials resemble sorbing ferric oxides (e.g., ferrihydrite and goethite) encountered in nature.

Also, even though the binding constants reported in a given study reflect directly the value taken for site density, there is no agreed upon method constraining this value. As Dzombak and Morel (1990) report, site densities and surface areas reported by studies in their compilation span a relatively broad range. Again, this lack of standardization makes data collected by different laboratories hard to integrate into a single database, let alone apply with confidence to a specific problem.

Nearly all of the data are collected at room temperature, furthermore, and there is little guidance how they might be corrected to apply at other temperatures. Far fewer data have been derived for sorption of anions than for cations. The theory does not account for the hysteresis commonly observed between the adsorption and desorption of a strongly bound ion. Finally, much work remains to be done before the results of laboratory experiments performed on simple mineral–water systems can be applied to the study of complex soils.

11
Three-Layer Complexation

Three-layer formulations of the surface complexation problem include the triple-layer and CD-MUSIC models, which comprise the subject of this chapter. The triple-layer model was developed by Davis *et al.* (1978) and Davis and Leckie (1980) to represent the behavior of outer-sphere complexes, those in which the mineral surface binds with the hydration sphere surrounding an ion.

The model was later generalized (e.g., Blesa *et al.*, 1984; Hayes and Leckie, 1987) to include inner-sphere complexes, where a surface binds to the ion itself. The CD-MUSIC model (Hiemstra and Van Riemsdijk, 1996) adapted the concept so that the electrical charge of a surface complex might be distributed across the interfacial region, rather than assigned as points of integer charge to specific positions therein.

In the various three-layer formulations, the 0-plane and d-plane mark the mineral surface and the innermost extent of the diffuse layer, as before, and a β-plane separates an inner from an outer Helmholtz layer (Fig. 11.1). The 0-plane generally protonates and deprotonates, and surface complexes may occupy the 0-plane, β-plane, and, in some cases, d-plane.

Three-layer formulations hold potential advantages for a geochemical modeler over the two-layer models discussed in Chapter 10. The former can be configured to better reflect the presumed structure of the interfacial region. Inner-sphere complexes, for example, might be assigned to the 0-plane and outer-sphere complexes placed along the β-plane.

In addition, the two-layer models distinguish between the background electrolyte, commonly a 1:1 salt, and whatever solutes of lesser concentration complex with the mineral surface. Three-layer models, in contrast, allow abundant ions in solution to sorb, generally on the β-plane, where they become more important and exert a stronger electrostatic effect as ionic strength increases (e.g., Hayes *et al.* 1988; 1991). The models, for this reason, may be preferred when working with fluids in the geosphere, where waters may be concentrated and in which strongly complexing ions can make up a significant portion of the background electrolyte (e.g., Sverjensky, 2006).

11.1 Triple-Layer Model

Uncomplexed sites >XOH in the triple-layer model can react with hydrogen ions to protonate,

$$>\mathrm{XOH_2^+} \rightleftarrows\ >\mathrm{XOH} + \mathrm{H^+}\,, \tag{11.1}$$

and deprotonate,

$$>\mathrm{XO^-} + \mathrm{H^+} \rightleftarrows\ >\mathrm{XOH}\,, \tag{11.2}$$

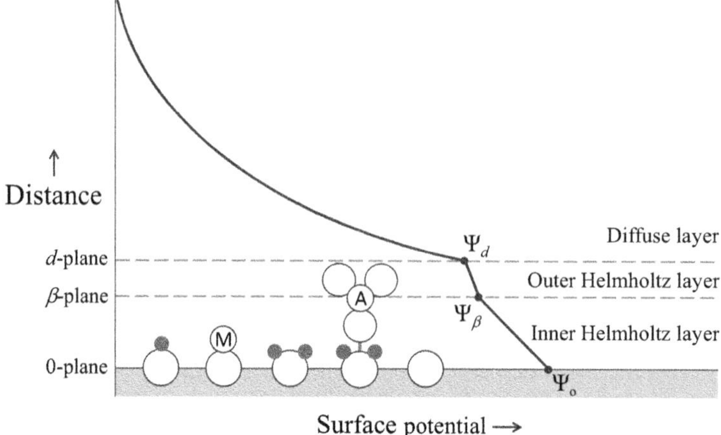

Figure 11.1 In a three-layer model, the 0-plane at the mineral surface is overlain by an inner Helmholtz layer extending to the β-plane. An outer Helmholtz layer then extends to the d-plane, which marks the innermost approach of the diffuse layer. Surface potential Ψ_o at the mineral surface decreases to Ψ_β at the β plane and Ψ_d at the d-plane, before dissipating with distance into the bulk fluid. Depicted are an uncomplexed hydroxyl site, a tightly bound metal cation, a protonated site, an oxyanion complex, and a deprotonated site.

on the 0-plane, just as they can in the two-layer formulations. In the model's original formulation (Davis *et al.*, 1978), other ions, including oxy- and hydroxy-ions, complexing with the mineral surface occupy the β-plane. In this way, the model diverges from the two-layer formulations, in which complexed ions lie along the 0-plane.

Davis and Leckie (1980), for example, considered that a sodium nitrate solution might react with surfaces of amorphous iron oxyhydroxide according to

$$>XO^-{:}Na^+ + H^+ \rightleftarrows >XOH + Na^+ , \tag{11.3}$$

and

$$>XOH_2^+{:}NO_3^- \rightleftarrows >XOH + NO_3^- + H^+ . \tag{11.4}$$

The notation ":" in the species' stoichiometries represents bonding within the surface complex between the 0-plane and β-plane.

More recent formulations have set strongly bound ions on the 0-plane while leaving weakly complexed ions on the β-plane. In modeling the surface complexation of inorganic lead, for example, Hayes and Leckie (1987) wrote reactions

$$>XOPb^+ + H^+ \rightleftarrows >XOH + Pb^{++} , \tag{11.5}$$

and

$$>XO^- : Pb^{++} + H^+ \rightleftarrows >XOH + Pb^{++} , \tag{11.6}$$

in which Pb^{++} lies on the 0-plane (Reaction 11.5) as well as the β-plane (Reaction 11.6). Surface chemists commonly consider the first case to be an inner-sphere complex and the second an outer-sphere complex.

Such classifications may be made largely by inference, however, since direct structural information is not necessarily taken into account when regressing a reaction's properties.

As we can see in the reactions above, the two positions in a surface complex carry electrical charge separately. Charge z^{o} on the 0-plane of $>\text{XO}^- : \text{Na}^+$ is -1, for example, and z^β for the complex is $+1$. As Reaction 11.3 proceeds from left to right, the changes in electrical charge Δz^{o} and Δz^β on the two planes are, respectively, $+1$ and -1, given that the uncomplexed site >XOH is electrically neutral. For complex $>\text{XOH}_2^+ : \text{NO}_3^-$, as a second example, z^{o} is $+1$ and z^β is -1, so Δz^{o} and Δz^β in Reaction 11.4 are -1 and $+1$.

11.2 CD-MUSIC Model

The CD-MUSIC model (Hiemstra and Van Riemsdijk, 1996) is a generalization of the triple-layer model in which electrical charge is distributed across the 0-plane, β-plane, and, in some cases, the d-plane of the mineral-water interface. Instead of assigning an integer charge to each plane in a surface complex, the CD-MUSIC technique works by taking into account the change in the planes' electrical charge when the complex forms. The changes can be, and commonly are, fractional.

Specifically, the method prescribes the changes in charge Δz_0, Δz_1, and Δz_2 at the 0-plane, β-plane, and d-plane, respectively, that arise from a complexation reaction. Our convention is to write reactions in terms of complex dissociation, rather than association. For this reason, and to maintain notation consistent with the other models, we denote the charge changes as $\Delta z^{\text{o}} = -\Delta z_0$, $\Delta z^\beta = -\Delta z_1$, and $\Delta z^d = -\Delta z_2$.

Weng *et al.* (2008), for example, wrote a reaction,

$$>\text{FeOHCa}^{+1.5} \rightleftarrows\ >\text{FeOH}^{-0.5} + \text{Ca}^{++}\,, \tag{11.7}$$

by which calcium ions complex with the goethite (FeOOH) surface. For the reaction, $\Delta z^{\text{o}} = -0.32$, $\Delta z^\beta = -1.68$, and $\Delta z^d = 0$. We know that charge on the uncomplexed site $>\text{FeOH}^{-0.5}$ lies on the 0-plane, so z^{o} on $>\text{FeOHCa}^{+1.5}$ must be -0.18, z^β for the complex needs to be $+1.68$, and z^d is zero. Weng *et al.* wrote a second reaction,

$$>\text{FeOHCaOH}^{+0.5} + \text{H}^+ \rightleftarrows\ >\text{FeOH}^{-0.5} + \text{Ca}^{++} + \text{H}_2\text{O}\,, \tag{11.8}$$

also for complexation with calcium ions. In this case, charge change Δz^β on the β-plane is -0.68, so z^β must be $+0.68$.

11.3 Governing Equations

In solving for speciation in the presence of a three-layer surface, we generally know the integral capacitance C_1 (farad m^{-2}) of the inner Helmholtz layer, as well as that C_2 of the outer layer, since the values commonly serve as adjustable parameters when regressing the experimental data underlying a reaction dataset (e.g., Sahai and Sverjensky, 1997). It remains to us to derive not only the distribution of mass in the system, but the surface charge densities, σ_{o}, σ_β, and σ_d (C m^{-2}), and surface potentials, Ψ_{o}, Ψ_β, and Ψ_d (V), at the three planes along the mineral–water interface (Fig. 11.1).

11.3.1 Mass Action Equation

As before (Eq. 10.26), a reaction for each surface complex A_q can be written

$$A_q = \nu_{wq} A_w + \sum_i \nu_{iq} A_i + \sum_k \nu_{kq} A_k + \sum_m \nu_{mq} A_m + \sum_p \nu_{pq} A_p \,, \tag{11.9}$$

in terms of the uncomplexed sites A_p and other basis species A_i, A_k, and A_m, where ν_{wq}, ν_{iq}, ν_{kq}, ν_{mq}, and ν_{pq} are the reaction coefficients.

The corresponding mass action equation,

$$m_q = \exp\left(\frac{F(\Delta z_q^{\mathrm{o}} \Psi_{\mathrm{o}} + \Delta z_q^{\beta} \Psi_{\beta} + \Delta z_q^{d} \Psi_d)}{R T_{\mathrm{K}}}\right) \frac{\mathrm{IAP}}{K_q} \,, \tag{11.10}$$

gives the molal concentration m_q of each surface complex. Here, F is the Faraday constant (96 485 C mol^{-1}), R is the gas constant (8.4143 V C mol^{-1} K^{-1}), IAP is the ion activity product for the reaction's right side,

$$\mathrm{IAP} = a_w^{\nu_{wq}} \prod^{i} (\gamma_i m_i)^{\nu_{iq}} \prod^{k} a_k^{\nu_{kq}} \prod^{m} f_m^{\nu_{mq}} \prod^{p} m_p^{\nu_{pq}} \,, \tag{11.11}$$

and K_q is the equilibrium constant. The relationship parallels Equation 10.29, with the form of the Boltzman factor expanded to reflect coulombic energies at each of the three planes.

In the triple-layer model, the Δz terms in the mass action equation for a reaction are given from the reaction coefficients and charge on each plane,

$$\Delta z_q^{\mathrm{o}} = \sum_p \nu_{pq} z_p^{\mathrm{o}} - z_q^{\mathrm{o}} \,, \quad \Delta z_q^{\beta} = -z_q^{\beta} \,, \quad \text{and} \quad \Delta z_q^{d} = 0 \,. \tag{11.12}$$

For the CD-MUSIC model, on the other hand, the Δz values are given directly.

11.3.2 Surface Charge Density

Surface charge density σ (C m^{-2}) at a given plane is determined from the charges z_p and z_q on the uncomplexed sites and surface complexes, respectively, and their molalites m_p and m_q. For the CD-MUSIC model, charges on the three planes are determined,

$$z_q^{\mathrm{o}} = \sum_p \nu_{pq} z_p^{\mathrm{o}} - \Delta z_q^{\mathrm{o}} \,, \quad z_q^{\beta} = -\Delta z_q^{\beta} \,, \quad \text{and} \quad z_q^{d} = -\Delta z_q^{d} \,, \tag{11.13}$$

from the reaction's Δz values, whereas charge at each plane is prescribed in the triple-layer method.

As before (Eq. 10.11), charge density on the 0-plane is

$$\sigma_{\mathrm{o}} = \frac{F n_w}{A_{\mathrm{sf}}} \left(\sum_p z_p^{\mathrm{o}} m_p + \sum_q z_q^{\mathrm{o}} m_q \right) \,, \tag{11.14}$$

where A_{sf} is sorbing surface area (m^2). The corresponding value at the β-plane,

$$\sigma_{\beta} = \frac{F n_w}{A_{\mathrm{sf}}} \sum_q z_q^{\beta} m_q \,, \tag{11.15}$$

follows from similar logic, given that uncomplexed sites carry no charge there. Finally, charge density at the d-layer,

$$\sigma_d = -\sqrt{8\,RT_K\,\varepsilon\varepsilon_o\,I \times 10^3}\,\sinh\left(\frac{z_\pm \Psi_d F}{2\,RT_K}\right), \tag{11.16}$$

is given by the Gouy–Chapman equation (Eq. 10.12).

11.3.3 Surface Potential

Surface potential at each of the three planes reflects the distribution of charge density and the integral capacitances C_1 and C_2 of the inner and outer Helmholtz layers. From the definition of capacitance,

$$C_1 = \frac{\sigma_o}{\Psi_o - \Psi_\beta}, \quad \text{and} \quad C_2 = \frac{\sigma_o + \sigma_\beta}{\Psi_\beta - \Psi_d}. \tag{11.17}$$

Rearranging gives

$$\sigma_o = C_1\left(\Psi_o - \Psi_\beta\right) \tag{11.18}$$

and

$$\sigma_\beta = C_2\left(\Psi_\beta - \Psi_d\right) - C_1\left(\Psi_o - \Psi_\beta\right). \tag{11.19}$$

Then, by charge balance,

$$\sigma_d = -C_2\left(\Psi_\beta - \Psi_d\right), \tag{11.20}$$

since $\sigma_o + \sigma_\beta + \sigma_d = 0$.

11.4 Numerical Solution

Determining the distribution of mass in a system holding a three-layer complexing surface largely parallels the procedure for a two-layer surface, described in the previous chapter (Chapter 10). The primary differences are use of an expanded mass action equation (Eq. 11.10) in place of Equation 10.29, and handling of the electrostatic terms. This section approaches the latter issue.

In the two-layer case we found need at each pass in the encompassing Newton–Raphson solution to iterate with Newton's method (see Section 3.2.1) for consistent values of charge density σ_o and surface potential Ψ_d. For three-layer surfaces, in contrast, we require consistency among three charge densities σ_o, σ_β, and σ_d, and three surface potentials Ψ_o, Ψ_β, and Ψ_d. For this reason, we need to nest a Newton–Raphson solution (Section 4.2.2) to the electrostatic problem within the outer Newton–Raphson iteration.

11.4.1 Residual Functions

The electrostatic iteration seeks the root of three residual functions, one for each plane in the mineral–water interface. Residual R_o is the inequality of Equations 11.18 and 11.14,

$$R_o = C_1\left(\Psi_o - \Psi_\beta\right) - \frac{F\,n_w}{A_{sf}}\left(\sum_p z_p^o m_p + \sum_q z_q^o m_q\right), \tag{11.21}$$

whereas R_β is the difference between Equations 11.19 and 11.15,

$$R_\beta = C_1\left(\Psi_\beta - \Psi_o\right) + C_2\left(\Psi_\beta - \Psi_d\right) - \frac{F\,n_w}{A_{sf}}\sum_q z_q^\beta m_q\,, \tag{11.22}$$

and R_d,

$$R_d = C_2\left(\Psi_d - \Psi_\beta\right) + \sqrt{8\,RT_K\,\varepsilon\varepsilon_o\,I \times 10^3}\,\sinh\left(\frac{z_\pm \Psi_d F}{2\,RT_K}\right), \tag{11.23}$$

reflects imbalance between Equations 11.20 and 11.16.

11.4.2 Jacobian Matrix

To find the roots of residual functions R_o, R_β, and R_d for the electrostatic problem, we need to derive the nine entries in the Jacobian matrix, composed of the partial derivatives of the residuals with respect to the independent variables, Ψ_o, Ψ_β, and Ψ_d. Noting that from differentiating Equation 11.10,

$$\frac{\partial m_q}{\partial \Psi_o} = \left(\frac{F\Delta z_q^o}{RT_K}\right) m_q\,,\quad \frac{\partial m_q}{\partial \Psi_\beta} = \left(\frac{F\Delta z_q^\beta}{RT_K}\right) m_q\,,\quad \text{and}\quad \frac{\partial m_q}{\partial \Psi_d} = \left(\frac{F\Delta z_q^d}{RT_K}\right) m_q\,, \tag{11.24}$$

we find the derivatives,

$$\frac{\partial R_o}{\partial \Psi_o} = C_1 - \frac{F^2 n_w}{RT_K A_{sf}}\sum_q \Delta z_q^o z_q^o m_q \tag{11.25}$$

$$\frac{\partial R_o}{\partial \Psi_\beta} = -C_1 - \frac{F^2 n_w}{RT_K A_{sf}}\sum_q \Delta z_q^\beta z_q^o m_q \tag{11.26}$$

$$\frac{\partial R_o}{\partial \Psi_d} = 0 \tag{11.27}$$

$$\frac{\partial R_\beta}{\partial \Psi_o} = -C_1 - \frac{F^2 n_w}{RT_K A_{sf}}\sum_q \Delta z_q^o z_q^\beta m_q \tag{11.28}$$

$$\frac{\partial R_\beta}{\partial \Psi_\beta} = C_1 + C_2 - \frac{F^2 n_w}{RT_K A_{sf}}\sum_q \Delta z_q^\beta z_q^\beta m_q \tag{11.29}$$

$$\frac{\partial R_\beta}{\partial \Psi_d} = -C_2 \tag{11.30}$$

$$\frac{\partial R_d}{\partial \Psi_o} = -\frac{F^2 n_w}{RT_K A_{sf}}\sum_q \Delta z_q^o z_q^d m_q \tag{11.31}$$

$$\frac{\partial R_d}{\partial \Psi_\beta} = -C_2 - \frac{F^2 n_w}{RT_K A_{sf}}\sum_q \Delta z_q^\beta z_q^d m_q \tag{11.32}$$

and

$$\frac{\partial R_d}{\partial \Psi_d} = C_2 + \frac{z_\pm F}{2\,RT_K}\sqrt{8\,RT_K\,\varepsilon\varepsilon_o\,I \times 10^3}\cosh\left(\frac{z_\pm \Psi_d F}{2\,RT_K}\right). \tag{11.33}$$

Writing the derivatives ensemble forms a 3×3 Jacobian matrix,

$$(\mathbf{J}) = \begin{pmatrix} \frac{\partial R_o}{\partial \Psi_o} & \frac{\partial R_o}{\partial \Psi_\beta} & \frac{\partial R_o}{\partial \Psi_d} \\ \frac{\partial R_\beta}{\partial \Psi_o} & \frac{\partial R_\beta}{\partial \Psi_\beta} & \frac{\partial R_\beta}{\partial \Psi_d} \\ \frac{\partial R_d}{\partial \Psi_o} & \frac{\partial R_d}{\partial \Psi_\beta} & \frac{\partial R_d}{\partial \Psi_d} \end{pmatrix}, \tag{11.34}$$

that allows us to find the unknown potentials by iteration, as described in Section 4.2.2.

11.5 Example Calculation

To illustrate behavior of the triple-layer model, we borrow the work of Hayes *et al.* (1988), as recorded in surface dataset `Goethite_Se.sdat`, to consider sorption of selenite (SeO_3^{--}) from a $NaNO_3$ solution onto a goethite (FeOOH) surface. In their compilation, the goethite surface has a specific surface area of 52 $m^2\,g^{-1}$ and holds seven >FeOH sites nm^{-2}; the integral capacitance C_1 is 1.1 farad m^{-2}, and C_2 is 0.2 farad m^{-2}.

In parallel to the SPECE8 procedure in Section 10.7, we read the corresponding surface dataset, specify that compositional constraints are to comprise sorbed as well as aqueous species, and set the mass of the sorbing mineral

```
read Goethite_Se.sdat
sorbate include
decouple Fe+++
swap Goethite for Fe+++
30 free gram Goethite
```

The system extent, then, is 30 g of goethite in contact with 1 kg of solvent water, the default, and the solutes therein. The fluid is a 0.1 molal $NaNO_3$ solution containing 100 μmolal Na_2SeO_3 At first, we set pH to 9

```
(cont'd)
NO3-    = 100    mmolal
SeO3-- =    0.1 mmolal
Na+     = 100.2 mmolal
balance on NO3-
pH = 9
go
```

and run the model, before repeating the calculation

```
(cont'd)
pH = 12
go
```

with pH set to 12. Table 11.1 summarizes results of the two calculations.

Table 11.1 *Calculated examples of triple-layer complexation*

	pH = 9		pH = 12	
Surface charge ($\mu C\ cm^{-2}$)				
0-plane, σ_o		−3.19		−18.2
β-plane, σ_β		2.91		17.8
d-plane, σ_d		0.28		0.4
Surface potential (mV)				
0-plane, Ψ_o		−46.8		−189.
β-plane, Ψ_β		−17.8		−24.1
d-plane, Ψ_d		−3.9		−5.3
SITE OCCUPANCY				
	mmolal	% of sites	mmolal	% of sites
$>FeOH$	17.07	94.12	15.10	83.29
$>FeO^- : Na^+$	0.677	3.73	2.91	16.04
$>FeOH_2^+ : NO_3^-$	0.200	1.10	0.0298	0.16
$>FeSeO_3^-$	0.0899	0.50	0.0003	0.00
$>FeOH_2^+$	0.0666	0.37	0.0151	0.08
$>FeO^-$	0.0219	0.12	0.0757	0.42
$>FeOH_2^+ : HSeO_3^-$	0.0086	0.05	0.0014	0.01
$>FeSeO_3^- : Na^+$	0.0014	0.01	0.0000	0.00
	18.14	100	18.14	100
SORBED FRACTIONS				
		% sorbed		% sorbed
NO_3^-		0.20		0.036
Na^+		0.68		2.90
SeO_3^{--}		99.89		1.65

We can check the predicted charge densities against the surface speciation shown in Table 11.1 using Equations 11.14–11.15. At pH 9, we add the product of each species' charge z_o or z_β with its concentration,

	z_o	mmolal	z_β	mmolal
$>FeOH$	$0\ \times$	17.07	$0\ \times$	17.07
$>FeO^-:Na^+$	$-1\ \times$	0.667	$+1\ \times$	0.667
$>FeOH_2^+:NO_3^-$	$+1\ \times$	0.200	$-1\ \times$	0.200
$>FeSeO_3^-$	$-1\ \times$	0.0899	$0\ \times$	0.0899
$>FeOH_2^+$	$+1\ \times$	0.0666	$0\ \times$	0.0666
$>FeO^-$	$-1\ \times$	0.0219	$0\ \times$	0.0219
$>FeOH_2^+:HSeO_3^-$	$+1\ \times$	0.0086	$-1\ \times$	0.0086
$>FeSeO_3^-:Na^+$	$-1\ \times$	0.0014	$+1\ \times$	0.0014
		−0.5153		0.4701

Carrying the resulting sums and recalling the mass of goethite and its specific surface area, the charge densities on the 0-plane and β-plane follow as

$$\begin{aligned}\sigma_o = {} & (96\,485\ \text{C mol}^{-1}) \times (1\ \text{kg H}_2\text{O}) \times (-0.5153 \times 10^{-3}\ \text{molal})/ \\ & (30\ \text{g}) \times (52\ \text{m}^2\ \text{g}^{-1}) = -0.0319\ \text{C m}^{-2} = -3.19\ \mu\text{C cm}^{-2}\end{aligned} \tag{11.35}$$

and

$$\begin{aligned}\sigma_\beta = {} & (96\,485\ \text{C mol}^{-1}) \times (1\ \text{kg H}_2\text{O}) \times (0.4701 \times 10^{-3}\ \text{molal})/ \\ & (30\ \text{g}) \times (52\ \text{m}^2\ \text{g}^{-1}) = 0.0291\ \text{C m}^{-2} = 2.91\ \mu\text{C cm}^{-2}.\end{aligned} \tag{11.36}$$

Electroneutrality requires $\sigma_d = -\sigma_o - \sigma_\beta$, so

$$\begin{aligned}\sigma_d = {} & (0.0319 - 0.0291)\ \text{C m}^{-2} = 0.0028\ \text{C m}^{-2} \\ & = 0.28\ \mu\text{C cm}^{-2}\end{aligned} \tag{11.37}$$

gives charge density on the d-plane.

We can confirm the calculated surface potentials by inserting the values predicted for Ψ_o, Ψ_β, and Ψ_d into Equations 11.18–11.20. Recalling that a farad is one coulomb per volt,

$$(1.1\ \text{C V}^{-1}\text{m}^{-2}) \times (-0.0468 + 0.0178)\ \text{V} = -0.0319\ \text{C m}^{-2} \tag{11.38}$$

$$\begin{aligned}&(0.2\ \text{C V}^{-1}\text{m}^{-2}) \times (-0.0178 + 0.0039)\ \text{V} - \\ &\quad (1.1\ \text{C V}^{-1}\text{m}^{-2}) \times (-0.0468 + 0.0178)\ \text{V} = 0.0291\ \text{C m}^{-2}\end{aligned} \tag{11.39}$$

and

$$-(0.2\ \text{C V}^{-1}\text{m}^{-2}) \times (-0.0178 + 0.0039)\ \text{V} = 0.0028\ \text{C m}^{-2}, \tag{11.40}$$

we arrive at the charge densities we derived from the surface speciation (Eqs. 11.35–11.37). The results determined at pH 12 might be verified in the same manner.

Comparing the two calculations (Table 11.1), we see the 0-plane carries a notably more negative electrical charge at pH 12 than pH 9, whereas charge on the β-plane becomes more positive. The shift in charge is due largely to the complexation of Na^+,

$$>\text{FeOH} + \text{Na}^+ \rightleftarrows\ >\text{FeO}^-\text{:Na}^+ + \text{H}^+, \tag{11.41}$$

with the goethite surface in response to the decreased activity of H^+. The effect of pH on selenite sorption is also dramatic. The oxyanion is nearly completely complexed with the goethite surface at pH 9, but the reaction

$$>FeSeO_3^- + H_2O \rightleftarrows >FeOH + SeO_3^{--} + H^+ \tag{11.42}$$

drives almost all of the selenite into solution by pH 12, again reflecting the decrease in H^+ activity.

11.6 Choice of Complexation Data

The surface complexation modeler has the good fortune of being able to take advantage of a broad array of data sources, ranging from individual published studies to sweeping syntheses. The latter category includes compilations by Dzombak and Morel (1987) for hydrous ferric oxide, as already noted, Mathur and Dzombak (2006) for goethite, and Karamalidis and Dzombak (2010) for gibbsite. The RES^3T online database (Brendler *et al.*, 2003) notably offers properties of more than 7000 complexation and sorption reactions, in which approximately 150 sorbates bind with almost 150 minerals (HZDR, 2021). Despite this wealth of information, however, assembling a complexation model entails more than simply compiling a list of reactions and their properties.

To describe surface complexation upon a given mineral, first of all, the geochemist is of course careful to select data cast in terms of a mutual formalism—1-pK_a or 2-pK_a; double-layer, triple-layer, or CD-MUSIC—and to include polydentate reactions derived in terms of a shared standard state, as discussed in Section 10.2.5. Other choices, such as the level of detail needed to describe a system, are more nuanced. Dzombak and Morel (1987) consider that UO_2^{++} binds directly with hydrous ferric oxide in only monodentate pairings, for example, whereas Turner and Sassman (1996) parameterized the uranyl–water–gibbsite system using monodentate and bidentate complexes of eleven stoichiometries, and still more in the presence of carbonate.

Thermodynamic properties of aqueous species and minerals were needed in the original study to regress binding constants and should therefore be carried into the calculation, as well, but not all publications cite this information. The modeler must further project the experimental realm onto a natural system. The ratio of solid to solution is small in most experimental studies, for example, but in nature can be quite large. And complexation data is almost invariably derived from experiments involving a single background electrolyte, such as a nitrate, perchlorate, or chloride salt, whereas natural waters are mixed electrolyte solutions.

As Lützenkirchen (2006) notes, finally, the choice of reaction data in a surface complexation model too often defaults to locating a dataset containing the chemical components of interest, rather than gleaning the data that best correspond to the conditions to be modeled. As the field continues to mature, we as modelers can aspire to increasingly choose the latter approach over the former.

12

Automatic Reaction Balancing

Conveniently, perhaps even miraculously, the equations developed in Chapter 5 to accomplish basis swaps can be used to balance chemical reactions automatically. Once the equations have been coded into a computer program, there is no need to balance reactions, compute equilibrium constants, or even determine equilibrium equations by hand. Instead, these procedures can be performed quickly and reliably on a small computer.

To balance a reaction, we first choose a species to appear on the reaction's left side, and express that species' composition in terms of a basis **B**. The basis might be a list of the elements in the species' stoichiometry, or an arbitrary list of species that combine to form the left-side species. Then we form a second basis **B**$'$ composed of species that we want to appear on the reaction's right side. To balance the reaction, we calculate the transformation matrix relating basis **B**$'$ to **B**, following the procedures in Chapter 5. The transformation matrix, in turn, gives the balanced reaction and its equilibrium constant.

12.1 Calculation Procedure

Two methods of balancing reactions are of interest. We can balance reactions in terms of the stoichiometries of the species considered. In this case, the existing basis **B** is a list of elements and, if charged species are involved, the electron e^-. Alternatively, we may use a dataset of balanced reactions, such as the LLNL database. Basis **B**, in this case, is the one used in the database to write reactions. We will consider each possibility in turn.

12.1.1 Using Species' Stoichiometries

A straightforward way to balance reactions is to use as the initial basis the stoichiometries of the species involved. If the species' free energies of formation are known, the reaction's equilibrium constant can be determined as well. In the stoichiometric approach, basis **B** is the list of elements that will appear in the reaction, plus the electron if needed. We write swap reactions and calculate a transformation matrix as described in Section 5.1. The equations in Sections 5.2 and 5.3 give the balanced reaction and associated equilibrium constant.

The process is best shown by example. Suppose that we wish to balance the reaction by which calcium clinoptilolite ($CaAl_2Si_{10}O_{24} \cdot 8H_2O$), a zeolite mineral, reacts to form muscovite [$KAl_3Si_3O_{10}(OH)_2$] and quartz (SiO_2). We choose to write the reaction in terms of the aqueous species Ca^{++}, K^+, and OH^-.

Reserving clinoptilolite for the reaction's left side, we write the stoichiometry of each remaining species in matrix form,

$$\begin{pmatrix} H_2O \\ Ca^{++} \\ K^+ \\ \text{Muscovite} \\ \text{Quartz} \\ OH^- \\ e^- \end{pmatrix} = \begin{pmatrix} 2 & 1 & 0 & 0 & 0 & 0 & 0 \\ 0 & 0 & 1 & 0 & 0 & 0 & -2 \\ 0 & 0 & 0 & 1 & 0 & 0 & -1 \\ 2 & 12 & 0 & 1 & 3 & 3 & 0 \\ 0 & 2 & 0 & 0 & 0 & 1 & 0 \\ 1 & 1 & 0 & 0 & 0 & 0 & 1 \\ 0 & 0 & 0 & 0 & 0 & 0 & 1 \end{pmatrix} \begin{pmatrix} H \\ O \\ Ca \\ K \\ Al \\ Si \\ e^- \end{pmatrix}. \tag{12.1}$$

Notice that we have added the electron to $\mathbf{B}$ and $\mathbf{B}'$ in order to account for the electrical charge on the aqueous species. This incorporation provides a convenient check: the electron's reaction coefficient must work out to zero in order for the reaction to be charge balanced.

We reverse the equation by computing the inverse to the coefficient matrix, giving

$$\begin{pmatrix} H \\ O \\ Ca \\ K \\ Al \\ Si \\ e^- \end{pmatrix} = \begin{pmatrix} 1 & 0 & 0 & 0 & 0 & -1 & 1 \\ -1 & 0 & 0 & 0 & 0 & 2 & -2 \\ 0 & 1 & 0 & 0 & 0 & 0 & 2 \\ 0 & 0 & 1 & 0 & 0 & 0 & 1 \\ 4/3 & 0 & -1/3 & 1/3 & -1 & -10/3 & 3 \\ 2 & 0 & 0 & 0 & 1 & -4 & 4 \\ 0 & 0 & 0 & 0 & 0 & 0 & 1 \end{pmatrix} \begin{pmatrix} H_2O \\ Ca^{++} \\ K^+ \\ \text{Muscovite} \\ \text{Quartz} \\ OH^- \\ e^- \end{pmatrix}. \tag{12.2}$$

The inverted matrix is the transformation matrix $(\boldsymbol{\beta})^{-1}$.

Now, we write the stoichiometry of the clinoptilolite and substitute the result above, giving the balanced reaction

$$\begin{aligned} \text{Ca-clinoptilolite} &= (16\ \ 32\ \ 1\ \ 0\ \ 2\ \ 10\ \ 0) \begin{pmatrix} H \\ O \\ Ca \\ K \\ Al \\ Si \\ e^- \end{pmatrix} \\ &= (16\ \ 32\ \ 1\ \ 0\ \ 2\ \ 10\ \ 0) \times \\ &\quad \begin{pmatrix} 1 & 0 & 0 & 0 & 0 & -1 & 1 \\ -1 & 0 & 0 & 0 & 0 & 2 & -2 \\ 0 & 1 & 0 & 0 & 0 & 0 & 2 \\ 0 & 0 & 1 & 0 & 0 & 0 & 1 \\ 4/3 & 0 & -1/3 & 1/3 & -1 & -10/3 & 3 \\ 2 & 0 & 0 & 0 & 1 & -4 & 4 \\ 0 & 0 & 0 & 0 & 0 & 0 & 1 \end{pmatrix} \begin{pmatrix} H_2O \\ Ca^{++} \\ K^+ \\ \text{Muscovite} \\ \text{Quartz} \\ OH^- \\ e^- \end{pmatrix} \end{aligned}$$

$$= (20/3 \;\; 1 \;\; -2/3 \;\; 2/3 \;\; 8 \;\; 4/3 \;\; 0)\begin{pmatrix} H_2O \\ Ca^{++} \\ K^+ \\ \text{Muscovite} \\ \text{Quartz} \\ OH^- \\ e^- \end{pmatrix} . \tag{12.3}$$

More simply,

$$\underset{\textit{Ca-clinoptilolite}}{CaAl_2Si_{10}O_{24} \cdot 8\,H_2O} + 2/3\,K^+ \rightleftarrows 20/3\,H_2O + Ca^{++} + 2/3 \underset{\textit{muscovite}}{KAl_3Si_3O_{10}(OH)_2} + 8 \underset{\textit{quartz}}{SiO_2} + 4/3\,OH^- , \tag{12.4}$$

which is the result we seek.

To calculate the reaction's equilibrium constant, we note that the free-energy change ΔG_{sw} of each of the swap reactions in Equation 12.1 is the negative free energy of formation from the elements of the corresponding species,

$$\Delta G^o_{sw} = -\Delta G^o_f \, . \tag{12.5}$$

The sign on the reaction free energies is reversed because the species appear on the left side of Equation 12.1. In other words, we are decomposing the species rather than forming them from the elements. We determine ΔG^o for the reaction by adding the values for ΔG^o_{sw}, just as we added log Ks in Section 2.1,

$$\Delta G^o = -\Delta G^o_f\,(\text{Ca-clinopt}) + (20/3 \;\; 1 \;\; -2/3 \;\; 2/3 \;\; 8 \;\; 4/3 \;\; 0)\begin{pmatrix} \Delta G^o_f\,(H_2O) \\ \Delta G^o_f\,(Ca^{++}) \\ \Delta G^o_f\,(K^+) \\ \Delta G^o_f\,(\text{Musc}) \\ \Delta G^o_f\,(\text{Qtz}) \\ \Delta G^o_f\,(OH^-) \\ \Delta G^o_f\,(e^-) \end{pmatrix} . \tag{12.6}$$

Substituting values of ΔG^o_f taken (in kJ mol^{-1}) from Robie *et al.* (1979) and the LLNL database, the equation,

$$\Delta G^o = 12\,764.67 + (20/3 \;\; 1 \;\; -2/3 \;\; 2/3 \;\; 8 \;\; 4/3 \;\; 0)\begin{pmatrix} -237.14 \\ -553.54 \\ -282.49 \\ -5590.76 \\ -856.29 \\ -157.29 \\ 0 \end{pmatrix} , \tag{12.7}$$

predicts a free-energy change of 31.32 kJ mol^{-1}. The value can be expressed as an equilibrium constant using the relation,

$$\log K = -\frac{\Delta G^o}{2.303\, R T_K} , \tag{12.8}$$

where R is the gas constant and T_K is absolute temperature, which gives a value for log K of –5.48.

12.1.2 Using a Reaction Database

A reaction dataset, such as the LLNL database, provides an alternative method for balancing reactions. Such a database contains reactions to form a number of aqueous species, minerals, and gases, together with the corresponding equilibrium constants. Reactions are written in terms of a generic basis set **B**, which probably does not correspond to set **B**′, our choice of species to appear in the reaction.

To balance a reaction, we write swap reactions relating **B**′ to **B**. Returning to the previous example, we wish to compute the reaction by which Ca-clinoptilolite transforms to muscovite and quartz. Reserving the clinoptilolite for the reaction's left side, we write the swap reactions for the basis transformation in matrix form. The reactions and associated equilibrium constants at 25 °C are

$$\begin{pmatrix} \mathrm{H_2O} \\ \mathrm{Ca^{++}} \\ \mathrm{K^+} \\ \text{Muscovite} \\ \text{Quartz} \\ \mathrm{OH^-} \end{pmatrix} = \begin{pmatrix} 1 & 0 & 0 & 0 & 0 & 0 \\ 0 & 1 & 0 & 0 & 0 & 0 \\ 0 & 0 & 1 & 0 & 0 & 0 \\ 6 & 0 & 1 & 3 & 3 & -10 \\ 0 & 0 & 0 & 0 & 1 & 0 \\ 1 & 0 & 0 & 0 & 0 & -1 \end{pmatrix} \begin{pmatrix} \mathrm{H_2O} \\ \mathrm{Ca^{++}} \\ \mathrm{K^+} \\ \mathrm{Al^{+++}} \\ \mathrm{SiO_2(aq)} \\ \mathrm{H^+} \end{pmatrix}$$
$$\log K^{\mathrm{sw}} = \begin{pmatrix} 0 \\ 0 \\ 0 \\ 14.56 \\ -4.00 \\ 13.99 \end{pmatrix}. \tag{12.9}$$

We reverse the equation by inverting the coefficient matrix to give

$$\begin{pmatrix} \mathrm{H_2O} \\ \mathrm{Ca^{++}} \\ \mathrm{K^+} \\ \mathrm{Al^{+++}} \\ \mathrm{SiO_2(aq)} \\ \mathrm{H^+} \end{pmatrix} = \begin{pmatrix} 1 & 0 & 0 & 0 & 0 & 0 \\ 0 & 1 & 0 & 0 & 0 & 0 \\ 0 & 0 & 1 & 0 & 0 & 0 \\ 4/3 & 0 & -1/3 & 1/3 & -1 & -10/3 \\ 0 & 0 & 0 & 0 & 1 & 0 \\ 1 & 0 & 0 & 0 & 0 & -1 \end{pmatrix} \begin{pmatrix} \mathrm{H_2O} \\ \mathrm{Ca^{++}} \\ \mathrm{K^+} \\ \text{Muscovite} \\ \text{Quartz} \\ \mathrm{OH^-} \end{pmatrix}. \tag{12.10}$$

The inverted matrix is the transformation matrix for the basis we have chosen.

The reaction in the LLNL database for Ca-clinoptilolite,

$$\text{Ca-clinoptilolite} = (12 \;\; 1 \;\; 0 \;\; 2 \;\; 10 \;\; -8) \begin{pmatrix} \mathrm{H_2O} \\ \mathrm{Ca^{++}} \\ \mathrm{K^+} \\ \mathrm{Al^{+++}} \\ \mathrm{SiO_2(aq)} \\ \mathrm{H^+} \end{pmatrix}, \tag{12.11}$$

has a log K at 25 °C of −9.12. Substituting the transformation matrix and multiplying through gives

$$\text{Ca-clinoptilolite} = (12\ \ 1\ \ 0\ \ 2\ \ 10\ \ -8) \times \begin{pmatrix} 1 & 0 & 0 & 0 & 0 & 0 \\ 0 & 1 & 0 & 0 & 0 & 0 \\ 0 & 0 & 1 & 0 & 0 & 0 \\ 4/3 & 0 & -1/3 & 1/3 & -1 & -10/3 \\ 0 & 0 & 0 & 0 & 1 & 0 \\ 1 & 0 & 0 & 0 & 0 & -1 \end{pmatrix} \begin{pmatrix} H_2O \\ Ca^{++} \\ K^+ \\ \text{Muscovite} \\ \text{Quartz} \\ OH^- \end{pmatrix}$$

$$= (20/3\ \ 1\ \ -2/3\ \ 2/3\ \ 8\ \ 4/3) \begin{pmatrix} H_2O \\ Ca^{++} \\ K^+ \\ Al^{+++} \\ SiO_2(aq) \\ H^+ \end{pmatrix}, \tag{12.12}$$

or, as before,

$$\underset{\textit{Ca-clinoptilolite}}{CaAl_2Si_{10}O_{24} \cdot 8\,H2O} + 2/3\,K^+ \rightleftarrows 20/3\,H_2O + Ca^{++} + 2/3 \underset{\textit{muscovite}}{KAl_3Si_3O_{10}(OH)_2} + 8 \underset{\textit{quartz}}{SiO2} + 4/3\,OH^- . \tag{12.13}$$

The formula in Section 5.3 gives the reaction's equilibrium constant,

$$\log K = -9.12 - (20/3\ \ 1\ \ -2/3\ \ 2/3\ \ 8\ \ 4/3) \begin{pmatrix} 0 \\ 0 \\ 0 \\ 14.56 \\ -4.00 \\ 13.99 \end{pmatrix}, \tag{12.14}$$

or log $K = -5.48$, as determined in the previous section.

The program RXN performs such calculations automatically. To follow the procedure above, we enter the commands

```
react Clinoptil-Ca
swap Muscovite for Al+++
swap Quartz for SiO2(aq)
swap OH- for H+
T = 25 C
go
```

giving the same result without hand calculation.

12.2 Dissolution of Pyrite

To further illustrate how the basis-swapping algorithm can be used to balance reactions, we consider several ways to represent the dissolution reaction of pyrite, FeS_2. Using the program RXN, we retrieve the reaction for pyrite as written in the LLNL database

```
react Pyrite
go
```

producing the result

$$\underset{pyrite}{FeS_2} + H_2O + 7/2\, O_2(aq) \rightleftarrows Fe^{++} + 2\, SO_4^{--} + 2\, H^+ . \tag{12.15}$$

By this reaction, sulfur from the pyrite oxidizes to form sulfate ions, liberating protons that acidify the solution.

The above reaction represents, in a simplified way, the origin of acid mine drainage. Streambeds in areas of acid drainage characteristically become coated with an orange layer of ferric precipitate. We can write a reaction representing the overall process by swapping ferric hydroxide in place of the ferrous ion:

(cont'd)
```
swap Fe(OH)3(ppd) for Fe++
go
```

The resulting reaction,

$$\underset{pyrite}{FeS_2} + 7/2\, H_2O + 15/4\, O_2(aq) \rightleftarrows \underset{ferric\ hydroxide}{Fe(OH)_3(ppd)} + 2\, SO_4^{--} + 4\, H^+ , \tag{12.16}$$

consumes more oxygen than the previous result, because not only sulfur but iron oxidizes.

Pyrite can dissolve into reducing as well as oxidizing solutions. To find the reaction by which the mineral dissolves to form H_2S, we swap this species into the basis in place of the sulfate ion

(cont'd)
```
unswap Fe++
swap H2S(aq) for SO4--
go
```

The result,

$$\underset{pyrite}{FeS_2} + H_2O + 2\, H^+ \rightleftarrows Fe^{++} + 2\, H_2S(aq) + 1/2\, O_2(aq) , \tag{12.17}$$

differs from the previous reactions in that it consumes protons and produces oxygen.

Is there a reaction by which pyrite can dissolve without changing the overall oxidation state of its sulfur? To see, we return the sulfate ion to the basis and swap H_2S for dissolved oxygen:

(cont'd)
```
unswap SO4--
swap H2S(aq) for O2(aq)
go
```

The reaction written in terms of the new basis is

$$\underset{pyrite}{FeS_2} + H_2O + 3/2\, H^+ \rightleftarrows Fe^{++} + 1/4\, SO_4^{--} + 7/4\, H_2S(aq)\,, \tag{12.18}$$

which neither consumes nor produces oxygen. Calculating the transformation matrices in these examples provides an interesting exercise, which is left to the reader.

12.3 Equilibrium Equations

A reaction's equilibrium equation is given directly from the form of the reaction and the value of the equilibrium constant. Hence, it is an easy matter to extend a reaction balancing program to report equilibrium lines. For example, the reaction

$$\underset{hematite}{Fe_2O_3} + 4\, H^+ \rightleftarrows 2\, Fe^{++} + 2\, H_2O + 1/2\, O_2(aq) \tag{12.19}$$

has an equilibrium constant at 25 °C of $10^{-16.9}$. From the mass action equation and definition of pH, the general equilibrium line is

$$\log K = 4\, \mathrm{pH} + 2\, \log a_{Fe^{++}} + 2\, \log a_{H_2O} + 1/2\, \log a_{O_2(aq)}\,, \tag{12.20}$$

since the activity of hematite is one. The specific equilibrium line at 25 °C for $a_{Fe^{++}} = 10^{-10}$ and $a_{H_2O} = 1$ is

$$\log a_{O_2(aq)} = 6.17 - 8\, \mathrm{pH}\,. \tag{12.21}$$

To use RXN to calculate the equilibrium lines above in general and specific forms, we type

```
react Hematite
pH = ?
long
go

T = 25
log a Fe++ = -10
a H2O = 1
go
```

The `long` command tells the program to show the reaction's equilibrium constant versus temperature and calculate its equilibrium equation; `pH = ?` causes the program to render the equation in terms of pH instead of $\log a_{H^+}$. To find the equilibrium lines written in terms of pe and Eh, we type

```
(cont'd)
swap e- for O2(aq)
pe = ?
go

Eh = ?
go
```

giving the results

$$pe = 23.03 - 3\,pH \qquad (12.22)$$

and

$$Eh = 1.363 - 0.178\,pH\,. \qquad (12.23)$$

12.3.1 Equilibrium Activity Ratio

Many mineralogic reactions involve exchange of cations or anions. Hence, geochemists commonly need to determine equilibrium lines in terms of activity ratios. Consider, for example, the reaction at 25 °C between the clay kaolinite [$Al_2Si_2O_5(OH)_4$] and the mica muscovite. The RXN commands

```
react Muscovite
swap Kaolinite for Al+++
T = 25
go
```

give the reaction

$$\underset{muscovite}{KAl_3Si_3O_{10}(OH)_2} + 1.5\,H_2O + H^+ \rightleftarrows K^+ + 1.5\,\underset{kaolinite}{Al_2Si_2O_5(OH)_4}\;, \qquad (12.24)$$

whose log K is 3.42. To calculate the equilibrium ratio a_{K^+}/a_{H^+}, assuming unit water activity, we type

```
(cont'd)
swap K+/H+ for H+
a H2O = 1
long
go
```

giving the result $a_{K^+}/a_{H^+} = 10^{3.42}$.

In our second example, we calculate the same ratio for the reaction between muscovite and potassium feldspar ($KAlSi_3O_8$; "maximum microcline" in the database) in the presence of quartz,

$$K^+ + 1/2\,\underset{muscovite}{KAl_3Si_3O_{10}(OH)_2} + 3\,\underset{quartz}{SiO_2} \rightleftarrows 3/2\,\underset{microcline}{KAlSi_3O_8} + H^+\,. \qquad (12.25)$$

The commands

```
T = 25
react "Maximum Microcline"
swap Muscovite for Al+++
swap Quartz for SiO2(aq)
swap K+/H+ for H+
long
factor 3/2
go
```

give the result $a_{K^+}/a_{H^+} = 10^{4.84}$. For convenience, the `factor` command above applies a multiplier of 1.5 to the reaction coefficients. We can quickly recalculate the equilibrium activity ratio for reaction in the presence of the silica polymorphs tridymite and amorphous silica:

(cont'd)
```
swap Tridymite for SiO2(aq)
go

swap Amrph^silica for SiO2(aq)
go
```

The resulting a_{K^+}/a_{H^+} values, respectively, are $10^{4.34}$ and $10^{0.98}$.

As a final example, we balance the reaction between the zeolite calcium clinoptilolite and the mica prehnite [$Ca_2Al_2Si_3O_{10}(OH)_2$] in the presence of quartz, and calculate at 200 °C the equilibrium activity ratio $a_{Ca^{++}}/a^2_{H^+}$. The commands

```
react Clinoptil-Ca
swap Prehnite for Al+++
swap Quartz for SiO2(aq)
T = 200
go
```

give the reaction

$$\underset{\text{Ca-clinoptilolite}}{CaAl_2Si_{10}O_{24}\cdot 8\,H_2O} + Ca^{++} \rightleftarrows 6\,H_2O + \underset{\text{prehnite}}{Ca_2Al_2Si_3O_{10}(OH)_2} + 7\underset{\text{quartz}}{SiO_2} + 2\,H^+ , \quad (12.26)$$

which has a log K of -10.23. To calculate the activity ratio, we type

```
swap Ca++/H+^2 for Ca++
a H2O = 1
long
go
```

which gives, as expected, the value calculated for log K.

12.3.2 Equilibrium Temperature

When the activity of each species in a reaction is known, we can determine the temperature (or temperatures) at which the reaction is in equilibrium. As an example, we calculate the temperature at which gypsum ($CaSO_4 \cdot 2\,H_2O$) dehydrates to form anhydrite ($CaSO_4$). The RXN commands

```
react Gypsum
swap Anhydrite for Ca++
long
go
```

give the reaction

$$\underset{\text{gypsum}}{CaSO_4 \cdot 2\,H2O} \rightleftarrows \underset{\text{anhydrite}}{CaSO_4} + 2\,H_2O\,, \quad (12.27)$$

and equilibrium equation

$$\log K = 2 \log a_{H_2O}\,. \quad (12.28)$$

The equilibrium temperature for any water activity is the temperature at which log K satisfies this equality. To find this value when the activity of water is one, we type

(cont'd)
```
a H2O = 1
go
```

The resulting equilibrium temperature, 43.7 °C, is the temperature at which the log K is zero. For a water activity of 0.7, the equilibrium temperature drops to 11.8 °C. Typing

(cont'd)
```
a H2O = ?
T = 25 C
```

we find that the two minerals are in equilibrium at 25 °C when the water activity is 0.815.

13
Uniqueness

A practical question that arises in quantitative modeling is whether the results of a modeling study are unique. In other words, is it possible to arrive at results that differ, at least slightly, from the original ones but nonetheless satisfy the governing equations and honor the input constraints?

In the broadest sense, of course, no model is unique (see, for example, Oreskes *et al.*, 1994). A geochemical modeler could conceptualize the problem differently, choose a different compilation of thermodynamic data, include more or fewer species and minerals in the calculation, or employ a different method of estimating activity coefficients. The modeler might allow a mineral to form at equilibrium with the fluid or require it to precipitate according to any of a number of published kinetic rate laws and rate constants, and so on. Since a model is a simplified version of reality that is useful as a tool (Chapter 2), it follows that there is no "correct" model, only a model that is most useful for a given purpose.

A more precise question (Bethke, 1992) is the subject of this chapter: in geochemical modeling is there but a single root to the set of governing equations that honors a given set of input constraints? We might call such a property mathematical uniqueness, to differentiate it from the broader aspects of uniqueness. The property of mathematical uniqueness is important because once the software has discovered a root to a problem, the modeler may abandon any search for further solutions. There is no concern that the choice of a starting point for iteration has affected the answer. In the absence of a demonstration of uniqueness, on the other hand, the modeler cannot be completely certain that another solution, perhaps a more realistic or useful one, remains undiscovered.

Geochemists, following early theoretical work in other fields, have long considered the multicomponent equilibrium problem (as defined in Chapter 3) to be mathematically unique. In fact, however, this assumption is not correct. Although relatively uncommon, there are examples of geochemical models in which more than one root of the governing equations satisfy the modeling constraints equally well. In this chapter, we consider the question of uniqueness and pose three simple problems in geochemical modeling that have nonunique solutions.

13.1 The Question of Uniqueness

As noted in Chapter 1, chemical modeling developed from efforts to calculate the thrust of a rocket fuel from its bulk composition. Mathematicians of the era (for example, Brinkley, 1947; White *et al.*, 1958; Boynton, 1960) analyzed the multicomponent equilibrium problem, including the uniqueness of its roots, in detail. Warga (1963) considered the problem of a thermodynamically ideal solution of known bulk composition. He showed that the free-energy surface representing the sum of the free energies of individual species, when traced along trajectories satisfying mass balance, is concave upward. The surface, hence, has a single minimum point and therefore a unique equilibrium state. Van Zeggeren and Storey (1970, pp. 31–32) cite several other studies that offer mathematical proofs of uniqueness.

Geochemists have generally believed that the uniqueness proofs hold in their field. Their proofs, however, are limited in two regards: they consider thermodynamically ideal solutions and assume that, as in the rocket fuel problem, the calculation is posed in terms of mass balance constraints. The first limitation is not known to be a serious problem in modeling speciation in aqueous fluids. The second limitation is important, however, because in geochemical modeling the bulk composition of the modeled system is seldom known completely. For this reason, geochemists generally construct models using a combination of mass balance and mass action constraints (see Chapter 4).

Mass balance constraints are specifications, when solving the governing equations (Eqs. 3.32–3.35), of the mole numbers M_w, M_i, and M_k of the components in the basis. Setting the bulk sodium content of a fluid, for example, represents a mass balance constraint. Mass action constraints include setting a species' molality m_i or activity a_i, the free mass n_k of a mineral, or the partial pressure P_m or fugacity f_m of a gas. In setting pH, the volume of quartz in a system, or the CO_2 partial pressure, the modeler poses a mass action constraint. By doing so, he or she violates an underlying assumption of the uniqueness proofs and opens the possibility of mathematical nonuniqueness.

13.2 Examples of Nonunique Solutions

To demonstrate nonuniqueness, we pose here three problems in geochemical modeling that each have two physically realistic solutions. In the first example, based on data from an aluminum solubility experiment, we assume equilibrium with an alumina mineral to fix the pH of a fluid of otherwise known composition. Setting pH by mineral equilibrium is a widespread practice in modeling the chemistry of deep groundwaters, and of fluids sampled from hydrothermal experiments, because it is difficult to directly measure the in situ pH of hot fluids. In this case, however, there are two possible solutions because many aluminous minerals, including hydroxides, clays, and micas, are amphoteric and hence equally soluble at low and high pH.

In SPECE8, we prepare our calculation by swapping boehmite (AlOOH) for H^+,

```
swap Boehmite for H+
1 free cm3 Boehmite
```

so the equilibrium with this mineral controls pH according to reactions such as

$$\underset{boehmite}{\mathrm{AlOOH}} + 2\,\mathrm{H^+} \rightleftarrows \mathrm{H_2O} + \mathrm{AlOH^{++}} \tag{13.1}$$

$$\underset{boehmite}{\mathrm{AlOOH}} + \mathrm{H^+} \rightleftarrows \mathrm{Al(OH)_2^+} \tag{13.2}$$

or

$$\underset{boehmite}{\mathrm{AlOOH}} + 2\,\mathrm{H_2O} \rightleftarrows \mathrm{Al(OH)_4^-} + \mathrm{H^+}\,, \tag{13.3}$$

depending on the predominant aluminum species in solution. In general, species $AlOH^{++}$ and $Al(OH)_2^+$ predominate under acidic conditions, whereas the hydroxy species $Al(OH)_4^-$ dominates under alkaline conditions (Fig. 13.1).

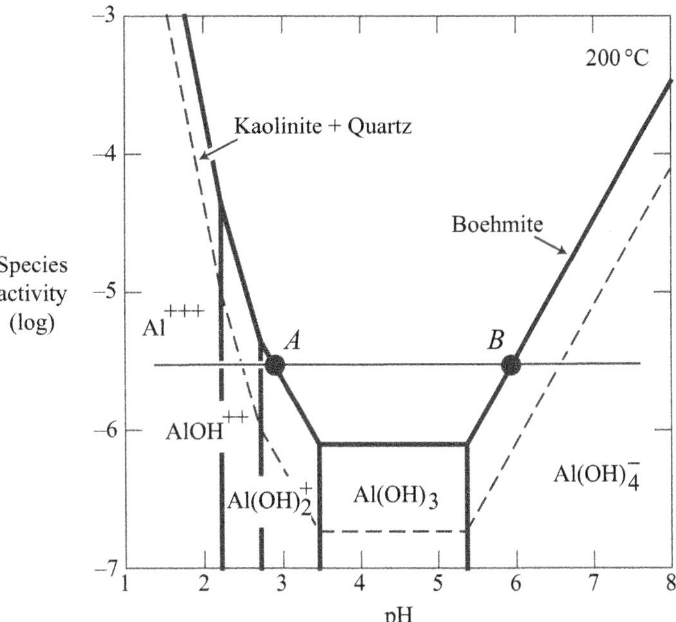

Figure 13.1 Solubility diagram for aluminum species in aqueous solution as a function of pH at 200 °C in the presence of boehmite (solid lines) and kaolinite plus quartz (dashed lines). Aluminum is soluble at a specific activity (horizontal line) either under acidic conditions as species $Al(OH)_2^+$, $AlOH^{++}$, or Al^{+++} (e.g., point *A*), or under alkaline conditions as $Al(OH)_4^-$ (point *B*).

We assume a 0.1 molal KCl solution containing hypothetical amounts of silica, aluminum, and carbonate. We set temperature to 200 °C and run the calculation

```
(cont'd)
K+         = 100 mmolal
Cl-        = 100 mmolal
SiO2(aq)  =   3 mmolal
Al+++      =  10 umolal
HCO3-      =  60 umolal

T = 200
print species = long
go
```

For simplicity, we do not allow supersaturated minerals to precipitate. The result is an acidic fluid in which $AlOH^{++}$ and $Al(OH)_2^+$ predominate among aluminum species.

We repeat the calculation, this time swapping $Al(OH)_4^-$ into the basis to represent dissolved aluminum

```
(cont'd)
swap Al(OH)4- for Al+++
go
```

The swap favors iteration toward a second root at alkaline pH where $Al(OH)_4^-$ is the predominant aluminum species in solution.

The chemistries corresponding to the two roots can be summarized as

	Root *A*	Root *B*
pH	3.1	6.3
total Al^{+++} (molal)	10.0×10^{-6}	10.0×10^{-6}
free $AlOH^{++}$	4.3×10^{-6}	1.5×10^{-12}
free $Al(OH)_2^+$	3.0×10^{-6}	1.8×10^{-9}
free $Al(OH)_4^-$	5.5×10^{-9}	9.2×10^{-6}

Repeating the calculations using minerals such as kaolinite [$Al_2Si_2O_5(OH)_4$] or muscovite [$KAl_3Si_3O_{10}(OH)_2$] to fix pH also produces nonunique results.

As a second example, we constrain a fluid's oxidation state by assuming equilibrium with pyrite (FeS_2). As before, direct information on this variable can be difficult to obtain, so it is not uncommon for modelers to use mineral equilibrium to fix a fluid's redox state. The choice of pyrite to buffer oxidation state, however, is perilous because pyrite sulfur, which is in the S^{-I} oxidation state, may dissolve by oxidation to sulfate (S^{6+}),

$$\underset{pyrite}{FeS_2} + H_2O + 7/2\, O_2(g) \rightleftarrows Fe^{++} + 2\, SO_4^{--} + 2\, H^+ \,, \tag{13.4}$$

or by reduction to H_2S (S^{-II}),

$$\underset{pyrite}{FeS_2} + H_2O + 2\, H^+ \rightleftarrows Fe^{++} + 2\, H_2S(aq) + 1/2\, O_2(g)\,. \tag{13.5}$$

As such, there are two redox states that satisfy the assumption of equilibrium with pyrite (Fig. 13.2).

In our example, we know the pH and iron and sulfur contents of a 1 molal NaCl solution at 100 °C. In SPECE8, we swap pyrite into the basis in place of $O_2(aq)$ and run the model

```
swap Pyrite for O2(aq)
1 free cm3 Pyrite

Na+    =  1 molal
Cl-    =  1 molal
Fe++   = 10 mmolal
SO4--  = 10 mmolal
pH = 4

T = 100
print species = long
go
```

In this case, the program converges to a relatively oxidized solution in which sulfur speciation is dominated by sulfate species.

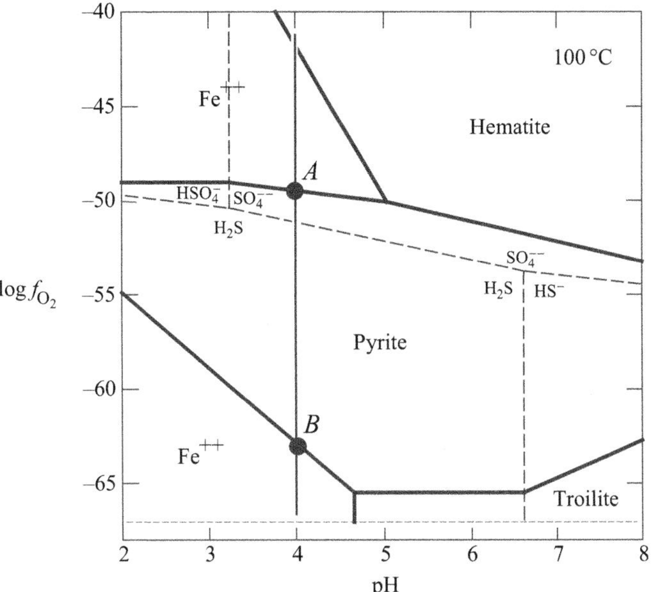

Figure 13.2 Redox–pH diagram for the Fe–S–H_2O system at 100 °C, showing speciation of sulfur (dashed line) and the stability fields of iron minerals (solid lines). Diagram is drawn assuming sulfur and iron species activities, respectively, of 10^{-3} and 10^{-4}. Broken line at bottom of diagram is the water stability limit at 100 atm total pressure. At pH 4, there are two oxidation states (points *A* and *B*) in equilibrium with pyrite under these conditions.

Swapping $H_2S(aq)$ into the basis in the place of SO_4^{--}, on the other hand,

(cont'd)
```
swap H2S(aq) for SO4--
go
```

changes the starting point of the iteration and causes the program to converge to a more reduced solution nearly devoid of sulfate species. The two roots are

	Root *A*	Root *B*
pH	4.0	4.0
log P_{O_2}	−50	−67
total Fe^{++} (molal)	0.010	0.010
total S	0.010	0.010
ΣSO_4^{--}	0.010	5.0×10^{-34}
$\Sigma H_2S(aq)$	3.3×10^{-7}	0.010

Here, the notations ΣSO_4^{--} and $\Sigma H_2S(aq)$ refer, respectively, to the sums of the molalities of the sulfate and sulfide sulfur species in solution.

Analogous examples of nonuniqueness can be constructed using any mineral or gas of intermediate oxidation state. Buffering the partial pressure of $N_2(g)$ or $SO_2(g)$, for example, would be a poor choice for constraining oxidation state, since the gases can either oxidize to NO_3^- and SO_4^{--}, respectively, or reduce to NH_4 and $H_2S(aq)$ species.

As a final example, we consider a fluid of known fluoride concentration whose calcium content is set by equilibrium with fluorite (CaF_2). The speciation of fluorine provides for two solutions to this problem. In dilute solutions, in which the free ion F^- dominates, the reaction

$$\underset{\textit{fluorite}}{CaF_2} \rightleftarrows Ca^{++} + 2\,F^- \tag{13.6}$$

requires that calcium content vary inversely with fluorine concentration. By this reaction, increasing the fluorine concentration leads to solutions that are less calcic, and vice versa.

As calcium content increases, especially at elevated temperature, the CaF^+ ion pair becomes predominant. The CaF^+ activity exceeds that of F^- at 200 °C whenever the activity of Ca^{++} is greater than about 10^{-3}; at 300 °C, the ion pair is favored at Ca^{++} activities as small as $10^{-4.8}$. Where CaF^+ dominates, the reaction

$$\underset{\textit{fluorite}}{CaF_2} + Ca^{++} \rightleftarrows 2\,CaF^+ \tag{13.7}$$

controls fluorite solubility. This reaction, in contrast to this previous one, requires that fluids become proportionally richer in calcium as their fluorine contents increase.

As shown in Figure 13.3, fluids of identical fluorine content but two distinct Ca^{++} activities can exist in equilibrium with fluorite. At 200 °C, setting the activity of dissolved fluorine to $10^{-3.5}$ allows two equilibrium activities of Ca^{++}: $10^{-4.3}$ and $10^{-1.6}$ (points *A* and *B*); the corresponding activities at 300 °C are $10^{-6.5}$ and $10^{-3.1}$.

To use REACT to discover the root at low Ca^{++} concentration, we swap fluorite for Ca^{++} and arbitrarily set a 1 molal NaCl solution of pH 5 that contains slightly less than 1 mmolal fluorine. We set temperature to 200 °C

```
swap Fluorite for Ca++
1 free cm3 Fluorite
F-  = 0.8 mmolal
Na+ = 1 molal
Cl- = 1 molal
pH  = 5

T = 200
print species = long
go
```

and run the model.

It is difficult to persuade REACT to locate the high Ca^{++} root directly. By experimentation we learn that if we gradually add 1.0165 moles of $CaCl_2$ to the fluid, the fluid initially becomes supersaturated with respect to fluorite but then returns to equilibrium.

```
(cont'd)
react 1 mol of Ca++
react 2 mol of Cl-
reactants times 1.0165
```

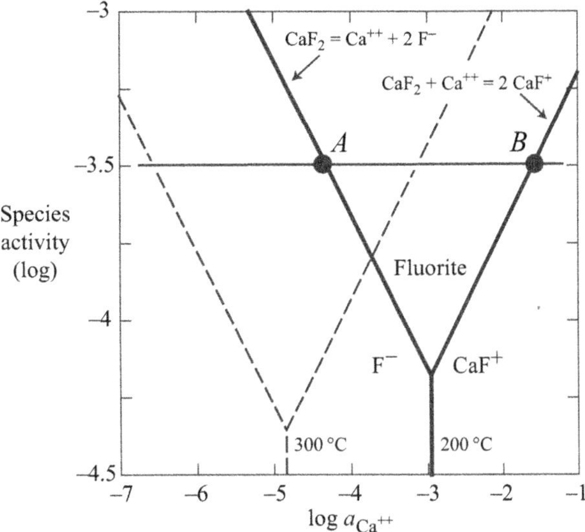

Figure 13.3 Solubility diagram for fluorite as a function of calcium ion activity at 200 °C (solid) and 300 °C (dashed lines). Fluorite is soluble at a specific activity (horizontal line) either as F^- at small Ca^{++} activity (point *A*) or as CaF^+ at high Ca^{++} activity (point *B*).

```
dump
fix pH
precip = off
go
```

Here, we set the `dump` option to remove excess fluorite from the system before beginning the path and fix pH to a constant value to prevent it from wandering as activity coefficients change with ionic strength.

The end point of the reaction path is the second root to the problem. At 300 °C, we find by trial and error that we need to add 0.08992 moles of $CaCl_2$ to reach the second root. The differences among the roots are summarized,

	200 °C		300 °C	
	Root *A*	Root *B*	Root *A*	Root *B*
pH	5.0	5.0	5.0	5.0
Ca^{++} (molal)	0.0026	1.0	0.00013	0.090
F^-	0.80×10^{-3}	0.80×10^{-3}	0.80×10^{-3}	0.80×10^{-3}
free F^-	0.34×10^{-3}	0.021×10^{-3}	0.73×10^{-3}	0.027×10^{-3}
free CaF^+	0.050×10^{-3}	0.75×10^{-3}	0.027×10^{-3}	0.77×10^{-3}
I (molal)	0.91	2.7	1.0	1.1

where I is ionic strength.

13.3 Coping With Nonuniqueness

The examples in the previous section demonstrate that nonunique solutions to the equilibrium problem can occur when the modeler constrains the calculation by assuming equilibrium between the fluid and a mineral or gas phase. In each example, the nonuniqueness arises from the nature of the multicomponent equilibrium problem and the variety of species distributions that can exist in an aqueous fluid. When more than one root exists, the iteration method and its starting point control which root the software locates.

In each of the cases, the dual roots differ from each other in terms of pH, sulfide content, or ionic strength, so that in a modeling study the "correct" root could readily be selected. The danger of nonuniqueness is that a modeler, having reached an inappropriate root, might not realize that a separate, more meaningful root to the problem exists.

Unfortunately, no software techniques exist currently to automatically search for additional roots. Instead, modelers must rely on their understanding of geochemistry to demonstrate uniqueness to their satisfaction. Activity–activity diagrams such as those presented in Figures 13.1–13.3 are the most useful tools for identifying additional roots.

PART II

REACTION PROCESSES

14

Mass Transfer

In previous chapters we have discussed the nature of the equilibrium state in geochemical systems: how we can define it mathematically, what numerical methods we can use to solve for it, and what it means conceptually. With this chapter we begin to consider questions of process rather than state. How does a fluid respond to changes in composition as minerals dissolve into it, or as it mixes with other fluids? How does a fluid evolve in response to changing temperature or variations in the partial pressure of a coexisting gas? In short, we begin to consider reaction modeling.

In this chapter we consider how to construct reaction paths that account for the effects of simple reactants, a name given to reactants that are added to or removed from a system at constant rates. We take on other types of mass transfer in later chapters. Chapter 15 treats the mass transfer implicit in setting a species' activity or gas's fugacity over a reaction path. In Chapter 17 we develop reaction models in which the rates of mineral precipitation and dissolution are governed by kinetic rate laws.

14.1 Simple Reactants

Simple reactants are those added to (or removed from) the system at constant rates over the reaction path. As noted in Chapter 2, we commonly refer to such a path as a titration model, because at each step in the process, much like in a laboratory titration, the model adds an aliquot of reactant mass to the system. Each reactant A_r is added at a rate n_r, expressed in moles per unit reaction progress, ξ. Negative values of n_r, of course, describe the removal rather than the addition of the reactant. Since ξ is unitless and varies from zero at the start of the path to one at the end, we can just as well think of n_r as the number of moles of the reactant to be added over the reaction path.

A simple reactant may be an aqueous species (including water), a mineral, a gas, or any entity of known composition. The only requirement is that we be able to form it by a reaction,

$$A_r \rightleftarrows \nu_{wr} A_w + \sum_i \nu_{ir} A_i + \sum_k \nu_{kr} A_k + \sum_m \nu_{mr} A_m \,, \tag{14.1}$$

among the basis entries. Significantly, we need not know the $\log K$ for this reaction; only the reaction coefficients (ν_{wr}, ν_{ir}, and so on) come into play in the calculation. We can therefore employ a substance as a reactant, even if we do not know its stability.

The calculation procedure for tracing a titration path is straightforward. We begin by calculating the equilibrium state of the initial system, as described in Chapter 4. Once we know the initial equilibrium state, we substitute the resulting bulk compositions (M_w, M_i, M_k, and M_m) for any free constraints. If we specified pH to set up the calculation, for example, we now take the value of M_{H^+} to constrain the governing equation for the H^+ component, leaving m_{H^+} as an unknown variable. Similarly, we take as a constraint the mole number M_k for each mineral component, replacing the mineral's free mass n_k. The exceptions to this process are species set at fixed or sliding activity and gases of fixed and sliding fugacity, as described in the next chapter; for these exceptions, we retain the free constraints.

Once a substitution of constraints is accomplished, the calculation consists of incrementally changing the system's bulk composition as a function of reaction progress and, after each increment, recalculating the equilibrium state. The bulk composition is given from the component masses (M_w^o, M_i^o, and M_k^o) present at $\xi = 0$, the reaction rates n_r, and the reactants' stoichiometric coefficients (ν_{wr}, and so on, from Reaction 14.1), according to

$$M_w(\xi) = M_w^o + \xi \sum_r \nu_{wr} n_r \tag{14.2}$$

$$M_i(\xi) = M_i^o + \xi \sum_r \nu_{ir} n_r \tag{14.3}$$

$$M_k(\xi) = M_k^o + \xi \sum_r \nu_{kr} n_r \,. \tag{14.4}$$

Alternatively, we can compute the new composition from the composition at the end ξ' of the previous step as

$$M_w(\xi) = M_w(\xi') + (\xi - \xi') \sum_r \nu_{wr} n_r \tag{14.5}$$

$$M_i(\xi) = M_i(\xi') + (\xi - \xi') \sum_r \nu_{ir} n_r \tag{14.6}$$

$$M_k(\xi) = M_k(\xi') + (\xi - \xi') \sum_r \nu_{kr} n_r \,. \tag{14.7}$$

The equations expressed in this stepwise manner are somewhat easier to integrate into certain reaction configurations, such as the flush or flow-through model described later in this chapter. We could also update M_m in this manner, but there is no need to do so. A gas species A_m appears in the basis only when its fugacity f_m is known, so the value of each M_m results from solving the governing equations, as described in Chapter 4.

To trace the reaction path, the model begins with the system at $\xi = 0$ and steps forward in reaction progress by setting ξ to $\Delta\xi$, where $\Delta\xi$ is the size of the reaction step. It then recomputes the bulk composition using Equations 14.5–14.7 and, honoring these values, iterates to the new equilibrium state. To start the iteration, the model takes the values of the unknown variables at the beginning of the step. The model then takes a further step forward in reaction progress, incrementing ξ to a value of $2\,\Delta\xi$, and recomputes the equilibrium state. The calculation continues in this fashion until it reaches $\xi = 1$.

We consider as an example the hydrolysis of potassium feldspar ($KAlSi_3O_8$), the first reaction path traced using a computer (Helgeson *et al.*, 1969). We specify the composition of a hypothetical water

```
Na+ =       5 mg/kg
K+ =        1 mg/kg
Ca++ =     15 mg/kg
Mg++ =      3 mg/kg
Al+++ =     1 ug/kg
SiO2(aq) = 3 mg/kg
Cl- =      30 mg/kg
```

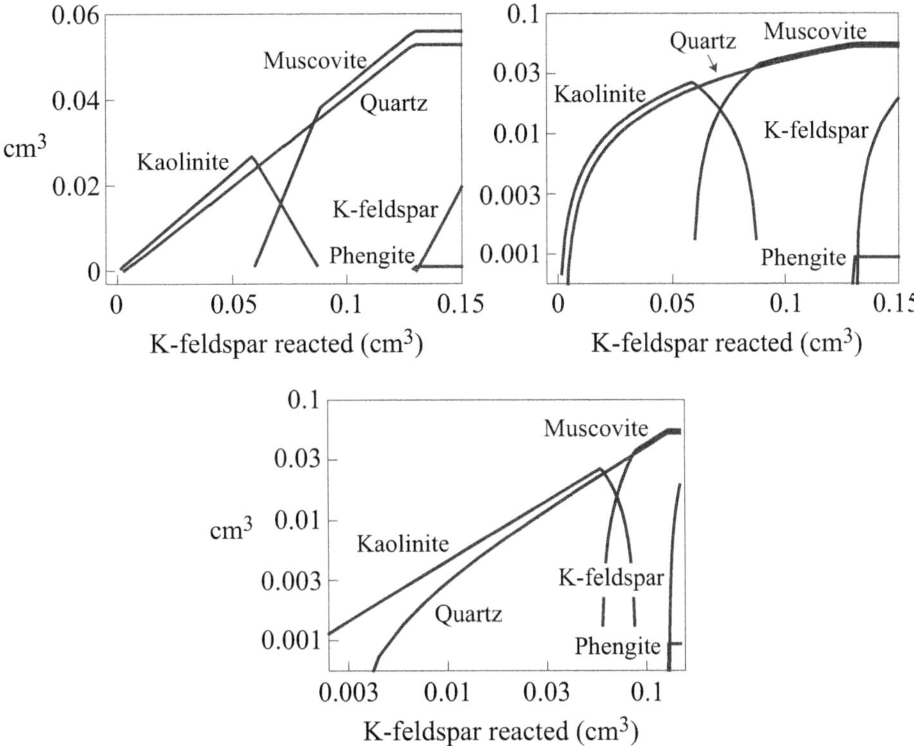

Figure 14.1 Mineralogical results of reacting potassium feldspar into a hypothetical water at 25 °C, plotted in linear, semilog (a "spaghetti diagram"), and log–log coordinates.

```
SO4-- =    8 mg/kg
HCO3- =   50 mg/kg
pH = 5
```

and then define a reaction path involving the addition of a small amount of feldspar

```
(cont'd)
react .15 cm3 of K-feldspar
```

Typing `go` triggers the calculation.

Figure 14.1 shows the mineralogical results of the calculation, plotted in linear, semilog, and log–log coordinates. Note that we have plotted each diagram in terms of the mass of feldspar reacted into the water. Common practice in reaction modeling is to present results plotted in terms of ξ, but this is an unfortunate convention. The reaction progress variable ξ has mathematical meaning within the modeling program, but its physical meaning vests only in terms of how the modeler sets the reaction rates n_r. By choosing a variable with physical meaning (such as reacted mass) when plotting results, we can present our calculation results in a more direct manner.

In the first segment of the reaction path (Fig. 14.1), the feldspar dissolves into solution, producing kaolinite [$Al_2Si_2O_5(OH)_4$] and quartz (SiO_2). The reaction gradually increases the activity ratio a_{K^+}/a_{H^+} until the solution reaches equilibrium with muscovite [$KAl_3Si_3O_{10}(OH)_2$]. At this point, the kaolinite begins to

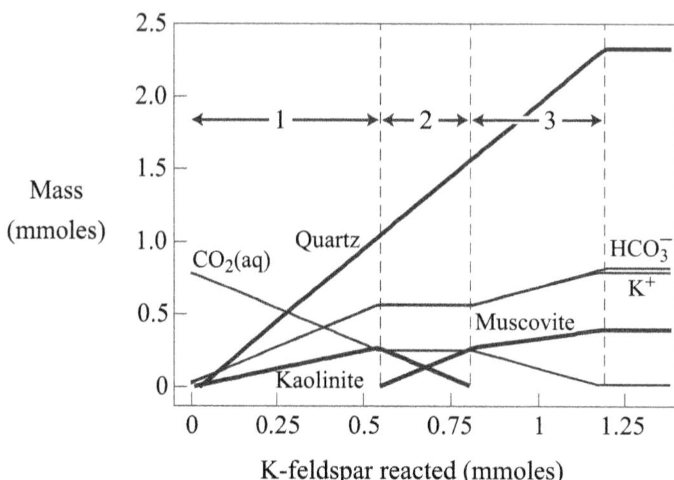

Figure 14.2 Results of the reaction path shown in Figure 14.1, plotted to allow the overall reaction to be extracted using the slopes-of-the-lines method.

dissolve, producing muscovite and quartz. The reaction continues after the kaolinite is consumed until the fluid reaches equilibrium with the feldspar [a small amount of phengite, $KAlMgSi_4O_{10}(OH)_2$, forms just before this point]. Once the fluid is in equilibrium with it, the feldspar simply accumulates as it is added to the system, and reaction with the fluid ceases. In the next section of this chapter, we consider how we can extract the chemical reactions occurring over each segment of the reaction path.

14.2 Extracting the Overall Reaction

The ultimate goal in reaction modeling is to discover the overall reaction that occurs within a system. Strangely, whereas the results of nearly every published study involving reaction modeling are presented in a "spaghetti diagram" (see, e.g., Fig. 14.1), few papers report the overall reaction. For this reason, some of the most important information in the results is obscured. Who could blame the reader for thinking, "Enough pasta, let's get to the meatballs"?

The procedure for determining the overall reaction, fortunately, is straightforward. The modeler plots the mole numbers of the species and minerals in the system against the mass of a reactant added to the system over the reaction path. The plot must be in linear coordinates with both axes in consistent units, such as moles or mmoles.

The slopes of the lines in the plot give the reaction coefficients for each species and mineral in the overall reaction. Species with negative slopes appear to the left of the reaction (with their coefficients set positive), and those with positive slopes are placed to the right. The reactant plotted on the horizontal axis appears to the left of the reaction with a coefficient of one. If there are additional reactants, these also appear on the reaction's left with coefficients equal to the ratios of their reaction rates n_r to that of the first reactant.

As an example, we consider the reaction path traced in the previous section (Fig. 14.1). To extract the overall reaction for each segment of the path, we construct a plot as described above. The result is shown in Figure 14.2. There are three segments in the reaction path: the precipitation of kaolinite, the transformation of kaolinite to muscovite, and the continued formation of muscovite once the kaolinite is exhausted. There is a distinct overall reaction for each segment.

From this plot (and one showing the mass of species H_2O, which does not fit on these axes), we can write down the slope of the line for each species and mineral. The results, compiled for each segment in the reaction, are

	Segment 1	Segment 2	Segment 3
$CO_2(aq)$	−1	0	−0.67
HCO_3^-	+1	0	+0.67
K^+	+1	0	+0.67
H_2O	−1.5	+1	−0.67
quartz	+2	+2	+2
kaolinite	+0.5	−1	–
muscovite	–	+1	+0.33

The values in the first column give the overall reaction for the first segment of the reaction path,

$$\underset{\textit{K-feldspar}}{KAlSi_3O_8} + 3/2\, H_2O + CO_2(aq) \rightarrow 2\ \underset{\textit{quartz}}{SiO_2} + 1/2\ \underset{\textit{kaolinite}}{Al_2Si_2O_5(OH)_4} + HCO_3^- + K^+ . \quad (14.8)$$

Similarly, the overall reactions for the second segment,

$$\underset{\textit{K-feldspar}}{KAlSi_3O_8} + \underset{\textit{kaolinite}}{Al_2Si_2O_5(OH)_4} \rightarrow 2\ \underset{\textit{quartz}}{SiO_2} + \underset{\textit{muscovite}}{KAl_3Si_3O_{10}(OH)_2} + H_2O , \quad (14.9)$$

and third segment,

$$\begin{aligned} &\underset{\textit{K-feldspar}}{KAlSi_3O_8} + 2/3\, CO_2(aq) + 2/3\, H_2O \rightarrow \\ &\qquad 2\ \underset{\textit{quartz}}{SiO_2} + 1/3\ \underset{\textit{muscovite}}{KAl_3Si_3O_{10}(OH)_2} + 2/3\, HCO_3^- + 2/3\, K^+ , \end{aligned} \quad (14.10)$$

are given by the coefficients in the second and third columns.

14.3 Special Configurations

In Chapter 2 we discussed three special configurations for tracing reaction paths: the dump, flow-through, and flush models. These models are special cases of mass transfer that can be implemented within the mathematical framework developed in this chapter.

In the dump configuration, the model discards the masses of any minerals present in the initial equilibrium system before beginning to trace the reaction path. To do so, the model updates the total composition to reflect the absence of minerals,

$$M_k = M_k - n_k , \quad (14.11)$$

and sets the mineral masses n_k to zero. Here, of course, we use the "=" to represent assignment of value, rather than algebraic equivalence. If the model considers surface complexation (Chapters 10–11), the mole numbers M_p of the surface components as well as the molalities m_p and m_q of the surface species must also be set to zero to reflect the disappearance of sorbing mineral surfaces.

The model then swaps aqueous species A_j into the basis locations A_k held by the mineral components. The technique for swapping the basis is explained in Chapter 5, and a method for selecting an appropriate species A_j to include in the basis is described in Chapter 4. When the procedure is complete, the equilibrium system contains only the original fluid.

In the flow-through model, any mineral mass present at the end of a reaction step is sequestered from the equilibrium system to avoid back-reaction. At the end of each step, the model eliminates the mineral mass (including any sorbed species) from the equilibrium system, keeping track of the total amount removed. To do so, it applies Equation 14.11 for each mineral component and sets each n_k to a vanishingly small number. It is best to avoid setting n_k to exactly zero in order to maintain the mineral entries A_k in the basis. The model then updates the system composition according to Equations 14.5–14.7 and takes another reaction step.

In a flush model, reactant fluid displaces existing fluid from the equilibrium system. It is simplest to implement this model by determining the mass of water entering the system over a step and eliminating an equal mass of water component and the solutes it contains from the system. In this case, we ignore any density differences between the fluids.

The model first determines (from the reaction rate n_r for water and the mass M_w of the water component) the fraction X_{disp} of fluid to be displaced from the system over a step. Typically, the model will limit the size $\Delta\xi$ of the reaction step to a value that will cause only a fraction (perhaps a tenth or a quarter) of the fluid present at the start of the step to be displaced, in the event that the modeler accidentally sets too large a step size. The formulae for determining the updated composition become

$$M_w(\xi) = \left(1 - X_{\text{disp}}\right) M_w(\xi') + \left(\xi - \xi'\right) \sum_r \nu_{wr} n_r \tag{14.12}$$

$$M_i(\xi) = \left(1 - X_{\text{disp}}\right) M_i(\xi') + \left(\xi - \xi'\right) \sum_r \nu_{ir} n_r \tag{14.13}$$

$$M_k(\xi) = \left(1 - X_{\text{disp}}\right) \left(M_k(\xi') - n_k\right) + n_k + \left(\xi - \xi'\right) \sum_r \nu_{kr} n_r\,. \tag{14.14}$$

These are Equations 14.5–14.7, modified to account for the loss of the displaced fluid.

15

Polythermal, Fixed, and Sliding Paths

In this chapter we consider how to construct reaction models that are somewhat more sophisticated than those discussed in the previous chapter, including reaction paths over which temperature varies and those in which species activities and gas fugacities are buffered. The latter cases involve the transfer of mass between the equilibrium system and an external buffer. Mass transfer in these cases occurs at rates implicit in solving the governing equations, rather than at rates set explicitly by the modeler. In Chapter 17 we consider the use of kinetic rate laws, a final method for defining mass transfer in reaction models.

15.1 Polythermal Reaction

Polythermal reaction paths are those in which temperature varies as a function of reaction progress, ξ. In the simplest case, the modeler prescribes the temperatures T_o and T_f at the beginning and end of the reaction path. The model then varies temperature linearly with reaction progress. This type of model is sometimes called a "sliding-temperature" path.

The calculation procedure for a sliding-temperature path is straightforward. In taking a reaction step, the model evaluates the temperature to be attained at the step's end. Since ξ varies from zero to one, temperature at any point ξ in reaction progress is given by

$$T(\xi) = T_o + \xi(T_f - T_o)\,, \tag{15.1}$$

as a function of the initial and final temperatures, as set by the modeler. The model then re-evaluates values for the reaction log Ks and the constants used to calculate activity coefficients (see Chapter 8). If reaction kinetics or isotopic fractionation is considered (Chapters 17 and 22), the model recalculates the reaction rate constants and isotopic fractionation factors.

A second type of polythermal path traces temperature as reactants mix into the equilibrium system. This case differs from a sliding-temperature path only in the manner in which temperature is determined. The modeler assigns a temperature T_o to the initial system, as before, and a distinct temperature T_{rct} to the reactants. By assuming that the heat capacities C_{P_f}, C_{P_k}, and C_{P_r} of the fluid, minerals, and reactants are constant over the temperature range of interest, we can calculate temperature $T(\xi)$ from energy balance and the temperature $T(\xi')$ at the onset of the step according to

$$T(\xi) = \frac{A\,T(\xi') + B\,T_{rct}}{A + B}$$

$$A = 1000\,(1 - X_{\text{disp}})\,C_{P_f} n_f + \sum_k C_{P_k} M_{W_k} n_k$$
$$B = (\xi - \xi') \sum_r C_{P_r} M_{W_r} n_r\,. \tag{15.2}$$

In this equation, ξ' and ξ are the values of reaction progress at the beginning and end of the step; n_f is the mass in kg of the fluid (equal to n_w, the water mass, plus the mass of the solutes); n_k is the mole number of each mineral; n_r is the reaction rate (moles) for each reactant; M_{W_k} is the mole weight (g mol^{-1}) of each mineral, and M_{W_r} is the mole weight for each reactant; and X_{disp} is the fraction of the fluid displaced over the reaction step in a flush model (X_{disp} is zero if a flush model is not invoked).

In an example of a sliding-temperature path, we consider the effects of cooling from 300 °C to 25 °C a system in which a 1 molal NaCl solution is in equilibrium with the feldspars albite ($NaAlSi_3O_8$) and microcline ($KAlSi_3O_8$), quartz (SiO_2), and muscovite [$KAl_3Si_3O_{10}(OH)_2$]. To set up the calculation, we enter the commands

```
swap Albite for Al+++
swap "Maximum Microcline" for K+
swap Muscovite for H+
swap Quartz for SiO2(aq)

1 molal Na+
1 molal Cl-
20 free cm3 Albite
10 free cm3 "Maximum Microcline"
5 free cm3 Muscovite
2 free cm3 Quartz

suppress "Albite low"
```

to define the fluid in equilibrium with the minerals, which are present in excess amounts. For simplicity, we suppress the entry "albite low," which is almost identical in stability to the entry "albite" in the thermodynamic database. We then set a polythermal path with the command

(cont'd)
```
T initial = 300, final = 25
```

and type `go` to start the calculation.

In the calculation results (Fig. 15.1), albite reacts to form microcline,

$$\underset{\textit{albite}}{NaAlSi_3O_8} + K^+ \rightarrow \underset{\textit{microcline}}{KAlSi_3O_8} + Na^+\,, \tag{15.3}$$

as the temperature decreases. The reaction is driven not by mass transfer but by variation in the stabilities of the minerals and the species in solution. The composition of the system (fluid plus minerals), in fact, remains constant over the reaction path. Chapters 26 and 27 give a number of further examples of the application of polythermal reaction paths.

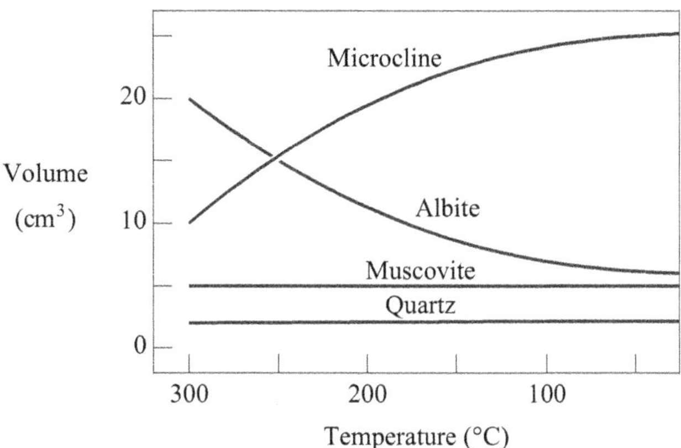

Figure 15.1 Mineralogical results of tracing a polythermal reaction path. In the calculation, a 1 molal NaCl solution in equilibrium with albite, microcline, muscovite, and quartz cools from 300 °C to 25 °C.

15.2 Fixed Activity and Fugacity

In a fixed-activity path, the activity of an aqueous species (or those of several species) maintains a constant value over the course of the reaction path. A fixed-fugacity path is similar, except that the model holds constant a gas fugacity instead of a species activity. Fixed-activity paths are useful in modeling laboratory experiments in which an aspect of a fluid's chemistry is maintained mechanically. In studying reaction kinetics, for example, it is common practice to hold constant the pH of a solution with a pH-stat. Fixed-activity paths are also convenient for calculating speciation diagrams, which by convention may be plotted at constant pH and oxygen activity. Fixed-fugacity paths are useful for tracing reaction paths in which a fluid remains in contact with a gas phase, such as the atmosphere.

To calculate a fixed-activity path, the model maintains within the basis each species A_i whose activity a_i is to be held constant. For each such species, the corresponding mass balance equation (Eq. 4.4) is reserved from the reduced basis, as described in Chapter 4, and the known value of a_i is used in evaluating the mass action equation (Eq. 4.7). Similarly, the model retains within the basis each gas A_m whose fugacity is to be fixed. We reserve the corresponding mass balance equation (Eq. 4.6) from the reduced basis and use the corresponding fugacity f_m in evaluating the mass action equation.

A complication to the calculation procedure for holding an aqueous species at fixed activity is the necessity of maintaining ionic charge balance over the reaction path. If the species is charged, the model must enforce charge balance at each step in the calculation by adjusting the concentration of a specified component, as discussed in Section 4.3. For example, if the pH is fixed over a path and the charge balance component is Cl^-, then the model will behave as if HCl were added to or removed from the system in the quantities needed to maintain a constant H^+ activity.

In an example of a fixed-fugacity path we model the dissolution of pyrite (FeS_2) at 25 °C. We start in REACT with a hypothetical water in equilibrium with hematite (Fe_2O_3) and oxygen in the atmosphere

```
swap O2(g) for O2(aq)
swap Hematite for Fe++

f O2(g) = .2
1 free mg Hematite
```

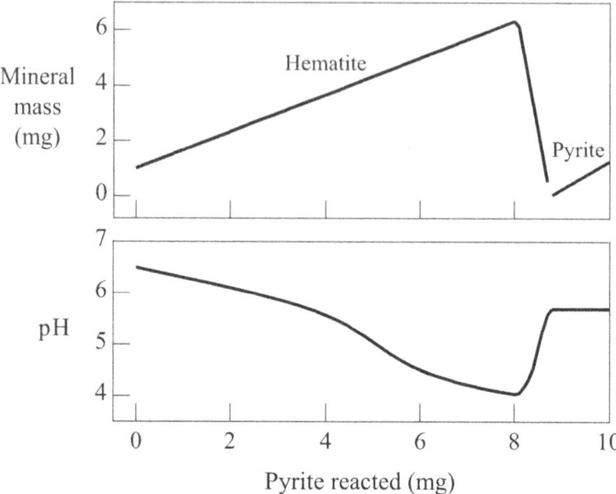

Figure 15.2 Mineralogical results (top) of reacting pyrite at 25 °C into a dilute water held closed to O_2, and variation in pH (bottom) over the reaction.

```
pH = 6.5

 4 mg/kg Ca++
 1 mg/kg Mg++
 2 mg/kg Na+
18 mg/kg HCO3-
 3 mg/kg SO4--
 5 mg/kg Cl-
```

into which we react a small amount of pyrite

(cont'd)

```
react 10 mg Pyrite
go
```

In this initial calculation, we do not fix the oxygen fugacity.

As shown in Figure 15.2, about 8 mg of pyrite dissolves into the water, producing hematite. The reaction drives pH from the initial value of 6.5 to about 4 before the water becomes reducing. At this point, the hematite redissolves and the fluid reaches equilibrium with pyrite, bringing the reaction to an end.

We can write the overall reaction by which hematite forms, using the slopes-of-the-lines method discussed in Chapter 14. Initially, the reaction proceeds as

$$\underset{pyrite}{FeS_2} + {}^{15}\!/_{4}\, O_2(aq) + 4\, HCO_3^- \rightarrow {}^{1}\!/_{2}\, \underset{hematite}{Fe_2O_3} + 2\, SO_4^{--} + 4\, CO_2(aq) + 2\, H_2O\,, \quad (15.4)$$

as can be seen from Figure 15.3. As the water becomes more acidic and the supply of HCO_3^- is depleted, a second reaction,

$$\underset{pyrite}{FeS_2} + {}^{15}\!/_{4}\, O_2(aq) + 2\, H_2O \rightarrow {}^{1}\!/_{2}\, \underset{hematite}{Fe_2O_3} + 2\, SO_4^{--} + 4\, H^+\,, \quad (15.5)$$

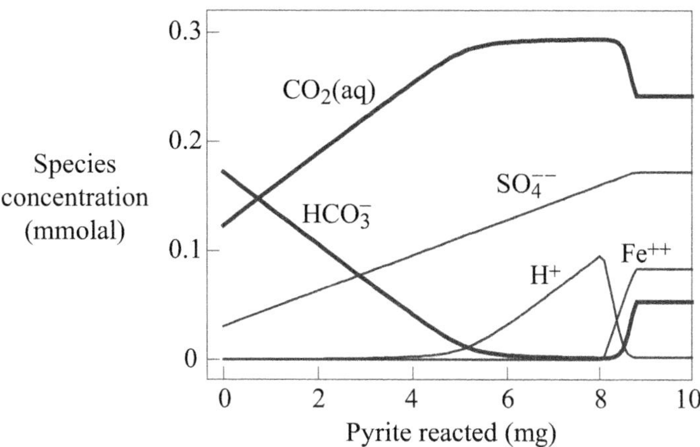

Figure 15.3 Variation in the concentrations of aqueous species involved in the dissolution reaction of pyrite, for the reaction path shown in Figure 15.2.

becomes dominant. Pyrite continues to dissolve until the available $O_2(aq)$ has been consumed.

How would the reaction have proceeded if the oxygen fugacity had been fixed by equilibrium with the atmosphere? To find out, we repeat the calculation, this time holding the oxygen fugacity constant

(cont'd)

```
fix f O2(g)
react 1000 mg Pyrite
go
```

and specifying a 100-fold increase in the supply of pyrite.

The fixed-fugacity path (Fig. 15.4) differs from the previous calculation (in which the fluid was closed to the addition of oxygen) in that pyrite dissolution continues indefinitely, since there is an unlimited supply of oxygen gas. Initially, the reaction proceeds as

$$\underset{pyrite}{FeS_2} + {}^{15}/_4\,O_2(g) + 2\,H_2O \rightarrow {}^{1}/_2\ \underset{hematite}{Fe_2O_3} + 2\,SO_4^{--} + 4\,H^+ , \tag{15.6}$$

as can be seen from Figure 15.5. Later, a second reaction that produces HSO_4^- instead of SO_4^{--},

$$\underset{pyrite}{FeS_2} + {}^{15}/_4\,O_2(aq) + 2\,H_2O \rightarrow {}^{1}/_2\ \underset{hematite}{Fe_2O_3} + 2HSO_4^- + 2\,H^+ , \tag{15.7}$$

becomes dominant. The H^+ produced by these reactions drives pH to values far more acidic than those in the closed-system case.

15.3 Sliding Activity and Fugacity

Sliding-activity and sliding-fugacity paths are similar to fixed-activity and fixed-fugacity paths, except that the model varies the buffered activity or fugacity over the reaction path rather than holding it constant. Once the equilibrium state of the initial system is known, the model stores the initial activity a_i^{o} or initial fugacity

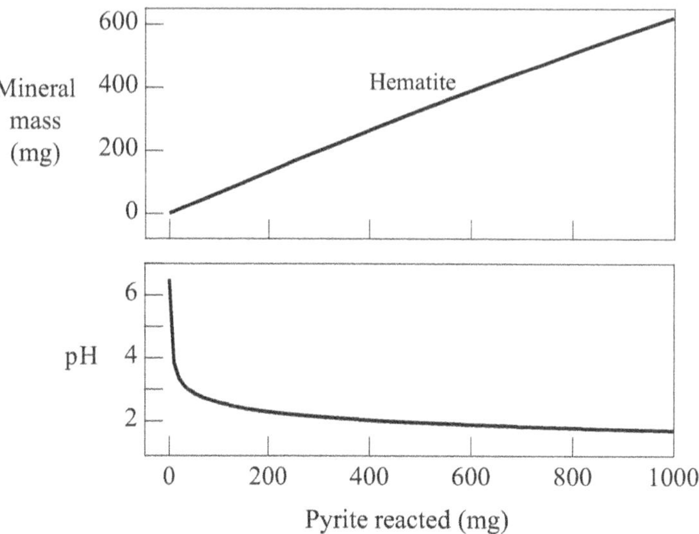

Figure 15.4 Mineralogical results (top) of a fixed-fugacity path in which pyrite dissolves at 25 °C into water held in equilibrium with O_2 in the atmosphere, and the variation in pH (bottom) over the path.

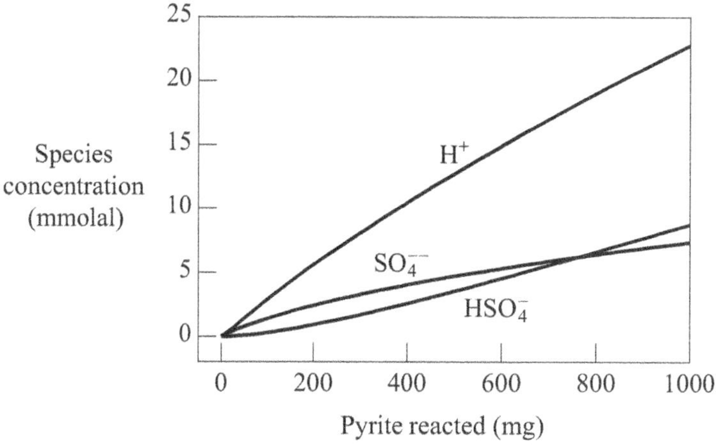

Figure 15.5 Concentrations of species involved in the dissolution of pyrite, for the fixed-fugacity path shown in Figure 15.4.

f_m^o of the buffered species or gas. (The modeler could set this value as a constraint on the initial system, but this is not necessary.)

The modeler supplies the final (or target) activity a_i^f or fugacity f_m^f, which will be achieved at the end of the reaction path, when $\xi = 1$. The modeler also specifies whether the activity or fugacity itself or the logarithm of activity or fugacity is to be varied. If the value is to be varied, the model determines it as a function of reaction progress according to

$$a_i(\xi) = a_i^o + \xi \left(a_i^f - a_i^o\right) \tag{15.8}$$

or

$$f_m(\xi) = f_m^o + \xi \left(f_m^f - f_m^o\right) . \tag{15.9}$$

If, on the other hand, the value's logarithm is to be varied, the model calculates the value according to

$$\log a_i(\xi) = \log a_i^o + \xi \left(\log a_i^f - \log a_i^o\right) \tag{15.10}$$

or

$$\log f_m(\xi) = \log f_m^o + \xi \left(\log f_m^f - \log f_m^o\right) . \tag{15.11}$$

To see the difference between sliding a value and sliding its logarithm, consider a path in which f_{O_2} varies from one (10^0) to 10^{-30}. If $\Delta\xi$ in this simple example is 0.2, the model will step along the reaction path in even steps of either f_{O_2} or $\log f_{O_2}$:

ξ	f_{O_2}	$\log f_{O_2}$
0	1.0	1
0.2	0.8	10^{-6}
0.4	0.6	10^{-12}
0.6	0.4	10^{-18}
0.8	0.2	10^{-24}
1.0	10^{-30} ($\simeq 0$)	10^{-30}

If we select the linear option (the center column), the path will stop at a series of oxidizing points followed by a single reducing point. If we choose the logarithmic option (right column), however, the path will visit a range of oxidation states.

In an example of a sliding-fugacity path, we calculate how CO_2 fugacity affects the solubility of calcite ($CaCO_3$). We begin by defining a dilute solution in equilibrium with calcite and the CO_2 fugacity of the atmosphere

```
swap Calcite for Ca++
swap CO2(g) for H+

10 mmolal Na+
10 mmolal Cl-
1/2 free cm3 Calcite
log f CO2(g) = -3.5
balance on HCO3-
go
```

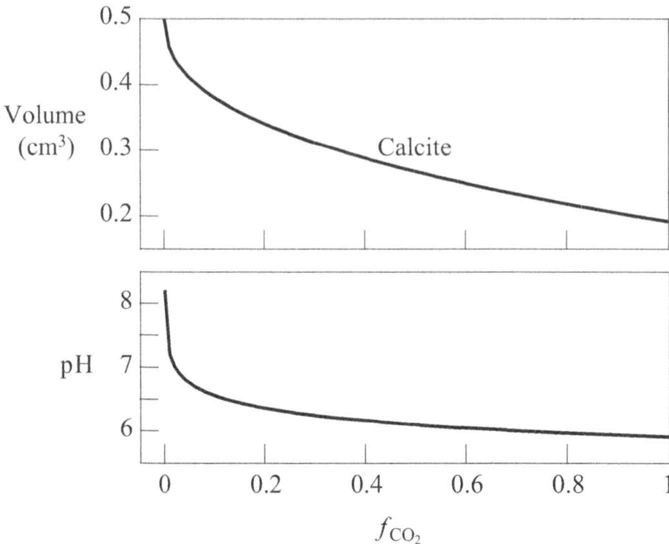

Figure 15.6 Effect of CO_2 fugacity on the solubility of calcite (top) and on pH (bottom), calculated at 25 °C using a sliding-fugacity path.

The resulting fluid has a pH of about 8.3. To vary the CO_2 fugacity, we set a path in which f_{CO_2} slides from the initial atmospheric value

(cont'd)
```
slide f CO2(g) to 1
go
```

to one.

In the calculation results (Fig. 15.6), increasing the CO_2 fugacity decreases the pH to about 6, causing calcite to dissolve into the fluid. The fugacity increase drives CO_2 from the buffer into the fluid, and most of the CO_2 (Fig. 15.7) becomes CO_2(aq). The nearly linear relationship between the concentration of CO_2(aq) and the fugacity of CO_2(g) results from the reaction

$$CO_2(g) \rightleftarrows CO_2(aq)\,, \tag{15.12}$$

which holds a_{CO_2} proportional to f_{CO_2}. Some of the gas, however, dissociates to produce HCO_3^- and H^+, and the resulting acid is largely consumed by dissolving calcite. The overall reaction (derived using the slopes-of-the-lines method) is approximately

$$5\,CO_2(g) + \underset{calcite}{CaCO_3} + H_2O \rightarrow 4\,CO_2(aq) + Ca^{++} + 2\,HCO_3^-\,. \tag{15.13}$$

In a second example, we calculate how pH affects sorption onto hydrous ferric oxide, expanding on our discussion (Section 10.5) of Dzombak and Morel's (1990) surface complexation model. We start as before, setting the dataset of surface reactions, suppressing the ferric minerals hematite (Fe_2O_3) and goethite (FeOOH), and specifying the amount of ferric oxide [represented in the calculation by $Fe(OH)_3$ precipitate] in the system

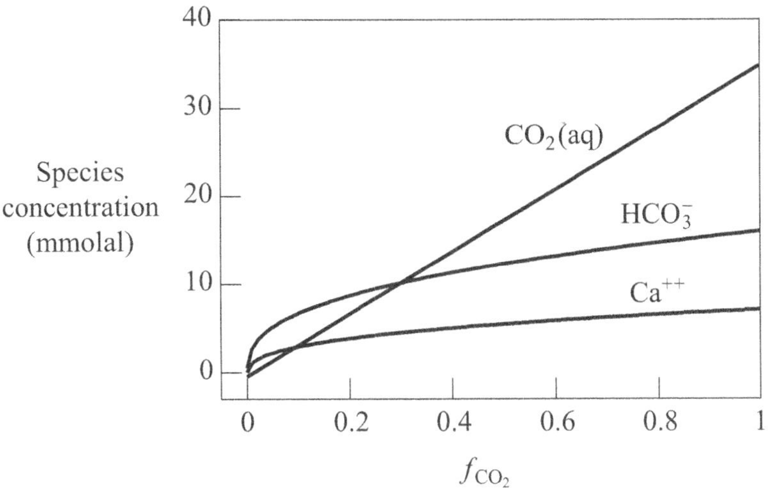

Figure 15.7 Species concentrations (mmolal) in the sliding-fugacity path shown in Figure 15.6.

```
read FeOH+.sdat
sorbate include
decouple Fe+++
suppress Hematite, Goethite
swap Fe(OH)3(ppd) for Fe+++
1 free gram Fe(OH)3(ppd)
```

We set a 0.1 molal NaCl solution and define a sliding-activity path in which pH varies from 4 to 12

```
(cont'd)
0.1 molal Na+
0.1 molal Cl-
pH = 4

slide pH to 12
precip = off
go
```

Figure 15.8 shows how the concentrations of the surface species vary with pH in the calculation results.

According to the complexation model, neither Na^+ nor Cl^- reacts with the surface, so the species consist entirely of surface sites (>(w)FeOH and >(s)FeOH) in their uncomplexed, protonated,

$$>\text{(w)FeOH} + \text{H}^+ \rightleftarrows >\text{(w)FeOH}_2^+ \,, \tag{15.14}$$

and deprotonated forms,

$$>\text{(w)FeOH} \rightleftarrows >\text{(w)FeO}^- + \text{H}^+ \,, \tag{15.15}$$

depending on the H^+ activity. Protonated sites dominate at low pH, resulting in a positive surface potential (Fig. 15.9), whereas the predominance of deprotonated sites at high pH yields a negative potential.

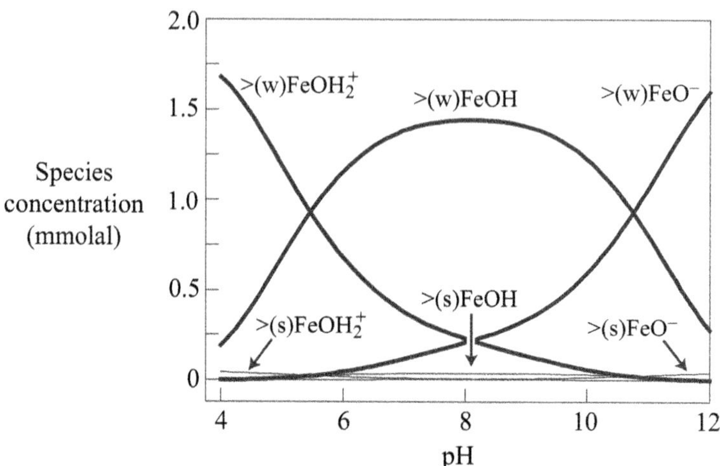

Figure 15.8 Concentrations (mmolal) of sites on a hydrous ferric oxide surface exposed at 25 °C to a 0.1 molal NaCl solution, calculated using a sliding-pH path.

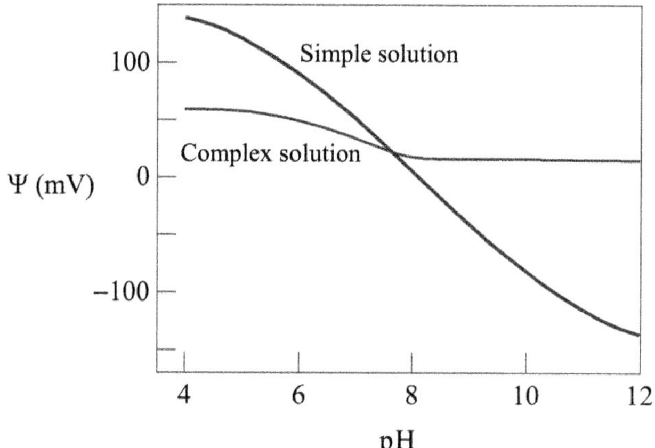

Figure 15.9 Variation of surface potential Ψ (mV) with pH for a hydrous ferric oxide surface in contact at 25 °C with a 0.1 molal NaCl solution (bold line) and a more complex solution (fine line) that also contains Ca, SO_4, Hg, Cr, As, and Zn.

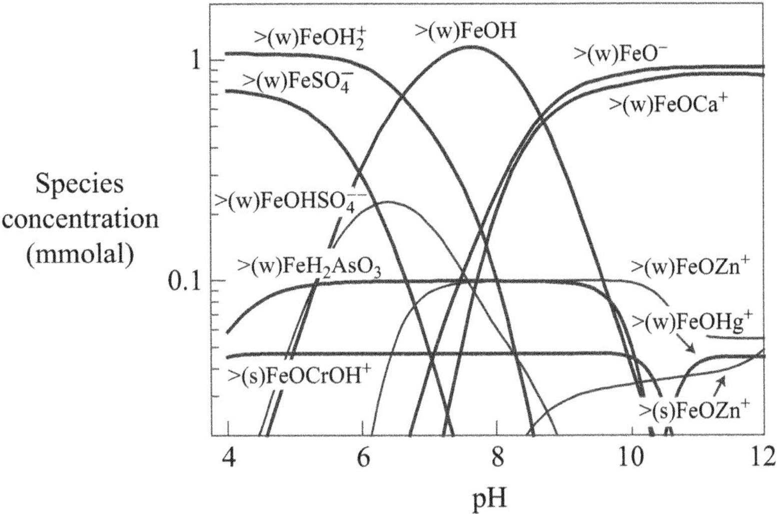

Figure 15.10 Concentrations (mmolal) of surface species on hydrous ferric oxide exposed at 25 °C to a solution containing Ca, SO_4, Hg, Cr, As, and Zn, calculated using a sliding-pH path.

To see how contact with a more complex solution affects the surface, we introduce to the fluid 10 mmoles of $CaSO_4$ and 100 μmoles each of Hg^{++}, Cr^{+++}, $As(OH)_4^-$, and Zn^{++}

(cont'd)
```
10 mmolal Ca++
10 mmolal SO4--
100 umolal Hg++
100 umolal Cr+++
100 umolal As(OH)4-
100 umolal Zn++
```

Typing `go` triggers the model, which again slides pH from 2 to 12.

In this case (Fig. 15.10), we observe a more complicated distribution of species. At low pH, the H^+ activity and positive surface potential drive SO_4^{--} to sorb according to the reaction

$$SO_4^{--} + {>}(w)FeOH + H^+ \rightleftarrows {>}(w)FeSO_4^- + H_2O\,. \tag{15.16}$$

Under alkaline conditions, on the other hand, reactions such as

$$Zn^{++} + {>}(w)FeOH \rightleftarrows {>}(w)FeOZn^+ + H^+ \tag{15.17}$$

$$Hg^{++} + {>}(w)FeOH \rightleftarrows {>}(w)FeOHg^+ + H^+ \tag{15.18}$$

and

$$Ca^{++} + {>}(w)FeOH \rightleftarrows {>}(w)FeOCa^+ + H^+\,, \tag{15.19}$$

promote the sorption of bivalent cations. These reactions produce negatively charged surface species at low pH and positively charged surface species at high pH, thereby reducing the magnitude of the surface charge under acidic as well as alkaline conditions. Hence, the surface is considerably less charged than it was in contact with the NaCl solution, as shown in Figure 15.9.

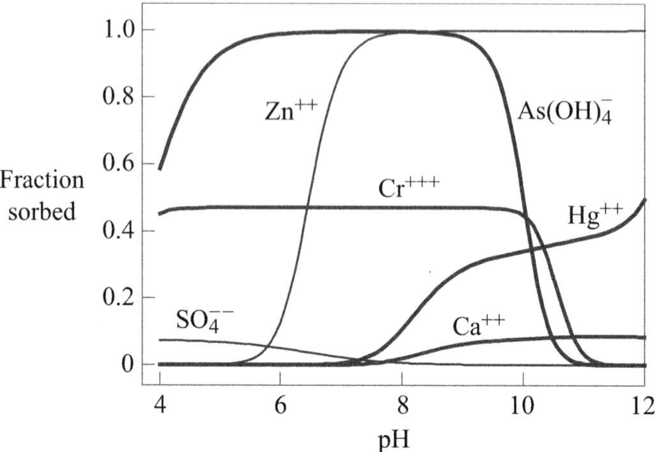

Figure 15.11 Concentrations (mmolal) of surface species on hydrous ferric oxide exposed at 25 °C to a solution containing Ca, SO_4, Hg, Cr, As, and Zn, calculated using a sliding-pH path.

The $As(OH)_4^-$ and Cr^{+++} components follow a pattern distinct from the other metals (Fig. 15.11), sorbing at only near-neutral pH. This pattern results from the manner in which the metals speciate in solution. As^{III} appears as $As(OH)_3$ when pH is less than 9, and as $As(OH)_4^-$, or AsO_2OH^{--} at higher pH. The sorption reactions for these species are

$$As(OH)_3 + {>}(w)FeOH \rightleftarrows {>}(w)FeH_2AsO_3 + H_2O \tag{15.20}$$

$$As(OH)_4^- + {>}(w)FeOH + H^+ \rightleftarrows {>}(w)FeH_2AsO_3 + 2\,H_2O \tag{15.21}$$

$$AsO_2OH^{--} + {>}(w)FeOH + 2\,H^+ \rightleftarrows {>}(w)FeH_2AsO_3 + H_2O\,. \tag{15.22}$$

There is no pH dependence to the reaction for $As(OH)_3$, which sorbs strongly when pH is near neutral. Under acidic conditions, however, high H^+ activities drive the reactions for protonation and the sorption of SO_4^{--}, which displace arsenic from the mineral surface. Arsenic sorbs poorly under alkaline conditions because low H^+ activities work against the complexation of $As(OH)_4^-$, or AsO_2OH^{--}, as shown in the reactions for these species.

The chromium follows a similar pattern. The component, present as Cr^{+++} at low pH, reacts successively to form $CrOH^{++}$, $Cr(OH)_2^+$, $Cr(OH)_3$, and $Cr(OH)_4^-$ as pH increases. The sorption reactions are

$$Cr^{+++} + {>}(s)FeOH + H_2O \rightleftarrows {>}(s)FeOCrOH^+ + 2\,H^+ \tag{15.23}$$

$$CrOH^{++} + {>}(s)FeOH \rightleftarrows {>}(s)FeOCrOH^+ + H^+ \tag{15.24}$$

$$Cr(OH)_2^+ + {>}(s)FeOH \rightleftarrows {>}(s)FeOCrOH^+ + H_2O \tag{15.25}$$

$$Cr(OH)_3 + {>}(s)FeOH + H^+ \rightleftarrows {>}(s)FeOCrOH^+ + 2\,H_2O \tag{15.26}$$

$$Cr(OH)_4^- + {>}(s)FeOH + 2\,H^+ \rightleftarrows {>}(s)FeOCrOH^+ + 3\,H_2O\,. \tag{15.27}$$

Significantly, the reactions for the species predominant at low pH favor the desorption of chromium when the H^+ activity is high, and those for the species predominant under alkaline conditions favor desorption when the H^+ activity is low. Hence Cr^{III}, like Ar^{III}, sorbs strongly only when pH is near neutral.

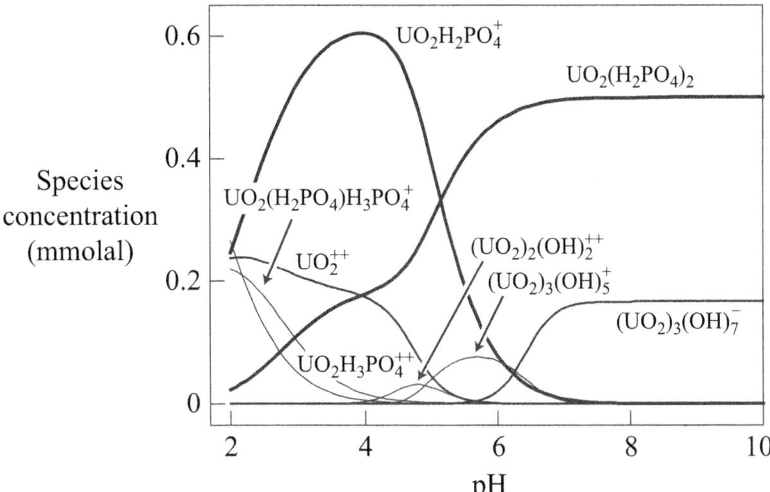

Figure 15.12 Speciation diagram at 25 °C for a 1 mmolal solution of hexavalent uranium containing 1 mmolal dissolved phosphate, calculated as a sliding-activity path.

In a final example of the use of a sliding-activity path, we calculate a speciation diagram, plotted versus pH, for hexavalent uranium in the presence of dissolved phosphate at 25 °C. We take a 10 mmolal NaCl solution containing 1 mmolal each of UO_2^{++}, the basis species for U^{VI}, and HPO_4^{--}

```
decouple UO2++
10 mmolal Na+
10 mmolal Cl-
1 mmolal UO2++
swap H3PO4 for HPO4--
1 mmolal H3PO4
```

Here we have swapped H_3PO_4 for HPO_4^{--} to help the program converge under acidic conditions. We specify the initial and final pH values for the calculation, set the program to avoid precipitating minerals (since we assume a fixed solution composition)

(cont'd)
```
pH = 2
slide pH to 10
precip = off
```

and type go to trigger the calculation. The resulting diagram (Fig. 15.12) shows the importance of complexing between U^{VI} and phosphate. We can, of course, make many variations on the calculation, such as setting different concentrations, including other components, allowing minerals to precipitate, and so on.

16

Geochemical Buffers

Buffers are reactions that at least temporarily resist change to some aspect of a fluid's chemistry. A pH buffer, for example, holds pH to an approximately constant value, opposing processes that would otherwise drive the solution acid or alkaline. The bicarbonate–CO_2 buffer,

$$HCO_3^- + H^+ \rightarrow CO_2(aq) + H_2O\,, \tag{16.1}$$

for example, consumes hydrogen ions when they are added to the system and produces them when they are consumed, thereby resisting variation in pH. The buffer operates until nearly all of the HCO_3^- is converted to $CO_2(aq)$, or vice versa. The thirsty reader might be interested to know that the concept of buffering (as well as the notation pH) was introduced by the brewing industry in Europe (Sørensen, 1909; see Rosing, 1993) as it sought to improve the flavor of beer.

Buffers such as the bicarbonate reaction are known as homogeneous buffers, because all of the constituents are found in the fluid phase. Many important buffering reactions in geochemical systems are termed heterogeneous (e.g., Rosing, 1993) because, in addition to the fluid, they involve minerals or a gas phase. Reducing minerals or oxygen in the atmosphere (examples of heterogeneous buffers) can control a fluid's oxidation state. Equilibrium with quartz fixes a fluid's silica content. Some buffers, such as those provided by assemblages of minerals, can be rather complex. Many reaction models, in fact, are designed to describe how buffers behave and how various buffering reactions interact.

In this chapter we construct models of buffering reactions, both homogeneous and heterogeneous. We concentrate on buffering reactions that are well known to geochemists, taking the opportunity to explore reaction modeling on familiar geochemical terrain. The methods discussed here, however, can be readily applied to more complicated situations, such as those involving multiple buffers or buffers involving a larger number of phases.

16.1 Buffers in Solution

We begin by considering the well-known pH buffer provided by the aqueous carbonate system and its effects on the ease with which a fluid can be acidified. We start with an alkaline NaCl solution containing a small amount of carbonate, and add 300 mmol of hydrochloric acid to it. The procedure in REACT is

```
pH = 12
HCO3- =   1 mmolal
Cl-   = 200 mmolal
balance on Na+
```

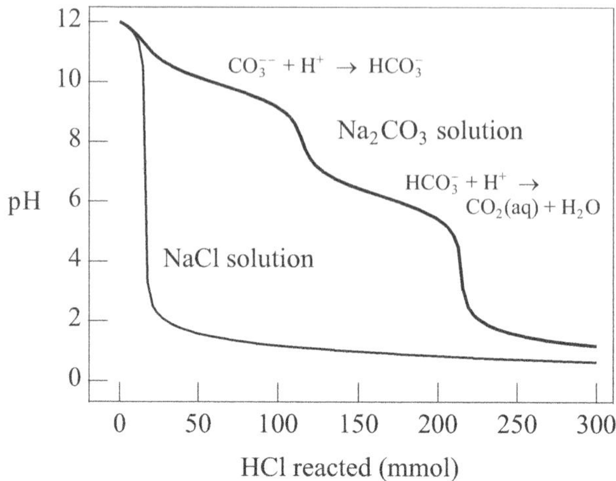

Figure 16.1 Calculated effects on pH of reacting hydrochloric acid into a 0.2 molal NaCl solution and a 0.1 molal Na_2CO_3 solution, as functions of the amount of HCl added. The two plateaus on the second curve represent the buffering reactions between CO_3^{--} and HCO_3^-, and between HCO_3^- and $CO_2(aq)$.

```
react 300 mmol HCl
go
```

As can be seen in Figure 16.1, the effect is to quickly drive the solution acidic. The only buffer is the presence of OH^- ions, which are quickly consumed by reaction with H^+ to produce water.

In a second experiment, we reverse the anion concentrations so that the fluid is dominantly an Na_2CO_3 solution:

```
(cont'd)
HCO3- = 100 mmolal
Cl-   =   1 mmolal
go
```

Since the HCO_3^- component is present mostly as the doubly charged species CO_3^{--}, its molality is half that of the balancing cation, Na^+. In this case (Fig. 16.1), the fluid resists acidification until more than 200 mmol of HCl have been added.

The buffering occurs in two stages, as shown in Figure 16.2: first the CO_3^{--} species in solution consume H^+ ions as they react to produce HCO_3^-, then the HCO_3^- reacts with H^+ to make $CO_2(aq)$. The two stages are represented in Figure 16.1 by nearly horizontal portions of the pH curve. When all of the CO_3^{--} and HCO_3^- species have been consumed, the solution quickly becomes acidic.

We can readily derive the overall reactions in the buffer from Figure 16.3 using the slopes-of-the-lines method, described in Chapter 14. In the first stage, the overall reaction is

$$0.83\, CO_3^{--} + 0.17\, NaCO_3^- + H^+ \rightarrow 0.83\, HCO_3^- + 0.17\, NaHCO_3\,. \tag{16.2}$$

The reaction for the second stage is

$$0.83\, HCO_3^- + 0.17\, NaHCO_3 + H^+ \rightarrow CO_2(aq) + 0.17\, Na^+ + H_2O\,. \tag{16.3}$$

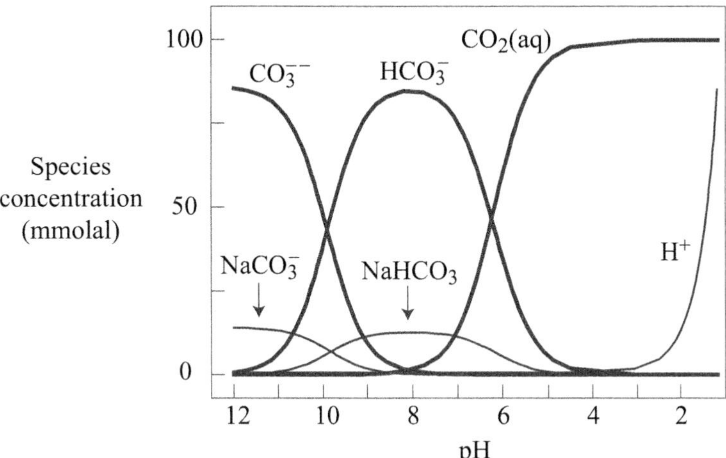

Figure 16.2 Concentrations of species in the carbonate buffer in a 0.1 molal Na_2CO_3 solution, plotted against pH.

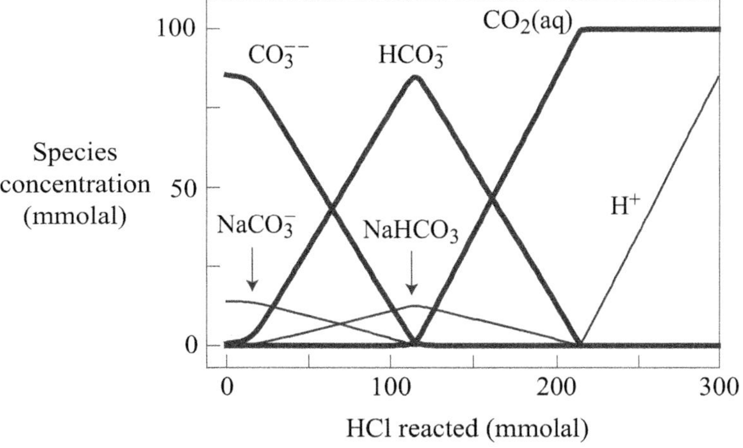

Figure 16.3 Species masses in the Na_2CO_3 solution plotted so that the overall reaction can be determined using the slopes-of-the-lines method.

It is common practice when writing overall reactions to omit mention of ion pairs whenever they are not considered important to the point being addressed. We could well write the reactions above as

$$CO_3^{--} + H^+ \rightarrow HCO_3^- \tag{16.4}$$

and

$$HCO_3^- + H^+ \rightarrow CO_2(aq) + H_2O\,. \tag{16.5}$$

The simplified form is not as exact but is less cluttered than the full form and shows more clearly the nature of the buffering reaction. In this book, we will often make simplifications of this sort.

In a practical example of the use of reaction modeling to trace buffering reactions, we consider the problem of interpreting the titration alkalinity of natural waters. Laboratories commonly report titration alkalinity rather than provide a direct analysis of a solution's carbonate content. Titration alkalinity is the solution's ability to neutralize strong acid (e.g., Snoeyink and Jenkins, 1980; Hem, 1985). The analyst titrates an acid such as H_2SO_4 into the solution until it reaches an endpoint pH, as indicated by the color change of an indicator such as methyl orange. The endpoint pH is generally in the range of 4.5 to 4.8.

The analyst reports the amount of acid required to reach the endpoint, generally expressed in terms of the number of mg of $CaCO_3$ that could be dissolved by the acid, per kg solution. Since the mole weight of $CaCO_3$ is 100.09 g and each mole of carbonate can neutralize two equivalents of acid, the conversion is

$$\frac{\text{meq acid}}{\text{kg solution}} \times \frac{50.05 \text{ mg CaCO}_3}{\text{meq acid}} = \frac{\text{mg CaCO}_3}{\text{kg solution}}\,. \tag{16.6}$$

Note that there are two equivalents of acid per mole of H_2SO_4.

At the titration endpoint, most carbonate in solution is present as $CO_2(aq)$. We can expect that each mmol of HCO_3^- originally present in solution will neutralize one meq of acid, according to the reaction

$$HCO_3^- + H^+ \rightarrow CO_2(aq) + H_2O\,. \tag{16.7}$$

Similarly, each mmol of CO_3^{--} originally present should neutralize two meq of acid,

$$CO_3^{--} + 2\,H^+ \rightarrow CO_2(aq) + H_2O\,. \tag{16.8}$$

Therefore, knowing from the initial pH the proportions of HCO_3^- and CO_3^{--} in the fluid, we can estimate the total carbonate content.

Unfortunately, such simple estimations can be in error. Hydroxyl, borate, silicate, ammonia, phosphate, and organic species can contribute to the solution's ability to buffer acid. For example, each of the reactions

$$HPO_4^{--} + H^+ \rightarrow H_2PO_4^- \tag{16.9}$$

$$B(OH)_4^- + H^+ \rightarrow B(OH)_3 + H_2O \tag{16.10}$$

$$NH_3 + H^+ \rightarrow NH_4^+ \tag{16.11}$$

can consume hydrogen ions during the titration. Other possible complications include the effects of activity coefficients and complex species.

A more rigorous method for interpreting an alkalinity measurement is to use a reaction model to reproduce the titration. The technique is to calculate the effects of adding acid to the original solution, assuming various carbonate contents. When we produce a model that reaches the endpoint pH after adding the acid, we have found the correct carbonate concentration.

Table 16.1 *Analysis of water from Mono Lake, California, USA (James Bischoff, personal communication)*

pH	9.68
Ca^{++} (mg kg^{-1})	4.6
Mg^{++}	42
Na^{+}	37 200
K^{+}	1 580
SO_4^{--}	12 074
Cl^{-}	20 100
$B(OH)_3$	2 760
F^{-}	54
HPO_4^{--}	120
Alkalinity, as $CaCO_3$	34 818
Dissolved solids	92 540

We now consider as an example an analysis (Table 16.1) of water from Mono Lake, California. The reported alkalinity of 34 818 mg kg^{-1} as $CaCO_3$ is equivalent to 700 meq of acid or 350 mmol of H_2SO_4. Since at this pH carbonate and bicarbonate species are present in roughly equal concentrations, we can quickly estimate the total carbonate concentration to be about 30 000 mg kg^{-1}. We take this value as a first guess and model the titration with REACT

```
TDS = 92540
pH = 9.68
Ca++    =    4.6 mg/kg
Mg++    =     42 mg/kg
Na+     = 37200 mg/kg
K+      =  1580 mg/kg
SO4--   = 12074 mg/kg
Cl-     = 20100 mg/kg
B(OH)3 =   2760 mg/kg
F-      =     54 mg/kg
HPO4-- =    120 mg/kg
HCO3-   = 30000 mg/kg

react 350 mmol of H2SO4
precip = off
go
```

Testing varying values for the total HCO_3^- concentration,

```
(cont'd)
HCO3-   = 25000 mg/kg
go
```

we find that 25 100 mg kg^{-1} gives a titration endpoint of 4.5, as shown in Figure 16.4. The result differs from

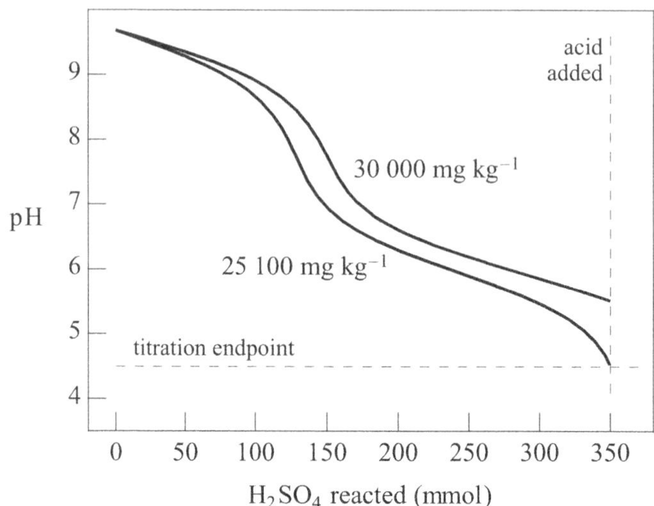

Figure 16.4 Use of reaction modeling to derive a fluid's carbonate concentration from its titration alkalinity, as applied to an analysis of Mono Lake water. When the correct HCO_3 total concentration (in this case, 25 100 mg kg^{-1}) is set, the final pH matches the titration endpoint.

our initial guess primarily because of the protonation of the $B(OH)_4^-$ and $NaB(OH)_4$ species to form $B(OH)_3$ plus water, as shown in Figure 16.5.

16.2 Minerals as Buffers

In a first example of how minerals can buffer a fluid's chemistry, we consider how a hypothetical groundwater that is initially in equilibrium with calcite ($CaCO_3$) at 25 °C might respond to the addition of an acid. In REACT, we enter the commands

```
swap Calcite for HCO3-
10 free cm3 Calcite

pH = 8
Na+ = 100 mmolal
Ca++ = 10 mmolal
balance on Cl-
```

to set the initial system containing a fluid of pH 8 in equilibrium with calcite. We then type

```
(cont'd)
dump
react 100 mmol of HCl
go
```

to define the reaction path and trigger the calculation. With the `dump` command, we specify that the calcite be separated from the fluid before the reaction path begins.

The only pH buffer in the calculation is the small concentration of carbonate species in solution. The buffer

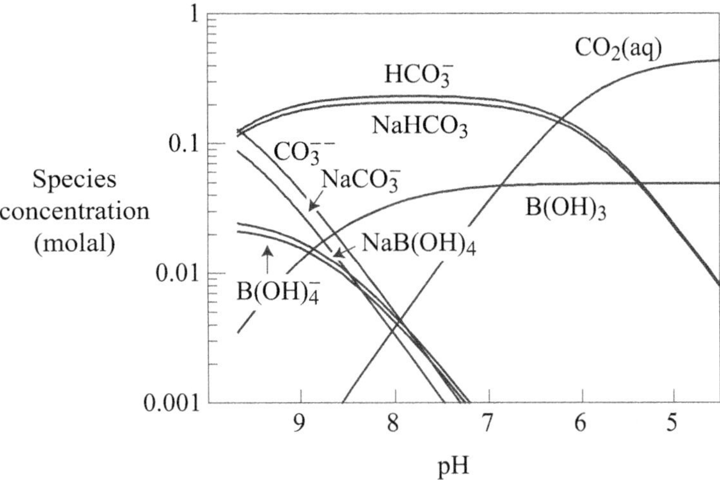

Figure 16.5 Concentrations of species in buffer reactions that contribute to the titration alkalinity of Mono Lake water, plotted against pH.

is quickly overwhelmed and the fluid shifts rapidly to acidic pH, as shown in Figure 16.6. The dominant reaction,

$$HCl \rightarrow H^+ + Cl^- , \tag{16.12}$$

is the dissociation of the HCl.

In a second calculation, we trace the same path while maintaining the fluid in equilibrium with the calcite. To do so, we enter the command

```
(cont'd)
dump = off
```

to disable the `dump` option and type `go` to start the path. In the calculation results (Fig. 16.6), pH decreases by only a small amount and the fluid becomes just mildly acidic. The CO_2 partial pressure rises steadily during the reaction, finally exceeding a value of one, at which point we would expect the gas to begin to effervesce against atmospheric pressure.

The overall reaction for the earliest portion of the reaction path, where HCO_3^- is the predominant carbonate species, is

$$2\,HCl + \underset{calcite}{CaCO_3} \rightarrow Ca^{++} + 2\,Cl^- + HCO_3^- + H^+ . \tag{16.13}$$

By this reaction, the fluid becomes more acidic with the addition of HCl. With decreasing pH, the $CO_2(aq)$ species quickly comes to dominate HCO_3^-. At this point, the principal reaction becomes

$$2\,HCl + \underset{calcite}{CaCO_3} \rightarrow Ca^{++} + 2\,Cl^- + CO_2(aq) + H_2O . \tag{16.14}$$

According to this reaction, adding HCl to the fluid no longer affects pH. Instead, calcite dissolves to neutralize the acid, leaving Ca^{++} and $CO_2(aq)$ in solution (Fig. 16.7).

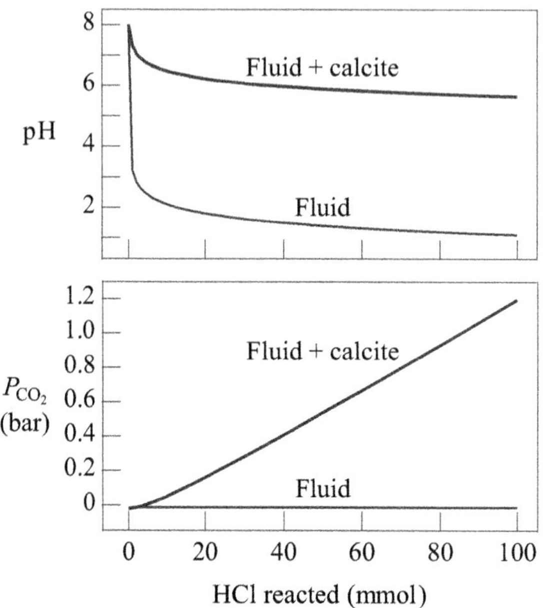

Figure 16.6 Effects on pH (top) and CO_2 partial pressure (bottom) of reacting HCl into a fluid not in contact with calcite (fine lines) and with the same fluid when it maintains equilibrium with calcite over the reaction path (bold lines).

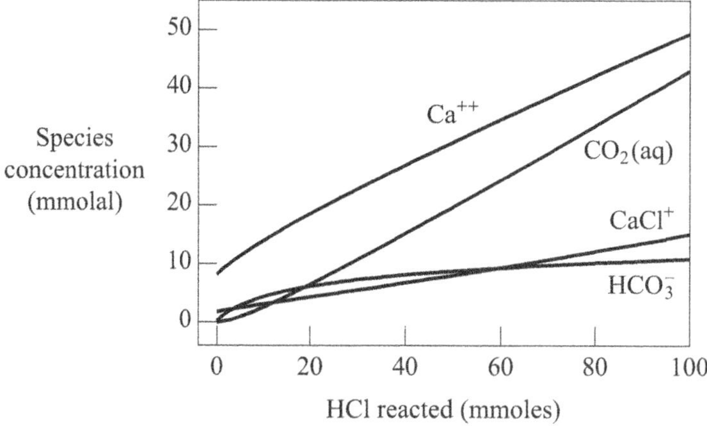

Figure 16.7 Concentrations (mmolal) of the predominant carbonate and calcium species over the course of a reaction path in which HCl is added to a fluid in contact with calcite.

In nature, the fluid would begin to exsolve $CO_2(g)$,

$$CO_2(aq) \rightarrow CO_2(g)\,, \tag{16.15}$$

once the $CO_2(aq)$ activity has built up sufficiently to drive the reaction forward. In this case, we can see that the acid introduced to the fluid is converted by reaction with calcite into CO_2 (or carbonic acid, equivalently) and then lost from the fluid by effervescence.

As a second example, we consider how the presence of pyrite (FeS_2) can serve to buffer a fluid's oxidation state. We set an initial system at 100 °C containing a 1 molal NaCl solution in equilibrium with 0.2 moles (about 5 cm^3) of pyrite and a small amount of hematite (Fe_2O_3). In REACT, the procedure is

```
T = 100
swap Pyrite for Fe++
swap Hematite for O2(aq)

200 free mmol Pyrite
  1 free mmol Hematite

pH = 8
Cl- = 1 molal
Na+ = 1 molal
SO4-- = 0.001 molal

go
```

The reaction between pyrite and hematite fixes the initial oxidation state to a reducing value ($\log f_{O_2} = -54$). We then set a reaction path in which we add oxygen to the fluid

(cont'd)

```
react 800 mmol of O2(aq)
go
```

simulating what might happen, for example, if O_2 were to diffuse into a reducing geologic formation.

Figure 16.8 shows the calculation results, and in Figure 16.9 the reaction path is projected onto an f_{O_2}–pH diagram drawn for the Fe–S–H_2O system. (To project the path onto the diagram, we complete the reaction path, start ACT2, enter the commands

```
T = 100
swap O2(g) for O2(aq)
diagram Fe++ on O2(g) vs pH
log a Fe++ = -6
log a SO4-- = -3
speciate SO4-- over X-Y

x from 0 to 9
y from -65 to 2
trace
```

and type go.) In the earliest portion of the path, the system responds to the addition of oxygen by shifting quickly toward low pH. The system's rapid acidification results from the production of H^+ according to the reaction,

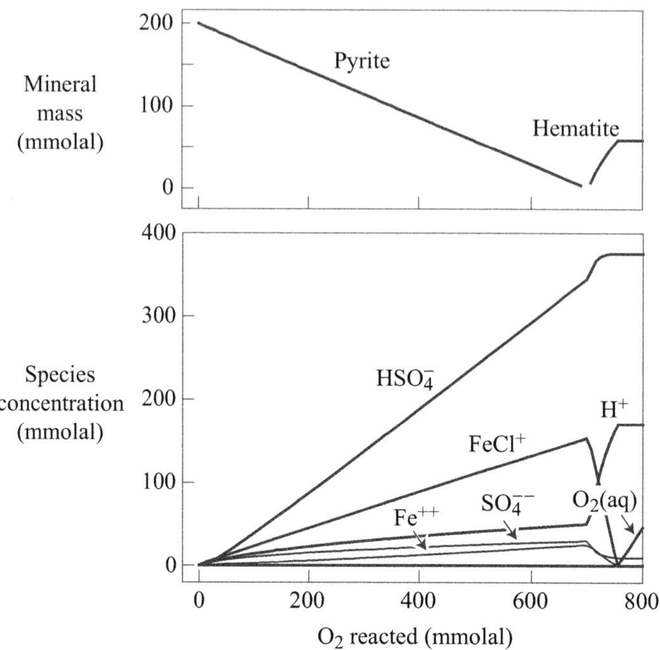

Figure 16.8 Results of reacting O_2 at 100 °C into a system containing pyrite. Pyrite dissolves (top) with addition of O_2. The reaction (bottom) produces bisulfate ions and ferric species ($FeCl^+$ and Fe^{++}), which in turn are consumed at the end of the path to form hematite.

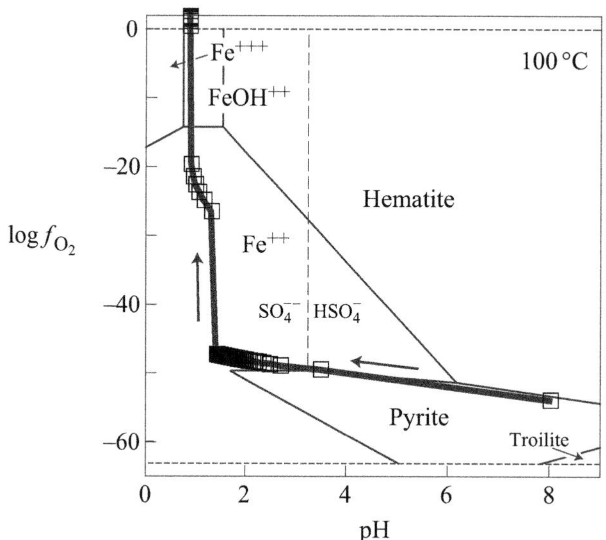

Figure 16.9 Projection (bold line) of reaction path shown in Figure 16.8 onto an f_{O_2}–pH diagram drawn for the Fe–S–H_2O system at 100 °C. Horizontal dashed lines are stability limits for water at 1 atm pressure. Vertical dashed line is equal-activity line for SO_4^{--} and HSO_4^-. Boxes show reaction steps taken in tracing the path; each step represents the addition of an 8 mmol aliquot of O_2.

$$\tfrac{7}{2}\,O_2(aq) + \underset{pyrite}{FeS_2} + H_2O \rightarrow Fe^{++} + 2\,SO_4^{--} + 2\,H^+\,, \tag{16.16}$$

(lumping $FeCl^+$ together with Fe^{++}). As the system moves to lower pH, the bisulfate species HSO_4^- becomes more abundant than SO_4^{--} because of protonation of the sulfate species,

$$SO_4^{--} + H^+ \rightarrow HSO_4^-\,. \tag{16.17}$$

At pH less than about 3, where bisulfate predominates, the dominant reaction is

$$\tfrac{7}{2}\,O_2(aq) + \underset{pyrite}{FeS_2} + H_2O \rightarrow Fe^{++} + 2\,HSO_4^-\,. \tag{16.18}$$

Since the reaction written in terms of HSO_4^- produces no H^+, the shift to lower pH slows and then ceases as the SO_4^{--} in solution is depleted.

At this point, the solution remains almost fixed in oxidation state and pH, accumulating Fe^{++} and HSO_4^- as the addition of oxygen causes pyrite to dissolve. When the pyrite is exhausted, the oxygen fugacity begins to rise rapidly. As oxygen is added, it is consumed in converting the dissolved ferrous iron to hematite,

$$\tfrac{1}{2}\,O_2(aq) + 2\,Fe^{++} + 2\,H_2O \rightarrow \underset{hematite}{Fe_2O_3} + 4\,H^+\,, \tag{16.19}$$

which further acidifies the solution. Only when the ferrous iron is exhausted does the fluid become fully oxidized.

If we eliminate the pyrite and retrace the calculation

```
(cont'd)
dump
react 1 mmol of O2(aq)
go
```

we find that just a small amount of O_2 is sufficient to oxidize the system. In this case, pH changes little over the reaction path.

16.3 Gas Buffers

With a final example, we consider how the presence of a gas phase can serve as a chemical buffer. A fluid, for example, might maintain equilibrium with the atmosphere, soil gas in the root zone, or natural gas reservoirs in deep strata. Gases such as O_2 and H_2 can fix oxidation state, H_2S can set the activity of dissolved sulfide, and CO_2 (as we demonstrate in this section) can buffer pH.

In this experiment, we take an acidic water in equilibrium with atmospheric CO_2 and titrate NaOH into it. In REACT, the commands

```
pH = 2
swap CO2(g) for HCO3-
log P CO2(g) = -3.5 bar
Na+ = .2 molal
balance on Cl-
```

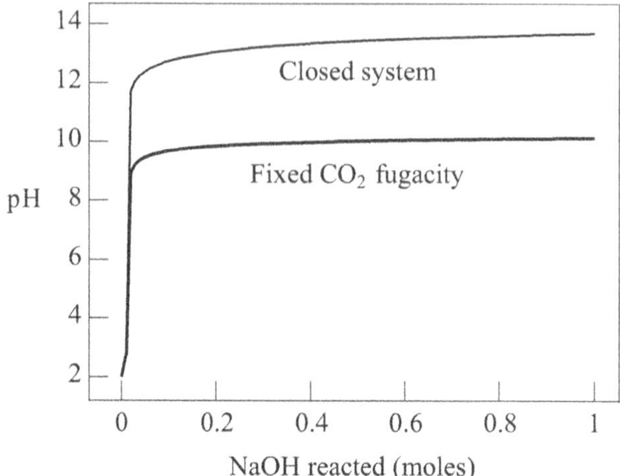

Figure 16.10 Calculated effects on pH of reacting sodium hydroxide into an initially acidic solution that is either closed to mass transfer (fine line) or in equilibrium with atmospheric CO_2 (bold line).

```
react 1 mole NaOH
go
```

set up the calculation, assuming the partial pressure of CO_2 in the atmosphere is $10^{-3.5}$ bar.

As shown in Figure 16.10, the fluid quickly becomes alkaline, approaching a pH of 14. Since the fluid's carbonate content is small, about 10 μmolal, little beyond the fluid's initial H^+ content,

$$NaOH + H^+ \rightarrow Na^+ + H_2O\,, \tag{16.20}$$

is available to buffer pH. Once the H^+ is exhausted, adding NaOH to the solution,

$$NaOH \rightarrow Na^+ + OH^-\,, \tag{16.21}$$

simply produces OH^-, driving the pH to high values.

Now we consider the same reaction occurring in a water that maintains equilibrium with atmospheric CO_2. With the REACT commands

```
(cont'd)
fix fugacity CO2(g)
go
```

we fix the CO_2 fugacity to its atmospheric value and retrace the path. In this case (Fig. 16.10), the pH rises initially but then levels off, approaching a value of 10.

The latter path differs from the closed system calculation because of the effect of $CO_2(g)$ dissolving into the fluid. In the initial part of the calculation, the $CO_2(aq)$ in solution reacts to form HCO_3^- in response to the changing pH. Since the fluid is in equilibrium with $CO_2(g)$ at a constant fugacity, however, the activity of $CO_2(aq)$ is fixed. To maintain this activity, the model transfers CO_2,

$$CO_2(g) \rightarrow CO_2(aq)\,, \tag{16.22}$$

from gas to fluid, replacing whatever $CO_2(aq)$ has reacted to form HCO_3^-.

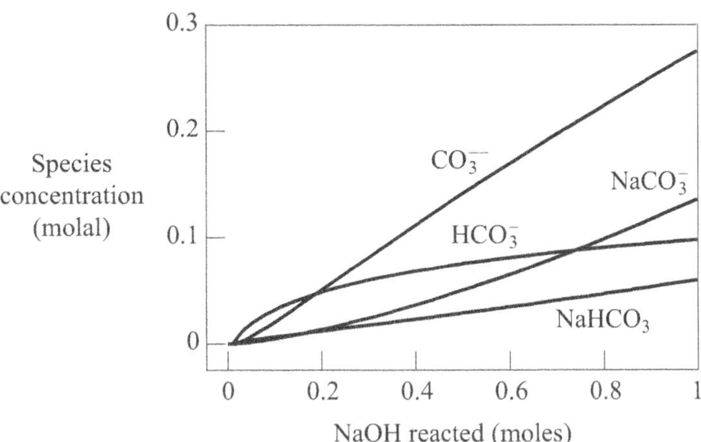

Figure 16.11 Concentrations (molal) of the predominant carbonate species over the course of a reaction path in which NaOH is added to a fluid that maintains equilibrium with CO_2 in the atmosphere.

The overall reaction for the earliest portion of the path, obtained by the slopes-of-the-lines method, is,

$$2\,NaOH + CO_2(g) \rightarrow 2\,Na^+ + HCO_3^- + OH^- \,. \tag{16.23}$$

In contrast to the unbuffered case (Reaction 16.21), two NaOHs are required to produce each OH^- ion. As pH continues to increase, the HCO_3^- reacts to produce CO_3^{--}, as shown in Figure 16.11. At this point, a second overall reaction,

$$2\,NaOH + CO_2(g) \rightarrow 2\,Na^+ + CO_3^{--} + H_2O\,, \tag{16.24}$$

becomes increasingly important. According to this reaction, adding NaOH increases the fluid's sodium and carbonate contents but does not affect pH. The CO_2, an acid gas, neutralizes the alkaline NaOH as fast as it is added. Once Reaction 16.24 comes to dominate Reaction 16.23, pH ceases to change. As long as CO_2 is available in the gas reservoir, the fluid maintains the buffered pH.

17

Kinetics of Dissolution and Precipitation

To this point we have measured reaction progress parametrically in terms of the reaction progress variable ξ, which is dimensionless. When in Chapter 14 we reacted feldspar with water, for example, we tied reaction progress to the amount of feldspar that had reacted and expressed our results along that coordinate. Studying reactions in this way is in many cases perfectly acceptable. But what if we want to know how much time it took to reach a certain point along the reaction path? Or, when modeling the reaction of granite with rainwater, how can we set the relative rates at which the various minerals in the granite dissolve? In such cases (e.g., Rimstidt, 2014), we need to incorporate reaction rate laws from the field of geochemical kinetics.

The differences between the study of thermodynamics and kinetics might be illustrated (e.g., Lasaga, 1981a) by the analogy of rainfall on a mountain. On the mountaintop, the rainwater contains a considerable amount of potential energy. With time, it flows downhill, losing energy (to be precise, losing hydraulic potential, the mechanical energy content of a unit mass of water; Hubbert, 1940), until it eventually reaches the ocean, its lowest possible energy level. The thermodynamic interpretation of the process is obvious: the water seeks to minimize its energy content.

But how long will it take for the rainfall to reach the ocean? The rain might enter a swift mountain stream, flow into a river, and soon reach the sea. It might infiltrate the subsurface and migrate slowly through deep aquifers until it discharges in a distant valley, thousands of years later. Or, perhaps it will find a faster route through a fracture network or flow through an open drill hole. There are many pathways, just as there are many mechanisms by which a chemical reaction can proceed. Clearly, the questions addressed by geochemical kinetics are more difficult to answer than are those posed in thermodynamics.

In geochemical kinetics, the rates at which reactions proceed are given (in units such as mol s^{-1} or mol yr^{-1}) by rate laws, as discussed in the next section. Kinetic theory can be applied to study reactions among the species in solution. We might, for example, study the rate at which the ferrous ion Fe^{++} oxidizes by reaction with O_2 to produce the ferric species Fe^{+++}. Since the reaction occurs within a single phase, it is termed *homogeneous*. Reactions involving more than one phase (including the reactions by which minerals precipitate and dissolve and those involving a catalyst) are called *heterogeneous*.

In this chapter we consider the problem of the kinetics of the heterogeneous reactions by which minerals dissolve and precipitate. This topic has received a considerable amount of attention in geochemistry, primarily because of the slow rates at which many minerals react and the resulting tendency of waters, especially at low temperature, to be out of equilibrium with the minerals they contact. We first discuss how rate laws for heterogeneous reactions can be integrated into reaction models and then calculate some simple kinetic reaction paths. In Chapter 30, we explore a number of examples in which we apply heterogeneous kinetics to problems of geochemical interest.

As discussed already in Chapter 7, redox reactions in many cases also proceed too slowly in the natural environment to attain equilibrium. The kinetics of redox reactions, both homogeneous and those catalyzed on a mineral surface are considered in detail in the next chapter, Chapter 18, and the role microbial life plays in catalyzing redox reactions is discussed in Chapter 19. To round out our discussion, Chapters 20 and 21 examine how fast aqueous and surface complexes associate and dissociate, and how rapidly gas species transfer into and out of an aqueous solution.

17.1 Kinetic Rate Laws

Despite the authority apparent in its name, no single "rate law" describes how quickly a mineral precipitates or dissolves. The mass action equation, which describes the equilibrium point of a mineral's dissolution reaction, is independent of reaction mechanism. A rate law, on the other hand, reflects our idea of how a reaction proceeds on a molecular scale. Rate laws, in fact, quantify the slowest or "rate-limiting" step in a hypothesized reaction mechanism.

Different reaction mechanisms can predominate in fluids of differing composition, since species in solution can serve to promote or inhibit the reaction mechanism. For this reason, there may be a number of valid rate laws that describe the reaction of a single mineral (e.g., Brady and Walther, 1989). It is not uncommon to find that one rate law applies under acidic conditions, another at neutral pH, and a third under alkaline conditions. We may discover, furthermore, that a rate law measured for reaction with deionized water fails to describe how a mineral reacts with electrolyte solutions.

In studying dissolution and precipitation, geochemists commonly consider that a reaction proceeds in five generalized steps:

1. diffusion of reactants from the bulk fluid to the mineral surface,
2. adsorption of the reactants onto reactive sites,
3. a chemical reaction involving the breaking and creation of bonds,
4. desorption of the reaction products, and
5. diffusion of the products from the mineral surface to the bulk fluid

(e.g., Brezonik, 1994). The adsorption and desorption processes (steps 2 and 4) are almost certainly rapid, so two classes of rate-limiting steps are possible (e.g., Lasaga, 1984). If the reaction rate depends on how quickly reactants can reach the surface by aqueous diffusion and the products can move away from it (steps 1 and 5), the reaction is said to be "transport controlled." If, on the other hand, the speed of the surface reaction (step 3) controls the rate, the reaction is termed "surface controlled." A reaction may be revealed to be transport controlled if its rate in the laboratory varies with stirring speed, or if a low value is found for its activation energy (defined later in this section).

Reactions for common minerals fall in both categories, but many important cases tend, except under acidic conditions, to be surface controlled (e.g., Aagaard and Helgeson, 1982; Stumm and Wollast, 1990). For this reason and because of their relative simplicity, we will consider in this chapter rate laws for surface-controlled reactions. The problem of integrating rate laws for transport-controlled reactions into reaction path calculations, nonetheless, is complex and interesting (Steefel and Lasaga, 1994), and warrants further attention.

Almost all published rate laws for surface-controlled reactions are presented in a form derived from transition state theory (Lasaga, 1981a, 1981b, 1984, 1998; Aagaard and Helgeson, 1982). According to the theory, a mineral dissolves by a mechanism involving the creation and subsequent decay of an *activated complex*, which is less stable (of higher free energy per mole) than either the bulk mineral or product species. The rate at which the activated complex decays controls how quickly the mineral dissolves.

The dissolution rate, according to the theory, does not depend on the mineral's saturation state. The precipitation rate, on the other hand, varies strongly with saturation, exceeding the dissolution rate only when the mineral is supersaturated. At the point of equilibrium, the dissolution rate matches the rate of precipitation so that the net rate of reaction is zero. There is, therefore, a strong conceptual link between the kinetic and thermodynamic interpretations: equilibrium is the state in which the forward and reverse rates of a reaction balance.

To formulate a kinetic reaction path, we consider one or more minerals $A_{\vec{k}}$ whose rates of dissolution and precipitation are to be controlled by kinetic rate laws. We wish to avoid assuming that the minerals $A_{\vec{k}}$ are in equilibrium with the system, so they do not appear in the basis (i.e., $A_{\vec{k}} \notin A_k$). We can write a reaction,

$$A_{\vec{k}} \rightleftarrows \nu_{w\vec{k}} A_w + \sum_i \nu_{i\vec{k}} A_i + \sum_k \nu_{k\vec{k}} A_k + \sum_m \nu_{m\vec{k}} A_m \,, \tag{17.1}$$

for $A_{\vec{k}}$ in terms of the current basis (A_w, A_i, A_k and A_m) and calculate the reaction's equilibrium constant $K_{\vec{k}}$.

Following transition state theory, we can write a rate law giving the dissolution rate $r_{\vec{k}}$ of mineral $A_{\vec{k}}$; the rate is the negative time rate of change of the mineral's mole number $n_{\vec{k}}$. The law takes the form

$$r_{\vec{k}} = -\frac{dn_{\vec{k}}}{dt} = (A_S\, k_+)_{\vec{k}} \prod^{\vec{j}} (m_{\vec{j}})^{p_{\vec{j}\vec{k}}} \left(1 - \frac{Q_{\vec{k}}}{K_{\vec{k}}}\right) . \tag{17.2}$$

Here, A_S is the mineral's surface area (cm^2) and k_+ is the intrinsic rate constant for the reaction. The concentrations of certain species $A_{\vec{j}}$, which make up the rate law's promoting and inhibiting species, are denoted $m_{\vec{j}}$, and $p_{\vec{j}\vec{k}}$ are those species' exponents, the values of which are derived empirically. $Q_{\vec{k}}$ is the activity product for Reaction 17.1 (Eq. 3.41). In the absence of promoting and inhibiting species, the units of the rate constant are mol cm^{-2} s^{-1}, and in any case these are the units of the product of k_+ and the Π term.

The promoting and inhibiting species $A_{\vec{j}}$ are most commonly aqueous species, but may also be mineral, gas, or surface species. For aqueous, mineral, and surface species, $m_{\vec{j}}$ is formally the volumetric concentration, in units such as mol cm^{-3} or mol l^{-1}, but in geochemical modeling we commonly carry this variable as the species' molality. Sometimes, especially when $A_{\vec{j}}$ is H^+ or OH^-, $m_{\vec{j}}$ in this equation is understood to stand for the species' activity rather than its molality. For a gas species, $m_{\vec{j}}$ represents partial pressure or fugacity.

There are three functional parts of Equation 17.2. The first grouping ($A_S\, k_+$) requires that the reaction proceed at a rate proportional to the surface area and the rate constant. The surface area of a sample can be measured by a nitrogen adsorption technique (the BET method) or estimated from geometric considerations, and the rate constant is determined experimentally. It is interesting to note that whenever A_S is zero, the reaction rate vanishes. A mineral that does not exist, therefore, cannot begin to precipitate until crystal nuclei form. Various theories have been suggested for describing the rate at which nuclei might develop spontaneously or on the surfaces of other minerals (e.g., Berner, 1980) and it is possible to integrate the theories into reaction models (Steefel and Van Cappellen, 1990). Considerable uncertainties exist in applying nucleation theory to practical cases, however, and we will not include the theory in the scope of our discussion.

The Π grouping in Equation 17.2 represents the role that species in solution play in promoting the reaction or inhibiting its progress. A species can promote the reaction by catalyzing formation of the activated complex (in which case the corresponding $p_{\vec{j}\vec{k}}$ is positive) or it can inhibit the reaction by impeding its formation ($p_{\vec{j}\vec{k}}$ is negative). The final grouping represents the thermodynamic drive for reaction. When mineral $A_{\vec{k}}$ is supersaturated, $Q_{\vec{k}} > K_{\vec{k}}$ and the mineral precipitates. When the mineral is undersaturated, it dissolves because $Q_{\vec{k}} < K_{\vec{k}}$.

The rate constant in Equation 17.2 can be related to temperature by the phenomenological Arrhenius equation,

$$k_+ = A\mathrm{e}^{-E_\mathrm{A}/RT_\mathrm{K}} \tag{17.3}$$

(e.g., Lasaga, 1981a). Here, A is the pre-exponential factor (mol cm^{-2} s^{-1}), E_A is the activation energy (J mol^{-1}), R is the gas constant (8.3143 J K^{-1} mol^{-1}), and T_K is absolute temperature (K). The values of A and E_A are determined for a given reaction by measuring k_+ at several temperatures and fitting the data in semilog coordinates.

In an example of the forms rate laws can take, we consider the reaction of albite ($NaAlSi_3O_8$), the dissolution of which was studied by Knauss and Wolery (1986). They found that the reaction proceeds according to different rate laws, depending on pH. From their results, we can write a rate law valid at pH values more acidic than about 1.5 as

$$r_\mathrm{alb} = A_\mathrm{S}\, k_+\, a_{\mathrm{H}^+} \left(1 - \frac{Q}{K}\right), \tag{17.4}$$

where at 70 °C, $k_+ = 10^{-12.2}$ mol cm^{-2} s^{-1}. In this case, just one species (H^+) appears in the Π terms of Equation 17.2; the corresponding value of the exponent p_{H^+} is one. In the pH range of about 1.5 to 8, a second reaction mechanism is predominant. In the corresponding rate law,

$$r_\mathrm{alb} = A_\mathrm{S}\, k_+ \left(1 - \frac{Q}{K}\right), \tag{17.5}$$

in which $k_+ = 10^{-15.1}$ mol cm^{-2} s^{-1} at 70 °C, there are no species in the Π terms. The reaction rate, therefore, is not affected by solution composition. A third law,

$$r_\mathrm{alb} = A_\mathrm{S}\, k_+\, a_{\mathrm{H}^+}^{-1/2} \left(1 - \frac{Q}{K}\right), \tag{17.6}$$

with $k_+ = 10^{-19.5}$ at 70 °C, describes the dominant mechanism at higher pH. The exponent p_{H^+} in this case is −0.5.

Some caveats about the form presented for the rate law (Eq. 17.2) are worth noting. First, although Equation 17.2 is linear in $Q_{\vec{k}}$, transition state theory does not demand that rate laws take such a form. There are nonlinear forms of the rate law that are equally valid (e.g., Merino *et al.*, 1993; Lasaga *et al.*, 1994) and that in some cases may be required to explain observations. Specifically, the term $Q_{\vec{k}}/K_{\vec{k}}$ can appear raised to an arbitrary (not necessarily integer) exponent, as can the entire $(1 - Q_{\vec{k}}/K_{\vec{k}})$ term (provided that its original sign is preserved). Such rate expressions have seldom been invoked in geochemistry, if only because most experiments have been designed to study the dissolution reaction under conditions far from equilibrium, where there is no basis for observing nonlinear effects. Nonlinear rate laws can, nonetheless, be readily incorporated into reaction models, as described in Appendix D.

Second, in deriving Equation 17.2 from transition state theory, it is necessary to assume that the overall reaction proceeds on a molecular scale as a single elementary reaction or a series of elementary reactions (e.g., Lasaga, 1984; Nagy *et al.*, 1991). In general, the elementary reactions that occur as a mineral dissolves and precipitates are not known. Thus, even though the form of Equation 17.2 is convenient and broadly applicable for explaining experimental results, it is not necessarily correct in the strictest sense.

17.2 From Laboratory to Application

The great value of kinetic theory is that it frees us from many of the constraints of the equilibrium model and its variants (partial equilibrium, local equilibrium, and so on; see Chapter 2). In early studies (e.g., Lasaga, 1984), geochemists were openly optimistic that the results of laboratory experiments could be applied directly to the study of natural systems. Transferring the laboratory results to field situations, however, has proved to be much more challenging than many first imagined.

Many minerals have been found to dissolve and precipitate in nature at dramatically different rates than they do in laboratory experiments. As first pointed out by Pačes (1983) and confirmed by subsequent studies, for example, albite weathers in the field much more slowly than predicted on the basis of reaction rates measured in the laboratory. The discrepancy can be as large as four orders of magnitude (Brantley, 1992, and references therein). As we calculate in Chapter 30, the measured reaction kinetics of quartz (SiO_2) suggest that water should quickly reach equilibrium with this mineral, even at low temperatures. Equilibrium between groundwater and quartz, however, is seldom observed, even in aquifers composed almost entirely of quartz sand. In a study of microbial activity and weathering on a regional scale, finally, Park *et al.* (2009) deduced rate constants six to seven orders of magnitude smaller than measured in laboratory experiments.

Geochemists (e.g., Aagaard and Helgeson, 1982) commonly attribute such discrepancies to difficulties in representing the surface area A_S of minerals in natural samples. In the laboratory, the mineral is fresh and any surface coatings have been removed. The same mineral in the field, however, may be shielded with oxide, hydroxide, or organic coatings. It may be occluded by contact with other materials, including reaction products, organic matter, and other grains. In addition, the aged surface of the natural sample is probably smoother than the laboratory material and hence contains fewer kinks and sharp edges, which are highly reactive.

Even where it is not occluded, the mineral surface may not be reactive. In the vadose zone, the surface may not be fully in contact with water or may contact water only intermittently. In the saturated zone, a mineral may touch virtually immobile water within isolated portions of the sediment's pore structure. Fluid chemistry in such microenvironments may bear little relationship to the bulk chemistry of the pore water. Since groundwater flow tends to be channeled through the most permeable portions of the subsurface, furthermore, fluids may bypass many or most of the mineral grains in a sediment or rock. The latter phenomenon is especially pronounced in fractured rocks, where only the mineral surfaces lining the fracture may be reactive.

There are other important factors beyond the state of the surface that may lead to discrepancies between laboratory and field studies. Measurement error in the laboratory, first of all, is considerable. Brantley (1992) notes that rate constants determined by different laboratories generally agree to within only a factor of about 30. Agreement to better than a factor of 5, she reasons, might not be an attainable goal.

There is no certainty, furthermore, that the reaction or reaction mechanism studied in the laboratory will predominate in nature. Data for reaction in deionized water, for example, might not apply if aqueous species present in nature promote a different reaction mechanism, or if they inhibit the mechanism that operated in the laboratory. Dove and Crerar (1990), for example, showed that quartz dissolves into dilute electrolyte solutions up to 30 times more quickly than it does in pure water. Laboratory experiments, furthermore, are nearly always conducted under conditions in which the fluid is far from equilibrium with the mineral, although reactions in nature proceed over a broad range of saturation states across which the laboratory results may not apply.

Further error is introduced if reactions distinct from those for which data is available affect the chemistry of a natural fluid. Consider as an example the problem of predicting the silica content of a fluid flowing through a quartz sand aquifer. There is little benefit in modeling the reaction rate for quartz if the more reactive minerals (such as clays and zeolites) in the aquifer control the silica concentration.

Finally, whereas most laboratory experiments have been conducted in largely abiotic environments, the action of bacteria may control reaction rates in nature (e.g., Chapelle, 2001). In the production of acid drainage (see Chapter 35), for example, bacteria such as *Thiobacillus ferrooxidans* control the rate at which pyrite (FeS_2) oxidizes (Taylor *et al.*, 1984; Okereke and Stevens, 1991). Laboratory observations of how quickly pyrite oxidizes in abiotic systems (e.g., Williamson and Rimstidt, 1994, and references therein), therefore, might poorly reflect the oxidation rate in the field.

17.3 Numerical Solution

The procedure for tracing a kinetic reaction path differs from the procedure for paths with simple reactants (Chapter 14) in two principal ways. First, progress in the simulation is measured in units of time t rather than by the reaction progress variable ξ. Second, the rates of mass transfer, instead of being set explicitly by the modeler (Eqs. 14.5–14.7), are computed over the course of the reaction path by a kinetic rate law (Eq. 17.2).

17.3.1 System Composition

From Equations 17.1 and 17.2, we can write the instantaneous rate of change in the system's bulk composition as

$$\frac{dM_w}{dt} = \sum_{\vec{k}} \nu_{w\vec{k}} r_{\vec{k}} \tag{17.7}$$

$$\frac{dM_i}{dt} = \sum_{\vec{k}} \nu_{i\vec{k}} r_{\vec{k}} \tag{17.8}$$

$$\frac{dM_k}{dt} = \sum_{\vec{k}} \nu_{k\vec{k}} r_{\vec{k}} \, . \tag{17.9}$$

In stepping forward from t' to a new point in time t, the instantaneous rate will change as the fluid's chemistry evolves. Rather than carrying the rate at t' over the step, it is more accurate (e.g., Richtmyer, 1957; Peaceman, 1977) to take the average of the rates at t' and t. In this case, the new bulk composition (at t) is given from its previous value (at t') and Equations 17.7–17.9 by

$$M_w(t) = M_w(t') + \frac{(t-t')}{2} \sum_{\vec{k}} \nu_{w\vec{k}} \left[r_{\vec{k}}(t') + r_{\vec{k}}(t) \right] \tag{17.10}$$

$$M_i(t) = M_i(t') + \frac{(t-t')}{2} \sum_{\vec{k}} \nu_{i\vec{k}} \left[r_{\vec{k}}(t') + r_{\vec{k}}(t) \right] \tag{17.11}$$

$$M_k(t) = M_k(t') + \frac{(t-t')}{2} \sum_{\vec{k}} \nu_{k\vec{k}} \left[r_{\vec{k}}(t') + r_{\vec{k}}(t) \right] . \tag{17.12}$$

We use these relations instead of Equations 14.5–14.7 when tracing a kinetic path.

17.3.2 Newton–Raphson Iteration

To solve for the chemical system at t, we use Newton–Raphson iteration to minimize a set of residual functions, as discussed in Chapter 4. For a kinetic path, the residual functions are derived by combining Equations 4.22–4.23 with Equations 17.10–17.11, giving

$$R_w = n_w \left(55.5 + \sum_j \nu_{wj} m_j \right) - M_w(t') - \frac{(t-t')}{2} \sum_{\vec{k}} \nu_{w\vec{k}} \left[r_{\vec{k}}(t') + r_{\vec{k}}(t) \right] \tag{17.13}$$

$$R_i = n_w \left(m_i + \sum_j \nu_{ij} m_j \right) - M_i(t') - \frac{(t-t')}{2} \sum_{\vec{k}} \nu_{i\vec{k}} \left[r_{\vec{k}}(t') + r_{\vec{k}}(t) \right] . \tag{17.14}$$

In order to derive the corresponding Jacobian matrix, we need to differentiate $r_{\vec{k}}$ (Eq. 17.2) with respect to n_w and m_i. Taking advantage of the relations,

$$\frac{\partial m_j}{\partial n_w} = 0 \quad \text{and} \quad \frac{\partial m_j}{\partial m_i} = \frac{\nu_{ij} m_j}{m_i} \tag{17.15}$$

(Eq. 4.25), and

$$\frac{\partial Q_{\vec{k}}}{\partial n_w} = 0 \quad \text{and} \quad \frac{\partial Q_{\vec{k}}}{\partial m_i} = \frac{\nu_{i\vec{k}} Q_{\vec{k}}}{m_i} \tag{17.16}$$

(following from the definition of the activity product; Eq. 3.41), we can show that $\partial r_{\vec{k}} / \partial n_w = 0$ and

$$\frac{\partial r_{\vec{k}}}{\partial m_i} = \frac{(A_S\, k_+)_{\vec{k}}}{m_i} \prod^{\vec{j}} (m_{\vec{j}})^{p_{\vec{j}\vec{k}}} \left[-\frac{\nu_{i\vec{k}} Q_{\vec{k}}}{K_{\vec{k}}} + \left(\sum_{\vec{j}} \nu_{i\vec{j}} p_{\vec{j}\vec{k}} \right) \left(1 - \frac{Q_{\vec{k}}}{K_{\vec{k}}} \right) \right] . \tag{17.17}$$

As discussed in Section 4.3, the entries in the Jacobian matrix are given by differentiating the residual functions (Eqs. 17.13–17.14) with respect to the independent variables n_w and m_i. The resulting entries are

$$J_{ww} = \frac{\partial R_w}{\partial n_w} = 55.5 + \sum_j \nu_{wj} m_j \tag{17.18}$$

$$J_{wi} = \frac{\partial R_w}{\partial m_i} = \frac{n_w}{m_i} \sum_j \nu_{wj} \nu_{ij} m_j - \frac{(t-t')}{2} \sum_{\vec{k}} \nu_{w\vec{k}} \frac{\partial r_{\vec{k}}}{\partial m_i} \tag{17.19}$$

$$J_{iw} = \frac{\partial R_i}{\partial n_w} = m_i + \sum_j \nu_{ij} m_j \tag{17.20}$$

$$J_{ii'} = \frac{\partial R_i}{\partial m_{i'}} = n_w \left(\delta_{ii'} + \frac{1}{m_{i'}} \sum_j \nu_{ij} \nu_{i'j} m_j \right) - \frac{(t-t')}{2} \sum_{\vec{k}} \nu_{i\vec{k}} \frac{\partial r_{\vec{k}}}{\partial m_{i'}} , \tag{17.21}$$

where the values $\partial r_{\vec{k}} / \partial m_i$ are given by Equation 17.17. Equations 17.18–17.21 take the place of Equations 4.26–4.29 in a kinetic model. As before (Eq. 4.30), $\delta_{ii'}$ is the Kronecker delta function.

The time-stepping proceeds as previously described (Chapter 14), with the slight complication that the surface areas A_S of the kinetic minerals must be evaluated after each iteration to account for changing mineral masses. For polythermal paths, each rate constant k_+ must be set before beginning a time step according to the Arrhenius equation (Eq. 17.3) to a value corresponding to the temperature at the new time level.

17.3.3 Time Weighting

The simple average of rates $r_{\vec{k}}(t')$ and $r_{\vec{k}}(t)$, those at the onset and conclusion of a time step, in Equations 17.10–17.12 can be profitably generalized,

$$\frac{1}{2}\left[r_{\vec{k}}(t') + r_{\vec{k}}(t)\right] \quad \rightarrow \quad (1-\theta)\, r_{\vec{k}}(t') + \theta\, r_{\vec{k}}(t)\,, \tag{17.22}$$

as a weighted average using the time weighting parameter $0 \leq \theta \leq 1$. The weighted average in this case needs to be extended into calculation of the residuals (Eqs. 17.13–17.14) and Jacobian entries (Eqs. 17.19 and 17.21).

When θ is zero, the calculation accounts for only the rate $r_{\vec{k}}(t')$ at a step's onset and is hence referred to as a forward-in-time solution. A value of one, on the other hand, recognizes only $r_{\vec{k}}(t)$; the case is then known as backward-in-time. Setting θ to one-half, which reverts weighting to a simple average, leads to a centered-in-time solution. In the context of integrating the rate law as an ordinary differential equation, the three cases—θ of zero, one-half, and one—amount to the Euler, trapezoidal, and backward Euler methods, respectively (e.g., Atkinson *et al.*, 2009).

The case $\theta = 1/2$, the simple average, is notable in that it minimizes the truncation error inherent in integrating a rate law numerically (see also Section 23.4.4). Adjusting θ toward one, however, works to improve stability of the numerical procedure and can allow for longer time steps. For this reason, the modeler may prefer to set a time weighting of perhaps 0.6 or 0.7.

Setting θ to zero, nonetheless, can be advantageous. Taking reaction rate into account at only the onset of a time step simplifies the calculations, especially computation of the Jacobian matrix, expediting code development. Also, rates need to be computed just once per time step, rather than upon each Newton–Raphson iteration, potentially speeding program execution, at least on a per-time-step basis.

Forward-in-time weighting, furthermore, may be favored in especially nonlinear cases, such as those sometimes encountered modeling microbial kinetics (see Chapter 19). A microbially catalyzed reaction proceeds at a rate that depends on biomass concentration, for example, yet the biomass grows and decays in response to reaction rate; the interplay can give rise to significant nonlinearity. Many reaction and reactive transport codes, for these reasons, employ forward-in-time weighting; the GWB programs allow the user to set time weighting arbitrarily.

17.4 Example Calculations

In an example of a kinetic reaction path, we calculate how quartz sand reacts at 100 °C with deionized water. According to Rimstidt and Barnes (1980), quartz reacts according to the rate law

$$r_{\text{qtz}} = A_S\, k_+ \left(1 - \frac{Q}{K}\right)\,, \tag{17.23}$$

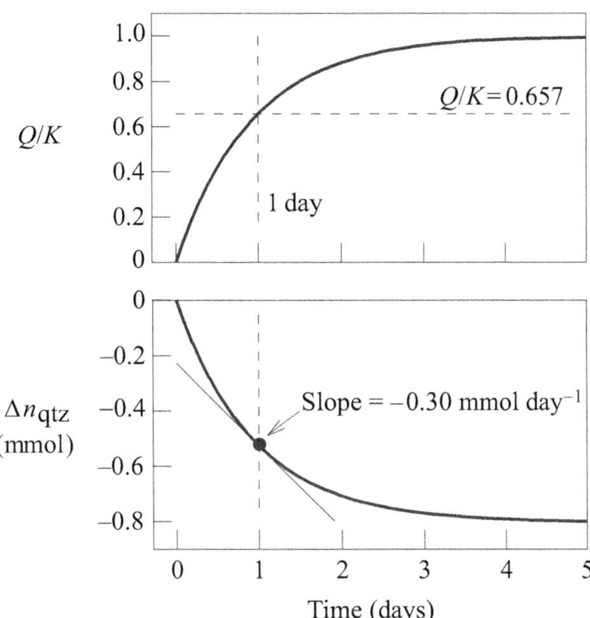

Figure 17.1 Results of reacting quartz sand at 100 °C with deionized water, calculated according to a kinetic rate law. Top diagram shows how the saturation state Q/K of quartz varies with time; bottom plot shows change in amount (mmol) of quartz in system (bold line). The slope of the tangent to the curve (fine line) is the instantaneous reaction rate, the negative of the dissolution rate, shown at 1 day of reaction.

with a rate constant k_+ at this temperature of about 2×10^{-15} mol cm^{-2} s^{-1}. From their data, we assume that the sand has a specific surface area of 1000 cm^2 g^{-1}.

In REACT, we set a system containing 1 kg of solvent water and 5 kg of quartz

```
time begin = 0 days, end = 5 days
T = 100
SiO2(aq) =  1 umolal
react 5000 g Quartz
kinetic Quartz rate_con = 2.0e-15  surface = 1000
go
```

and allow it to react for 5 days. Figure 17.1 shows the calculation results. Quartz dissolves with time, adding a small amount of SiO_2(aq) to solution. The increasing silica content of the solution causes the saturation state Q/K of quartz to increase, slowing the reaction rate. After 5 days, the reaction has approached equilibrium ($Q/K \simeq 1$) and the reaction nearly ceases.

We can calculate the reaction rate according to Equation 17.23 and compare it with the calculation results. The required quantities at 1 day of reaction are

$$\begin{aligned} A_S &= 5000 \text{ g} \times 1000 \text{ cm}^2 \text{ g}^{-1} = 5 \times 10^6 \text{ cm}^2 \\ k_+ &= 2 \times 10^{-15} \text{ mol cm}^{-2} \text{ s}^{-1} \\ Q/K &= 10^{-0.182} = 0.657\,. \end{aligned} \tag{17.24}$$

Entering these values into the rate equation (Eq. 17.23) gives the dissolution rate

$$\begin{aligned} r_{qtz} &= (5 \times 10^6 \text{ cm}^2) \times (2 \times 10^{-15} \text{ mol cm}^{-2} \text{ s}^{-1}) \times (1 - 0.657) \\ &= 3.43 \times 10^{-9} \text{ mol s}^{-1} = 0.30 \text{ mmol day}^{-1} . \end{aligned} \tag{17.25}$$

We can confirm that on a plot of the mole number n_{qtz} for quartz versus time (Fig. 17.1), this value is the slope of the tangent line and hence the dissolution rate $-dn_{qtz}/dt$ we expect.

In a slightly more complicated example, we calculate the rate at which albite dissolves at 70 °C into an acidic NaCl solution. We use the rate law shown in Equation 17.4, which differs from Equation 17.23 by the inclusion of the a_{H^+} term, and a rate constant of 6.3×10^{-13} mol cm^{-2} s^{-1} determined for this temperature by Knauss and Wolery (1986). We set pH to a constant value of 1.5, as though this value were maintained by an internal buffer or an external control such as a pH-stat, and allow 250 g of the mineral to react with 1 kg of solvent water for 30 days.

The procedure in REACT is

```
time begin = 0 days, end = 30 days
T = 70

pH = 1.5
0.1 molal Cl-
0.1 molal Na+
1 umolal SiO2(aq)
1 umolal Al+++

react 250 grams of "Albite low"
kinetic "Albite low" rate_con = 6.3e-13, apower(H+) = 1, surface = 1000
fix pH

go
```

We can quickly verify the calculation results (Fig. 17.2). Choosing day 15 of the reaction,

$$\begin{aligned} A_S &= 250 \text{ g} \times 1000 \text{ cm}^2 \text{ g}^{-1} = 2.5 \times 10^5 \text{ cm}^2 \\ k_+ &= 6.3 \times 10^{-13} \text{ mol cm}^{-2} \text{ s}^{-1} \\ a_{H^+} &= 10^{-1.5} = 3.16 \times 10^{-2} \\ (1 - Q/K) &= (1 - 10^{-10.1}) \simeq 1 . \end{aligned} \tag{17.26}$$

Substituting into Equation 17.4,

$$\begin{aligned} r_{alb} &= (2.5 \times 10^5 \text{ cm}^2) \times (6.3 \times 10^{-13} \text{ mol cm}^{-2} \text{ s}^{-1}) \times (3.16 \times 10^{-2}) \times (1) \\ &= 5.0 \times 10^{-9} \text{ mol s}^{-1} = 0.43 \text{ mmol day}^{-1} , \end{aligned} \tag{17.27}$$

we find the dissolution rate $-dn_{alb}/dt$ shown in Figure 17.2. Note that the fluid in this example remains so undersaturated with respect to albite that the value of Q/K is nearly zero and hence has virtually no influence on the reaction rate. For this reason, the reaction rate remains nearly constant over the course of the calculation.

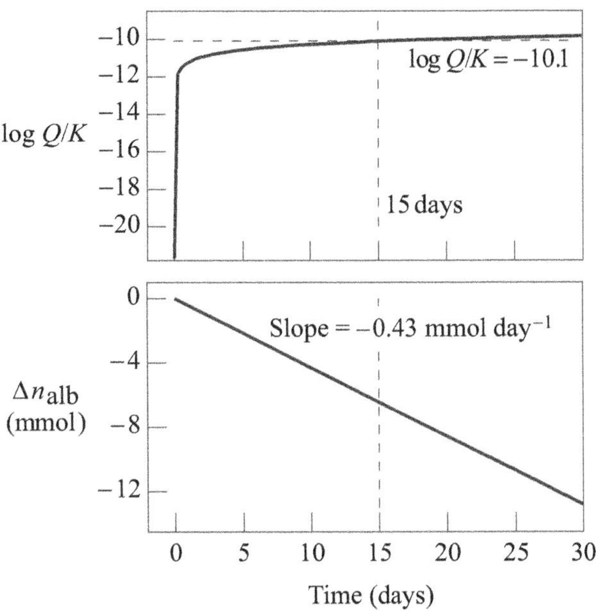

Figure 17.2 Results of reacting albite at 70 °C with an NaCl solution maintained at pH 1.5, calculated as a kinetic reaction path. Top diagram shows how the saturation index of albite varies with time; bottom plot shows change in amount (mmol) of albite.

17.5 Modeling Strategy

A practical consideration in reaction modeling is choosing the extent to which reaction kinetics should be integrated into the calculations. On the one hand, kinetic theory is an important generalization of the equilibrium model that lets us account for the fact that fluids and minerals do not necessarily coexist at equilibrium. On the other hand, the theory can add considerable complexity to developing and evaluating a reaction model.

We might take a purist's approach and attempt to use kinetic theory to describe the dissolution and precipitation of each mineral that might appear in the calculation. Such an approach, although appealing and conceptually correct, is seldom practical. The database required to support the calculation would have to include rate laws for every possible reaction mechanism for each of perhaps hundreds of minerals. Even unstable minerals that can be neglected in equilibrium models would have to be included in the database, since they might well form in a kinetic model (see Section 30.4, Ostwald's Step Rule). If we are to allow new minerals to form, furthermore, it will be necessary to describe how quickly each mineral can nucleate on each possible substrate.

The modeling software would have to trace a number of reactions occurring at broadly different rates. Although certainly feasible, such a calculation can present practical difficulties, especially at the onset of a reaction path, if the software must take very small steps to accurately trace the progress of the faster reactions. For each calculation, furthermore, we would need to be able to set initial conditions, requiring knowledge of hydrologic factors (flow rates, residence times, and so on) that in real life are not always available.

A worthwhile strategy for conceptualizing kinetic reaction paths is to divide mineral reactions into three groups. The first group contains the reactions that proceed quickly over the time span of the calculation.

We can safely assume that these minerals remain in equilibrium with the fluid. A second group consists of minerals that react negligibly over the calculation and hence may be ignored or "suppressed." The reactions for the remaining minerals fall in the third group, for which we need to account for reaction kinetics. Our time is best spent attempting to define the rates at which minerals in the latter group react.

18

Redox Kinetics

Reaction kinetics enter into a geochemical model, as we noted in the previous chapter, whenever a reaction proceeds quickly enough to affect the distribution of mass, but not so quickly that it reaches the point of thermodynamic equilibrium. In Part I of this book, we considered two broad classes of reactions that in geochemistry commonly deviate from equilibrium.

The first class, discussed in detail in Chapter 6, was reaction between a fluid and the minerals it contacts. The kinetics of the reactions by which minerals dissolve and precipitate was the subject of the preceding chapter (Chapter 17). The second class of reactions commonly observed to be in disequilibrium in natural waters, as discussed in Chapter 7, is redox reactions. The subject of this chapter is modeling the rates at which redox reactions proceed within the aqueous solution, or when catalyzed on a mineral surface or by the action of an enzyme. In the following chapter (Chapter 19), we consider the related question of how rapidly redox reactions proceed when catalyzed in the geosphere by the action of microbial life.

Kinetic redox reactions are simulated within the context of a redox disequilibrium model, a geochemical model constructed to account for disequilibrium among species of differing redox state, as described in Chapter 7. In such models, one or more redox coupling reactions are disabled. It is worth noting that the development here, although cast in terms of redox reactions, can be applied equally well to describe the kinetics of other reaction types, such as the formation and decomposition of complex species. If a complex species is set in the thermodynamic database to result from a coupling reaction, that couple can be disabled and the species' creation or destruction in a simulation controlled by a kinetic rate law, even though the reaction in fact involves neither oxidation nor reduction.

18.1 Rate Laws for Oxidation and Reduction

The rate law for a redox reaction predicts the rate at which a species is transformed by a specific reaction mechanism, or combination of mechanisms, from one oxidation state to another. For example, Wehrli and Stumm (1988, 1989) studied the oxidation of vanadyl (V^{IV}) to vanadate (V^{V}) by reaction with dissolved O_2.

They showed this process can proceed in aqueous solution by the oxidation of the hydroxy complex $VO(OH)^+$, according to the reaction

$$VO(OH)^+ + {}^1\!/_4\, O_2(aq) + {}^3\!/_2\, H_2O \rightarrow H_2VO_4^- + 2\,H^+ \,, \tag{18.1}$$

and the oxidation rate is given by the rate law

$$\begin{aligned} r^+ &= -\left(\frac{dM_{V^{IV}}}{dt}\right)^+ = \left(\frac{dM_{V^V}}{dt}\right)^+ \\ &= n_w \; k_+ \; m_{H^+}^{-1} \; m_{VO^{++}} \; m_{O_2(aq)} \, , \end{aligned} \tag{18.2}$$

where VO^{++} is the predominant vanadyl species at pH < 6. In this equation, r^+ is the rate (mol s^{-1}) at which the reaction proceeds in the forward direction, n_w is the number of kg of solvent water, k_+ is the rate constant (s^{-1}, in this case), M is a component mole number, and m represents a species' molal concentration.

A set of concentration factors appears for the promoting and inhibiting species: H^+, VO^{++}, and O_2(aq), in this case. Promoting and inhibiting species may, but do not necessarily, appear in the redox reaction. As noted in the previous chapter, m is formally set in volumetric units, but in geochemical modeling is commonly taken as the species' molality. The rate constant k_+ has units such that multiplying it by the molalities of the promoting and inhibiting species gives a value in molal s^{-1}.

To trace a reaction path incorporating redox kinetics, we first set a model in redox disequilibrium by disabling one or more redox couples, then specify the reaction in question and the rate law by which it proceeds. To model the progress of Reaction 18.1, for example, we would disable the redox couple between vanadyl and vanadate species. In a model of the oxidation of Fe^{++} by manganite (MnOOH), we would likely disable the couples for both iron and manganese.

The redox reaction may include solvent water, basis and secondary aqueous species, mineral and gas species, and uncomplexed and complexed surface sites. Each such reaction, denoted $\vec{r}$, can be written in a generalized form as

$$0 \rightarrow \nu_{w\vec{r}} A_w + \sum_i \nu_{i\vec{r}} A_i + \sum_j \nu_{j\vec{r}} A_j + \sum_l \nu_{l\vec{r}} A_l + \sum_n \nu_{n\vec{r}} A_n + \sum_p \nu_{p\vec{r}} A_p + \sum_q \nu_{q\vec{r}} A_q \, . \tag{18.3}$$

As before, species with negative reaction coefficients ν are reactant species, which in a less abstract rendering would appear with positive coefficients on the left side of the reaction, and those with positive coefficients are product species.

The reaction has an associated equilibrium constant $K_{\vec{r}}$, which is understood to carry Boltzman terms accounting for electrostatic effects of any surface complexation reactions embedded within the redox reaction. The reaction's activity product at any point in the simulation is given by

$$Q_{\vec{r}} = a_w^{\nu_{w\vec{r}}} \prod^i (\gamma_i m_i)^{\nu_{i\vec{r}}} \prod^j (\gamma_j m_j)^{\nu_{j\vec{r}}} \prod^l a_l^{\nu_{l\vec{r}}} \prod^n f_n^{\nu_{n\vec{r}}} \prod^p m_p^{\nu_{p\vec{r}}} \prod^q m_q^{\nu_{q\vec{r}}} \, , \tag{18.4}$$

from the activities, fugacities, and molalities of the various reactant and product species. The activity a_l of a mineral species of fixed composition is unity, and hence the product function over l is carried here as a formality.

Experimental studies of reaction kinetics are commonly carried out under conditions far from equilibrium, where the thermodynamic drive is strong and forward reaction overwhelms reverse. Rate laws reported in the literature, such as Equation 18.2, therefore, generally represent the forward reaction rate. The laws can be expressed in the general form

$$r_{\vec{r}}^+ = n_w \; (k_+)_{\vec{r}} \prod^{\vec{j}} (m_{\vec{j}})^{P_{\vec{j}\vec{r}}} \, , \tag{18.5}$$

where $\vec{j}$ indexes the promoting and inhibiting species, described in the previous chapter (Chapter 17). Rate here is expressed in moles of the turnover of Reaction 18.3, per second.

In geochemical modeling, we prefer to use rate laws that predict the net rather than the forward reaction rate, to avoid the possibility of a reaction running past the point of equilibrium and continuing in a simulation, impossibly, against the thermodynamic drive. The net reaction rate $r_{\vec{r}}$ is the difference between the forward rate, given by the rate law above, and the rate at which reaction proceeds in the reverse direction,

$$r_{\vec{r}} = r_{\vec{r}}^{+} - r_{\vec{r}}^{-} \,. \tag{18.6}$$

An important result of nonequilibrium thermodynamics (Boudart, 1976) is that the ratio of the forward and reverse rates of a chemical reaction varies with the reaction's free-energy change according to

$$\frac{r_{\vec{r}}^{+}}{r_{\vec{r}}^{-}} = \mathrm{e}^{-(\omega\,\Delta G_{\vec{r}}/RT_{\mathrm{K}})} \,. \tag{18.7}$$

Here, $\Delta G_{\vec{r}}$ is the free-energy change of Reaction 18.3 (kJ mol^{-1}), R is the gas constant (8.3143 J K^{-1} mol^{-1}), and T_{K} is absolute temperature (K). Factor ω is the reciprocal of the *average stoichiometric number*, which can be taken as the number of times the rate-determining step in Reaction 18.3 occurs per turnover of the reaction (Jin and Bethke, 2005).

According to this equation, a reaction proceeds forward where its free-energy change is negative, and backward where the change is positive. At its equilibrium point, where the free-energy change is zero, the reaction proceeds forward and backward at equal rates. This is a statement of Tolman's *principle of microscopic reversibility* (see Brezonik, 1994; Lasaga, 1998).

Since, from the definitions of the ion activity product and equilibrium constant (Chapter 3),

$$\Delta G_{\vec{r}} = RT_{\mathrm{K}} \ln \frac{Q_{\vec{r}}}{K_{\vec{r}}} \,, \tag{18.8}$$

we can combine Equations 18.6 and 18.7 with the forward rate law 18.5 to give

$$r_{\vec{r}} = n_w \, (k_+)_{\vec{r}} \prod^{\vec{j}} (m_{\vec{j}})^{P_{\vec{j}\vec{r}}} \left[1 - \left(\frac{Q_{\vec{r}}}{K_{\vec{r}}} \right)^{\omega} \right] . \tag{18.9}$$

We see the addition of a thermodynamic term to Equation 18.5 gives the net reaction rate, and hence a rate law of more general use in geochemical modeling than a law describing only forward reaction.

It is worth noting that the values of both k_+ and ω in Equation 18.9 depend on how the kinetic reaction (Reaction 18.3) is written. If we were to arbitrarily double each of the reaction's coefficients, the value of the rate constant k_+ would be cut in half, because twice as many of the reactant species would be consumed, and twice as many product species produced, per reaction turnover. The rate determining step, furthermore, would occur twice as often per reaction turnover, doubling the average stoichiometric number and requiring ω to be halved as well.

18.2 Heterogeneous Catalysis

Contact with a mineral surface can in many cases allow a redox reaction to proceed at a rate considerably greater than attainable within an aqueous solution itself. The catalyzing mineral sorbs the electron donating and accepting species, then, within its structure or along its surface, conducts electrons from one to the other.

Where electron transfer by this pathway proceeds more rapidly than via a direct transfer in solution between colliding molecules, the redox reaction proceeds preferentially by heterogeneous catalysis.

Catalysis by such a mechanism can be accounted for in a kinetic rate equation by including as a factor the catalyzing mineral's surface area. Sung and Morgan (1981), for example, in studying the oxidation on Mn^{II} by dissolved dioxygen,

$$Mn^{++} + 1/4\, O_2(aq) + 3/2\, H_2O \rightarrow MnOOH(s) + 2\, H^+ \,, \tag{18.10}$$

in the presence of a ferric oxide surface, suggested a rate law

$$r^+ = -\left(\frac{dM_{Mn^{II}}}{dt}\right)^+ = A_S\, k_+\, m^2_{OH^-}\, m_{O_2(aq)}\, m_{Mn^{++}} \,, \tag{18.11}$$

where A_S is the ferric surface area (cm^2).

Following development in the previous section, rate laws of this type can be written in a general form,

$$r_{\vec{r}} = (A_S\, k_+)_{\vec{r}} \prod^{\vec{j}}_{j} (m_{\vec{j}})^{p_{\vec{j}\vec{r}}} \left[1 - \left(\frac{Q_{\vec{r}}}{K_{\vec{r}}}\right)^{\omega}\right] , \tag{18.12}$$

where the rate constant k_+ has units of mol cm^{-2} s^{-1}, divided by the units of the product term. As was the case for Equation 18.9, this equation predicts the net rate of reaction, the difference between the forward and reverse rates.

To incorporate into a geochemical model a rate law of this form, it is common practice to specify the specific surface area (cm^2 g^{-1}) of the catalyzing mineral. The mineral's surface area A_S, then, is determined over the course of the simulation as the product of the specific surface area and the mineral's mass.

A more robust way to write a rate law for a catalytically promoted reaction is to include the concentrations of one or more surface complexes, in place of the surface area A_S. In this case, the simulation can account not only for the catalyzing surface area, since the mass of a surface complex varies with the area of the sorbing surface, but the effects of pH, competing ions, and so on.

Liger *et al.* (1999), for example, studied the reduction of uranyl (U^{VI}) by dissolved ferrous iron,

$$UO_2^{++} + 2\, Fe^{++} + 2\, H_2O \rightarrow U(OH)_4 + 2\, Fe^{+++} \,, \tag{18.13}$$

in the presence of hematite (Fe_2O_3). They found that the reaction rate in their experiments varied directly with the masses of the hydroxy species UO_2OH^+ and $FeOH^+$ sorbed to the hematite surface, according to the rate law

$$r^+ = -\left(\frac{dM_{U^{VI}}}{dt}\right)^+ = n_w\, k_+\, m_{>FeO\cdot UO_2OH}\, m_{>FeO\cdot FeOH} \,, \tag{18.14}$$

where $>FeO \cdot UO_2OH$ and $>FeO \cdot FeOH$ are the species bonded to a deprotonated surface site, $>FeO^-$. The uranyl, in fact, may bond to a bidentate site as $(>FeO)_2 \cdot UO_2$, but using the unidentate site provides an equally good fit to the experimental results.

The rate law is of the form of Equation 18.5 in the previous section, and the equivalent law giving the net reaction rate is Equation 18.9. We can, therefore, account for the effect of catalysis on a redox reaction using the same formulation as the case of homogeneous reaction, if we include surface complexes among the promoting and inhibiting species. In Chapter 32, we consider in detail how this law can be integrated into a reaction path simulation.

The example of uranyl reduction shows the utility of this approach. The concentrations of the two surface complexes vary strongly with pH, and this variation explains the observed effect of pH on reaction rate, using a single value for the rate constant k_+. If we had chosen to let the catalytic rate vary with surface area, according to 18.12, we could not reproduce the pH effect, even using H^+ and OH^- as promoting and inhibiting species (since the concentration of a surface species depends not only on fluid composition, but the number of surface sites available). We would in this case need to set a separate value for the rate constant at each pH considered, which would be inconvenient.

18.3 Enzymes

An *enzyme* is an organic molecule, specifically a protein, that catalyzes a chemical reaction. Or, more simply, an enzyme is a biological catalyst. The enzyme binds with a reactant species, known as the *substrate*, to form a reactive intermediate species. While bound, the substrate can transform to a reaction product that would not form, or would form very slowly, in the absence of the enzyme.

It is in principle possible for a free enzyme to promote reaction in a geochemical system, but enzyme kinetics are invoked in geochemical modeling most commonly to describe the effect of microbial metabolism. Microbes are sometimes described from a geochemical perspective as self-replicating enzymes. This is of course a considerable simplification of reality, as we will discuss in the following chapter (Chapter 19), since even the simplest metabolic pathway involves a series of enzymes.

Nonetheless, if the microbial population is steady and geochemical conditions such as pH are controlled, an enzymatic model can be appropriate. Bekins *et al.* (1998), for example, considered how the mineralization of phenol,

$$\underset{\textit{phenol}}{C_6H_5OH(aq)} + {}^{13}\!/_2\, H_2O \rightarrow {}^{7}\!/_2\, \underset{\textit{methane}}{CH_4(aq)} + {}^{5}\!/_2\, HCO_3^- + {}^{5}\!/_2\, H^+ , \tag{18.15}$$

by methanogens in a laboratory experiment might be cast as an enzymatically promoted reaction, according to the rate law

$$r^+ = n_w\, r_{\max} \frac{m_{C_6H_5OH(aq)}}{m_{C_6H_5OH(aq)} + K_A} . \tag{18.16}$$

They used a value of 4.1×10^{-11} molal s^{-1} for $r_{\max}$, the maximum reaction rate, and 1.4×10^{-5} molal for K_A, the half-saturation constant. We consider application of this kinetic law in detail in Chapter 32.

Enzymatic catalysis can be represented more generally by the reaction between a substrate A and enzyme E,

$$A + E \underset{k_2}{\overset{k_1}{\rightleftarrows}} AE \overset{k_3}{\rightarrow} P + E , \tag{18.17}$$

to give the reactive intermediate AE, which can react to give the product species P, releasing the enzyme unchanged. Here k_1, k_2, and k_3 are the rate constants for the constituent reactions in the direction shown, assuming the reactions behave as though they are elementary.

The total enzyme concentration is the sum of the concentrations of the free and bound forms, E and AE, and the ratio of the latter values depends only on the substrate concentration and the three rate constants in 18.17. From these observations, it follows that for an enzymatically promoted kinetic reaction $\vec{r}$,

$$r^+_{\vec{r}} = n_w\, (k_+)_{\vec{r}}\, m_E \frac{m_A}{m_A + K_{A\vec{r}}} \tag{18.18}$$

gives the forward reaction rate (e.g., Brezonik, 1994; Lasaga, 1998). Here, m_E is the total enzyme concentration, the sum of the concentrations of E and AE, and $K_{A\vec{r}}$ is the half-saturation constant (molal^{-1}) describing binding between enzyme and substrate.

This relation is the broadly known *Michaelis–Menten equation*. The effect of substrate concentration m_A on the rate predicted by this equation follows a characteristic pattern. Where substrate concentration is considerably smaller than the half-saturation constant ($m_A \ll K_{A\vec{r}}$), most of the enzyme is present in its free form E and the concentration of the reactive intermediate EA depends on the availability of the substrate A. In this case, $(m_A + K_{A\vec{r}}) \simeq K_{A\vec{r}}$ and reaction rate r^+ given by 18.18 is proportional to m_A. For the opposite case, ($m_A \gg K_{A\vec{r}}$), little free enzyme E is available to complex with A. Now, $(m_A + K_{A\vec{r}}) \simeq m_A$ and reaction rate is invariant with respect to substrate concentration.

The enzyme concentration m_E in a geochemical application is likely to represent in a general way the amount of biomass in the system. It is not common to attempt to determine values for k_+ and m_E separately; instead, the product r_{max} of the two variables is generally reported, as was the case for Equation 18.16. As well as this, the value of $K_{A\vec{r}}$, formally given from the rate constants as $(k_2 + k_3)/k_1$, is observed by laboratory experiment.

In geochemical modeling, as already noted, we prefer rate laws giving the net rather than forward reaction rate. Writing Reaction 18.17 to account for reverse reaction,

$$\mathrm{A} + \mathrm{E} \underset{k_2}{\overset{k_1}{\rightleftarrows}} \mathrm{AE} \underset{k_4}{\overset{k_3}{\rightleftarrows}} \mathrm{P} + \mathrm{E}\,, \tag{18.19}$$

the net reaction rate is given by

$$r_{\vec{r}} = n_w\,(k_+)_{\vec{r}}\; m_E \frac{\dfrac{m_A}{K_{A\vec{r}}}}{1 + \dfrac{m_A}{K_{A\vec{r}}} + \dfrac{m_P}{K_{P\vec{r}}}} \left[1 - \left(\frac{Q_{\vec{r}}}{K_{\vec{r}}}\right)^{\omega}\right] \tag{18.20}$$

(e.g., Cox, 1994), where $K_{P\vec{r}}$ (molal^{-1}) is the half-saturation constant for the reverse reaction, equal to $(k_2 + k_3)/k_4$. For consistency with previous development, we can include the possibility of promoting and inhibiting species, giving the rate law for an enzymatic reaction,

$$r_{\vec{r}} = n_w\,(k_+)_{\vec{r}} \prod^{\vec{j}} (m_{\vec{j}})^{P_{\vec{j}\vec{r}}}\; m_E \frac{\dfrac{m_A}{K_{A\vec{r}}}}{1 + \dfrac{m_A}{K_{A\vec{r}}} + \dfrac{m_P}{K_{P\vec{r}}}} \left[1 - \left(\frac{Q_{\vec{r}}}{K_{\vec{r}}}\right)^{\omega}\right], \tag{18.21}$$

in final form.

18.4 Numerical Solution

Tracing a kinetic redox path is a matter of redistributing mass among the basis entries, adding mass, for example, to oxidized basis entries at the expense of reduced entries. The stoichiometry of the mass transfer is given by the kinetic reaction 18.3, and the transfer rate $r_{\vec{r}}$ is determined by the associated rate law (Eq. 18.9, 18.12, or 18.21).

A set of mass transfer coefficients $\vec{\nu}$ describes how many of each of the basis entries are created or destroyed (in which case the coefficient is negative) per turnover of Reaction 18.3. The coefficients are given by

$$\begin{aligned}\vec{\nu}_{w\vec{r}} = \nu_{w\vec{r}} + \sum_j \nu_{j\vec{r}}\nu_{wj} + \sum_l \nu_{l\vec{r}}\nu_{wl} + \\ \sum_n \nu_{n\vec{r}}\nu_{wn} + \sum_p \nu_{p\vec{r}}\nu_{wp} + \sum_q \nu_{q\vec{r}}\nu_{wq}\end{aligned} \tag{18.22}$$

$$\begin{aligned}\vec{\nu}_{i\vec{r}} = \nu_{i\vec{r}} + \sum_j \nu_{j\vec{r}}\nu_{ij} + \sum_l \nu_{l\vec{r}}\nu_{il} + \\ \sum_n \nu_{n\vec{r}}\nu_{in} + \sum_p \nu_{p\vec{r}}\nu_{ip} + \sum_q \nu_{q\vec{r}}\nu_{iq}\end{aligned} \tag{18.23}$$

$$\begin{aligned}\vec{\nu}_{k\vec{r}} = \sum_j \nu_{j\vec{r}}\nu_{kj} + \sum_l \nu_{l\vec{r}}\nu_{kl} + \\ \sum_n \nu_{n\vec{r}}\nu_{kn} + \sum_p \nu_{p\vec{r}}\nu_{kp} + \sum_q \nu_{q\vec{r}}\nu_{kq}\,.\end{aligned} \tag{18.24}$$

As described in Chapter 3, ν_{wj} and so on are the reaction coefficients by which species are made up from the current basis entries. Mass transfer coefficients are not needed for gases in the basis, because no accounting of mass balance is maintained on the external buffer, and the coefficients for the mole numbers M_p of the surface sites are invariably zero, since sites are neither created nor destroyed by a properly balanced reaction.

Comparing the development here to the accounting for the kinetics of mineral precipitation and dissolution presented in the previous chapter (Chapter 17), we see that the mass transfer coefficients $\vec{\nu}_{w\vec{r}}$ and so on serve a function parallel to the coefficients $\nu_{w\vec{r}}$, etc., in Reaction 17.1. The rates of change in the mole number of each basis entry, accounting for the effect of each kinetic redox reaction carried in the simulation, for example,

$$\frac{dM_w}{dt} = \sum_{\vec{r}} \vec{\nu}_{w\vec{r}}\, r_{\vec{r}} \tag{18.25}$$

$$\frac{dM_i}{dt} = \sum_{\vec{r}} \vec{\nu}_{i\vec{r}}\, r_{\vec{r}} \tag{18.26}$$

$$\frac{dM_k}{dt} = \sum_{\vec{r}} \vec{\nu}_{k\vec{r}} r_{\vec{r}}\,, \tag{18.27}$$

are given by substituting the mass transfer coefficients into Equations 17.7–17.9, in place of the reaction coefficients.

Similar substitution into Equations 17.10–17.12 gives masses of the basis entries at the end of a time step, Equations 17.13–17.14 yield the residual functions, and Equations 17.18–17.21 give the entries in the Jacobian matrix. In evaluating the Jacobian, the derivatives $\partial r_{\vec{r}}/\partial n_w$ and $\partial r_{\vec{r}}/\partial m_i$ can be obtained by differentiating the appropriate rate law (Eq. 18.9, 18.12, or 18.21), as discussed in Appendix D, or their values determined just as efficiently by finite differences.

18.5 Example Calculation

As an example, we consider the reduction by dihydrogen sulfide (H_2S) of hexavalent chromium (Cr^{VI}) to its trivalent form (Cr^{III}). Kim *et al.* (2001) report that when pH is in the range 6.5–10, this process can occur via a reaction,

$$2\,\mathrm{CrO_4^{--}} + 3\,\mathrm{H_2S} + 4\,\mathrm{H^+} \rightarrow 2\,\mathrm{Cr(OH)_3(s)} + 3\,\mathrm{S(s)} + 2\,\mathrm{H_2O}\,, \tag{18.28}$$

that produces a Cr^{III} hydroxide precipitate and native sulfur. The reaction proceeds according to a second-order rate law,

$$r^+ = n_w\, k_+\, m_{\mathrm{H_2S(aq)}}\, m_{\mathrm{CrO_4^{--}}}\,, \tag{18.29}$$

which is first order with respect to both $H_2S(aq)$ and CrO_4^{--}. The rate constant k_+ they found, expressed here accounting for the reduction of two chromiums per reaction turnover, is 0.32 $\mathrm{molal^{-1}\ s^{-1}}$.

As pH increases over the range studied, $H_2S(aq)$ reacts to form HS^-. The reaction rate observed, and that predicted by Equation 18.29, decreases sharply. Thermodynamics strongly favors forward progress of the reaction, so reverse reaction is insignificant and $r \simeq r^+$. The $(1 - Q/K)$ term in the overall rate law (Eq. 18.9), although formally appended to Equation 18.29 and carried in the calculation, therefore, remains very close to unity and the effect of thermodynamics on the reaction rate in this case is negligible.

In setting up a reaction path, we find there is no entry in the "thermo.tdat" database for $Cr(OH)_3(s)$. To write the kinetic reaction, we can use the mineral Cr_2O_3 as a proxy, since it is the dehydrated form of the hydroxide phase. This substitution alters the reaction's free-energy yield, but forward progress is favored so strongly that the reaction rate predicted is not affected. If this were not the case, we would need to add to the database a mineral $Cr(OH)_3(s)$ of appropriate stability.

A second complication is that we would like to decouple zero-valent sulfur from the element's other redox states, since Reaction 18.28 produces native sulfur, but the database does not include such a coupling reaction. Situations of this nature are not uncommon, occurring when an element in a certain oxidation state is stable as a solid, but no corresponding aqueous species occurs under geochemical conditions. To work the problem, we invent a fictitious zero-valent species S(aq) with an arbitrarily low stability. Setting $\log K$ for the reaction

$$\mathrm{S(s)} \rightleftarrows \mathrm{S(aq)} \tag{18.30}$$

to −9, the species will occur in the simulation at only nmolal levels. Combining this reaction with the reaction for native sulfur ("Sulfur-rhmb" in the database),

$$\mathrm{S(s)} + 3/2\,\mathrm{O_2(aq)} + \mathrm{H_2O} \rightleftarrows \mathrm{SO_4^{--}} + 2\,\mathrm{H^+}\,, \tag{18.31}$$

which has a $\log K$ of 93.23 at 25 °C, gives

$$\mathrm{S(aq)} + 3/2\,\mathrm{O_2(aq)} + \mathrm{H_2O} \rightleftarrows \mathrm{SO_4^{--}} + 2\,\mathrm{H^+}\,, \tag{18.32}$$

for which $\log K = 102.23$. We include Reaction 18.32 in the database as a redox couple and change the reaction for "Sulfur-rhmb" to Reaction 18.30, saving the resulting file as "thermo+S0.tdat".

Using the modified thermodynamic database, we simulate reaction over 300 minutes in a fluid buffered to a pH of 7. We prescribe a redox disequilibrium model by disabling redox couples for chromium and sulfur. We set 10 mmolal NaCl as the background electrolyte, initial concentrations of 200 µmolal for Cr^{VI} and 800 µmolal for H_2S, and small initial masses of Cr_2O_3 and S(aq). Finally, we set Equation 18.29 as the rate law and specify that pH be held constant over the simulation.

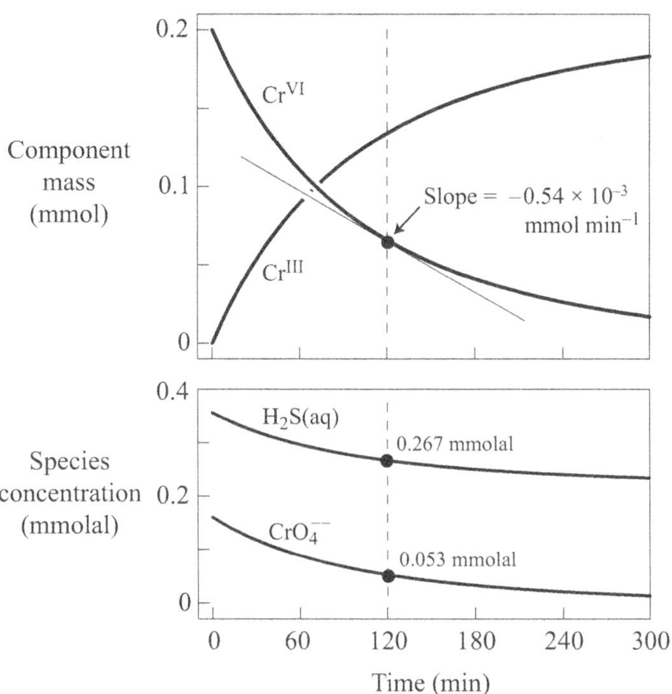

Figure 18.1 Results of simulating the reduction of hexavalent chromium by H_2S, as a function of time. The slope of the tangent to the curve (fine line) is the rate of Cr^{VI} reduction at 120 minutes of reaction.

The procedure in REACT is

```
read thermo+S0.tdat
time end = 300 min
pH = 7

decouple CrO4--
decouple HS-
decouple S(aq)

10 mmolal Na+
10 mmolal Cl-
.2 mmolal CrO4--
.8 mmolal HS-

swap Cr2O3 for Cr+++
.001 free mg Cr2O3
10^-6 mmolal S(aq)

kinetic redox-1 rxn = \
  "2 CrO4-- + 4 H+ + 3 H2S(aq) -> Cr2O3 + 5 H2O + 3 Sulfur-Rhmb"\
  rate_con = .32, mpower(H2S(aq)) = 1, mpower(CrO4--) = 1
fix pH
```

Typing go triggers the simulation.

Figure 18.1 shows the calculation results. The mass of Cr^{VI} decreases at a rate mirroring the increase in Cr^{III} mass, which is twice the rate at which Reaction 18.28 proceeds. Dissolved sulfide in the simulation is divided approximately evenly between HS^- and $H_2S(aq)$, since pH is held to 7. The reaction consumes $H_2S(aq)$ as well as CrO_4^{--}, causing the concentration of each to decline. Since the two concentrations appear as first-order terms in the rate law, reaction rate also decreases with time.

We can quickly verify correct accounting of the reaction kinetics at any point in the simulation. At 120 minutes, for example,

$$\begin{aligned}
\frac{\mathrm{d}M_{\mathrm{Cr^{VI}}}}{\mathrm{d}t} &= -\frac{\mathrm{d}M_{\mathrm{Cr^{III}}}}{\mathrm{d}t} \\
&= -2\, n_w\, k_+\, m_{\mathrm{H_2S}}\, m_{\mathrm{CrO_4^{--}}} \\
&= -2 \times 1\ \mathrm{kg} \times (0.32\ \mathrm{molal^{-1}\ s^{-1}}) \times (0.267 \times 10^{-3}\ \mathrm{molal}) \\
&\qquad \times (0.053 \times 10^{-3}\ \mathrm{molal}) \\
&= -9.05 \times 10^{-9}\ \mathrm{mol\ s^{-1}} \\
&= -0.543 \times 10^{-3}\ \mathrm{mmol\ min^{-1}}\,,
\end{aligned} \tag{18.33}$$

which is the slope of the line in Figure 18.1 representing Cr^{VI} mass versus time.

Chapter 32 includes a number of examples in which redox kinetics are incorporated into reaction path simulations.

19

Microbial Kinetics

Redox reactions in the geochemical environment, as discussed in previous chapters (Chapters 7 and 18), are commonly in disequilibrium at low temperature, their progress described by kinetic rate laws. The reactions may proceed in solution homogeneously or be catalyzed on the surface of minerals or organic matter. In a great many cases, however, they are promoted by the enzymes of the ambient microbial community.

In this chapter, we consider how the microbial community catalyzes redox reactions, perhaps changing in size and composition as it does. The kinetics of such reactions are of special interest, because of the close relationship between geochemical conditions and microbial ecology. The microbes promote reactions that change geochemical conditions, many times significantly, and the geochemistry controls the nature of the microbial community that can exist in a given environment.

From the geochemist's perspective, a microbe can be thought of as a self-replicating bundle of enzymes. Microbes use their collections of enzymes to catalyze redox reactions, harvesting some of the energy released for their own purposes. Since microbial growth increases the ability of the community to catalyze redox reactions, and catalyzing the reactions provides microbes the energy they need to grow, a microbially mediated reaction is by nature autocatalytic.

19.1 Microbial Respiration and Fermentation

Chemosynthetic microorganisms derive the energy they need to live and grow from chemical species in their environment, reaping the benefits of the redox disequilibrium characteristic of geochemical environments (see Chapter 7). The microbes catalyze redox reactions, harvesting some of the energy released and storing it as *adenosine triphosphate*, or ATP. This is an unstable molecule and hence rich in energy; it serves as the cell's energy currency, available for purposes ranging from cell maintenance to growth and reproduction.

The microbes use two general strategies to synthesize ATP: *respiration* and *fermentation*. A respiring microbe captures the energy released when electrons are transferred from a reduced species in the environment to an oxidized species (Fig. 19.1). The reduced species, the electron donor, sorbs to a complex of redox enzymes, or a series of such complexes, located in the cell membrane. The complex strips from the donor one or more electrons, which cascade through a series of enzymes and coenzymes that make up the *electron transport chain* to a terminal enzyme complex, also within the cell membrane.

The oxidized species, the electron acceptor, sorbs to the terminal complex and takes up the electrons, becoming reduced. The newly oxidized donating species and the accepting species, now reduced, desorb from the redox complexes and move back into the environment. The energy released by the electrons passing through the transport chain does the critical work, driving hydrogen ions from within the cell's cytoplasm across the cell membrane, a process known as *proton translocation*. The hydrogen ions, under the drive

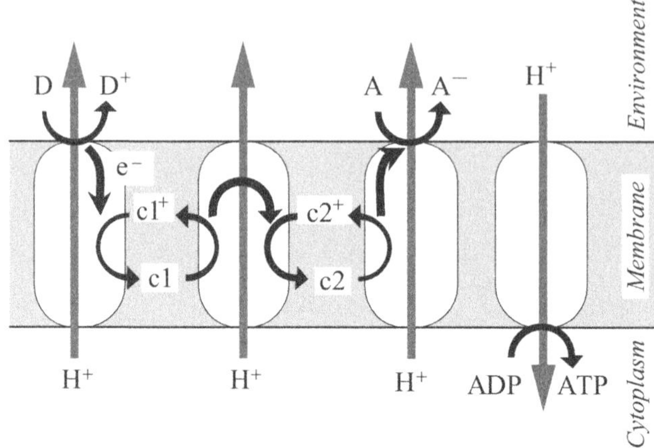

Figure 19.1 Generalized depiction of the process of microbial respiration, after Jin and Bethke (2003). An electron donating species D sorbs to a redox complex in the cell membrane (left oval), from either outside or inside the cell. The sorbed species gives up one or more electrons e^-, leaving the species in oxidized form D^+ to desorb and return to the environment. Electrons pass along an electron transport chain composed of various enzyme complexes (ovals) and coenzymes (c1, $c1^+$, c2, $c2^+$), driving the translocation of hydrogen ions H^+ from the cell's cytoplasm across the cell membrane. An electron accepting species A sorbs to a terminal complex and takes up electrons from the transport chain, thereby being reduced to A^-, which desorbs and returns to the environment. The translocated hydrogen ions H^+ re-enter the cell through ATP synthase (oval on right), creating ATP from ADP and orthophosphate ions.

of strong gradients in electrical and chemical potential, re-enter the cell through *ATP synthase*, a special membrane-bound enzyme. The ATP synthase, activated by movement of the hydrogen ions, binds together *adenosine diphosphate* (ADP) and orthophosphate ions in the cell's cytoplasm to form ATP.

This process of creating ATP, known as *electron transport phosphorylation,* then, involves two half-cell reactions, one at the electron donation site and the other where the electrons are accepted from the transport chain. Taking aerobic sulfide oxidation as an example, the donating species $H_2S(aq)$ gives up electrons, two at a time, to a series of redox complexes. With the loss of each pair of electrons, the sulfide oxidizes first to S^0, then thiosulfate, sulfite, and finally sulfate.

The electron donation half-reaction, the sum of the four donation steps, is

$$H_2S(aq) + 4\,H_2O \rightarrow SO_4^{--} + 10\,H^+ + 8\,e^- . \tag{19.1}$$

The electrons, having passed through the transport chain to the terminal enzyme, are taken up by the reduction of dioxygen,

$$O_2(aq) + 4\,H^+ + 4\,e^- \rightarrow 2\,H_2O\,, \tag{19.2}$$

to water. Adding together the half-reactions gives

$$H_2S(aq) + 2\,O_2(aq) \rightarrow SO_4^{--} + 2\,H^+\,, \tag{19.3}$$

which is the redox reaction the microbes exploit.

In environments lacking a suitable external electron acceptor—such as dioxygen, sulfate, or ferric iron—respiration is not possible. Here, many organic compounds may be metabolized by fermenting microorganisms. Microbes of this class may create ATP by a direct coupling mechanism, using a process known as *substrate level phosphorylation*, SLP; with an ion translocation mechanism like that employed by respirers, as already described; or by a combination of SLP and ion translocation.[1]

A fermenter in general *dismutates* its substrate into two compounds, one more oxidized than the substrate and the other more reduced. Fermenting microbes, for example, are commonly responsible for breaking down complex organic matter in clays and shales into simpler, more oxidized organic molecules and $H_2(aq)$. Acetoclastic methanogens, as a second example, dismutate acetate to methane and carbonate,

$$CH_3COO^- + H_2O \rightarrow CH_4(aq) + HCO_3^- , \tag{19.4}$$

using an ion translocation mechanism. The methanogens work by cleaving the carbon–carbon bond in the acetate ion (Zehnder *et al.*, 1980). Stoichiometrically, the dismutation reaction is an electron donating half-reaction,

$$1/2\ CH_3COO^- + 2\ H_2O \rightarrow HCO_3^- + 9/2\ H^+ + 4\ e^- , \tag{19.5}$$

combined with an accepting half-reaction,

$$1/2\ CH_3COO^- + 9/2\ H^+ + 4\ e^- \rightarrow CH_4(aq) + H_2O\,. \tag{19.6}$$

The substrate, as we can see in this example, serves in a fermentation as both the electron donating and electron accepting species.

19.2 Monod Equation

The *Monod equation* is the relation most commonly applied to describe the rate at which a microbe metabolizes its substrate (e.g., Panikov, 1995). Taking acetotrophic sulfate reduction as an example, the redox reaction

$$CH_3COO^- + SO_4^{--} \rightarrow 2\ HCO_3^- + HS^- \tag{19.7}$$

provides the energy the microbes need to live and reproduce. If acetate, the electron donor, is the limiting reactant, the Monod equation gives reaction rate as

$$r^+ = n_w\ k_+\ [X]\ \frac{m_{\mathrm{Ac}}}{m_{\mathrm{Ac}} + K'_{\mathrm{S}}} , \tag{19.8}$$

where r^+ is the forward reaction rate (mol s^{-1}), n_w is solvent mass (kg), k_+ is the rate constant (mol mg^{-1} s^{-1}), $[X]$ is biomass concentration (mg kg^{-1}), m_{Ac} is acetate concentration (molal), and K'_{S} is the half-saturation constant (molal) for the substrate, acetate.

If sulfate rather than acetate is limiting, the equation is written

$$r^+ = n_w\ k_+\ [X]\ \frac{m_{\mathrm{SO_4}}}{m_{\mathrm{SO_4}} + K'_{\mathrm{S}}} , \tag{19.9}$$

in terms of the electron acceptor. Here, $m_{\mathrm{SO_4}}$ is sulfate concentration (molal) and K'_{S} is the corresponding half-saturation constant (molal). Writing the rate law in the form

$$r^+ = n_w\ k_+\ [X]\ \frac{m_{\mathrm{Ac}}}{m_{\mathrm{Ac}} + K'_{\mathrm{D}}}\ \frac{m_{\mathrm{SO_4}}}{m_{\mathrm{SO_4}} + K'_{\mathrm{A}}} , \tag{19.10}$$

where K'_{D} and K'_{A} are half-saturation constants (molal) for the donating and accepting half-reactions, provides

[1] For our purposes, fermentation refers to microbial growth on a single substrate in the absence of an external electron acceptor besides hydrogen ions, without regard for the mechanism of phosphorylation.

for the possibility of the reaction being limited by either acetate or sulfate, or both; this expression is the *dual Monod equation.*

As substrate concentration varies, the Monod equation behaves like the Michaelis–Menten equation discussed in the previous chapter (Eq. 18.18). If m_{Ac} is small relative to K'_S, the ratio $m_{Ac}/(m_{Ac} + K'_S)$ in Equation 19.8 is close to m_{Ac}/K'_S, so reaction rate is proportional to substrate concentration. Where m_{Ac} exceeds K'_S, however, $m_{Ac}/(m_{Ac} + K'_S) \simeq 1$ and the reaction rate approaches its maximum value r_{max}, where $r_{max} = n_w\, k_+\, [X]$.

The Monod equation differs from the Michaelis–Menten equation in that it includes as a factor biomass concentration $[X]$, which can change with time. A microbe as it catalyzes a redox reaction harvests some of the energy liberated, which it uses to grow and reproduce, increasing $[X]$. At the same time, some microbes in the population decay or are lost to predation. The time rate of change in biomass,

$$\frac{d[X]}{dt} = \frac{Y\, r}{n_w} - D\, [X]\,, \tag{19.11}$$

is given from the reaction rate r, where Y is the growth yield (mg mol^{-1}), the amount of biomass created per mole of reaction turnover, and D is the decay constant (s^{-1}), the fraction of the biomass expected to be lost in an interval of time.

19.3 Thermodynamically Consistent Rate Laws

The Monod equation predicts the forward rate r^+ of reaction, rather than the net rate r, which is the difference $r^+ - r^-$ between the forward and reverse rates. This distinction is of little practical significance for reactions that liberate significant amounts of free energy, because for those cases the reverse rate is negligible. The aerobic oxidation of organic compounds, for example, is so strongly favored thermodynamically that, in the presence of detectable levels of O_2(aq) and the organic substrate, the forward and net rates are indistinguishable.

Metabolisms such as methanogenesis and bacterial sulfate reduction, conversely, liberate sufficiently little energy that in common geochemical environments the forward and reverse rates can be of similar magnitude. In constructing a geochemical model, furthermore, we wish to avoid allowing simulations in which a microbially catalyzed reaction might proceed, impossibly, beyond its point of equilibrium. We should, therefore, account for the effect of thermodynamics on the net rate of reaction (Jin and Bethke, 2007).

In considering the energetics of a microbially catalyzed reaction, it is important to recall that progress of the redox reaction (e.g., Reaction 19.7) is coupled to synthesis of ATP within the cell, so the overall reaction is the redox reaction combined with ATP synthesis. The free energy liberated by the overall reaction is the energy liberated by the redox reaction, less that consumed to make ATP. The overall reaction's equilibrium point is where this difference vanishes; at this point, the forward and reverse rates balance and there is no net reaction. If the redox reaction were to supply less energy than needed to create ATP, the overall reaction would run backward. In this case, ATP would be expended and the cell would need to shut down its electron transport chain, to avoid expending its energy stores.

To account for reverse as well as forward reaction, the Monod (and dual Monod) equation can be modified by appending to it a thermodynamic potential factor, as shown by Jin and Bethke (2005), in which case the equation predicts the net rate of reaction. The thermodynamic factor F_T, which can vary from zero to one, is given by

$$F_{\mathrm{T}} = 1 - \exp\left(\frac{\Delta G_{\mathrm{r}} + m\Delta G_{\mathrm{P}}}{\chi R T_{\mathrm{K}}}\right), \tag{19.12}$$

where ΔG_{r} is the free-energy change (J mol^{-1}) of the redox reaction, ΔG_{P} is the free-energy change (J mol^{-1}) under cellular conditions of the phosphorylation reaction,

$$\mathrm{ADP} + \mathrm{PO}_4^{3-} \rightarrow \mathrm{ATP} + \mathrm{H_2O}, \tag{19.13}$$

of ADP to produce ATP, m is the number of ATPs produced per reaction turnover, χ is the average stoichiometric number of the overall reaction, R is the gas constant (8.3143 J mol^{-1} K^{-1}), and T_{K} is absolute temperature (K). The value of ΔG_{P} under typical physiological conditions is in the range of about 40–50 kJ mol^{-1} (e.g., White, 2007).

The free-energy change of the redox reaction is given by

$$\Delta G_{\mathrm{r}} = R T_{\mathrm{K}} \ln \frac{Q}{K}, \tag{19.14}$$

from the reaction's ion activity product Q and its equilibrium constant K. The average stoichiometric number χ is the number of times the rate controlling step in the overall reaction, commonly translocation of hydrogen ions across the cell membrane, occurs per turnover of the overall reaction. For this reason, the value of χ depends on how the reaction is written. If the reaction coefficients were doubled, for example, χ would increase by a factor of two, as would the value of m.

Appending F_{T} to the Monod equation, we write a thermodynamically consistent rate law,

$$r_{\vec{m}} = n_w\,(k_+)_{\vec{m}}\,[X]\,\frac{m_{\mathrm{S}}}{m_{\mathrm{S}} + K'_{\mathrm{S}\vec{m}}}\,F_{\mathrm{T}}, \tag{19.15}$$

for a microbial reaction $\vec{m}$. Here, $r_{\vec{m}}$ is the reaction's net rate (mol s^{-1}), m_{S} is substrate concentration (molal), and $K'_{\mathrm{S}\vec{m}}$ is the substrate's half-saturation constant (molal). Similarly, the dual Monod equation in thermodynamically consistent form becomes

$$r_{\vec{m}} = n_w\,(k_+)_{\vec{m}}\,[X]\,\frac{m_{\mathrm{D}}}{m_{\mathrm{D}} + K'_{\mathrm{D}\vec{m}}}\,\frac{m_{\mathrm{A}}}{m_{\mathrm{A}} + K'_{\mathrm{A}\vec{m}}}\,F_{\mathrm{T}}, \tag{19.16}$$

where m_{D} and m_{A} are molal concentration of the electron donating and accepting species, and $K'_{\mathrm{D}\vec{m}}$ and $K'_{\mathrm{A}\vec{m}}$ are the corresponding half-saturation constants (molal). In each case, F_{T} is given by Equation 19.12.

19.4 General Kinetic Model

In a series of papers, Jin and Bethke (2002; 2003; 2005; 2007) and Jin (2007) derived a generalized rate expression describing microbial respiration and fermentation. They account in their rate model for an electron donating half-cell reaction,

$$\sum_{\mathrm{D}} \nu_{\mathrm{D}}\mathrm{D} \rightarrow \sum_{\mathrm{D}^+} \nu_{\mathrm{D}^+}\mathrm{D}^+ + n\,\mathrm{e}^-, \tag{19.17}$$

an electron accepting half-reaction,

$$\sum_{\mathrm{A}} \nu_{\mathrm{A}}\mathrm{A} + n\,\mathrm{e}^- \rightarrow \sum_{\mathrm{A}^-} \nu_{\mathrm{A}^-}\mathrm{A}^-, \tag{19.18}$$

and an intracellular reaction,

$$m \text{ ADP} + m \text{ PO}_4^{3-} \rightarrow m \text{ ATP} + m \text{ H}_2\text{O}\,, \tag{19.19}$$

to synthesize ATP.

The redox reaction that supplies the microbe with energy from the environment,

$$\sum_{\text{D}} \nu_{\text{D}}\text{D} + \sum_{\text{A}} \nu_{\text{A}}\text{A} \rightarrow \sum_{\text{D+}} \nu_{\text{D+}}\text{D}^{+} + \sum_{\text{A}^-} \nu_{\text{A}^-}\text{A}^{-}\,, \tag{19.20}$$

is the sum of the first two reactions, and the overall reaction,

$$\begin{aligned} &\sum_{\text{D}} \nu_{\text{D}}\text{D} + \sum_{\text{A}} \nu_{\text{A}}\text{A} + m \text{ ADP} + m \text{ PO}_4^{3-} \rightarrow \\ &\qquad \sum_{\text{D+}} \nu_{\text{D+}}\text{D}^{+} + \sum_{\text{A}^-} \nu_{\text{A}^-}\text{A}^{-} + m \text{ ATP} + m \text{ H}_2\text{O}\,, \end{aligned} \tag{19.21}$$

is the combination of all three. The resulting rate expression is

$$r_{\vec{m}} = n_w \,(k_+)_{\vec{m}}\, [X]\, F_{\text{D}}\, F_{\text{A}}\, F_{\text{T}}\,, \tag{19.22}$$

where F_{D} and F_{A} are unitless kinetic factors related to the electron donating and accepting half-reactions and, as before (Eq. 19.12), F_{T} is the thermodynamic potential factor.

The kinetic factors, which can vary from zero to one, are given by

$$F_{\text{D}} = \frac{\prod^{\text{D}} m_{\text{D}}^{\beta_{\text{D}}}}{\prod^{\text{D}} m_{\text{D}}^{\beta_{\text{D}}} + K_{\text{D}} \prod^{\text{D+}} m_{\text{D+}}^{\beta_{\text{D+}}}} \tag{19.23}$$

and

$$F_{\text{A}} = \frac{\prod^{\text{A}} m_{\text{A}}^{\beta_{\text{A}}}}{\prod^{\text{A}} m_{\text{A}}^{\beta_{\text{A}}} + K_{\text{A}} \prod^{\text{A}^-} m_{\text{A}^-}^{\beta_{\text{A}^-}}}\,, \tag{19.24}$$

where K_{D} and K_{A} are kinetic constants for the electron donating and accepting half-reactions, and β_{D}, $\beta_{\text{D+}}$, and so on, are exponents defined for each species in the redox reaction.

The β values can be determined formally only through careful experiments. Significantly, however, β values are needed only for species whose concentrations change under the conditions of interest; for other species, the quantities m^{β} can be folded into the corresponding kinetic constant, or the kinetic constant and rate constant. For species whose concentrations are likely to change, such as the reaction substrates, β is commonly taken to be one, in the absence of contradictory information.

If the concentrations of only the electron donor and acceptor are considered to vary, each $m_{\text{D+}}$ is invariant and the term $K_{\text{D}} \prod m_{\text{D+}}^{\beta_{\text{D+}}}$ in Equation 19.23 reverts to a half-saturation constant K'_{D}. Similarly, the corresponding term in Equation 19.24 may be represented by K'_{A}. Now, we see that the dual Monod equation (Eq. 19.16) is a specific simplification of the general rate law (Eq. 19.22).

The kinetic factors F_{D} and F_{A} and thermodynamic potential factor F_{T} are largest where the electron donor and acceptor are abundant, and the reaction products are not. If under such conditions all three factors are equal to one, as is not uncommon, the reaction rate predicted by Equation 19.22 reaches its maximum value, $r_{\text{max}} = n_w\, k_+\, [X]$. As the substrates are depleted with reaction progress, and reaction products accumulate, the factors eventually decrease toward zero, slowing the reaction to a near stop.

If one of the kinetic factors approaches zero first, the reaction rate is said to be kinetically controlled; if the thermodynamic factor falls first, the control is thermodynamic. The aerobic consumption of organic species and highly reduced compounds such as H_2(aq) and H_2S(aq) invariably shows kinetic controls, because the thermodynamic drive for the oxidation of these compounds is quite large. Since methanogens and sulfate reducers operate under a considerably smaller thermodynamic drive, in contrast, it is not uncommon for their reaction rates to be controlled thermodynamically.

19.5 Example Calculation

We consider as an example the growth of sulfate-reducing bacteria on acetate. A solution, initially sterile, contains 500 mmolal NaCl, 20 mmolal $CaSO_4$, 2 mmolal $FeCO_3$, and 1 mmolal $NaCH_3COO$; its pH is buffered to 7.2. At $t = 0$, the solution is inoculated with enough of the sulfate reducers to bring the initial biomass, expressed in terms of the dry weight of the cells, to 0.1 mg kg^{-1}. The solution is incubated for 3 weeks.

The bacteria in our example promote Reaction 19.7 at a rate given by Equation 19.15, the thermodynamically consistent form of the Monod equation,

$$r = n_w \, k_+ \, [X] \, \frac{m_{\text{Ac}}}{m_{\text{Ac}} + K'_{\text{S}}} \, F_{\text{T}} \,, \tag{19.25}$$

and grow according to Equation 19.11. From a study of the kinetics of *Desulfobacter postgatei* by Ingvorsen *et al.* (1984), we take a rate constant k_+ of 10^{-9} mol mg^{-1} s^{-1}, a half-saturation constant K'_{S} of 70 μmolal, and a growth yield of 4300 mg mol^{-1}; we ignore decay given the brief duration of the experiment, and in the presumed absence of predation.

Taking the rate-limiting step in the electron transport chain to be trans-membrane proton translocation, which occurs about five times per sulfate consumed (Rabus *et al.*, 2006), the average stoichiometric number χ (entered into REACT as $\omega = 1/\chi$) for Reaction 19.7 is five. Sulfate reducers conserve about 45 kJ mol^{-1} of sulfate consumed (Qusheng Jin, unpublished data), so we set ΔG_{P} to this value and m to one. From Equations 19.12 and 19.14, then, we can write

$$F_{\text{T}} = 1 - \left(\frac{Q}{K}\right)^{1/5} \exp\left(\frac{45 \text{ kJ mol}^{-1}}{5 \, RT_{\text{K}}}\right), \tag{19.26}$$

which is the thermodynamic potential factor for this case.

We can expect the sulfide produced by the bacteria to react with the iron in solution to form mackinawite (FeS), a precursor to pyrite (FeS_2). The mineral is not in the default thermodynamic database, so we add to the file the reaction

$$\underset{\textit{mackinawite}}{\text{FeS}} + \text{H}^+ \rightleftarrows \text{Fe}^{++} + \text{HS}^- \,, \tag{19.27}$$

with log $K = -3.6$ at 25 °C and save the result as "thermo+Mackinawite.tdat".

To run the simulation, we decouple acetate from carbonate, and sulfide from sulfate, and suppress the iron sulfide minerals pyrite and troilite (FeS), which are more stable than mackinawite, but unlikely to form. We set the fluid composition, including an amount of HS^- small enough to avoid significantly supersaturating mackinawite, and define the rate law for the sulfate reducers.

The procedure in REACT is

```
read thermo+Mackinawite.tdat
time end = 21 days
decouple CH3COO-
decouple HS-
suppress Pyrite, Troilite

Na+      = 501.   mmolal
Ca++     =  20.   mmolal
Fe++     =   2.   mmolal
Cl-      = 500.   mmolal
SO4--    =  20.   mmolal
HCO3-    =   2.   mmolal
CH3COO-  =   1.   mmolal
HS-      =    .3  umolal
pH = 7.2

kinetic microbe-SRB \
  rxn = "CH3COO- + SO4-- -> 2*HCO3- + HS-", \
  biomass = .1, growth_yield = 4300, \
  ATP_energy = -45, ATP_number = 1, order1 = 1/5, \
  mpower(CH3COO-) = 1, mpowerD(CH3COO-) = 1, \
  rate_con = 10^-9, KD = 70.e-6
fix pH
go
```

Figure 19.2 shows the calculation results.

Reaction in the simulation begins slowly, but proceeds more rapidly as biomass accumulates, reflecting the reaction's autocatalytic nature. The reaction rate continues to increase until most of the acetate is consumed, at which point it slows abruptly to a near stop. In the simulation, dissolved ferrous iron is present in excess amount. The bisulfide produced as a result of bacterial sulfate reduction reacts with the iron,

$$Fe^{++} + HS^- \rightarrow \underset{\textit{mackinawite}}{FeS} + H^+ \,, \tag{19.28}$$

to form mackinawite, holding sulfide concentration low and maintaining a strong thermodynamic drive. The kinetic factor F_D falls below the thermodynamic factor F_T throughout the simulation, and hence controls the reaction rate. The reaction rate, in other words, is kinetically controlled.

We can verify the model results by hand calculation. At $t = 7$ days, the kinetic factor F_D in the rate law (Eq. 19.25) is given by

$$\begin{aligned} F_D &= m_{Ac}/(m_{Ac} + K'_S) \\ &= 0.6663 \text{ mmolal}/(0.6663 \text{ mmolal} + 0.070 \text{ mmolal}) = 0.9049 \,, \end{aligned} \tag{19.29}$$

where m_{Ac} is molal concentration of the free acetate ion, not the total concentration of the acetate component. To figure F_T according to Equation 19.26, we calculate the log Q for Reaction 19.7,

$$\begin{aligned} \log Q &= 2 \, \log a_{HCO_3^-} + \log a_{HS^-} - \log a_{Ac} - \log a_{SO_4^{--}} \\ &= 2\,(-3.293) - 6.672 + 3.334 + 2.638 \\ &= -7.286 \,. \end{aligned} \tag{19.30}$$

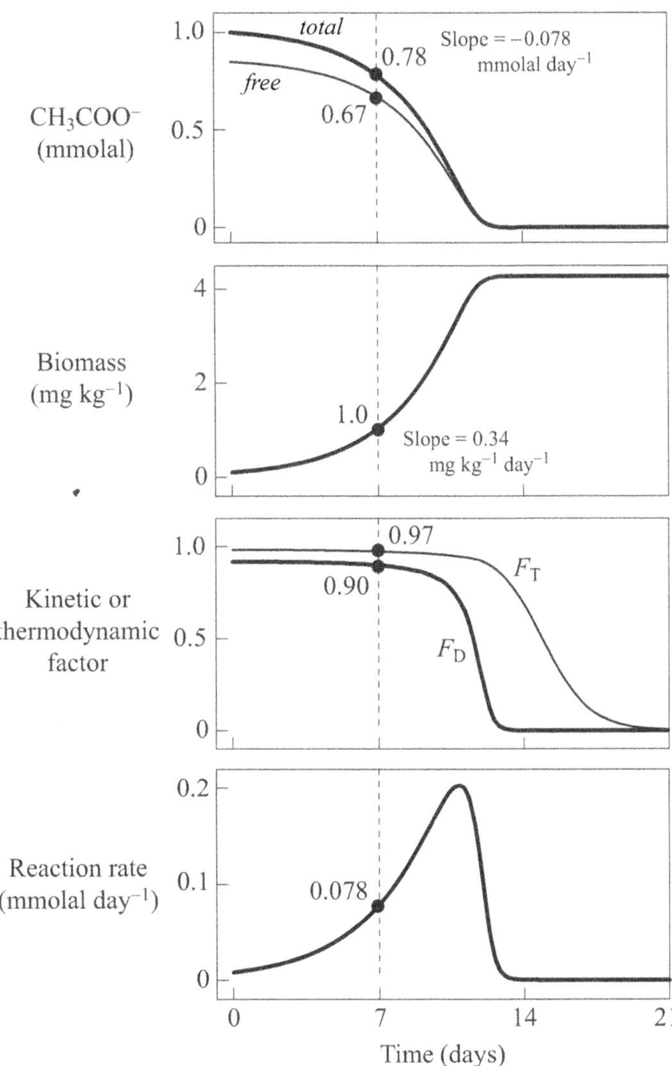

Figure 19.2 Results of modeling at 25 °C bacterial sulfate reduction using acetate as the electron donor, according to a thermodynamically consistent form of the Monod equation. Labels identify values and line slopes after 7 days of reaction.

Then, noting log K for the reaction at 25 °C is 8.404, we compute the Q/K term,

$$(Q/K)^{1/5} = 10^{(-7.286-8.404)/5} = 10^{-1.961} = 0.0007278\,, \tag{19.31}$$

as well as the ATP phosphorylation term,

$$\begin{aligned}\frac{\Delta G_{\mathrm{P}}}{5\,RT_{\mathrm{K}}} &= (45\,000\ \mathrm{J\ mol^{-1}})/(5)(8.3143\ \mathrm{J\ mol^{-1}\ K^{-1}}) \\ &\qquad \times (298.15\ \mathrm{K}) = 3.631\,,\end{aligned} \tag{19.32}$$

giving the thermodynamic potential factor,

$$F_{\mathrm{T}} = 1 - (0.0007278)\,\mathrm{e}^{(3.631)} = 0.9725 \tag{19.33}$$

for the sulfate reduction reaction.

The reaction rate, expressed per kg solvent, then, is

$$\begin{aligned}\frac{r}{n_w} &= -\frac{1}{n_w}\frac{\mathrm{d}M_{\mathrm{Ac}}}{\mathrm{d}t} \\ &= k_+\,[X]\,F_{\mathrm{D}}F_{\mathrm{T}} \\ &= (10^{-9}\ \mathrm{mol\ mg^{-1}\ s^{-1}})\ (1.032\ \mathrm{mg\ kg^{-1}})(0.9049)\ (0.9725) \\ &= 0.9082\times10^{-9}\ \mathrm{molal\ s^{-1}} \\ &= 0.07847\ \mathrm{mmolal\ day^{-1}}\,,\end{aligned} \tag{19.34}$$

which is the negative slope of the concentration of the acetate component, plotted against time (Fig. 19.2). The growth rate is given by

$$\begin{aligned}\frac{\mathrm{d}[X]}{\mathrm{d}t} &= \frac{rY}{n_w} \\ &= (0.07847\ \mathrm{mmolal\ day^{-1}})\ (4.3\ \mathrm{mg\ mmol^{-1}}) \\ &= 0.3374\ \mathrm{mg\ kg^{-1}\ day^{-1}}\,,\end{aligned} \tag{19.35}$$

which matches the slope of biomass plotted versus time.

20
Association and Dissociation Kinetics

Tracing the rates at which aqueous species form complexes with other species in solution, or with the reactive sites on mineral surfaces, and conversely how rapidly such complexes break apart, follows closely from our discussion in Chapter 17 of the kinetics of mineral precipitation and dissolution. Casey and Swaddle (2003) and Casey and Rustad (2007), in fact, point to a conceptual continuum extending from complexing in solution through reaction with aqueous clusters and macromolecules to the attachment to and detachment from mineral surfaces. As such, we should be capable of building complexation kinetics into our modeling as readily as we account for the rates at which minerals dissolve and precipitate.

Applied to surface chemistry, incorporating association and dissociation kinetics into geochemical modeling holds significant promise (Jenne, 1998; Chorover and Brusseau, 2008). Reactive transport studies (see Chapter 24), as applied currently, commonly account for the kinetics of mineral dissolution and precipitation. In seeming contradiction, however, the studies almost invariantly assume that sorption reactions can be modeled using equilibrium isotherms, as discussed in Chapters 9–11. The equilibrium assumption in such cases is notably vulnerable. Surface chemists, furthermore, observe hystereses in the isotherms they measure, depending on whether sorbate concentration is increasing or decreasing; the phenomenon cannot be explained using equilibrium concepts, but may be productively approached using kinetic theory (e.g., Limousin *et al.*, 2007).

Complexation kinetics can factor into the construction of a geochemical model for a variety of reasons and over a range of time spans, although perhaps more commonly on the human than the geologic spectrum. A number of industrial processes and laboratory techniques rely on complexing agents to mobilize and release metals; designing such procedures is at least in part a kinetic problem. In the environment, rate laws might be needed to describe the breakdown of ion pairs when a geothermal water is quenched and the reactions arising when a migrating fluid encounters natural organic matter, or when dissimilar fluids mix. Appelo and Postma (2005, p. 529) describe a case in which EDTA in effluent from a sewage treatment plant reacts slowly with the major ions in river water to form Ca–EDTA and Zn–EDTA complexes. Rate laws can be used to prescribe sorption–desorption hystereses in a reaction model, and water flowing through an aquifer may deviate significantly from equilibrium with mineral surfaces in the formation, requiring a kinetic formulation of the surface chemistry there. The sections in the chapter describe how to incorporate complexation kinetics into multicomponent models.

20.1 Kinetic Rate Laws

As an example of the dissociation of an aqueous complex, cobaltic hexammine breaks down somewhat slowly at room temperature, at a rate that varies with pH (Yoneda, 1958). Under alkaline conditions, we might expect the reaction

$$\mathrm{Co(NH_3)_6^{+++}} \rightarrow \mathrm{Co^{+++}} + 6\,\mathrm{NH_3} \tag{20.1}$$

to proceed at a rate proportional to the complex's molal concentration,

$$r^+ = -n_w \frac{dm_{\mathrm{Co(NH_3)_6^{+++}}}}{dt} = n_w k_+ \, m_{\mathrm{Co(NH_3)_6^{+++}}} \,, \tag{20.2}$$

where r^+ (mol s^{-1}) is the forward rate of reaction, n_w is solvent mass (kg), and k_+ (s^{-1}) is the corresponding rate constant.

The reverse rate of reaction r^- might reasonably be ignored in this case, in light of the ammine complex's instability. We can nonetheless cast the rate law in a thermodynamically consistent form, as discussed in Chapters 18 and 19. In this case, the net rate of reaction $r = r^+ - r^-$ is given,

$$r = n_w k_+ \, m_{\mathrm{Co(NH_3)_6^{+++}}} \left(1 - \frac{Q}{K}\right) , \tag{20.3}$$

where Q and K are the reaction's ion activity product and equilibrium constant. (As noted in Appendix D, the thermodynamic term may assume a nonlinear form.)

The ratio of Q to K for the reaction,

$$\frac{Q}{K} = \frac{m^{\mathrm{eq}}_{\mathrm{Co(NH_3)_6^{+++}}}}{m_{\mathrm{Co(NH_3)_6^{+++}}}} \,, \tag{20.4}$$

is the quotient of the complex's equilibrium to actual concentration, taking ionic strength to be invariant, since,

$$Q = \frac{a_{\mathrm{Co^{+++}}} \, a^6_{\mathrm{NH_3}}}{\gamma_{\mathrm{Co(NH_3)_6^{+++}}} \, m_{\mathrm{Co(NH_3)_6^{+++}}}} \tag{20.5}$$

and

$$K = \frac{a_{\mathrm{Co^{+++}}} \, a^6_{\mathrm{NH_3}}}{\gamma_{\mathrm{Co(NH_3)_6^{+++}}} \, m^{\mathrm{eq}}_{\mathrm{Co(NH_3)_6^{+++}}}} \,. \tag{20.6}$$

Substituting, the thermodynamically consistent law (Eq. 20.3),

$$r = n_w k_+ \left(m_{\mathrm{Co(NH_3)_6^{+++}}} - m^{\mathrm{eq}}_{\mathrm{Co(NH_3)_6^{+++}}} \right) , \tag{20.7}$$

holds that dissociation proceeds at a rate proportional to the difference between the actual and equilibrium concentrations. Where the complex's equilibrium concentration is small relative to its actual concentration, the rate law reverts to the original form (Eq. 20.2).

We take as a second example formation of the cupric fluoride ion pair. By our convention, complex species appear to the left, so association drives the reaction,

$$\mathrm{CuF^+} \leftarrow \mathrm{Cu^{++}} + \mathrm{F^-} \,, \tag{20.8}$$

from right to left. The association rate r^- (mol s^{-1}), the rate the reaction proceeds backward, might be given by a law of the form

$$r^- = n_w \frac{dm_{\mathrm{CuF^+}}}{dt} = n_w \, k_- \, m_{\mathrm{Cu^{++}}} \, m_{\mathrm{F^-}} \,, \tag{20.9}$$

where k_- is the intrinsic rate constant (molal^{-1} s^{-1}) in the reverse direction. If r is the net rate at which the reaction moves left to right, the net association rate, $-r = r^- - r^+$, is given,

$$-r = n_w\, k_-\, m_{\mathrm{Cu}^{++}}\, m_{\mathrm{F}^-} \left(1 - \frac{K}{Q}\right), \tag{20.10}$$

or, substituting, as before,

$$-r = n_w\, k_-\left[m_{\mathrm{Cu}^{++}}\, m_{\mathrm{F}^-} - (m_{\mathrm{Cu}^{++}}\, m_{\mathrm{F}^-})^{\mathrm{eq}}\right], \tag{20.11}$$

when we honor thermodynamic consistency.

To model the rate at which a solid phase releases its surface acidity (see Chapters 10–11) into its environs,

$$>\mathrm{XOH}_2^+ \rightarrow\ >\mathrm{XOH} + \mathrm{H}^+ , \tag{20.12}$$

as a final example, we might write a rate law

$$r = -n_w \frac{\mathrm{d}m_{>\mathrm{XOH}_2^+}}{\mathrm{d}t} = n_w k_+\, m_{>\mathrm{XOH}_2^+} \left(1 - \frac{Q}{K}\right). \tag{20.13}$$

Replacing Q/K as before (Eq. 20.4), the law becomes

$$r = n_w k_+ \left(m_{>\mathrm{XOH}_2^+} - m^{\mathrm{eq}}_{>\mathrm{XOH}_2^+}\right), \tag{20.14}$$

where $m^{\mathrm{eq}}_{>\mathrm{XOH}_2^+}$ is the equilibrium concentration of the protonated surface sites.

For comparison, the semi-empirical pseudo-first order law commonly applied in sorption kinetic studies (e.g., Gupta and Bhattacharyya, 2011; Tan and Hameed, 2017) may be expressed

$$-\frac{\mathrm{d}q}{\mathrm{d}t} = k_1\,(q - q^{\mathrm{eq}}) , \tag{20.15}$$

where q is adsorption capacity (mg g^{-1}), the mass sorbed per g sorbent; k_1 is a rate constant (s^{-1}); and q^{eq} is the capacity at equilibrium. Since q for a given sorbate is proportional to the molal concentration m of its surface complex,

$$q = \frac{1000\, M_{\mathrm{W}}\, n_w}{n_{\mathrm{sorb}}}\, m , \tag{20.16}$$

where M_{W} is the sorbate's mole weight (g mol^{-1}) and n_{sorb} is sorbent mass (g), the value of k_1 may be carried into Equation 20.14 directly as k_+.

20.2 Numerical Solution

To build kinetic complexation into a reaction model, we need to segregate species into two groups: equilibrium species, whose concentrations are controlled by mass action equations (e.g., Eqs. 4.7 and 10.29), and kinetic species, the concentrations of which vary in time according to rate laws. The solution procedures for aqueous and surface complexes are broadly similar, as described below.

20.2.1 Kinetics of Aqueous Complexation

In modeling the formation and decay of aqueous complexes in a multicomponent system, we designate one or more secondary species $A_{\vec{j}} \in A_j$ to serve as kinetic species. As a secondary species, a kinetic species cannot appear in the basis ($A_{\vec{j}} \notin A_i$). Each kinetic species dissociates to the basis by a kinetic reaction,

$$A_{\vec{j}} \rightleftarrows \nu_{w\vec{j}} A_w + \sum_i \nu_{i\vec{j}} A_i + \sum_k \nu_{k\vec{j}} A_k + \sum_m \nu_{m\vec{j}} A_m \,, \tag{20.17}$$

in the same form as an equilibrium reaction (i.e., Reaction 3.22); the reaction's equilibrium constant is $K_{\vec{j}}$.

Reaction Rate

A law of form familiar from previous chapters prescribes the rate $r_{\vec{j}}$ (mol s^{-1}),

$$r_{\vec{j}} = -n_w \frac{dm_{\vec{j}}}{dt} = n_w k_+ \prod^{j'} (m_{j'})^{p_{j'\vec{j}}} \left(1 - \frac{Q_{\vec{j}}}{K_{\vec{j}}}\right) , \tag{20.18}$$

at which the reaction proceeds: positive when $A_{\vec{j}}$ is consumed and negative when it is produced. Here, $m_{\vec{j}}$ is the kinetic species' molal concentration, $m_{j'}$ and $p_{j'\vec{j}}$ are concentrations and powers of the promoting and inhibiting species, and $Q_{\vec{j}}$ is the reaction's ion activity product. In the absence of promoting and inhibiting species, the intrinsic rate constant k_+ has units of molal $^{-1}$ s^{-1}.

From the definitions of equilibrium constant and ion activity product (Equations 3.26 and 3.41), and taking ionic strength to be constant, we note

$$\frac{Q_{\vec{j}}}{K_{\vec{j}}} = \frac{m^{\text{eq}}_{\vec{j}}}{m_{\vec{j}}} , \tag{20.19}$$

where the equilibrium concentration $m^{\text{eq}}_{\vec{j}}$ of kinetic species $A_{\vec{j}}$ is given,

$$m^{\text{eq}}_{\vec{j}} = \frac{a_w^{\nu_{w\vec{j}}} \prod^i (\gamma_i m_i)^{\nu_{i\vec{j}}} \prod^k a_k^{\nu_{k\vec{j}}} \prod^m f_m^{\nu_{m\vec{j}}}}{\gamma_{\vec{j}} K_{\vec{j}}} , \tag{20.20}$$

by Equation 4.7.

Mass Balance Equations

To integrate the rate expression (Eq. 20.18), we march forward along time in a series of discrete steps. Advancing from t' to t, the kinetic species' concentration $m_{\vec{j}}(t)$ at the end of a step is given,

$$m_{\vec{j}}(t) = m_{\vec{j}}(t') - \frac{(t - t')}{2 n_w} \left[r_{\vec{j}}(t') + r_{\vec{j}}(t) \right] , \tag{20.21}$$

from concentration $m_{\vec{j}}(t')$ at the onset, the step length $t - t'$, and the average of reaction rate at the onset $r_{\vec{j}}(t')$ and end $r_{\vec{j}}(t)$ of the step. As before (Eq. 17.22), the rate term can be profitably expressed as a weighted average, to allow adjustable time weighting in the solution procedure.

The mass balance equations (Eqs. 3.28–3.29) for the system now need to be expanded,

$$M_w = n_w \left(55.5 + \sum_{j \neq \vec{j}} \nu_{wj} m_j + \sum_{\vec{j}} \nu_{w\vec{j}} m_{\vec{j}}(t) \right) \tag{20.22}$$

$$M_i = n_w \left(m_i + \sum_{j \neq \vec{j}} \nu_{ij} m_j + \sum_{\vec{j}} \nu_{i\vec{j}} m_{\vec{j}}(t) \right) , \tag{20.23}$$

to account for the contributions of equilibrium species and kinetic complexes to the system's bulk composition separately.

Residual Functions

The residual functions follow as the inequality between the right and left sides of the mass balance equations,

$$R_w = n_w \left(55.5 + \sum_{j \neq \vec{j}} \nu_{wj} m_j + \sum_{\vec{j}} \nu_{w\vec{j}} m_{\vec{j}}(t') \right) - \frac{(t-t')}{2} \sum_{\vec{j}} \nu_{w\vec{j}} \left[r_{\vec{j}}(t') + r_{\vec{j}}(t) \right] - M_w \tag{20.24}$$

$$R_i = n_w \left(m_i + \sum_{j \neq \vec{j}} \nu_{ij} m_j + \sum_{\vec{j}} \nu_{i\vec{j}} m_{\vec{j}}(t') \right) - \frac{(t-t')}{2} \sum_{\vec{j}} \nu_{i\vec{j}} \left[r_{\vec{j}}(t') + r_{\vec{j}}(t) \right] - M_i , \tag{20.25}$$

once Equation 20.21 has been substituted for the concentrations $m_{\vec{j}}(t)$ of the kinetic species. The functions correspond to Equations 4.24–4.25 in our discussion of the multicomponent equilibrium problem.

Jacobian Matrix

The Jacobian matrix, as before, holds the partial derivatives of the residual functions (R_w, R_i) with respect to the unknown values (n_w, m_i). Noting that

$$\frac{\partial m_j}{\partial m_i} = \frac{\nu_{ij} m_j}{m_i} \quad \text{and} \quad \frac{\partial}{\partial m_i} \left(\frac{Q_{\vec{j}}}{K_{\vec{j}}} \right) = \frac{\nu_{i\vec{j}}}{m_i} \frac{Q_{\vec{j}}}{K_{\vec{j}}} , \tag{20.26}$$

we can write the derivatives of the kinetic rates with respect to the unknowns,

$$\frac{\partial r_{\vec{j}}}{\partial n_w} = k_+ \prod^{j'} m_{j'}^{p_{j'\vec{j}}} \left(1 - \frac{Q_{\vec{j}}}{K_{\vec{j}}} \right) \tag{20.27}$$

$$\frac{\partial r_{\vec{j}}}{\partial m_i} = \frac{n_w k_+}{m_i} \prod^{j'} m_{j'}^{p_{j'\vec{j}}} \left[\left(\sum_{j'} \nu_{ij'} p_{j'\vec{j}} \right) \left(1 - \frac{Q_{\vec{j}}}{K_{\vec{j}}} \right) - \frac{\nu_{i\vec{j}} Q_{\vec{j}}}{K_{\vec{j}}} \right] . \tag{20.28}$$

The entries in the Jacobian matrix follow as

$$J_{ww} = 55.5 + \sum_{j \neq \vec{j}} \nu_{wj} m_j + \sum_{\vec{j}} \nu_{w\vec{j}} m_{\vec{j}}(t') - \frac{(t-t')}{2} \sum_{\vec{j}} \nu_{w\vec{j}} \frac{\partial r_{\vec{j}}}{\partial n_w} \tag{20.29}$$

$$J_{wi} = \frac{n_w}{m_i} \sum_{j \neq \vec{j}} \nu_{wj} \nu_{ij} m_j - \frac{(t-t')}{2} \sum_{\vec{j}} \nu_{w\vec{j}} \frac{\partial r_{\vec{j}}}{\partial m_i} \tag{20.30}$$

$$J_{iw} = m_i + \sum_{j \neq \vec{j}} \nu_{ij} m_j + \sum_{\vec{j}} \nu_{i\vec{j}} m_{\vec{j}}(t') - \frac{(t-t')}{2} \sum_{\vec{j}} \nu_{i\vec{j}} \frac{\partial r_{\vec{j}}}{\partial n_w} \tag{20.31}$$

and

$$J_{ii'} = n_w \left(\delta_{ii'} + \frac{1}{m_{i'}} \sum_{j \neq \vec{j}} \nu_{ij} \nu_{i'j} m_j \right) - \frac{(t-t')}{2} \sum_{\vec{j}} \nu_{i\vec{j}} \frac{\partial r_{\vec{j}}}{\partial m_{i'}} . \tag{20.32}$$

These relations correspond to Equations 4.26–4.29 in our solution to the equilibrium problem. At this point, we can determine the distribution of species at each step in a time marching procedure using Newton–Raphson iteration, as described in Section 4.2.

20.2.2 Sorption Kinetics

To model the kinetics of sorption and surface complexation, we take the concentrations of an arbitrary number of the sorbed species or surface complexes $A_{\vec{q}} \in A_q$ to vary according to kinetic rate laws. Each kinetic complex decomposes to the basis according to a reaction

$$A_{\vec{q}} \rightleftarrows \nu_{w\vec{q}} A_w + \sum_i \nu_{i\vec{q}} A_i + \sum_k \nu_{k\vec{q}} A_k + \sum_m \nu_{m\vec{q}} A_m + \nu_{p\vec{q}} A_p , \tag{20.33}$$

in the form of Reaction 10.26; the corresponding equilibrium constant is $K_{\vec{q}}$.

Reaction Rate

The concentration of each kinetic surface species varies in time according to a rate law

$$r_{\vec{q}} = -n_w \frac{\mathrm{d}m_{\vec{q}}}{\mathrm{d}t} = n_w k_+ \prod^{j'} (m_{j'})^{p_{j'\vec{q}}} \left(1 - \frac{Q_{\vec{q}}}{K_{\vec{q}}} \right) . \tag{20.34}$$

Here, $m_{\vec{q}}$ is molality of the kinetic species $A_{\vec{q}}$ in question, variables $m_{j'}$ and $p_{j'\vec{q}}$ represent concentrations and powers of the promoting and inhibiting species, and $Q_{\vec{q}}$ is the reaction's ion activity product; the rate constant k_+ has units of molal s^{-1}, in the absence of promotion and inhibition.

The reaction rate $r_{\vec{q}}$ (mol s^{-1}) is positive when the species desorbs from the surface, and negative when it sorbs. As before (Eq. 20.19), the ratio of activity product to equilibrium constant is given,

$$\frac{Q_{\vec{q}}}{K_{\vec{q}}} = \frac{m_{\vec{q}}^{\mathrm{eq}}}{m_{\vec{q}}} , \tag{20.35}$$

where $m_{\vec{q}}^{\mathrm{eq}}$ is the species' equilibrium concentration, as calculated by the corresponding mass action equation (e.g., Eq. 10.29).

Mass Balance Equations

As we integrate the kinetic rate law, concentration $m_{\vec{q}}(t)$ at the completion of a time step is given,

$$m_{\vec{q}}(t) = m_{\vec{q}}(t') - \frac{(t-t')}{2n_w} \left[r_{\vec{q}}(t') + r_{\vec{q}}(t) \right] , \tag{20.36}$$

from concentration $m_{\vec{q}}(t')$ at the step's onset, its duration $t - t'$, and the reaction rates calculated (Eq. 20.34) for t' and t.

The mass balance relationships for the kinetic case are

$$M_w = n_w \left(55.5 + \sum_j \nu_{wj} m_j + \sum_{q \neq \vec{q}} \nu_{wq} m_q + \sum_{\vec{q}} \nu_{w\vec{q}} m_{\vec{q}}(t) \right) \tag{20.37}$$

$$M_i = n_w \left(m_i + \sum_j \nu_{ij} m_j + \sum_{q \neq \vec{q}} \nu_{iq} m_q + \sum_{\vec{q}} \nu_{i\vec{q}} m_{\vec{q}}(t) \right) \tag{20.38}$$

$$M_p = n_w \left(m_p + \sum_{q \neq \vec{q}} \nu_{pq} m_q + \sum_{\vec{q}} \nu_{p\vec{q}} m_{\vec{q}}(t) \right) , \tag{20.39}$$

where, as with aqueous complexation, we sum over kinetic as well as equilibrium surface species.

Residual Functions

The residual functions reflecting inequality in the mass balance equations,

$$\begin{aligned} R_w = n_w & \left(55.5 + \sum_j \nu_{wj} m_j + \sum_{q \neq \vec{q}} \nu_{wq} m_q + \sum_{\vec{q}} \nu_{w\vec{q}} m_{\vec{q}}(t') \right) \\ & - \frac{(t - t')}{2} \sum_{\vec{q}} \nu_{w\vec{q}} \left[r_{\vec{q}}(t') + r_{\vec{q}}(t) \right] - M_w \end{aligned} \tag{20.40}$$

$$\begin{aligned} R_i = n_w & \left(m_i + \sum_j \nu_{ij} m_j + \sum_{q \neq \vec{q}} \nu_{iq} m_q + \sum_{\vec{q}} \nu_{i\vec{q}} m_{\vec{q}}(t') \right) \\ & - \frac{(t - t')}{2} \sum_{\vec{q}} \nu_{i\vec{q}} \left[r_{\vec{q}}(t') + r_{\vec{q}}(t) \right] - M_i \end{aligned} \tag{20.41}$$

$$\begin{aligned} R_p = n_w & \left(m_p + \sum_{q \neq \vec{q}} \nu_{pq} m_q + \sum_{\vec{q}} \nu_{p\vec{q}} m_{\vec{q}}(t') \right) \\ & - \frac{(t - t')}{2} \sum_{\vec{q}} \nu_{p\vec{q}} \left[r_{\vec{q}}(t') + r_{\vec{q}}(t) \right] - M_p , \end{aligned} \tag{20.42}$$

arise from substituting Equation 20.36 into Equations 20.37–20.39.

Jacobian Matrix

The entries in the Jacobian matrix represent partial derivatives of the residual functions (R_w, R_i, R_p) with respect to the unknowns (n_w, m_i, m_q),

$$J_{ww} = 55.5 + \sum_j \nu_{wj} m_j + \sum_{q \neq \vec{q}} \nu_{wq} m_q + \sum_{\vec{q}} \nu_{w\vec{q}} m_{\vec{q}}(t') - \frac{(t - t')}{2} \sum_{\vec{q}} \nu_{w\vec{q}} \frac{\partial r_{\vec{q}}}{\partial n_w} \tag{20.43}$$

$$J_{wi} = n_w \left(\frac{1}{m_i} \sum_j \nu_{wj} \nu_{ij} m_j + \sum_{q \neq \vec{q}} \nu_{wq} \frac{\partial m_q}{\partial m_i} \right) - \frac{(t - t')}{2} \sum_{\vec{q}} \nu_{w\vec{q}} \frac{\partial r_{\vec{q}}}{\partial m_i} \tag{20.44}$$

$$J_{wp} = n_w \sum_{q \neq \vec{q}} \nu_{wq} \frac{\partial m_q}{\partial m_p} - \frac{(t - t')}{2} \sum_{\vec{q}} \nu_{w\vec{q}} \frac{\partial r_{\vec{q}}}{\partial m_p} \tag{20.45}$$

$$J_{iw} = m_i + \sum_j \nu_{ij} m_j + \sum_{q \neq \vec{q}} \nu_{iq} m_q + \sum_{\vec{q}} \nu_{i\vec{q}} m_{\vec{q}}(t') - \frac{(t - t')}{2} \sum_{\vec{q}} \nu_{i\vec{q}} \frac{\partial r_{\vec{q}}}{\partial n_w} \tag{20.46}$$

$$J_{ii'} = n_w \left(\delta_{ii'} + \frac{1}{m_{i'}} \sum_j \nu_{ij} \nu_{i'j} m_j + \sum_{q \neq \vec{q}} \nu_{iq} \frac{\partial m_q}{\partial m_{i'}} \right) - \frac{(t - t')}{2} \sum_{\vec{q}} \nu_{i\vec{q}} \frac{\partial r_{\vec{q}}}{\partial m_{i'}} \tag{20.47}$$

$$J_{ip} = n_w \sum_{q \neq \vec{q}} \nu_{iq} \frac{\partial m_q}{\partial m_p} - \frac{(t - t')}{2} \sum_{\vec{q}} \nu_{i\vec{q}} \frac{\partial r_{\vec{q}}}{\partial m_p} \tag{20.48}$$

$$J_{pw} = m_p + \sum_{q \neq \vec{q}} \nu_{pq} m_q + \sum_{\vec{q}} \nu_{p\vec{q}} m_{\vec{q}}(t') - \frac{(t - t')}{2} \sum_{\vec{q}} \nu_{p\vec{q}} \frac{\partial r_{\vec{q}}}{\partial n_w} \tag{20.49}$$

$$J_{pi} = n_w \sum_{q \neq \vec{q}} \nu_{pq} \frac{\partial m_q}{\partial m_i} - \frac{(t - t')}{2} \sum_{\vec{q}} \nu_{p\vec{q}} \frac{\partial r_{\vec{q}}}{\partial m_i} \tag{20.50}$$

$$J_{pp'} = n_w \left(\delta_{pp'} + \sum_{q \neq \vec{q}} \nu_{pq} \frac{\partial m_q}{\partial m_{p'}} \right) - \frac{(t - t')}{2} \sum_{\vec{q}} \nu_{p\vec{q}} \frac{\partial r_{\vec{q}}}{\partial m_{p'}} . \tag{20.51}$$

The rate derivatives in the Jacobian entries are perhaps most conveniently evaluated using finite differences, given the variety of formalisms in which sorption and surface complexation models (see Chapters 9–11) can be cast.

20.3 Aluminum Fluoride Complexation

High levels of dissolved aluminum are detrimental to fish and other biota in surface waters affected by acid rain, where the element's harmfulness depends on water chemistry (e.g., Santore *et al.*, 2018). Ligands such as fluoride play contradictory roles in controlling aluminum toxicity in natural waters (e.g., Berger *et al.*, 2015). They serve to reduce the availability of the most toxic forms of the element, the monomeric aquo- and hydroxo-species, by binding them into complexes; at the same time, the complexing promotes the solubility of whatever aluminous minerals may be in contact with the water, potentially increasing the total aluminum load.

Given that fluoride complexes commonly dominate the aluminum speciation in natural waters (e.g., Driscoll *et al.*, 1980), environmental geochemists are interested in knowing how rapidly the complexes form. Plankey *et al.* (1986) studied the rate at which fluoride ions pair with dissolved aluminum species from pH 1 to 5, bracketing the range observed in acid rain, to form AlF^{++} complexes. In their experiments, they added

Table 20.1 *Apparent rate constants* k_1 *(molal*$^{-1}$ *s*$^{-1}$*) and* k_2 *(molal*$^{-2}$ *s*$^{-1}$*) for the ion pairing of aluminum and fluoride at differing pH values (Plankey et al., 1986)*

pH	k_1	k_2
4.85	726	2.74×10^8
4.35	262	4.77×10^6
3.85	110	1.45×10^6
3.36	61	4.69×10^5
2.88	65	—
1.62	88	—
1.42	117	—
1.12	199	—

aliquots of $Al(NO_3)_3$ to pH-buffered, fluoride-bearing NaCl solutions and observed the initial rate of F^- uptake using an ion-selective electrode.

From the equilibrium speciation of aluminum and fluoride in this pH range, we expect the AlF^{++} complex to form by several parallel reactions, especially

$$AlF^{++} \leftarrow Al^{+++} + F^- \tag{20.52}$$

$$AlF^{++} + H_2O \leftarrow AlOH^{++} + F^- + H^+ \tag{20.53}$$

$$AlF^{++} + H^+ \leftarrow Al^{+++} + HF \tag{20.54}$$

$$AlF^{++} + H_2O \leftarrow AlOH^{++} + HF\,. \tag{20.55}$$

Again, we follow our convention of placing the complex species to the left. Regressing their results, Plankey *et al.* proposed that the initial association rate r_o^- (mol s^{-1}) for AlF^{++} can be described by

$$r_o^- = n_w \frac{dm_{AlF^{++}}}{dt} = k_1 m_{Al^{III}} m_{F^{-I}} + k_2 m_{Al^{III}} m_{F^{-I}}^2\,, \tag{20.56}$$

at a specific pH. Here, $m_{Al^{III}}$ and $m_{F^{-I}}$ are the total dissolved concentrations (molal) of aluminum and fluoride, the sums, respectively, of the concentrations of the individual aluminum and fluoride species in solution. Table 20.1 shows how the apparent rate constants k_1 (molal^{-1} s^{-1}) and k_2 (molal^{-2} s^{-1}) in the rate law vary with pH. Significantly, since the rate law is cast in terms of bulk composition rather than the concentrations of individual species, it is able to predict only the initial rate of reaction.

To calculate the initial association rate for an experiment containing 0.2 mmolal aluminum and 0.015 mmolal fluoride at pH 2.88, for example, we start REACT and enter the commands

```
Al+++ =    .20  mmolal
NO3-  =    .60  mmolal
F-    =    .015 mmolal
Cl-   = 100.    mmolal
Na+   = 100.    mmolal
pH    =   2.88
suppress AlF2+, AlF3, AlF4-, AlF5--
```

to set fluid composition. The final command suppresses the higher-order aluminum fluoride complexes; otherwise, the code would render them at equilibrium concentration. Next, we set a rate law

```
(cont'd)
kinetic AlF++ 1 nmolal, reverse, rate_law = \
   'Wmass * 65 * totmolal("Al+++") * totmolal("F-")'
go initial
```

representing Equation 20.56 and trigger the model to calculate the initial system. The first command segregates AlF^{++} as a kinetic species $A_{\bar{j}}$, sets it at near-zero concentration, and prescribes the rate law giving the complex's initial association rate r_0^-. The resulting rate of 1.95×10^{-7} molal s^{-1} matches the measured value of 1.98×10^{-7}, in these units. Repeating the calculation for 0.4 mmolal Al^{+++} and 0.02 mmolal F^-

```
(cont'd)
Al+++ =    .40 mmolal
NO3-  =   1.2  mmolal
F-    =    .02 mmolal
go initial
```

predicts a rate of 5.20×10^{-7} molal s^{-1}, which compares to the measured value of 5.24×10^{-7} molal s^{-1}.

To plot how the initial association rate r_0^- varies with pH, we note the data in Table 20.1 can be represented by the correlations

$$\log k_1 = 3.38 - 1.25\,\text{pH} + 0.236\,\text{pH}^2 \tag{20.57}$$

$$\log k_2 = 1.78\,\text{pH} - 0.551\,. \tag{20.58}$$

Starting anew in REACT, we define the solution's electrolyte composition

```
Al+++ =    .20  mmolal
NO3-  =    .60  mmolal
F-    =    .015 mmolal
Cl-   = 100.    mmolal
Na+   = 100.    mmolal
suppress AlF2+, AlF3, AlF4-, AlF5--
```

as before, and set a rate law

```
(cont'd)
kinetic AlF++ 1 nmolal, reverse, rate_law = {
   k1 = 10.0^(3.38 - 1.25 * pH + 0.236 * pH^2)
   k2 = 10.0^(1.78 * pH - 0.551)
   r  = k1 * totmolal("Al+++") * totmolal("F-") + \
        k2 * totmolal("Al+++") * totmolal("F-")^2
   RETURN Wmass * r
}
```

representing Equations 20.56–20.58. We then specify that pH in the simulation is to progress from 1 to 5, set a small enough time span that only the initial reaction rate will be considered, disable precipitation to prevent aluminum hydroxide minerals from forming

```
(cont'd)
pH = 1.0; slide pH to 5.0
time end 0.001 sec
```

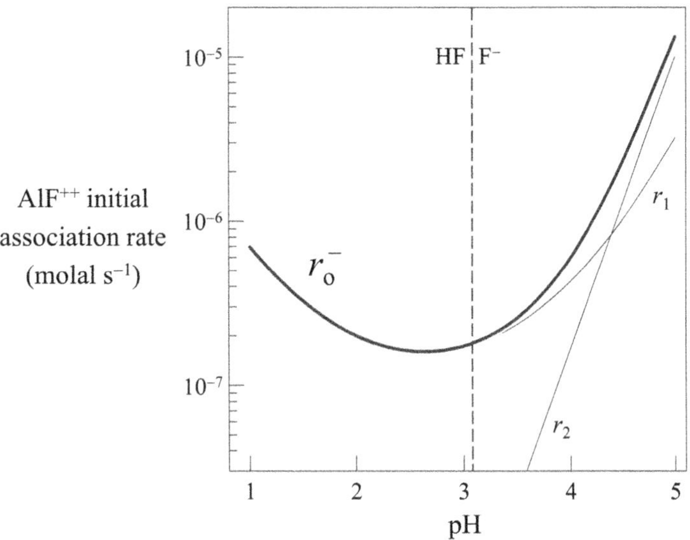

Figure 20.1 Predicted initial association rate for the AlF^{++} ion pair at 25 °C in a 100 mmolal NaCl solution containing 0.015 mmolal NaF into which sufficient $Al(NO_3)_3$ is introduced to provide a final Al concentration of 0.2 mmolal. Plot shows r_o^- versus pH, as given by the rate law of Plankey *et al.* (1986; Eq. 20.56). Line r_1 is the first term in the rate law; r_2, the second.

```
precip off
go
```

and trigger the calculation.

In the calculation results (Fig. 20.1), reaction is slowest near pH 2.5. Here, most aluminum is present as Al^{+++} and much of the fluoride is found as free F^-; each ion is protected by a strong hydration sphere. At lesser pH, F^- gives way to the more reactive HF ion pair, allowing reaction rate to increase. Reaction rate also increases at higher pH, where $AlOH^{++}$ progressively replaces Al^{+++} in solution. Above about pH 4, the second term in Equation 20.56 becomes important, causing r_o^- to rise sharply. The term likely reflects the increasing importance of polymerized aluminum species (e.g., Zhao *et al.*, 2009) in the experiments, as pH approaches neutrality.

21
Kinetics of Gas Transfer

The movement of gas species across the air–water interface is a central aspect of biogeochemical cycling and plays a critical role in controlling not only the composition of the atmosphere, but the chemistry of aquatic and marine systems. Uptake by seawater is a primary sink of a number of climatically active gases, including $CO_2(g)$ and $SO_2(g)$, whereas release of $CH_4(g)$ constitutes a net source to the atmosphere. The oceans take up $^{14}CO_2(g)$ released by historic bomb testing and absorb other anthropogenic contaminants, such as chlorofluorocarbons.

In lacustrine settings, influx of $O_2(g)$ from the air supports respiration by microbes and animal life; the supply of atmospheric $CO_2(g)$ for photosynthesis is in some cases a limiting factor on primary productivity. Surface waters absorb volatile organic compounds from the air, and various organic pollutants in groundwater volatilize within the vadose zone into soil gases and, ultimately, the atmosphere.

Gas stripping techniques, in which a stream of air or a specific gas percolates through still or moving water, are widely utilized in laboratory techniques, water treatment, and environmental remediation. Field geochemists employ gas stripping to measure trace levels of dissolved gases such as $H_2(g)$ in natural waters, and treatment plants strip on an industrial scale to remove $CO_2(g)$ and other undesired gases from drinking water. Air sparging is an in situ stripping technique used by environmental engineers to remove volatile organic chemicals from groundwater and soil.

Developing an understanding of gas transfer rates, then, lies at the core of a range of applications, from unraveling the complexities of global warming and ocean acidification to the effective design of industrial processes and environmental remedies. In this chapter, we consider how to incorporate gas transfer kinetics into our geochemical modeling.

21.1 Kinetic Rate Laws

A gas species crosses the interface between a fluid and a gas reservoir at a rate that reflects the extent to which its free concentration in solution deviates from its saturated concentration. The free concentration is that of the molecular gas species—$O_2(aq)$, $N_2(aq)$, or $CO_2(aq)$, for example—and saturated concentration is free concentration at equilibrium with the external reservoir. Total concentration, in contrast, would reflect the amount of a gas dissolved in the fluid. The two concentrations are equivalent for inert gases, but the distinction is important when considering a reactive gas. When HCl(g) dissolves in water, for example, most of it ionizes to H^+ and Cl^-; only a small fraction remains as the free ion pair HCl(aq). $CO_2(g)$, as a second example, dissolves to not only $CO_2(aq)$, but HCO_3^- and CO_3^{--}, in proportions that depend on pH.

Expressed quantitatively, the rate $r_{\vec{n}}$ (mol s^{-1}) at which a kinetic gas $A_{\vec{n}}$ crosses the interface can be expressed by the law

$$r_{\vec{n}} = A_c K_1 \left(C_{\vec{n}}^{\text{sat}} - C_{\vec{n}}^{\text{free}} \right), \tag{21.1}$$

where the rate is positive for uptake by the fluid (e.g., Brezonik, 1994; Stumm and Morgan, 1996). Here, A_c is the interfacial contact area (cm^2), and $C_{\vec{n}}^{\text{sat}}$ and $C_{\vec{n}}^{\text{free}}$ are the saturated and free concentrations, respectively, in mol cm^{-3}. A transfer coefficient K_1 (cm s^{-1}), also known as the transfer velocity, describes ease of transport across the interface. The coefficient depends on environmental factors such as wind, waves, bubbles, and temperature, as well as the nature of the gas in question (e.g., Liss, 1983; Esters *et al.*, 2017).

The saturated and free concentrations are given by Henry's law,

$$C_{\vec{n}}^{\text{sat}} = K_{\text{H}} P_{\vec{n}}^{\text{ext}} \quad \text{and} \quad C_{\vec{n}}^{\text{free}} = K_{\text{H}} P_{\vec{n}}, \tag{21.2}$$

in terms of the Henry's law constant K_{H} (mol cm^{-3} bar^{-1}) for a gas and its partial pressures $P_{\vec{n}}^{\text{ext}}$ and $P_{\vec{n}}$ (bar) in the external reservoir and fluid, respectively. A Henry's law constant is specific to the gas in question and varies with temperature and ionic strength.

Defining a rate constant $(k_+)_{\vec{n}}$ (mol cm^{-2} s^{-1} bar^{-1}) for gas transfer as the product $K_1 K_{\text{H}}$ and substituting Equation 21.2 into 21.1, we arrive at a rate law in the useful form

$$r_{\vec{n}} = n_w A_{\text{sp}} (k_+)_{\vec{n}} \left(P_{\vec{n}}^{\text{ext}} - P_{\vec{n}} \right). \tag{21.3}$$

Here, n_w is the mass of solvent water (kg), as before, and A_{sp} (cm^2 kg^{-1}) is specific contact area, the interfacial area per kg of solvent. Many geochemical modeling codes, including the GWB programs, use the gas's fugacity $f_{\vec{n}}$ as a proxy for its partial pressure when calculating gas transfer rates, so the rate law takes the form

$$r_{\vec{n}} = n_w A_{\text{sp}} (k_+)_{\vec{n}} \left(f_{\vec{n}}^{\text{ext}} - f_{\vec{n}} \right). \tag{21.4}$$

In this case, the rate constant $(k_+)_{\vec{n}}$ is given by $K_1 K_{\text{H}} / \varphi_{\vec{n}}$, where $\varphi_{\vec{n}}$ is the fugacity coefficient, as described in Section 3.3.8, and takes on units of mol cm^{-2} s^{-1}.

21.2 Numerical Solution

To trace kinetic gas transfer in a reaction model, we identify one or more gas species $A_{\vec{n}}$, each of which passes between the fluid and an external gas reservoir at a rate prescribed by a kinetic law. A kinetic gas reacts with the basis species according to Reaction 3.43,

$$A_{\vec{n}} \rightleftarrows \nu_{w\vec{n}} A_w + \sum_i \nu_{i\vec{n}} A_i + \sum_k \nu_{k\vec{n}} A_k + \sum_m \nu_{m\vec{n}} A_m, \tag{21.5}$$

where the corresponding equilibrium constant is $K_{\vec{n}}$.

The gas's fugacity $f_{\vec{n}}^{\text{ext}}$ in equilibrium with the external reservoir is prescribed as a boundary condition in the simulation, whereas fugacity $f_{\vec{n}}$ in the fluid is given by Equation 3.44,

$$f_{\vec{n}} = \frac{a_w^{\nu_{w\vec{n}}} \prod^{i} (\gamma_i m_i)^{\nu_{i\vec{n}}} \prod^{k} a_k^{\nu_{k\vec{n}}} \prod^{m} f_m^{\nu_{m\vec{n}}}}{K_{\vec{n}}}. \tag{21.6}$$

The rate at which Reaction 21.5 proceeds from left to right, the rate the gas is taken up by the fluid, is given by a rate law,

$$r_{\vec{n}} = n_w A_{\text{sp}}(k_+)_{\vec{n}} \prod^{j'} (m_{j'})^{p_{j'\vec{n}}} \left(f_{\vec{n}}^{\text{ext}} - f_{\vec{n}} \right), \tag{21.7}$$

of general form. For consistency with our development to this point, we allow the possibility of promoting and inhibiting species, where $m_{j'}$ is the molality of such species and $p_{j'\vec{n}}$ is the corresponding exponent.

When kinetic gases cross the interface, the mole numbers of the basis entries change,

$$\frac{\mathrm{d}M_w}{\mathrm{d}t} = \sum_{\vec{n}} \nu_{w\vec{n}} r_{\vec{n}}, \quad \frac{\mathrm{d}M_i}{\mathrm{d}t} = \sum_{\vec{n}} \nu_{i\vec{n}} r_{\vec{n}}, \quad \text{and} \quad \frac{\mathrm{d}M_k}{\mathrm{d}t} = \sum_{\vec{n}} \nu_{k\vec{n}} r_{\vec{n}}, \tag{21.8}$$

at rates that reflect the transport rates $r_{\vec{n}}$ (Eq. 21.7). Moving forward from time t' to t, then, the mole numbers after the time step,

$$M_w(t) = M_w(t') + \frac{(t-t')}{2} \sum_{\vec{n}} \nu_{w\vec{n}} \left[r_{\vec{n}}(t') + r_{\vec{n}}(t) \right] \tag{21.9}$$

$$M_i(t) = M_i(t') + \frac{(t-t')}{2} \sum_{\vec{n}} \nu_{i\vec{n}} \left[r_{\vec{n}}(t') + r_{\vec{n}}(t) \right] \tag{21.10}$$

$$M_k(t) = M_k(t') + \frac{(t-t')}{2} \sum_{\vec{n}} \nu_{k\vec{n}} \left[r_{\vec{n}}(t') + r_{\vec{n}}(t) \right], \tag{21.11}$$

follow from the corresponding values at the step's onset and the transport rates $r_{\vec{n}}$ figured at t' and t. The rate terms in these equations may be recast as weighted averages, as we did in Equation 17.22, to allow adjustable time weighting in the solution procedure.

Substituting Equations 21.9–21.10 into Equations 4.22–4.23, we arrive at a pair of residual functions,

$$R_w = n_w \left(55.5 + \sum_j \nu_{wj} m_j \right) - M_w(t') - \frac{(t-t')}{2} \sum_{\vec{n}} \nu_{w\vec{n}} \left[r_{\vec{n}}(t') + r_{\vec{n}}(t) \right] \tag{21.12}$$

$$R_i = n_w \left(55.5 + \sum_j \nu_{ij} m_j \right) - M_i(t') - \frac{(t-t')}{2} \sum_{\vec{n}} \nu_{i\vec{n}} \left[r_{\vec{n}}(t') + r_{\vec{n}}(t) \right]. \tag{21.13}$$

Noting that

$$\frac{\partial m_j}{\partial n_w} = 0, \quad \frac{\partial m_j}{\partial m_i} = \frac{\nu_{ij} m_j}{m_i}, \quad \frac{\partial f_{\vec{n}}}{\partial n_w} = 0, \quad \text{and} \quad \frac{\partial f_{\vec{n}}}{\partial m_i} = \frac{\nu_{i\vec{n}} f_{\vec{n}}}{m_i}, \tag{21.14}$$

the partial derivatives of the rate law (Eq. 21.7) with respect to the independent variables are $\partial r_{\vec{n}} / \partial n_w = 0$ and

$$\frac{\partial r_{\vec{n}}}{\partial m_i} = \frac{n_w A_{sp}(k_+)_{\vec{n}}}{m_i} \prod^{j'} m_{j'}^{p_{j'\vec{n}}} \left[\left(\sum_{j'} \nu_{ij'} p_{j'\vec{n}} \right) \left(f_{\vec{n}}^{\text{ext}} - f_{\vec{n}} \right) - \nu_{i\vec{n}} f_{\vec{n}} \right]. \tag{21.15}$$

Entries in the Jacobian matrix follow from differentiating the residual functions

$$J_{ww} = 55.5 + \sum_j \nu_{wj} m_j \tag{21.16}$$

$$J_{wi} = n_w \left(\frac{1}{m_i} \sum_j \nu_{wj} \nu_{ij} m_j + \frac{(t - t')}{2} \sum_{\vec{n}} \nu_{w\vec{n}} \frac{\partial r_{\vec{n}}}{\partial m_i} \right) \tag{21.17}$$

$$J_{iw} = m_i + \sum_j \nu_{ij} m_j \tag{21.18}$$

and

$$J_{ii'} = n_w \left(\delta_{ii'} + \frac{1}{m_{i'}} \sum_j \nu_{ij} \nu_{i'j} m_j - \frac{(t - t')}{2} \sum_{\vec{n}} \nu_{i\vec{n}} \frac{\partial r_{\vec{n}}}{\partial m_{i'}} \right), \tag{21.19}$$

where the rate derivatives are given by Equation 21.15. Iterating using the Newton–Raphson method (see Section 4.2) leads to $M_w(t)$ and $M_i(t)$; component masses $M_k(t)$ at the end of the time step are then given directly by Equation 21.11.

21.3 CO_2 Efflux From a Small Lake

Gas transfer across the surface of a freshwater lake constitutes in most cases a net source of $CO_2(g)$ to the atmosphere (e.g., Pace and Prairie, 2005). The efflux represents escape of inorganic carbon derived from some combination of supply by recharge waters and biological activity within the lake. Even a lake that emits carbon dioxide across its surface, nonetheless, may serve as a net carbon sink on a watershed scale, by segregating and burying some fraction of the allochthonous carbon carried into it by recharge waters, windfall, and so on (e.g., Webb *et al.*, 2019).

Biological activity within a lake can serve as either a net source or sink of dissolved carbon dioxide. A lake is net autotrophic if photosynthesis takes up more $CO_2(aq)$ than respiring organisms produce, and net heterotrophic in the contrary case. The net ecosystem production, or NEP, is the difference between the primary production of organic carbon compounds by photosynthesis and their oxidation by respiring organisms. The NEP, expressed in units such as g C m^{-2} day^{-1}, is therefore positive in a net autotrophic, and negative in a net heterotrophic lake (e.g., Bogard and del Giorgio, 2016).

To construct an example calculation, we consider a small, well-mixed lake that is recharged by base flow and discharges as streamwater. Groundwater influx is sufficiently continuous that the lakewater chemistry is invariant. Knowing the lake is of h = 300 cm average depth, and holds a Ca–HCO_3 water with an alkalinity of 40 mg kg^{-1} as $CaCO_3$ and a pH of 6, we wish to determine the $CO_2(g)$ flux between lake and atmosphere, assuming a transfer coefficient K_1 of 1 cm hr^{-1}, or 2.78×10^{-4} cm s^{-1}.

To begin, we need to compute the distribution of species in the lake water so we can calculate the appropriate Henry's law constant K_H for $CO_2(g)$ and the corresponding rate constant $(k_+)_{\vec{n}}$. In REACT, running the commands

```
HCO3- = 40 mg/kg_as_CaCO3
pH    = 6
balance on Ca++
go
```

predicts that the free $CO_2(aq)$ concentration is 1.78 mmolal, or 1.78×10^{-6} mol cm^{-3}, since fluid density ρ is 0.996 g cm^{-3}. The calculated P_{CO_2} in the water is 0.0508 bar and the fugacity coefficient φ_{CO_2} is 0.995. The TDS is only 144 mg kg^{-1}, so solution mass is functionally equivalent to solvent mass for our purposes.

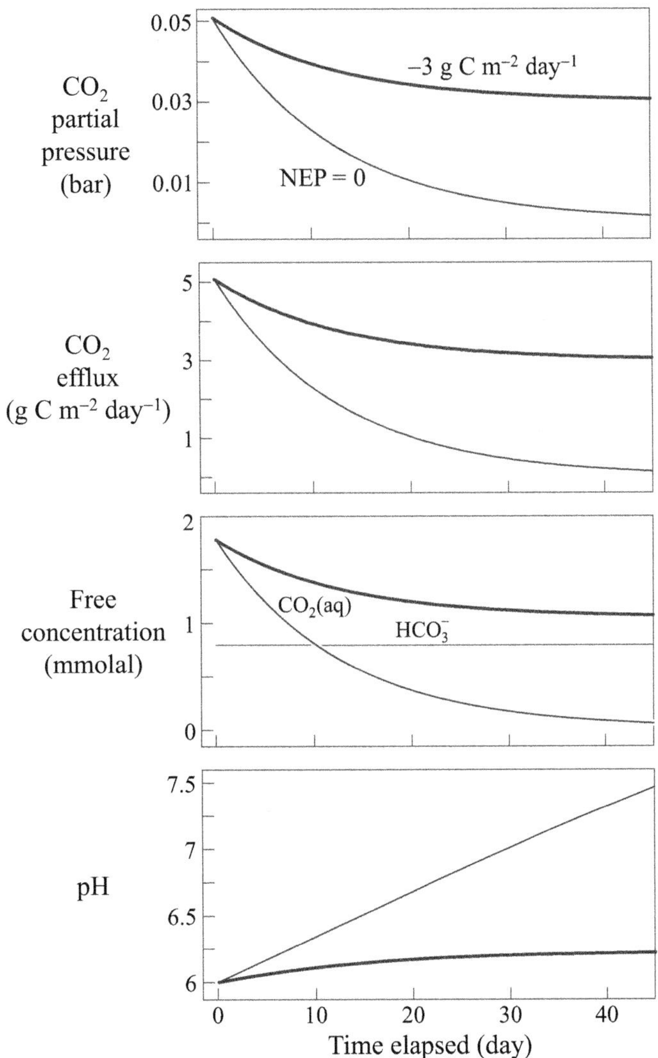

Figure 21.1 Change in lakewater chemistry with time as CO_2(g) diffuses into the atmosphere. Medium lines show calculation in which respiration and photosynthesis are in balance, so that net ecosystem production (NEP) is zero; bold lines show case of negative NEP, in which respiration produces more CO_2(aq) than photosynthesis consumes. The free concentration of HCO_3^- (fine line) is similar in each case.

By Equation 21.2, K_H is the ratio of free concentration to partial pressure, or 3.50×10^{-5} mol cm^{-3} bar^{-1}. REACT uses a rate law in the form of Equation 21.4, so the rate constant $(k_+)_{\vec{n}}$ is given as $K_1 K_H / \varphi_{CO_2}$, which is 9.78×10^{-9} mol cm^{-2} s^{-1}. By geometry, the specific contact area A_{sp} equals $1000 / \rho h$, or 3.35 cm^2 kg^{-1}. Continuing in REACT, we prescribe the rate law for gas transfer

(cont'd)
```
kinetic CO2(g) rate_con = 9.78e-9, contact = 3.35, P_ext = 10^-3.5 atm
go initial
```

in terms of the rate constant and contact area. We further set a value of $10^{-3.5}$ atm for atmospheric P_{CO_2}, which the program converts and carries internally as f_{CO_2}. The calculation predicts a transfer rate $r_{\vec{n}} / n_w$ of -1.65×10^{-9} molal s^{-1}. Dividing by the specific contact area A_{sp} and inverting the sign gives a $CO_2(g)$ flux from lake to atmosphere of 4.93×10^{-10} mol cm^{-2} s^{-1}, or 5.11 g C m^{-2} day^{-1}.

As a variation on the calculation, we assume base flow into the same lake is seasonal and that the given alkalinity and pH were observed at a point in time after recharge ceased. Taking the NEP to be zero, the case when photosynthesis and respiration are in balance, we would now like to know how the lake chemistry and gas flux evolve. The commands

(cont'd)
```
time end 45 days
go
```

launch a transient simulation in REACT, using the existing configuration as the initial condition.

In the calculation results (Fig. 21.1), escape of carbon dioxide to the atmosphere progressively diminishes $CO_2(aq)$ in the lake water and thereby drives down P_{CO_2}. The P_{CO_2} decrease, in turn, suppresses gas loss, so that a month into the simulation the efflux proceeds at less than a tenth of its original rate. Depleting $CO_2(aq)$ further serves to consume HCO_3^- and H^+ from the lake water by driving forward the reaction

$$HCO_3^- + H^+ \rightarrow CO_2(aq) + H_2O\,. \tag{21.20}$$

After 45 days, the reaction has shifted by 0.0030 mmolal, the amount of $CO_2(aq)$ degassed less the change observed in the species' free mass. The shift is small relative to the amount of free HCO_3^- in solution, but considerably in excess of the free H^+ concentration, 0.0010 mmolal, in the initial system.

Mass of the HCO_3^- species, as a result, holds nearly steady over the course of the simulation, whereas the H^+ concentration falls sharply, leading pH in the lake water to rise. In response, HCO_3^- and H_2O in solution deprotonate to liberate CO_3^{--} and OH^-, respectively, and H^+. The latter reactions make up for the 0.0020 mmolal deficit in hydrogen ions needed to account for the observed shift in Reaction 21.20.

To investigate how biological activity might affect the calculation results, we consider the case of a net heterotrophic lake. At an NEP of -3.0 g C m^{-2} day^{-1}, $CO_2(aq)$ is introduced to the lake water at the rate of 0.97×10^{-9} molal s^{-1}. To account in the REACT simulation for the internal source, we append the commands

(cont'd)
```
react 0.97e-9 molal/s CO2(aq)
go
```

to the configuration. In the calculation results (Fig. 21.1), we see that the biological activity acts to replace some of the carbon dioxide lost to the atmosphere. For this reason, P_{CO_2} and the efflux rate plateau at a higher level than in the biologically neutral simulation, and the lake chemistry is less strongly affected.

22

Stable Isotopes

Stable isotopes serve as naturally occurring tracers that can provide much information about how chemical reactions proceed in nature, such as which reactants are consumed and at what temperature reactions occur. The stable isotopes of several of the lighter elements are sufficiently abundant and fractionate strongly enough to be of special usefulness. Foremost in importance are hydrogen, carbon, oxygen, and sulfur.

The strong conceptual link between stable isotopes and chemical reaction makes it possible to integrate isotope fractionation into reaction modeling, allowing us to predict not only the mineralogical and chemical consequences of a reaction process, but also the isotopic compositions of the reaction products. By tracing the distribution of isotopes in our calculations, we can better test our reaction models against observation and perhaps better understand how isotopes fractionate in nature.

Bowers and Taylor (1985) were the first to incorporate isotope fractionation into a reaction model. They used a modified version of EQ3/EQ6 (Wolery, 1979) to study the convection of hydrothermal fluids through the oceanic crust, along midocean ridges. Their calculation method is based on evaluating mass balance equations, as described in this chapter.

As originally derived, however, the mass balance model has an important (and well acknowledged) limitation: implicit in its formulation is the assumption that fluid and minerals in the modeled system remain in isotopic equilibrium over the reaction path. This assumption is equivalent to assuming that isotope exchange between fluid and minerals occurs rapidly enough to maintain equilibrium compositions.

We know, however, that isotope exchange in nature tends to be a slow process, especially at low temperature (e.g., O'Neil, 1987). This knowledge comes from experimental study (e.g., Cole and Ohmoto, 1986) as well as from the simple observation that, unless they have reacted together, groundwaters and minerals are seldom observed to be in isotopic equilibrium with each other. In fact, if exchange were a rapid process, it would be very difficult to interpret the origin of geologic materials from their isotopic compositions: the information would literally diffuse away.

Lee and Bethke (1996) presented an alternative technique, also based on mass balance equations, in which the reaction modeler can segregate minerals from isotopic exchange. By segregating the minerals, the model traces the effects of the isotope fractionation that would result from dissolution and precipitation reactions alone. Not unexpectedly, segregated models differ broadly in their results from reaction models that assume isotopic equilibrium.

The segregated model works by defining a subset of the chemical equilibrium system called the isotope equilibrium system. The isotope system contains (1) the fluid, (2) any minerals not held segregated from isotope exchange, (3) any segregated minerals that dissolve over the current reaction step, and (4) the increments in mass of any segregated minerals that precipitate over the step. By holding the components of just the isotope system in equilibrium, the calculation procedure assures that the compositions of segregated minerals

Table 22.1 *The stable isotopes of hydrogen, carbon, oxygen, and sulfur, and their approximate terrestrial abundances (Wedepohl* et al., *1978)*

Isotope	Abundance (%)	Isotope	Abundance (%)
Hydrogen		*Carbon*	
^{1}H	99.984	^{12}C	98.89
^{2}H	0.016	^{13}C	1.11
Oxygen		*Sulfur*	
^{16}O	99.763	^{32}S	94.94
^{17}O	0.0375	^{33}S	0.77
^{18}O	0.1995	^{34}S	4.27
		^{36}S	0.02

change only when new mass precipitates. Conversely, the segregated minerals affect the composition of the fluid only when they dissolve. In this way, chemical reaction is the only force driving isotope fractionation.

These ideas might be further developed to model situations in which exchange among aqueous species occurs slowly. The oxygen in SO_4^{--} species might not be allowed to exchange with water, or H_2S sulfur might be held segregated from sulfur in SO_4^{--}. To date, however, no description of isotope disequilibrium within the fluid phase has been implemented within a reaction model.

In this chapter, we develop a mass balance model of the fractionation in reacting systems of the stable isotopes of hydrogen, carbon, oxygen, and sulfur. We then demonstrate application of the model by simulating the isotopic effects of the dolomitization reaction of calcite.

22.1 Isotope Fractionation

Each species and phase of an element with two or more stable isotopes consists of light and heavy isotopes in proportions that can be measured by using a mass spectrometer (e.g., Faure, 1986). The isotope ratio R is the quotient of the number of moles of a heavy isotope (such as ^{18}O) to the number of moles of a light isotope (such as ^{16}O).

It is not especially practical, however, to express isotopic composition in terms of R. One isotope (e.g., ^{16}O) greatly dominates the others in each case (Table 22.1), so values of R are small, rather inconvenient numbers. In addition, mass spectrometers can measure the difference in isotopic composition between two samples much more accurately than they can determine an absolute ratio R. For these reasons, isotope geochemists express isotopic composition in a δ notation,

$$\delta = \left[\frac{R_{\text{Sample}} - R_{\text{Standard}}}{R_{\text{Standard}}}\right] \times 1000\,, \tag{22.1}$$

as the permil (‰) deviation from a standard. For example, the δ value for ^{18}O in a sample is

$$\delta^{18}\text{O} = \left[\frac{(^{18}\text{O}/^{16}\text{O})_{\text{Sample}} - (^{18}\text{O}/^{16}\text{O})_{\text{SMOW}}}{(^{18}\text{O}/^{16}\text{O})_{\text{SMOW}}}\right] \times 1000\,, \tag{22.2}$$

expressed relative to the SMOW standard; SMOW is the composition of "standard mean ocean water," the usual standard for this element. In this notation, a sample with a positive $\delta^{18}O$ is enriched in the heavy isotope ^{18}O relative to the standard, whereas a negative value shows that the sample is depleted in the isotope.

When species or phases are in isotopic equilibrium, their isotopic ratios differ from one another by predictable amounts. The segregation of heavier isotopes into one species and light isotopes into the other is called isotope fractionation. The fractionation among species is represented by a fractionation factor α, which is determined empirically. The fractionation factor between species A and B is the ratio

$$\alpha_{A-B} = \frac{R_A}{R_B} = \frac{\delta_A + 1000}{\delta_B + 1000} \,. \tag{22.3}$$

By rearranging this equation, we can express the isotopic composition of species A,

$$\delta_A = \alpha_{A-B}(1000 + \delta_B) - 1000 \,, \tag{22.4}$$

from the fractionation factor and the composition of B. The useful expression

$$1000 \ln \alpha_{A-B} \simeq \delta_A - \delta_B \tag{22.5}$$

follows from the approximation $\ln x \simeq x - 1$ for $x \to 1$. Since this form gives the difference in δ values between A and B directly, fractionation factors are commonly compiled expressed as $1000 \ln \alpha$.

As a rule, isotopes fractionate more strongly at low temperatures than they do at high temperatures. Variation of the fractionation factor with absolute temperature T_{K} can be fit to a polynomial, such as

$$1000 \ln \alpha_{A-B} = a + bT_{\text{K}} + c\left(\frac{10^3}{T_{\text{K}}}\right) + d\left(\frac{10^6}{T_{\text{K}}^2}\right) + e\left(\frac{10^9}{T_{\text{K}}^3}\right) + f\left(\frac{10^{12}}{T_{\text{K}}^4}\right) . \tag{22.6}$$

Here, coefficients a through f are determined by fitting measurements to the polynomial. Equations of this form fit the data well in most cases, but may fail to describe hydrogen fractionation at high temperature in hydrated minerals or in minerals containing hydroxyl groups.

In REACT, dataset "isotope.dat" contains polynomial coefficients that define temperature functions for the fractionation factors of species, minerals, and gases. The factors describe fractionation relative to a reference species chosen for each element. The reference species for oxygen and hydrogen is solvent water, H_2O. CO_2 and H_2S, in either aqueous or gaseous form, serve as reference species for carbon and sulfur.

22.2 Mass Balance Equations

The key to tracing isotope fractionation over a reaction path, as shown by Bowers and Taylor (1985), is writing a mass balance equation for each fractionating element. We begin by specifying a part of the chemical system to be held in isotopic equilibrium, as already described. The isotope system, of course, excludes segregated minerals. At each step over the course of the reaction path, we apply the mass balance equations to determine, from the bulk composition of the isotope system, the isotopic composition of each species and phase. In this section, we derive the mass balance equation for each element; in the next we show how the equations can be integrated into reaction modeling.

We begin with oxygen. The total number of moles $n^{\text{T}}_{^{18}\text{O}}$ of ^{18}O in the isotope system is the sum of the mole numbers for this isotope in (1) the solvent, $n^{w}_{^{18}\text{O}}$, (2) the aqueous species, $n^{j}_{^{18}\text{O}}$, and (3) whatever mineral mass appears in the isotope system, $n^{k'}_{^{18}\text{O}}$. Expressed mathematically,

$$n^{\text{T}}_{^{18}\text{O}} = n^{w}_{^{18}\text{O}} + \sum_j n^{j}_{^{18}\text{O}} + \sum_{k'} n^{k'}_{^{18}\text{O}} \,. \tag{22.7}$$

A parallel expression,

$$n^{\mathrm{T}}_{^{16}\mathrm{O}} = n^{w}_{^{16}\mathrm{O}} + \sum_{j} n^{j}_{^{16}\mathrm{O}} + \sum_{k'} n^{k'}_{^{16}\mathrm{O}}\,, \tag{22.8}$$

applies to ^{16}O. Note that we use the notation k' to identify the mineral mass carried in the isotope system; as before, superscript k denotes minerals in the chemical system, whether in isotopic equilibrium or not.

Using these expressions and Equation 22.2, and recognizing that the mole ratio $n_{^{16}\mathrm{O}}/n_{\mathrm{O}}$ of ^{16}O to elemental oxygen is nearly constant, we can show that the total (or bulk) isotopic composition of the isotope system,

$$\delta^{18}\mathrm{O}_{\mathrm{T}} \simeq \frac{1}{n^{\mathrm{T}}_{\mathrm{O}}} \left[n^{w}_{\mathrm{O}}\, \delta^{18}\mathrm{O}_{w} + \sum_{j} n^{j}_{\mathrm{O}}\, \delta^{18}\mathrm{O}_{j} + \sum_{k'} n^{k'}_{\mathrm{O}}\, \delta^{18}\mathrm{O}_{k'} \right]. \tag{22.9}$$

can be calculated from the δ values of solvent ($\delta^{18}\mathrm{O}_{w}$), species ($\delta^{18}\mathrm{O}_{j}$), and minerals ($\delta^{18}\mathrm{O}_{k'}$).

This equation can be expanded by expressing the compositions of species and minerals in terms of their fractionation factors and the composition $\delta^{18}\mathrm{O}_{w}$ of solvent water, the reference species. From Equation 22.4,

$$\delta^{18}\mathrm{O}_{j} = \alpha_{j-w} \left(1000 + \delta^{18}\mathrm{O}_{w}\right) - 1000 \tag{22.10}$$

and

$$\delta^{18}\mathrm{O}_{k'} = \alpha_{k'-w} \left(1000 + \delta^{18}\mathrm{O}_{w}\right) - 1000\,. \tag{22.11}$$

By substituting these relations,

$$\begin{aligned} n^{\mathrm{T}}_{\mathrm{O}}\, \delta^{18}\mathrm{O}_{\mathrm{T}} = n^{w}_{\mathrm{O}}\, \delta^{18}\mathrm{O}_{w} &+ \sum_{j} n^{j}_{\mathrm{O}} \left[\alpha_{j-w} \left(1000 + \delta^{18}\mathrm{O}_{w}\right) - 1000\right] \\ &+ \sum_{k'} n^{k'}_{\mathrm{O}} \left[\alpha_{k'-w} \left(1000 + \delta^{18}\mathrm{O}_{w}\right) - 1000\right], \end{aligned} \tag{22.12}$$

and solving for the composition of solvent water, we arrive at

$$\delta^{18}\mathrm{O}_{w} = \frac{n^{\mathrm{T}}_{\mathrm{O}}\delta^{18}\mathrm{O}_{\mathrm{T}} - 1000 \left[\sum_{j} n^{j}_{\mathrm{O}} \left(\alpha_{j-w} - 1\right) + \sum_{k'} n^{k'}_{\mathrm{O}} \left(\alpha_{k'-w} - 1\right) \right]}{n^{w}_{\mathrm{O}} + \sum_{j} n^{j}_{\mathrm{O}}\alpha_{j-w} + \sum_{k'} n^{k'}_{\mathrm{O}}\alpha_{k'-w}}, \tag{22.13}$$

which is the mass balance equation for oxygen.

Equation 22.13 is useful because it allows us to use the system's total isotopic composition, $\delta^{18}\mathrm{O}_{\mathrm{T}}$, to determine the compositions of the solvent and each species and mineral. The calculation proceeds in two steps. First, given $\delta^{18}\mathrm{O}_{\mathrm{T}}$ and the fractionation factors α_{j-w} and $\alpha_{k'-w}$ for the various species and minerals, we compute the composition of the solvent, applying Equation 22.13. Second, we use this result to calculate the composition of each species and mineral directly, according to Equations 22.10 and 22.11.

The equations for the isotope pairs ^{2}H/^{1}H, ^{13}C/^{12}C, and ^{34}S/^{32}S parallel the relations for ^{18}O/^{16}O, except that the reference species for carbon and sulfur are CO_2 and H_2S, rather than solvent water. Carbon and sulfur compositions are commonly reported with respect to the PDB (Pee Dee belemnite) and CDT (Canyon Diablo troilite) standards, instead of SMOW. It makes little difference which standard we choose in applying these equations, however, as long as we carry a single standard for each element through the calculation.

Fractionation of hydrogen, carbon, and sulfur isotopes among the aqueous species is set by the relations

$$\delta^2\mathrm{H}_j = \alpha_{j-w}\left(1000 + \delta^2\mathrm{H}_w\right) - 1000 \tag{22.14}$$
$$\delta^{13}\mathrm{C}_j = \alpha_{j-\mathrm{CO_2}}\left(1000 + \delta^{13}\mathrm{C}_{\mathrm{CO_2}}\right) - 1000 \tag{22.15}$$
$$\delta^{34}\mathrm{S}_j = \alpha_{j-\mathrm{H_2S}}\left(1000 + \delta^{34}\mathrm{S}_{\mathrm{H_2S}}\right) - 1000\,. \tag{22.16}$$

Similarly, the equations

$$\delta^2\mathrm{H}_{k'} = \alpha_{k'-w}\left(1000 + \delta^2\mathrm{H}_w\right) - 1000 \tag{22.17}$$
$$\delta^{13}\mathrm{C}_{k'} = \alpha_{k'-\mathrm{CO_2}}\left(1000 + \delta^{13}\mathrm{C}_{\mathrm{CO_2}}\right) - 1000 \tag{22.18}$$
$$\delta^{34}\mathrm{S}_{k'} = \alpha_{k'-\mathrm{H_2S}}\left(1000 + \delta^{34}\mathrm{S}_{\mathrm{H_2S}}\right) - 1000 \tag{22.19}$$

define fractionation among minerals.

Compositions of the reference species are given by

$$\delta^2\mathrm{H}_w = \frac{n_\mathrm{H}^\mathrm{T}\,\delta^2\mathrm{H}_\mathrm{T} - 1000\left[\sum_j n_\mathrm{H}^j\left(\alpha_{j-w} - 1\right) + \sum_{k'} n_\mathrm{H}^{k'}\left(\alpha_{k'-w} - 1\right)\right]}{n_\mathrm{H}^w + \sum_j n_\mathrm{H}^j\,\alpha_{j-w} + \sum_{k'} n_\mathrm{H}^{k'}\,\alpha_{k'-w}} \tag{22.20}$$

$$\delta^{13}\mathrm{C}_{\mathrm{CO_2}} = \frac{n_\mathrm{C}^\mathrm{T}\delta^{13}\mathrm{C}_\mathrm{T} - 1000\left[\sum_j n_\mathrm{C}^j\left(\alpha_{j-\mathrm{CO_2}} - 1\right) + \sum_{k'} n_\mathrm{C}^{k'}\left(\alpha_{k'-\mathrm{CO_2}} - 1\right)\right]}{\sum_j n_\mathrm{C}^j\,\alpha_{j-\mathrm{CO_2}} + \sum_{k'} n_\mathrm{C}^{k'}\,\alpha_{k'-\mathrm{CO_2}}} \tag{22.21}$$

$$\delta^{34}\mathrm{S}_{\mathrm{H_2S}} = \frac{n_\mathrm{S}^\mathrm{T}\,\delta^{34}\mathrm{S}_\mathrm{T} - 1000\left[\sum_j n_\mathrm{S}^j\left(\alpha_{j-\mathrm{H_2S}} - 1\right) + \sum_{k'} n_\mathrm{S}^{k'}\left(\alpha_{k'-\mathrm{H_2S}} - 1\right)\right]}{\sum_j n_\mathrm{S}^j\,\alpha_{j-\mathrm{H_2S}} + \sum_{k'} n_\mathrm{S}^{k'}\,\alpha_{k'-\mathrm{H_2S}}}\,, \tag{22.22}$$

which constitute the mass balance equations for these elements.

22.3 Fractionation in Reacting Systems

Integrating isotope fractionation into the reaction path calculation is a matter of applying the mass balance equations while tracing over the course of the reaction path the system's total isotopic composition. Much of the effort in programming an isotope model consists of devising a careful accounting of the mass of each isotope.

The calculation begins with an initial fluid of specified isotopic composition. The model, by mass balance, assigns the initial compositions of the solvent and each aqueous species. The model then sets the composition of each unsegregated mineral to be in equilibrium with the fluid. The modeler specifies the composition of each segregated mineral as well as that of each reactant to be added to the system.

The model traces the reaction path by taking a series of steps along reaction progress, moving forward each step from ξ_1 to ξ_2. Over a step, the system's isotopic composition can change in two ways: reactants can be added or removed, and segregated minerals can dissolve.

Using ^{18}O as an example, we can calculate the composition $\delta^{18}\mathrm{O_T}(\xi_2)$ at the end of the step according to the equation

$$\delta^{18}\mathrm{O_T}(\xi_2) = \frac{1}{n_\mathrm{O}^\mathrm{T}(\xi_2)}\Big[n_\mathrm{O}^\mathrm{T}\,\delta^{18}\mathrm{O}_T(\xi_1) + \sum_r \Delta n_r\, n_\mathrm{O}^r\,\delta^{18}\mathrm{O}_r - \sum_{k\neq k'} \Delta n_k^{(-)}\, n_\mathrm{O}^k\,\delta^{18}\mathrm{O}_k\Big]. \tag{22.23}$$

Here, $\delta^{18}\mathrm{O_T}(\xi_1)$ is the composition at the beginning of the reaction step. Δn_r is the number of moles of reactant r added over the step (its value is negative for a reactant being removed from the system) and n_O^r is the number of oxygen atoms in the reactant's stoichiometry. For each segregated mineral $k \neq k'$ that dissolves over the step, $\Delta n_k^{(-)}$ is the mineral's change in mass (in moles) and n_O^k is its stoichiometric oxygen content. $\delta^{18}\mathrm{O}_r$ and $\delta^{18}\mathrm{O}_k$ are the isotopic compositions of reactant and segregated mineral. The mole number $n_\mathrm{O}^\mathrm{T}(\xi_2)$ of oxygen at the end of the step,

$$n_\mathrm{O}^\mathrm{T}(\xi_2) = n_\mathrm{O}^\mathrm{T}(\xi_1) + \sum_r \Delta n_r\, n_\mathrm{O}^r - \sum_{k\neq k'} \Delta n_k^{(-)}\, n_\mathrm{O}^k\,, \tag{22.24}$$

is given by summing over the contributions of reactants and dissolving segregated minerals.

The parallel equations for hydrogen, carbon, and sulfur isotopes are

$$\delta^{2}\mathrm{H_T}(\xi_2) = \frac{1}{n_\mathrm{H}^\mathrm{T}(\xi_2)}\Big[n_\mathrm{H}^\mathrm{T}\,\delta^{2}\mathrm{H_T}(\xi_1) + \sum_r \Delta n_r\, n_\mathrm{H}^r\,\delta^{2}\mathrm{H}_r - \sum_{k\neq k'} \Delta n_k^{(-)}\, n_\mathrm{H}^k\,\delta^{2}\mathrm{H}_k\Big] \tag{22.25}$$

$$n_\mathrm{H}^\mathrm{T}(\xi_2) = n_\mathrm{H}^\mathrm{T}(\xi_1) + \sum_r \Delta n_r\, n_\mathrm{H}^r - \sum_{k\neq k'} \Delta n_k^{(-)}\, n_\mathrm{H}^k \tag{22.26}$$

$$\delta^{13}\mathrm{C_T}(\xi_2) = \frac{1}{n_\mathrm{C}^\mathrm{T}(\xi_2)}\Big[n_\mathrm{C}^\mathrm{T}\,\delta^{13}\mathrm{C_T}(\xi_1) + \sum_r \Delta n_r\, n_\mathrm{C}^r\,\delta^{13}\mathrm{C}_r - \sum_{k\neq k'} \Delta n_k^{(-)}\, n_\mathrm{C}^k\,\delta^{13}\mathrm{C}_k\Big] \tag{22.27}$$

$$n_\mathrm{C}^\mathrm{T}(\xi_2) = n_\mathrm{C}^\mathrm{T}(\xi_1) + \sum_r \Delta n_r\, n_\mathrm{C}^r - \sum_{k\neq k'} \Delta n_k^{(-)}\, n_\mathrm{C}^k \tag{22.28}$$

$$\delta^{34}\mathrm{S_T}(\xi_2) = \frac{1}{n_\mathrm{S}^\mathrm{T}(\xi_2)}\Big[n_\mathrm{S}^\mathrm{T}\,\delta^{34}\mathrm{S_T}(\xi_1) + \sum_r \Delta n_r\, n_\mathrm{S}^r\,\delta^{34}\mathrm{S}_r - \sum_{k\neq k'} \Delta n_k^{(-)}\, n_\mathrm{S}^k\,\delta^{34}\mathrm{S}_k\Big] \tag{22.29}$$

$$n_\mathrm{S}^\mathrm{T}(\xi_2) = n_\mathrm{S}^\mathrm{T}(\xi_1) + \sum_r \Delta n_r\, n_\mathrm{S}^r - \sum_{k\neq k'} \Delta n_k^{(-)}\, n_\mathrm{S}^k\,. \tag{22.30}$$

Once we have computed the total isotopic compositions, we calculate the compositions of the reference species using the mass balance equations (Eqs. 22.13, 22.20, 22.21, 22.22). We can then use the isotopic

Table 22.2 *Assigning isotopic compositions in a reaction model (Lee and Bethke, 1996)*

Initial system	
Water & aqueous species	In equilibrium with each other, reflecting fluid's bulk composition, as set by modeler.
Unsegregated minerals & gases	In equilibrium with fluid.
Segregated minerals	Set by modeler.
Reactants being added to system	
Species, minerals, & gases	Set by modeler.
Buffered gases	In equilibrium with initial system.
Reactants being removed from system	
Species, minerals, gases, & buffered gases	In equilibrium with system at beginning of current step.
Chemical system, over course of reaction path	
Water, species, & unsegregrated minerals	In equilibrium with each other, reflecting system's bulk composition.
Segregated minerals	Increment precipitated over reaction step forms in equilibrium with fluid; composition of pre-existing mass is unaffected.
Gases	In equilibrium with current system.

compositions of the reference species to calculate the compositions of the other species (Eqs. 22.10, 22.14, 22.15, 22.16) and the unsegregated minerals (Eqs. 22.11, 22.17, 22.18, 22.19).

To update the composition of each segregated mineral, we average the composition (e.g., $\delta^{18}\mathrm{O}_{k'}$) of the mass that precipitated over the reaction step and the composition (e.g., $\delta^{18}\mathrm{O}_k(\xi_1)$) of whatever mass was present at the onset. The averaging equation for oxygen, for example, is

$$\delta^{18}\mathrm{O}_k(\xi_2) = \frac{n_k\,\delta^{18}\mathrm{O}_k(\xi_1) + \Delta n_k^{(+)}\,\delta^{18}\mathrm{O}_{k'}}{n_k(\xi_1) + \Delta n_k^{(+)}}\,, \tag{22.31}$$

where n_k is the mole number of mineral k and $\Delta n_k^{(+)}$ is the increase in this value over the reaction step. The compositions of segregated minerals that dissolve over the step are unchanged.

The precise manner in which we apply the mass transfer equations (Eqs. 22.23–22.30) depends on how we have configured the reaction path. Table 22.2 provides an overview of the process of assigning isotopic compositions to the various constituents of the model. When a simple reactant is added to the system (see Chapter 14), the increment Δn_r added is the reaction rate n_r multiplied by the step length, $\xi_2 - \xi_1$. The modeler explicitly prescribes the reactant's isotopic composition ($\delta^{18}\mathrm{O}_r$, and so on). When a simple reactant is removed from the system, the value of Δn_r is negative and the reactant's isotopic composition is the value in equilibrium with the system at the beginning of the step. A mineral reactant that precipitates and dissolves according to a kinetic rate law (Chapter 17) is treated in the same fashion, except that the model must calculate the increment Δn_r (which is positive when the mineral dissolves and negative when it precipitates) from the rate law.

In fixed- and sliding-fugacity paths, the model transfers gas into and out of an external buffer to obtain the fugacity desired at each step along the path (see Chapter 15). The increment Δn_r is the change in the total mole number M_m of the gas component as it passes to and from the buffer (see Chapter 3). When gas passes from buffer to system (Δn_r is positive), it is probably most logical to take its isotopic composition as the value in equilibrium with the initial system, at the start of the reaction path. Gas passing from system to buffer (i.e., Δn_r is negative) should be in isotopic equilibrium with the system at the start of the current reaction step. For polythermal paths (Chapter 15), it is necessary to update the fractionation factors α at each step, according to Equation 22.6, before evaluating the fractionation and mass balance equations (Eqs. 22.10, 22.14, 22.15, and 22.16; 22.11, 22.17, 22.18, and 22.19; and 22.13, 22.20, 22.21, and 22.22).

22.4 Dolomitization of a Limestone

As an example of how we might integrate isotope fractionation into reaction modeling (borrowing from Lee and Bethke, 1996), we consider the dolomitization of a limestone as it reacts after burial with a migrating pore fluid. When dolomite [$CaMg(CO_3)_2$] forms by alteration of a carbonate mineral, geochemists commonly assume that the dolomite reflects the isotopic composition of carbon in the precursor mineral (e.g., Mattes and Mountjoy, 1980; Meyers and Lohmann, 1985). This assumption seems logical since the reservoir of carbon in the precursor mineral, in this case calcite ($CaCO_3$), is likely to exceed that available from dissolved carbonate.

Some groundwaters in sedimentary basins, on the other hand, are charged with CO_2. In this case, if dolomite forms by reaction of a limestone with large volumes of migrating groundwater, the reservoir of dissolved carbon might be considerable. As we have noted, furthermore, the dolomitization reaction is best considered to occur in the presence of the carbon provided by dissolving calcite, not the entire reservoir of carbon present in the rock. Hence, it is interesting to use reaction modeling to investigate the factors controlling the isotopic composition of authigenic dolomite.

In our example, we test the consequences of reacting an isotopically light (i.e., nonmarine) limestone at 60 °C with an isotopically heavier groundwater that is relatively rich in magnesium. We start by defining the composition of a hypothetical groundwater that is of known CO_2 partial pressure (we initially set P_{CO_2} to 1 bar) and in equilibrium with dolomite:

```
T = 60
swap CO2(g) for H+
swap Dolomite-ord for HCO3-
P CO2(g) = 1 bar
1 free mol Dolomite-ord

Na+ = .1 molal
Cl- = .1 molal
Ca++ = .01 molal
Mg++ = .01 molal

go
```

The resulting fluid has an activity ratio $a_{Mg^{++}}/a_{Ca^{++}}$ of 1.26. By the reaction

$$\underset{\textit{dolomite}}{CaMg(CO_3)_2} + Ca^{++} \rightleftarrows 2\ \underset{\textit{calcite}}{CaCO_3} + Mg^{++}\,, \tag{22.32}$$

a fluid in equilibrium with calcite and dolomite has a ratio of 0.08 (as we can quickly show with program RXN). Hence, the groundwater is undersaturated with respect to calcite.

We then "pick up" the fluid from the previous step as a reactant and define a system representing the limestone and its pore fluid. We specify that the rock contains 3000 cm^3 of calcite, implying a porosity of about 25% since the extent of the system is 1 kg (about 1 liter) of solvent water. The pore fluid is similar to the reactant fluid, except that it contains less magnesium. The procedure is

(cont'd)
```
pickup reactants = fluid

swap Calcite for HCO3-
3000 free cm3 Calcite

pH = 6
Ca++ = .02  molal
Na+ =  .1   molal
Cl- =  .1   molal
Mg++ = .001 molal

reactants times 10
flush
```

The final commands define a reaction path in which 10 kg of reactant fluid gradually migrate through the system, displacing the existing (reacted) pore fluid. Typing `go` triggers the calculation.

As expected, the fluid, as it migrates through the limestone, converts calcite into dolomite. For each kg of fluid reacted, about 0.65 cm^3 (17.5 mmol) of calcite is consumed and 0.56 cm^3 (8.7 mmol) of dolomite forms.

To trace isotope fractionation over the reaction, we need to set the composition of the carbon in the initial system and the reactant fluid. We set $\delta^{13}C_{PDB}$ of the initial fluid to -10‰. By equilibrium with this value, the calcite has a composition of -6.1‰, as might be observed in a nonmarine limestone. We then set the composition of dissolved carbonate (HCO_3^- is the carbon-bearing component, as we can verify by typing `show`) in the reactant fluid to 0‰, a value typical for marine carbonate rocks. Finally, we specify that calcite and dolomite be held segregated from isotopic exchange. The procedure for tracing fractionation at low water–rock ratios is

(cont'd)
```
carbon initial = -10, HCO3- = 0
segregate Calcite, Dolomite-ord
go
```

To carry the calculation to high water–rock ratios, we enter the commands

(cont'd)
```
reactants times 4700
go
```

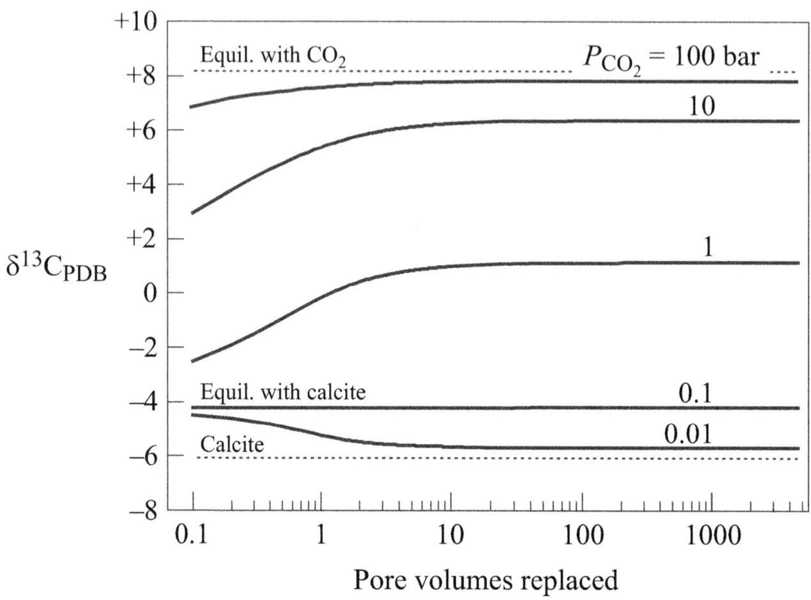

Figure 22.1 Carbon isotopic composition of dolomite formed by reaction between a limestone and migrating groundwater, calculated by holding minerals segregated from isotope exchange. In the calculation, calcite in the limestone reacts at 60 °C with the migrating groundwater, producing dolomite. Solid lines show results calculated assuming differing CO_2 partial pressures (from 0.01 to 100 bar) for the migrating fluid, plotted against the number of times the pore fluid was displaced. Horizontal lines show compositions of calcite in the limestone, and of dolomite in isotopic equilibrium with the calcite and with CO_2 in the migrating groundwater.

where the multiplier is set large enough for all the limestone in the system to transform into dolomite. We can then repeat the entire procedure (taking care each time to first type `reset`) for differing CO_2 partial pressures.

The calculation results (Fig. 22.1) show that the isotopic composition of the product dolomite depends strongly on the value assumed for the partial pressure of CO_2. At low pressure ($P_{CO_2} < 0.1$ bar), the dolomite forms at compositions similar to the δ^{13}C of the limestone, consistent with common interpretations in sedimentary geochemistry. At moderate to high CO_2 pressure ($P_{CO_2} > 1$ bar), however, the dolomite δ^{13}C more closely reflects the heavier carbon in the migrating pore fluid. Because of the fractionation of carbon between dolomite and CO_2 ($1000 \ln \alpha \simeq 4.6$), the isotopic composition of dolomite formed from groundwater rich in CO_2 can be as much as about 4‰ heavier than the groundwater δ^{13}C.

For each CO_2 pressure chosen, the dolomite composition during the reaction approaches a steady state, as can be seen in Figure 22.1. The steady state reflects the isotopic composition that balances sources (the unreacted fluid that enters the system and the calcite that dissolves over a reaction step) and sinks (the fluid displaced and dolomite precipitated over the step) of carbon to the isotope system. When the CO_2 pressure is low, the carbon source is isotopically light (being dominated by the dissolving calcite), and so are the sinks. At high partial pressure, conversely, the dominant carbon source is the heavy carbon in the migrating fluid. The pore fluid in the reacting system and the mineral mass that precipitates from it are therefore isotopically heavy.

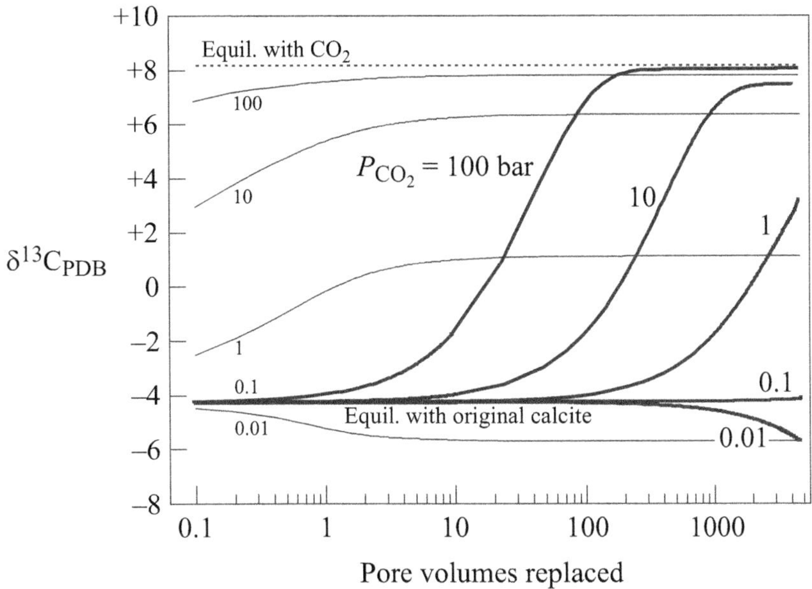

Figure 22.2 Isotopic composition (bold lines) of dolomite formed by reaction between a limestone and migrating groundwater, assuming that minerals maintain isotopic equilibrium over the simulation, determined assuming different CO_2 partial pressures. Fine lines show results of simulation holding minerals segregated from isotopic exchange, as already presented (Fig. 22.1).

How would the results differ if we had assumed isotopic equilibrium among minerals instead of holding them segregated from isotopic exchange? To find out, we enter the commands

```
(cont'd)
unsegregate ALL
reactants times 10
go
```

Then, to carry the run to high water–rock ratios, we type

```
(cont'd)
reactants times 4700
go
```

In the unsegregated case (Fig. 22.2), a family of curves, one for each CO_2 pressure chosen, represents dolomite composition. At low water–rock ratios, regardless of CO_2 pressure, the product dolomite forms at the composition in isotopic equilibrium with the original calcite. Because of the fractionation between the minerals, the dolomite is about 1.8‰ lighter than the calcite. As the reaction proceeds, the compositions of each component of the system (fluid, calcite, and dolomite) become isotopically heavier, reflecting the introduction of heavy carbon by the migrating fluid. Transition from light to heavy compositions occurs after a small number of pore volumes have been displaced by a fluid of high CO_2 pressure, or a large quantity of fluid of low CO_2 pressure.

These predictions differ qualitatively from the results of the segregated model (Fig. 22.1), as can be seen by comparing the bold and fine lines in Figure 22.2. In contrast to the segregated model, the initial and final $\delta^{13}C$

values of the dolomite in the unsegregated model are independent of P_{CO_2}, and the final value does not depend on the original composition of the calcite. Dolomite compositions predicted by the two models coincide only when CO_2 partial pressure is quite high ($P_{CO_2} \simeq 100$ bar), and then only at high water–rock ratios. Under other conditions, the predictions diverge. Since the unsegregated model depends on the unrealistic premise of maintaining isotopic equilibrium at low temperature, these results argue strongly for the importance of holding minerals segregated from isotope exchange when modeling many types of reaction processes.

23

Transport in Flowing Groundwater

To this point, we have considered relatively simple descriptions of mass transfer: reactants added to or removed from a chemical system at prescribed rates, the exchange of gases with external reservoirs, and kinetic rate laws. A general model of reactions occurring in a system open to groundwater flow would account at each location in the flow for the rate at which the flowing groundwater delivers reactants, and how fast it carries away reaction products. The model would consider not only the solutes carried by the flowing groundwater, but their movement by molecular diffusion and hydrodynamic dispersion, the mixing within the flow.

Such models are known as reactive transport models and are the subject of the next chapter (Chapter 24). We treat the preliminaries in this chapter, introducing the subjects of groundwater flow and mass transport, how flow and transport are described mathematically, and how transport can be modeled in a quantitative sense. We formalize our discussion for the most part in two dimensions, keeping in mind that the equations we use can be simplified quickly to account for transport in one dimension, or generalized to three dimensions.

23.1 Groundwater Flow

Transport in flowing groundwater is controlled primarily by the pattern and rate of flow, which are described by Darcy's law. Darcy's law says that groundwater migrates from high hydraulic potential to low, according to

$$\begin{aligned} q_x &= -\frac{k_x}{\mu}\frac{\partial \Phi}{\partial x} \\ q_y &= -\frac{k_y}{\mu}\frac{\partial \Phi}{\partial y} \end{aligned} \tag{23.1}$$

(Hubbert, 1940; Freeze and Cherry, 1979). Here, q_x and q_y are specific discharge (cm^3 cm^{-2} s^{-1}) in the x and y directions, k_x and k_y are permeability (cm^2) along those coordinates, μ is the fluid viscosity (poise, or g cm^{-1} s^{-1}), and Φ is hydraulic potential (g cm^{-1} s^{-2}), the mechanical energy contained in a unit volume of groundwater.

Hydraulic potential is the sum of the $V\mathrm{d}P$ work done on the water and its potential energy. The quantity is given by

$$\Phi = P + \rho g Z \,, \tag{23.2}$$

where P is fluid pressure (g cm^{-1} s^{-2}), ρ is fluid density (g cm^{-3}), g the acceleration of gravity (cm s^{-2}), and Z elevation (cm) relative to sea level or another datum. Specific discharge in an arbitrary direction of flow is given as

$$q = \sqrt{q_x^2 + q_y^2}\,, \tag{23.3}$$

from the components of the discharge vector.

Darcy's law can be written in an alternative form in terms of hydraulic head h (cm), the height to which water would rise above sea level if free to flow into a well, and hydraulic conductivity K_x, K_y (cm s^{-1}) of the sediment. In this case, groundwater flows from high head to low, at a discharge given by

$$\begin{aligned} q_x &= -K_x \frac{\partial h}{\partial x} \\ q_y &= -K_y \frac{\partial h}{\partial y}\,. \end{aligned} \tag{23.4}$$

Changes in hydraulic head reflect variation in the hydraulic potential, according to $d\Phi = \rho g\, dh$, and hydraulic conductivity is proportional to permeability,

$$K_x = \frac{\rho g k_x}{\mu} \qquad K_y = \frac{\rho g k_y}{\mu}\,, \tag{23.5}$$

for a fluid of constant viscosity; hence the forms are equivalent where μ does not vary, or where the conductivity is corrected for the variation.

Neither of these forms of Darcy's law is correct where fluid density varies, such as in convecting flows. In this case, the discharge is given by

$$\begin{aligned} q_x &= -\frac{k_x}{\mu}\left(\frac{\partial P}{\partial x} + \rho g \frac{\partial x}{\partial Z}\right) \\ q_y &= -\frac{k_y}{\mu}\left(\frac{\partial P}{\partial y} + \rho g \frac{\partial y}{\partial Z}\right)\,. \end{aligned} \tag{23.6}$$

The flow pattern in this case reflects the distribution of fluid pressure and fluid density; it cannot be determined from the gradient of any single potential function.

Specific discharge has units such as cm^3 of water per cm^2 of the saturated medium per second, so nominally it can be expressed in velocity units, like cm s^{-1}. The quantity is, however, properly considered a volume flux and not the groundwater velocity, which is the average rate at which water molecules translate through space. Groundwater can move only through the pore space of a sediment or rock, so the velocity depends on the porosity ϕ, which is the unitless ratio of pore volume to bulk volume, according to

$$v_x = \frac{q_x}{\phi} \qquad v_y = \frac{q_y}{\phi}\,. \tag{23.7}$$

Here v_x and v_y are the components of the velocity vector in the coordinate directions. Velocity v in an arbitrary direction is

$$v = \frac{q}{\phi} \tag{23.8}$$

or

$$v = \sqrt{v_x^2 + v_y^2}\,. \tag{23.9}$$

The value of groundwater velocity within a rock or sediment, then, invariably exceeds that of specific discharge.

23.2 Mass Transport

Chemical mass is redistributed within a groundwater flow regime as a result of three principal transport processes: advection, hydrodynamic dispersion, and molecular diffusion (e.g., Bear, 1972; Freeze and Cherry, 1979). Collectively, they are referred to as *mass transport*. The nature of these processes and how each can be accommodated within a transport model for a multicomponent chemical system are described in the following sections.

23.2.1 Advection

Advective transport, or simply *advection*, refers to movement of chemical mass within a flowing fluid or gas. For our purposes, it is most commonly migration of aqueous species along with groundwater. In constructing a transport model, we prefer to consider how much each of the thermodynamic components—the total masses of the basis entries A_w, A_i, A_k, and A_m—moves, rather than track migration of the free masses of each individual aqueous species.

Taking C_w, C_i, C_k, and C_m as the volumetric concentrations (mol cm^{-3}) of basis entries mobile in the groundwater, the advective fluxes (mol cm^{-2} s^{-1}) of the components are given from the specific discharge (q_x, q_y) as

$$q^{\mathrm{A}}_{x_w} = q_x C_w \qquad q^{\mathrm{A}}_{y_w} = q_y C_w \tag{23.10}$$

$$q^{\mathrm{A}}_{x_i} = q_x C_i \qquad q^{\mathrm{A}}_{y_i} = q_y C_i \tag{23.11}$$

$$q^{\mathrm{A}}_{x_k} = q_x C_k \qquad q^{\mathrm{A}}_{y_k} = q_y C_k \tag{23.12}$$

$$q^{\mathrm{A}}_{x_m} = q_x C_m \qquad q^{\mathrm{A}}_{y_m} = q_y C_m \,. \tag{23.13}$$

Here, q^{A}_x is the mass of a component carried across a unit plane oriented normal to the x direction, per unit time, and q^{A}_y is the same for a plane normal to y.

Volumetric concentration is the mobile fraction of a component's mass at the nodal block in question, divided by fluid volume there. The concentrations are given from the component masses (M_w, M_i, M_k, and M_m), adjusted for the immobile species, and the fluid volume V as

$$C_w = \frac{1}{V}\left(M_w - n_w \sum_q \nu_{wq} m_q\right) \tag{23.14}$$

$$C_i = \frac{1}{V}\left(M_i - n_w \sum_q \nu_{iq} m_q\right) \tag{23.15}$$

$$C_k = \frac{1}{V}\left(M_k - n_k - n_w \sum_q \nu_{kq} m_q\right) \tag{23.16}$$

$$C_m = \frac{1}{V}\left(M_m - n_w \sum_q \nu_{mq} m_q\right) . \tag{23.17}$$

As before, m_q are the molal concentrations of the sorbing species A_q, and ν_{wq}, etc., are the coefficients of the reaction forming A_q from the basis. In these equations, we have taken minerals and sorbed species as being immobile, although this assumption might be relaxed to account, for example, for the migration of colloids or suspended sediment.

In treating sorbing surfaces that carry an electrical charge, recall that a diffuse layer of counterions separates the surface from the mobile fluid. The layer serves to balance charge on the surface and hence needs to be held immobile if the advective flux is to be electrically neutral. For this reason, we must include the masses of the counterions when evaluating the summations over the sorbed species A_q in Equations 23.14–23.17. The various sorption and surface complexation theories provide little guidance in predicting the composition of the diffuse layer, so the modeler may need to construct it by ad hoc means, such as from the most abundant ion of opposing charge, or as a mixture of counterions in proportions reflecting the fluid's bulk composition.

23.2.2 Hydrodynamic Dispersion

The water molecules and dissolved species making up flowing groundwater do not pass through the subsurface in an orderly fashion. Instead of following simple trajectories, they branch continually into threads, moving around grains, into and out of areas of high conductivity, or along fractures. The threads may combine with threads of distant origin, or recombine with those from which they have previously split. Some threads move ahead of the average flow, others are retarded relative to it; a thread may follow the centerline of the flow or stray far to the side.

In this way, the groundwater continually mixes by mechanical means, an effect known as *hydrodynamic dispersion* (e.g., Freeze and Cherry, 1979). *Microscopic dispersion*, as the name suggests, arises at the level of the pores and pore throats in the sediment or rock. It accounts for the branching of the flow around grains and through microscopic fractures, and the more rapid migration of molecules in the center of pore throats than along the edges.

Macroscopic dispersion, on the other hand, arises from heterogeneities on a scale larger than individual pores and grains. Such heterogeneities include laminae, layers, and formations of contrasting permeability; fractures larger than the microscopic ones already considered; and karst channels, joints and faults. Dispersion of this sort is sometimes referred to as *differential advection*.

Hydrodynamic dispersion is in many cases taken to be a "Fickian" process, one whose transport law takes the form of Fick's law of molecular diffusion. If flow is along x only, so that $v_x = v$ and $v_y = 0$, the dispersive fluxes (mol cm^{-2} s^{-1}) along x and y for a component i are given by

$$\begin{aligned} q^{\mathrm{D}}_{x_i} &= -\phi D_{\mathrm{L}} \frac{\partial C_i}{\partial x} \\ q^{\mathrm{D}}_{y_i} &= -\phi D_{\mathrm{T}} \frac{\partial C_i}{\partial y} \,. \end{aligned} \tag{23.18}$$

Here ϕ is the porosity of the medium and D_{L} and D_{T} are the coefficients of hydrodynamic dispersion (cm^2 s^{-1}) in the longitudinal (along the flow) and transverse (across the flow) directions. Parallel equations are written for components w, k, and m, in terms of C_w, C_k, and C_m, as defined in the previous section. By these equations, we see that dispersion transports a component from areas of high to low concentration, working to smooth out the component's distribution.

The dispersion coefficients for the case of flow along a coordinate axis are calculated from the absolute value of the velocity v as

$$\begin{aligned} D_{\mathrm{L}} &= D^* + \alpha_{\mathrm{L}} \left| v \right| \\ D_{\mathrm{T}} &= D^* + \alpha_{\mathrm{T}} \left| v \right| \,, \end{aligned} \tag{23.19}$$

where D^* is the coefficient of molecular diffusion (cm^2 s^{-1}) in the medium, and α_L and α_T are the dispersivity (cm) of the medium in the longitudinal and transverse directions.

Dispersivity is a property that depends on the nature of the sediment or rock in question, as well as the scale on which dispersion is observed. There is no typical value: a dispersivity of 1 cm might be observed in a laboratory experiment, whereas a value of 100 m (10 000 cm) might be found to apply in a field study. Dispersion is generally more rapid along the direction of flow than across it, so $\alpha_L > \alpha_T$. Typical values of the diffusion coefficient D^* in porous media are in the range 10^{-7} to 10^{-6} cm^2 s^{-1}.

Two points in Equation 23.19 are apparent. First, through inclusion of the diffusion coefficient, the dispersion coefficient is defined to encompass not only mechanical mixing but molecular diffusion, taking advantage of the Fickian form assumed for the transport equation. There are limitations to lumping dispersion and diffusion together, as we will discuss. Second, the mechanical component of the dispersion coefficient, represented by the dispersivity, is proportional to the velocity of groundwater flow. This is because the rate of mechanical mixing would be expected to vary directly with flow rate—where flow is most rapid, mixing should be strongest.

Where groundwater is allowed to flow in an arbitrary direction instead of along one of the coordinate axes, the dispersion coefficient assumes a tensor rather than vector form. In this case, the dispersive fluxes are given by

$$\begin{aligned} q^{\mathrm{D}}_{x_i} &= -\phi\left(D_{xx}\frac{\partial C_i}{\partial x} + D_{xy}\frac{\partial C_i}{\partial y}\right) \\ q^{\mathrm{D}}_{y_i} &= -\phi\left(D_{yx}\frac{\partial C_i}{\partial x} + D_{yy}\frac{\partial C_i}{\partial y}\right), \end{aligned} \tag{23.20}$$

where the entries in the dispersion tensor are

$$\begin{aligned} D_{xx} &= D^* + \alpha_L\frac{v_x^2}{|v|} + \alpha_T\frac{v_y^2}{|v|} \\ D_{yy} &= D^* + \alpha_L\frac{v_y^2}{|v|} + \alpha_T\frac{v_x^2}{|v|} \\ D_{xy} &= D_{yx} = (\alpha_L - \alpha_T)\,\frac{v_x v_y}{|v|} \end{aligned} \tag{23.21}$$

(e.g., Bear, 1972; 1979). We see that Equation 23.19 is a special form of the dispersion tensor for the case $v_y = 0$.

The dispersivity of natural sediments and rocks, as we have noted, is a property notable for its tendency to scale strongly with the scale on which it is observed (e.g., Neuman, 1990). Dispersivities observed in field studies, for example, are almost invariably larger than those observed on a smaller scale of study in the laboratory. And, in a given field study, dispersivities observed over long distances are commonly larger than those observed locally.

There are two general strategies for accounting in mass transport simulations for dispersion, given the strong scaling of dispersivity. The most straightforward is to represent dispersion at all the scales considered as a Fickian process, using one value for longitudinal dispersivity, and a second for transverse. The dispersivities assumed are those appropriate to the domain taken as a whole, and the resulting dispersion coefficients (Eqs. 23.19 and 23.21) reflect the sum of microscopic and macroscopic effects. The principal disadvantage to this approach is that it applies the same dispersivity to transport over small areas of the domain—those covering just a few nodal blocks—as it does to the domain as a whole, in contradiction to the observed scaling effect (e.g., Gelhar, 1986).

A more detailed and potentially more realistic approach is to divide dispersion into two components: one present on the scale of the individual nodal blocks in the numerical solution, and a second arising from differential advection through the nodes. The domain is taken to be heterogeneous, with the permeability at individual nodal blocks assigned according to a stochastic model, using a random number generator. The first component is represented, as before, as a Fickian process, using dispersivity values appropriate to the size of the nodal blocks. The second component, in contrast, accounts for the dispersion observed at larger scales; it arises from the preferred flow of groundwater along high-permeability pathways through the assemblage of nodal blocks. Transport is most rapid along those pathways, giving rise in the simulation to differential advection and hence macrodispersion.

23.2.3 Molecular Diffusion

Molecular diffusion (or *self-diffusion*) is the process by which molecules show a net migration, most commonly from areas of high to low concentration, as a result of their thermal vibration, or Brownian motion. The majority of reactive transport models are designed to simulate the distribution of reactions in groundwater flows and, as such, the accounting for molecular diffusion is lumped with hydrodynamic dispersion, in the definition of the dispersivity.

The accounting for diffusion in these models, in fact, is in many cases a formality. This is because, as can be seen from Equations 23.19 and 23.21, the contribution of the diffusion coefficient D^* to the coefficient of hydrodynamic dispersion D is likely to be small, compared to the effect of dispersion. If we assume a dispersivity α of 100 cm, for example, then the product αv representing dispersion will be larger than a diffusion coefficient of 10^{-7}–10^{-6} cm^2 s^{-1} wherever groundwater velocity v exceeds 10^{-9}–10^{-8} cm s^{-1}, or just 0.03–0.3 cm yr^{-1}.

In light of this argument, it is common practice in groundwater transport modeling to apply a single value for the diffusion coefficient to all aqueous species. The practice neglects the fact that diffusion coefficients vary from species to species, reflecting in large part variation in hydrated radius, and hence is not valid everywhere. Groundwater, for example, may flow quite slowly within shale aquitards, intact blocks of igneous and metamorphic rock, or backfill at waste repositories. In such environments, diffusion may be more important than dispersion, and accurate values for the diffusion coefficients may be needed.

A straightforward response to this issue is to remove D^* from Equations 23.19 and 23.21 and then calculate the diffusive flux of each aqueous species (indexed j) individually, according to

$$\begin{aligned} q_{x_j}^{\text{Diff}} &= -\phi D_j^* \frac{\partial C_j}{\partial x} \\ q_{y_j}^{\text{Diff}} &= -\phi D_j^* \frac{\partial C_j}{\partial y} \, . \end{aligned} \tag{23.22}$$

Here, D_j^* is the species' diffusion coefficient, and C_j its volumetric concentration. This strategy is fully valid only for uncharged species. If the species are electrically charged, cations will begin diffusing more readily than anions, or vice versa, creating an electrical potential, which will in turn affect the rate at which the ions diffuse. Newman and Thomas-Alyea (2004) describe how to model diffusion in ionic systems in the presence of an electrical potential. If the electrical potential is ignored in the simulation, local charge balance will be lost.

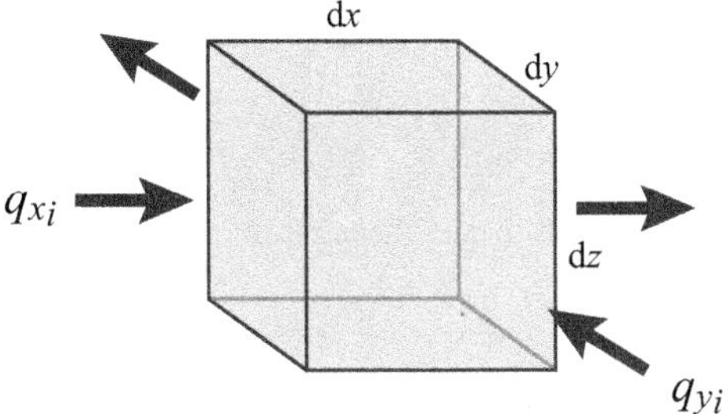

Figure 23.1 Control volume of an aquifer, showing the origin of the divergence principle. Volume dimensions are $dx \times dy \times dz$. The rate at which a chemical component i accumulates within the volume depends on the divergence of the mass fluxes; i.e., the rate at which the component's mass is transported into the volume along x and y, less the rate it is transported out.

23.3 Advection–Dispersion Equation

The effect of advection and dispersion on the distribution of a chemical component within flowing groundwater is described concisely by the *advection–dispersion equation.* This partial differential equation can be solved subject to boundary and initial conditions to give the component's concentration as a function of position and time.

23.3.1 Derivation

The advection–dispersion equation follows directly from the transport laws already presented in this chapter, and the *divergence principle.* The latter states that the time rate of change in the concentration of a component depends on how rapidly the advective and dispersive fluxes change with distance. If, for example, more of component i passes into the control volume shown in Figure 23.1 across its left and front faces than moves out across its right and back, the component is accumulating in the control volume and its concentration there increasing. The time rate of change in concentration is given by

$$\frac{\partial C_i}{\partial t} = -\frac{\partial}{\partial x}\left(q^{\mathrm{D}}_{x_i} + q^{\mathrm{A}}_{x_i}\right) - \frac{\partial}{\partial y}\left(q^{\mathrm{D}}_{y_i} + q^{\mathrm{A}}_{y_i}\right) \tag{23.23}$$

(Bear, 1972; 1979; Freeze and Cherry, 1979), which is the negative divergence of the fluxes. This equation is a statement of the divergence principle, applied to solute mass.

Substituting the transport laws for advection and dispersion (Eqs. 23.11 and 23.20), and noting that groundwater velocity v is related to specific discharge q according to Equation 23.7, gives

$$\begin{aligned}\frac{\partial C_i}{\partial t} = \frac{\partial}{\partial x}\left(D_{xx}\frac{\partial C_i}{\partial x}\right) + \frac{\partial}{\partial x}\left(D_{xy}\frac{\partial C_i}{\partial y}\right) + \frac{\partial}{\partial y}\left(D_{yx}\frac{\partial C_i}{\partial x}\right) + \\ \frac{\partial}{\partial y}\left(D_{yy}\frac{\partial C_i}{\partial y}\right) - \frac{\partial}{\partial x}\left(v_x C_i\right) - \frac{\partial}{\partial y}\left(v_y C_i\right),\end{aligned} \tag{23.24}$$

which is the advection–dispersion equation in two dimensions. In deriving this relation, we take porosity ϕ as invariant; in the next chapter (Chapter 24), we will relax this assumption. Parallel forms of the equation, of course, can be written for any component, in terms of C_w, C_k, and C_m. Considering flow and transport along only x leaves

$$\frac{\partial C_i}{\partial t} = \frac{\partial}{\partial x}\left(D_L \frac{\partial C_i}{\partial x}\right) - \frac{\partial}{\partial x}(v_x C_i) , \tag{23.25}$$

which is the advection–dispersion equation in one dimension.

23.3.2 Péclet Number

We see in Equations 23.24 and 23.25 that two types of terms contribute to the evolution of solute concentration: those representing the effects of advection, and those accounting for diffusion and dispersion. The nondimensional *Péclet number*[1] represents the importance of these processes, relative to one another.

In an interval of time Δt, advection carries a nonreacting solute across a distance $L = v_x \Delta t$, and the solute is dispersed over a characteristic distance $L \simeq \sqrt{D_L \Delta t}$ (e.g., Crank, 1975). For any given Δt, then, advection and dispersion observed at a length scale L are equally important wherever

$$D_L = L v_x . \tag{23.26}$$

Rearranging gives the Péclet number,

$$\text{Pe} = \frac{L v_x}{D_L} , \tag{23.27}$$

which is a measure of the effect of advection relative to diffusion and dispersion, at a given scale of observation.

Where the Péclet number has a value near one, advection and dispersion are of comparable importance. Values much greater than one signify the dominance of advection, and those less than one indicate that diffusion or dispersion dominates. In the presence of groundwater flowing at any appreciable rate, $D_L \simeq \alpha_L v_x$ (from Eq. 23.19), as already discussed, and the Péclet number becomes

$$\text{Pe} = \frac{L}{\alpha_L} . \tag{23.28}$$

Now, Pe depends only on the magnitude of the dispersivity relative to the scale of observation. The Péclet number of flowing groundwater is generally greater than one, reflecting the dominance of advection, since dispersivity is invariably found to be smaller than the scale on which it is observed (e.g., Neuman, 1990).

23.4 Numerical Solution

Equation 23.24 can be solved analytically to give closed-formed answers to simple problems. Many mass transport problems, however, including all but the most straightforward in reactive transport, require the equation to be evaluated numerically (e.g., Phillips, 1991). There are a variety of methods for doing so, including finite difference, finite element, particle tracking, and so on (e.g., Zheng and Bennett, 2002). Most numerical models of multicomponent reactive transport are based on the finite difference method, which we describe in this section.

[1] We might properly refer to this value as the "apparent Péclet number," because by many formal definitions the Péclet number accounts for the relative importance of advection and molecular diffusion, without mention of hydrodynamic dispersion.

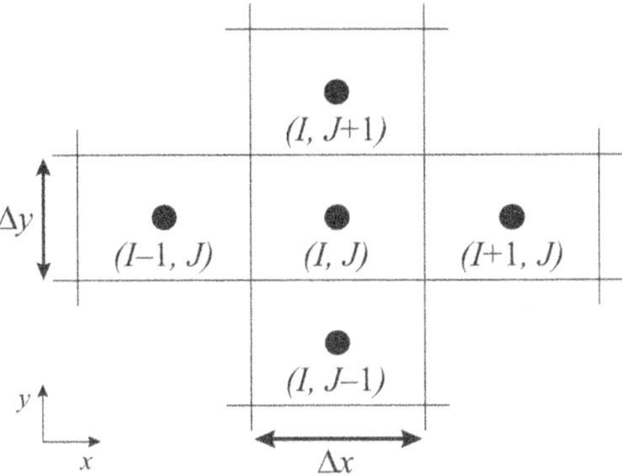

Figure 23.2 Indexing of nodal blocks in a finite difference grid, showing point (I, J) and its nearest neighbors. Properties of the nodal blocks are projected onto nodal points located at the center of each block.

23.4.1 Gridding

The first step in obtaining a finite difference solution is to divide the domain into nodal blocks, as shown in Figure 2.11. This process is known as *discretization*, or more simply as *gridding*. In a one-dimensional simulation, the domain is divided into N_x nodal blocks, so for a domain of length L, each nodal block is of length $\Delta x = L/N_x$. In two dimensions, there are $N_x \times N_y$ blocks, each of dimensions $\Delta x \times \Delta y$, where for a domain of width W, $\Delta y = W/N_y$.

The nodal blocks are indexed along x by I, and along y by J, where the indices may be taken to vary from 1 to N_x and 1 to N_y. Some recent codes use the "pointer and offset" indexing convention of modern programming languages, in which case I varies from 0 to $N_x - 1$ and J from 0 to $N_y - 1$. Either way, the neighbor to the left of (I, J) is $(I-1, J)$, the underlying node is $(I, J-1)$, and so on, as shown in Figure 23.2.

Each nodal block represents a distinct system, as we have defined it (Fig. 2.1). Conceptually, the properties of the entire block are projected onto a nodal point at the block's center (Fig. 23.2). A single value for any variable is carried per node in a transport or reactive transport simulation. There is one Ca^{++} concentration, one pH, one porosity, and so on. In other words, there is no accounting in the finite difference method for the extent to which the properties of a groundwater or the aquifer through which it flows might vary across the finite dimensions of a nodal block. All such variation exists among the nodal blocks, not within them.

23.4.2 Finite Difference Approximation

In a finite difference model, the differential equation representing mass transport (Eq. 23.24 or 23.25) is converted into an approximate, algebraic form that can readily be evaluated using a computer. A derivative of concentration in space evaluated between nodal points (I, J) and $(I+1, J)$, for example, can be written,

$$\left.\frac{\partial C_i}{\partial x}\right|^{(I+1/2,J)} \simeq \frac{C_i^{(I+1,J)} - C_i^{(I,J)}}{\Delta x}. \tag{23.29}$$

Here, $(I+1/2, J)$ represents the region centered between the nodal blocks in question, (I, J) and $(I+1, J)$, and Δx is the block spacing along x. The first derivative of concentration, in other words, is taken as the amount the variable changes between two nodal blocks, divided by the block spacing. As Δx becomes smaller, the approximation more closely reflects the actual derivative.

Rendering a transport equation in finite difference form is a straightforward and well known process (e.g., Peaceman, 1977; Smith, 1986). The derivatives of C_i in time and space are replaced with finite difference equivalents. The resulting difference equation written at a nodal block (I, J) now includes concentration at that block and its four neighbors: $C_i^{(I,J)}$, $C_i^{(I-1,J)}$, $C_i^{(I+1,J)}$, $C_i^{(I,J-1)}$, and $C_i^{(I,J+1)}$.

The difference equation may be rendered with the spatial derivative evaluated at the beginning or the end of the time step, or an average of the two, and with the first derivatives in space (the advection terms) shifted toward or away from the direction of groundwater flow. Depending on these choices, the resulting equation may be referred to as either forward- or backward-in-time, and forward- or backward-in-distance. In this book, we will consider only difference equations that are forward-in-time (the explicit method) and backward-in-distance (the upstream weighting method). We defer full details of the derivation until the following chapter (Chapter 24), where we construct a model of the transport of reacting solutes.

A transport model is evaluated using a time marching procedure. Concentration along the perimeter of the domain is known from the boundary conditions prescribed, and the initial conditions fix $C_i^{(I,J)}$ at each nodal block, at the onset of the simulation. Using the difference equation, the model calculates new values for concentration, as they vary in response to transport over an interval Δt. Now, $C_i^{(I,J)}$ is known at the end of the first step, $t = \Delta t$. The model steps forward again in time, calculating concentration at $t = 2\,\Delta t$, $t = 3\,\Delta t$, and so on, until it reaches the targeted end point of the simulation.

23.4.3 Stability and Courant Number

In following the time marching procedure, there are specific limits to the length of the time step that may be taken. If the time step exceeds the limits, the solution will become unstable: rather than following a smooth trajectory, small errors will be amplified until the solution oscillates wildly (e.g., Peaceman, 1977; Smith, 1986). It is important, therefore, for the modeler to choose a suitable time step, or for the model to impose one.

In the case of diffusion in the absence of advection, the stability of a forward-in-time solution in one dimension is given by the *von Neuman criterion*,

$$\left(\frac{2D_\mathrm{L}}{\Delta x^2}\right)\Delta t \le 1\,. \tag{23.30}$$

By this relation, the limiting time step varies with the square of the nodal block spacing. If the number of blocks in a domain is doubled, the spacing is cut in half and the limiting time step must be reduced by three-quarters.

In a backward-in-distance solution for advective transport in the absence of dispersion or diffusion, the *Courant criterion* limits the time step. In one dimension, the *grid Courant number* is the number of nodal blocks the fluid traverses over a time step. By the Courant criterion, the Courant number Co must not exceed one, or

$$\mathrm{Co} = \left(\frac{|v_x|}{\Delta x}\right)\Delta t \le 1\,, \tag{23.31}$$

when modeling transport of an unreactive solute. In other words, the solution is stable only when a time step is chosen so the fluid moves in a time step no farther than the length of a single nodal block.

Considering dispersion and advection simultaneously, the stability criterion for the forward-in-time, backward-in-distance procedure becomes

$$\left(\frac{|v_x|}{\Delta x} + \frac{2D_{\mathrm{L}}}{\Delta x^2}\right)\Delta t \leq 1\,. \tag{23.32}$$

In two dimensions, the grid Courant number is

$$\mathrm{Co} = \left(\frac{|v_x|}{\Delta x} + \frac{|v_y|}{\Delta y}\right)\Delta t\,, \tag{23.33}$$

and the stability criterion is given by

$$\left(\frac{|v_x|}{\Delta x} + \frac{|v_y|}{\Delta y} + \frac{2D_{xx}}{\Delta x^2} + \frac{2D_{yy}}{\Delta y^2} + \frac{4D_{xy}}{\Delta x \Delta y}\right)\Delta t \leq 1\,. \tag{23.34}$$

23.4.4 Numerical Dispersion

By replacing the derivatives in the transport equation with finite difference approximations, we introduce specific inaccuracies. To see this, we write a Taylor series expanding the difference between $C_i^{(I+1,J)}$ and $C_i^{(I,J)}$,

$$C_i^{(I+1,J)} - C_i^{(I,J)} = (\Delta x)\left.\frac{\partial C_i}{\partial x}\right|^{(I,J)} + \frac{(\Delta x)^2}{2!}\left.\frac{\partial^2 C_i}{\partial x^2}\right|^{(I,J)} + \frac{(\Delta x)^3}{3!}\left.\frac{\partial^3 C_i}{\partial x^3}\right|^{(I,J)} + \cdots\,. \tag{23.35}$$

Comparing this equation to a finite difference approximation (Eq. 23.29), we see that in the numerical solution we carry only the first term in the series, the $\partial C_i/\partial x$ term, omitting the higher-order entries. The Taylor series is truncated, then, and the resulting error called *truncation error*.

Truncation error arises from approximating each of the various space and time derivatives in the transport equation. The error resulting from the derivative in the advection term is especially notable and has its own name. It is known as *numerical dispersion* because it manifests itself in the calculation results in the same way as hydrodynamic dispersion.

To see why numerical dispersion arises, consider solute passing into a nodal block, across its upstream face. Over a time step, the solute might traverse only a fraction of the block's length. In the numerical solution, however, solute is distributed evenly within the block. At the end of the time step, some of it has in effect flowed across the entire nodal block and is in position to be carried into the next block downstream, in the subsequent time step. In this way, the numerical procedure advances some of the solute relative to the mean groundwater flow, much as hydrodynamic dispersion does.

For the transport of a nonreacting solute in one dimension, a coefficient of numerical dispersion can be defined,

$$D_{\mathrm{num}} = \frac{1}{2}\, v_x \Delta x\,(1 - \mathrm{Co})\;, \tag{23.36}$$

assuming the forward-in-time, backward-in-distance scheme we have been discussing (Peaceman, 1977; Smith, 1986). As before, $\mathrm{Co} = v\Delta t/\Delta x$ is the grid Courant number. The results of the finite difference solution resemble the exact solution to the advection–dispersion equation obtained by setting the dispersion coefficient to not D_{L}, but $D_{\mathrm{L}} + D_{\mathrm{num}}$.

Where $D_{\rm num}$ dominates $D_{\rm L}$, which is not uncommon, the calculation result reflects numerical more than actual dispersion. Substituting the grid Péclet number,

$$\mathrm{Pe_g} = \frac{\Delta x\ v_x}{D_{\rm L}}\,, \tag{23.37}$$

into Equation 23.36 gives

$$\frac{D_{\rm num}}{D_{\rm L}} = \frac{1}{2}\,\mathrm{Pe_g}\,(1-\mathrm{Co})\ . \tag{23.38}$$

From this result, we can be assured a solution reflects hydrodynamic rather than numerical dispersion—that is, $D_{\rm L} > D_{\rm num}$—wherever $\mathrm{Pe_g} < 2$.

Numerical dispersion can be minimized in several ways. The nodal block spacing Δx can be set small by dividing the domain into as many blocks as practical. The value specified for $D_{\rm L}$ can be reduced to account for the anticipated numerical dispersion. And a time step can be chosen to give a grid Courant number Co as close to one as allowed by the stability criterion (Eq. 23.32).

None of these options is completely satisfactory. Refining the grid sufficiently to force $D_{\rm num}$ smaller than $D_{\rm L}$ is commonly possible in one dimension, but may not be practical in two or three, since decreasing nodal block size not only increases the amount of computing required at each time step, but the number of time steps. It may be possible in a one-dimensional simulation to decrease $D_{\rm L}$ by the value of $D_{\rm num}$, but the effective value of $D_{\rm num}$ differs between nonreacting and reacting solutes, because the Courant number for a reacting solute is generally less than for a nonreacting species. And choosing a time step bringing Co as close to one as possible is most effective for the nonreacting solutes in a simulation, since they have larger Courant numbers than the reacting solutes.

23.5 Example Calculation

We trace as an example a model in one dimension of the migration of inorganic lead ion through an aquifer, according to Equation 23.25, assuming the Pb^{++} undergoes no reaction. The aquifer is 1 km long (L = 1000 m) and, taking discharge q_x = 30 m^3 m^{-2} yr^{-1} and porosity ϕ = 30%, groundwater flows through it at a velocity $v_x = q_x/\phi$ of 100 m yr^{-1}. The groundwater is virtually devoid of the solute at the onset, but starting at t = 0 and continuing for 2 years, it is present in the recharge at 0.1 mmolal concentration; after this interval, concentration in the recharge returns to near zero.

We consider two cases, one with a higher Péclet number than the other. Dispersivity $\alpha_{\rm L}$ in the first case is set to 0.03 m; in the second, it is 3 m. The procedure in X1T is

```
interval start   =  0 yr, fluid = contaminated
interval elution =  2 yr, fluid = ambient
interval end     = 10 yr

length = 1000 m
Nx = 400
discharge 30 m3/m2/yr; porosity = 30%

scope initial
   Na+  = 10    mmolal
   Pb++ = 1e-12 mmolal
   balance on Cl-
```

```
scope contaminated = initial
   Pb++ = 0.1   mmolal
scope ambient = initial

dispersivity = 0.03 m
go
```

followed by

```
(cont'd)
dispersivity = 3 m
go
```

Since Pe $\simeq L/\alpha_L$, the two cases correspond, respectively, to Péclet numbers on the scale of the aquifer ($L = 1000$ m) of 33 000 and 330.

In the calculation results (Fig. 23.3), the solute migrates through the aquifer at the velocity of the groundwater. In the higher Péclet number case, the sudden appearance and then disappearance of the solute in the recharge gives rise to a pulse that passes through the aquifer much like a square wave. For the lower Péclet number, however, the sharp concentration fronts attenuate quickly as groundwater moves away from the recharge area, in response to hydrodynamic dispersion. In this case, the pulse broadens considerably, as some solute moves ahead of the average groundwater flow, and some lags behind it.

Javandel *et al.* (1984) provide a closed form solution to Equation 23.25 that we can evaluate for these boundary and initial conditions. The exact results coincide closely but not precisely with the numerical approximation, as shown in Figure 23.3. Specifically, the numerical results seem to reflect a slightly lower Péclet number, or a somewhat larger dispersivity, than the exact solution. The divergence is attributable primarily to the effect of numerical dispersion, as described in Section 23.4.4.

To explore the phenomenon of numerical dispersion, we can set dispersivity α_L in the simulation to zero, so that by Equation 23.19 the coefficient of hydrodynamic dispersion falls back to the value of the diffusion coefficient D^*. The fact that D^*, taken here as 10^{-10} m^2 s^{-1}, is a small value implies a large Pe. As such, in the absence of numerical dispersion we expect to see the contaminant pulse migrate through the aquifer as a square wave.

By Equation 23.36, we know that numerical dispersion varies with the nodal block spacing $\Delta x = L/N_x$, so we first test the case of $N_x = 40$

```
(cont'd)
dispersivity = 0
Nx = 40
go
```

and then repeat the calculation for 100

```
(cont'd)
Nx = 100
go
```

and then 400 and 1000 nodes along x, yielding a series of simulations with progressively finer griddings. We see that computing effort increases sharply with larger settings for N_x, since the calculation procedure not only needs to span more nodal blocks, but must take smaller time steps to maintain numerical stability, as described in Section 23.4.3.

Figure 23.4 shows the effect of numerical dispersion on the simulation results. When the domain is gridded

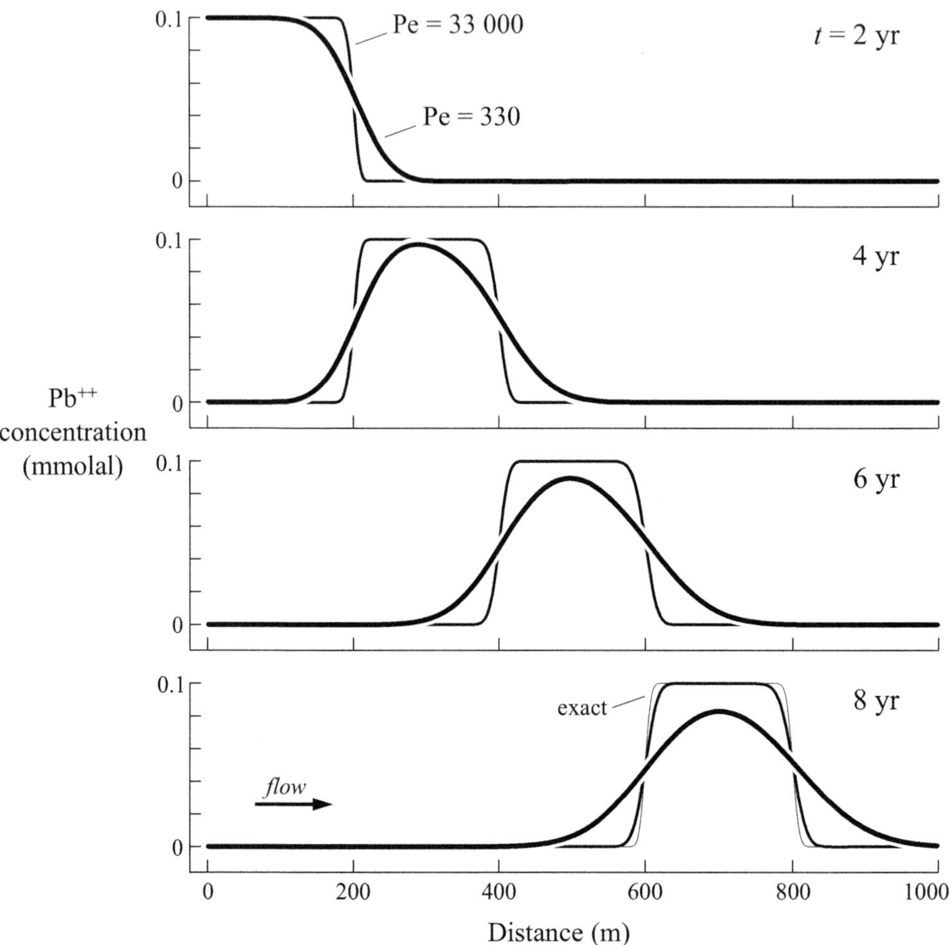

Figure 23.3 Results of a numerical solution ($N_x = 400$) for the nonreactive transport of inorganic lead through an aquifer, calculated for two Péclet numbers, Pe. Contaminant is virtually absent initially, but from $t = 0$ to $t = 2$ years recharge at the left boundary contains Pb^{++} at concentration 0.1 mmolal. After this imbibition, contaminant in recharge returns to near zero concentration. Fine line shows result for dispersivity α_L of 0.03 m, corresponding to a Péclet number on the scale of the aquifer of 33 000; bold line shows results for $\alpha_L = 3$ m, or Pe = 330. Very fine line at $t = 8$ years shows closed form solution to the problem for the high Péclet number case.

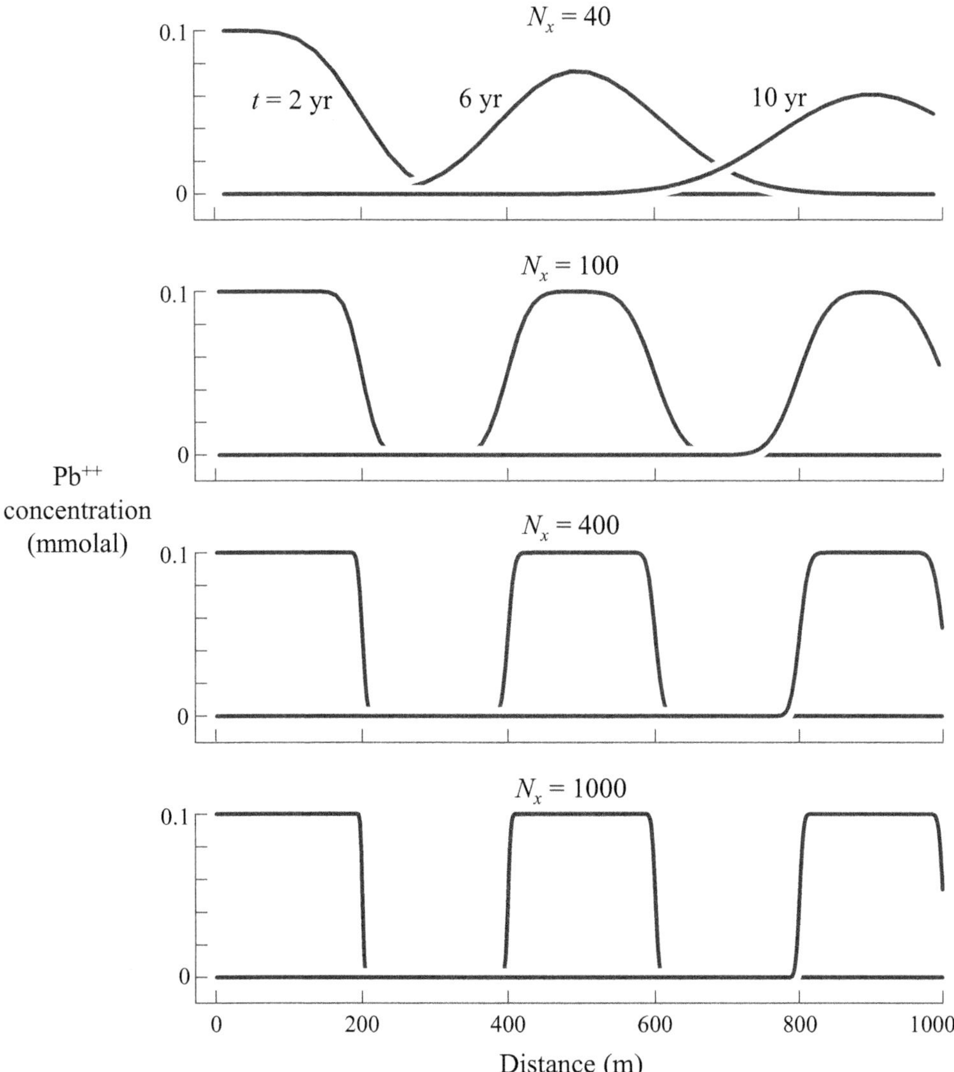

Figure 23.4 Effect of numerical dispersion on a simulation in which a pulse of inorganic lead contamination invades an aquifer, shown for different grid spacings. Dispersivity in the simulation is set to zero, so the result in the absence of numerical error would closely resemble a square wave. Simulations with large numbers of nodal blocks and hence fine grid spacings most closely approach the expected result.

coarsely, using few nodal blocks, numerical dispersion is a dominant feature. The contaminant pulse in this case spreads out broadly as it flows through the domain. Numerical dispersion diminishes as the grid is refined, however, leading to sharper pulses that progressively take on the form of a square wave.

24
Reactive Transport

Many times in geochemical modeling we want to understand not only what reactions proceed in an open chemical system, but where they take place (e.g., Steefel *et al.*, 2005). In a problem of groundwater contamination, for example, we may wish to know not only the extent to which a contaminant might sorb, precipitate, or degrade, but how far it will migrate before doing so.

To this point, we have discussed how to model the reactions occurring within a single representative volume, as shown in Figure 2.1. Such configurations are sometimes called *lumped parameter models*, because the properties of the entire system are represented by a single set of values. There is one pH for the entire volume, one ionic strength, a single mass for each chemical component, and so on. In studying reaction within flowing groundwater, in contrast, we may want to build a *distributed parameter model*, a model in which the properties vary across the system.

To construct models of this sort, we combine reaction analysis with transport modeling, the description of the movement of chemical species within flowing groundwater, as discussed in the previous chapter (Chapter 23). The combination is known as *reactive transport modeling*, or, in contaminant hydrology, *fate and transport modeling*.

24.1 Mathematical Model

A reactive transport model, as the name implies, is reaction modeling implemented within a transport simulation. It may be thought of as a reaction model distributed over a groundwater flow. In other words, we seek to trace the chemical reactions that occur at each point in space, accounting for the movement of reactants to that point, and reaction products away from it.

We formalize the discussion below in terms of C_i, the mobile concentration of a species component A_i, as defined by Equation 23.15. We bear in mind that, even though we will not write them explicitly, parallel equations exist for components A_w, A_k, and A_m, written in terms of C_w, C_k, and C_m (Eqs. 23.14, 23.16, and 23.17).

24.1.1 Governing Equation

In the previous chapter (Section 23.3), we showed the equation describing transport of a nonreacting solute in flowing groundwater (Eq. 23.24) arises from the divergence principle and the transport laws. By this equation, the time rate of change in the dissolved concentration of a chemical component at any point in the domain depends on the net rate the component accumulates or is depleted by transport. The net rate is the rate the component moves into a control volume, less the rate it moves out.

For a reacting solute, the net rate of accumulation is the rate due to transport, plus the rate chemical reactions add the component to the groundwater, or less the rate they remove it. Including the effects of reaction in Equation 23.24, and allowing porosity ϕ to vary with position and time, gives,

$$\begin{aligned}\frac{\partial(\phi C_i)}{\partial t} = & \frac{\partial}{\partial x}\left(\phi D_{xx}\frac{\partial C_i}{\partial x}\right) + \frac{\partial}{\partial x}\left(\phi D_{xy}\frac{\partial C_i}{\partial y}\right) + \\ & \frac{\partial}{\partial y}\left(\phi D_{yx}\frac{\partial C_i}{\partial x}\right) + \frac{\partial}{\partial y}\left(\phi D_{yy}\frac{\partial C_i}{\partial y}\right) - \\ & \frac{\partial}{\partial x}(\phi v_x C_i) - \frac{\partial}{\partial y}(\phi v_y C_i) + \phi R_i \, .\end{aligned} \tag{24.1}$$

Here, R_i is reaction rate (mol cm^{-3} s^{-1}), the net rate at which chemical reactions add component i to solution, expressed per unit volume of water. As before, C_i is the component's dissolved concentration (Eqs. 23.14–23.17), D_{xx} and so on are the entries in the dispersion tensor, and (v_x, v_y) is the groundwater velocity vector. For transport in a single direction x, the equation simplifies to,

$$\frac{\partial(\phi C_i)}{\partial t} = \frac{\partial}{\partial x}\left(\phi D_{\mathrm{L}}\frac{\partial C_i}{\partial x}\right) - \frac{\partial}{\partial x}(\phi v_x C_i) + \phi R_i \, , \tag{24.2}$$

where D_{L} is the coefficient of longitudinal dispersion.

The reaction rate R_i in these equations is a catch-all for the many chemical and physical processes by which a component can be added to or removed from solution in a geochemical model. It is the sum of the effects of equilibrium reactions, such as dissolution and precipitation of buffer minerals and the sorption and desorption of species on mineral surfaces; the kinetics of mineral dissolution and precipitation reactions, redox reactions, microbial activity, and so on; and diffusion into and out of stagnant zones (see Chapter 25).

If pH at a point in the domain drops, for example, an aluminosilicate mineral in contact with the fluid will likely dissolve, giving a positive reaction rate for the components making up the mineral. Where contaminated water infiltrates an aquifer, the contaminant may sorb, leading to a negative R_i. If clean water later enters, the contaminant may desorb, in which case R_i for the contaminant will be positive. For a species that oxidizes or is reduced in a redox disequilibrium model, or a substrate consumed by microbial activity, R_i is negative.

24.1.2 Attenuation and Retardation

In environmental hydrology, *attenuation* means a decrease in a contaminant's concentration or toxicity. Groundwater pollutants may attenuate by precipitating, being degraded by microorganisms, sorbing to aquifer solids, or being oxidized or chemically reduced to a less toxic form. Radioisotopes attenuate by radioactive decay. And attenuation results when contaminated water mixes with clean water, by dilution.

Retardation refers to the process by which chemical reactions slow the transport of a contaminant plume through the subsurface, relative to the average groundwater flow. Retardation commonly results from sorption of the contaminant onto aquifer solids, or from precipitation. A sorbing contaminant introduced to an aquifer, for example, traverses the aquifer more slowly that the flowing groundwater, because some of the contaminant is continually removed from solution. The contaminant will arrive at a point along the aquifer some time after the water that originally contained it. If clean water is introduced to a contaminated area, conversely, a sorbing contaminant is flushed from the aquifer more slowly than water migrates along it. Even under ideal circumstances, groundwater in an aquifer containing a sorbing pollutant may need to be replaced many times before the aquifer is clean.

In simple cases, the mobility in the subsurface of a sorbing contaminant can be described by a *retardation factor*. Where contaminated water passes into a clean aquifer, a reaction front develops. The front separates clean, or nearly clean water downstream from fully contaminated water upstream. Along the front, the sorption reaction removes the contaminant from solution. The retardation factor describes how rapidly the front moves through the aquifer, relative to the groundwater. A retardation factor of two means the front, and hence the contamination, will take twice as long as the groundwater to traverse a given distance.

For a species affected only by an equilibrium sorption reaction, if the species' sorbed concentration C_{S_i} depends directly on its dissolved concentration C_i, the reaction rate is

$$R_i = -\frac{\partial C_{S_i}}{\partial t} = -\frac{\partial C_{S_i}}{\partial C_i}\frac{\partial C_i}{\partial t} \,. \tag{24.3}$$

Where sorption can be described by the reaction K_d approach (Section 9.1),

$$\frac{\partial C_{S_i}}{\partial C_i} = 1000\,\frac{(1-\phi)}{\phi}\,\frac{\rho_s}{\rho}\,\gamma_i K'_\text{d} \,, \tag{24.4}$$

as can be seen by differentiating Equation 9.3. Here ρ_s and ρ are the densities of dry sediment and water (g cm^{-3}), γ_i is the species' activity coefficient, and K'_d is the distribution coefficient. By convention, a value of 2.65 g cm^{-3} is used for ρ_s.

Substituting these relations into the reactive transport equation (Eq. 24.2), and taking porosity to be constant, gives

$$R_\text{F}\frac{\partial C_i}{\partial t} = \frac{\partial}{\partial x}\left(D_\text{L}\frac{\partial C_i}{\partial x}\right) - \frac{\partial}{\partial x}\left(v_x C_i\right) \,, \tag{24.5}$$

where R_F,

$$R_\text{F} = 1 + 1000\,\frac{(1-\phi)}{\phi}\,\frac{\rho_s}{\rho}\,\gamma_j K'_\text{d} \,, \tag{24.6}$$

is the retardation factor. For a weakly sorbing solute, K'_d is small, R_F is close to one, and there is little retardation. A large K'_d, conversely, leads to significant retardation and a large retardation factor.

By similar logic, the retardation factor for a solute that sorbs according to a Freundlich isotherm is

$$R_\text{F} = 1 + 1000\,n_\text{f}\,\frac{(1-\phi)}{\phi}\,\frac{\rho_s}{\rho}\,(\gamma_i)^{n_\text{f}}(C_i)^{n_\text{f}-1}K'_\text{d} \,, \tag{24.7}$$

as can be seen by differentiating Equation 9.15; here, n_f is the Freundlich coefficient. Likewise, the retardation coefficient corresponding to a Langmuir isotherm (Eq. 9.20) is

$$R_\text{F} = 1 + \frac{K\,C_\text{T}}{(K + C_i)^2} \,, \tag{24.8}$$

where K is the equilibrium constant for the Langmuir reaction and C_T is the total concentration of sorbing sites, whether complexed or not. The retardation factor in these two cases varies with solute concentration and hence with position and time. Only where sorption is described by a constant distribution coefficient is retardation a property of an aquifer and its sorbing surfaces alone.

Retardation also arises when a fluid undersaturated or supersaturated with respect to a mineral invades an aquifer, if the mineral dissolves or precipitates according to a kinetic rate law. When the fluid enters the aquifer, a reaction front, which may be sharp or diffuse, develops and passes along the aquifer at a rate less than the average groundwater velocity. Lichtner (1988) has derived equations describing the retardation arising from dissolution and precipitation for a variety of reactive transport problems of this sort.

24.1.3 Damköhler Number

In a simple situation of the type just described, water undersaturated or supersaturated with respect to a mineral invades an aquifer, where the mineral dissolves or precipitates according to a kinetic rate law, adding or removing a species to or from solution. With time, the species' concentration along the aquifer approaches a steady state distribution, trending from the unreacted value at the inlet toward the equilibrium concentration.

Where dissolution or precipitation is sufficiently rapid, the species' concentration quickly approaches the equilibrium value as water migrates along the aquifer; the system is said to be *reaction controlled.* Alternatively, given rapid enough flow, water passes along the aquifer too quickly for the species concentration to be affected significantly by chemical reaction. The system in this case is *transport controlled.* The relative importance of reaction and transport is described formally by the nondimensional *Damköhler number*, written Da.

In Chapter 17, we wrote rate laws for simple dissolution and precipitation reactions, such as those for the silica minerals forming from $SiO_2(aq)$. Rewriting Equation 17.23 in terms of volumetric concentration C_i, assuming the activity coefficient γ_i does not vary over the reaction, gives the rate law

$$\frac{\partial C_i}{\partial t} = \frac{(A_S/V)\,k_+}{C_{eq}}\left(C_{eq} - C_i\right)\,, \tag{24.9}$$

where (A_S/V) is mineral surface area per volume of fluid (cm^2 cm^{-3}), k_+ is the rate constant (mol cm^{-2} s^{-1}), and C_{eq} is the equilibrium species concentration (mol cm^{-3}).

The reaction's characteristic time τ_{rxn} is the interval that would be required for the reaction to reach equilibrium, if the instantaneous rate were held constant. In other words,

$$\tau_{rxn}\frac{\partial C_i}{\partial t} = \left(C_{eq} - C_i\right)\,, \tag{24.10}$$

or, from Equation 24.9,

$$\tau_{rxn} = \frac{C_{eq}}{(A_S/V)\,k_+}\,. \tag{24.11}$$

Now, if the system is observed over a distance scale L, the characteristic time τ_{trans} associated with advective transport is simply the time L/v_x required for groundwater to traverse distance L. The Damköhler number, the ratio of the two characteristic times,

$$\mathrm{Da} = \frac{\tau_{trans}}{\tau_{rxn}} = \frac{(A_S/V)\,k_+\,L}{v_x\,C_{eq}}\,, \tag{24.12}$$

is a measure of the importance of reaction relative to transport, as observed at scale L (Knapp, 1989). Where $\mathrm{Da} \gg 1$, reaction dominates transport, and where $\mathrm{Da} \ll 1$, vice versa.

A notable aspect of this equation is that L appears within it as prominently as the rate constant k_+ or the groundwater velocity v_x, indicating the balance between the effects of reaction and transport depends on the scale at which it is observed. Transport might control fluid composition where unreacted water enters the aquifer, in the immediate vicinity of the inlet. The small scale of observation L would lead to a small Damköhler number, reflecting the lack of contact time there between fluid and aquifer. Observed in its entirety, on the other hand, the aquifer might be reaction controlled, if the fluid within it has sufficient time to react toward equilibrium. In this case, L and hence Da take on larger values than they do near the inlet.

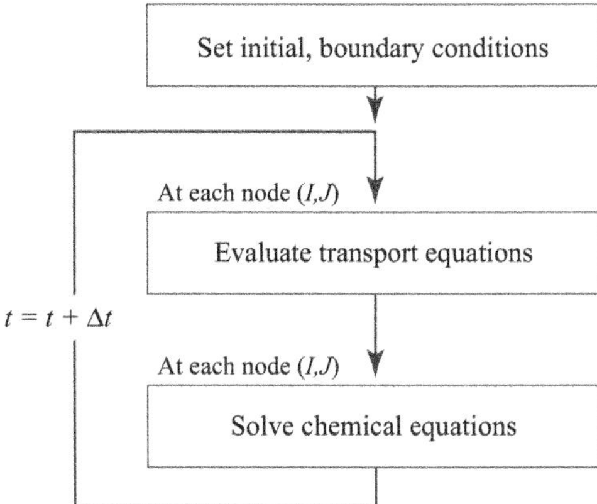

Figure 24.1 Operator splitting method for tracing a reactive transport simulation. To step forward from $t = 0$, the initial condition, to $t = \Delta t$, evaluate transport of the chemical components into and out of each nodal block, using the current distribution of mass. The net transport is the amount of component mass accumulating in a block over the time step. Once the updated component masses are known, evaluate the chemical equations to give a revised distribution of mass at each block. Repeat procedure, stepping to $t = 2\Delta t$, $t = 3\Delta t$, and so on, until the simulation endpoint is reached.

24.2 Numerical Solution

In the simplest cases of reactive transport, a species sorbs according to a linear isotherm (Chapter 9), or reacts kinetically by a zero-order or first-order rate law. There is a single reacting species, and only one reaction is considered. In these cases, the governing equation (Eq. 24.1 or 24.2) can be solved analytically or numerically, using methods parallel to those established to solve the groundwater transport problem, as described in the previous chapter (Chapter 23).

A reactive transport model in a more general sense treats a multicomponent system in which a number of equilibrium and perhaps kinetic reactions occur at the same time. This problem requires more specialized solution techniques, a variety of which have been proposed and implemented (e.g., Yeh and Tripathi, 1989; Steefel and MacQuarrie, 1996). Of the techniques, the *operator splitting method* is best known and most commonly used.

In this method, we write the governing equation (Eq. 24.1 or 24.2) for reactive transport in the conceptual form

$$\frac{\partial C_i}{\partial t} = \mathcal{O}_{\mathrm{T}}(C_i) + \mathcal{O}_{\mathrm{R}}(C_i)\,, \qquad (24.13)$$

where $\mathcal{O}_{\mathrm{T}}$ is the transport operator, and $\mathcal{O}_{\mathrm{R}}$ the reaction operator. The transport operator applied to C_i returns the terms in the governing equation representing dispersion and advection, and $\mathcal{O}_{\mathrm{R}}(C_i)$ simply represents the reaction rate R_i, accounting for whatever reactions are considered in the model.

At each time step, instead of evaluating the entire governing equation at once, as we did for the transport equation, we treat first the transport terms, then the reaction terms. The process is shown in Figure 24.1. We have split each step in the time marching procedure into two parts, according to operator; hence the method's name.

24.2.1 Operator Splitting Method

To march forward in time according to the operator splitting method, we divide the procedure for advancing the time step into four substeps:

1. Calculate the concentration $C_i^{(I,J)}$ of the mobile fraction of component i at each block from the current values of $M_i^{(I,J)}$, using Equation 23.15. For water, mineral, and gas components (components w, k, and m), use Equations 23.14, 23.16, and 23.17.
2. From the concentrations $C_i^{(I,J)}$, evaluate the mass fluxes of component i into and out of each nodal block, due to advection and dispersion, as well as diffusion into and out of stagnant zones (Chapter 25).
3. Update the component mass $M_i^{(I,J)}$ from the mass at the beginning of the time step, the net rate of mass accumulation, and the length of the time step Δt. To do so, calculate the rate at which i enters the block, less the rate it leaves. The difference is the net rate at which i is accumulating or being depleted, in mol s^{-1}.
4. Using the new values of $M_i^{(I,J)}$, evaluate the equations describing the chemistry of the updated system, as described in Chapter 4. Here, we solve the equations describing the distribution of mass derived in Chapter 3, accounting as required for redox disequilibrium (Chapter 7), sorption and surface complexation (Chapters 9–11), and chemical kinetics (Chapters 17–21).

In the first two substeps we compute the rates of transport, whereas in the last two we account for the chemical reactions. Details of carrying out substeps (ii) and (iii) are given below.

24.2.2 Mass Fluxes

To carry out substep (ii), we need to compute the mass fluxes across left and right faces of the control volume (the low and high x sides; see Fig. 23.1), and the front and back faces (low and high y). The faces are denoted $(I-1/2, J)$, $(I+1/2, J)$, $(I, J-1/2)$, and $(I, J+1/2)$, respectively.

The advective flux (mol cm^{-2} s^{-1}) of component i (Eq. 23.11) across the left side of the control volume is calculated as

$$q_{x_i}^{A}\Big|^{(I-1/2,J)} = \begin{cases} q_x^{(I-1/2,J)} C_i^{(I-1,J)} & \text{if } q_x^{(I-1/2,J)} > 0 \\ q_x^{(I-1/2,J)} C_i^{(I,J)} & \text{if } q_x^{(I-1/2,J)} < 0 \end{cases}, \tag{24.14}$$

where $q_x^{(I-1/2,J)}$ is the specific discharge (cm^3 cm^{-2} s^{-1}) from block $(I-1, J)$ to (I, J). Variable C_i in Equation 23.11 is represented by concentration (mol cm^{-3}) at one of the two blocks, depending on whether groundwater flows in the positive or negative x direction; this formality reflects the choice of a backward-in-space (or upstream weighted) procedure. The parallel equations for the other faces are

$$q_{x_i}^{A}\Big|^{(I+1/2,J)} = \begin{cases} q_x^{(I+1/2,J)} C_i^{(I,J)} & \text{if } q_x^{(I+1/2,J)} > 0 \\ q_x^{(I+1/2,J)} C_i^{(I+1,J)} & \text{if } q_x^{(I+1/2,J)} < 0 \end{cases} \tag{24.15}$$

$$q_{y_i}^{A}\Big|^{(I,J-1/2)} = \begin{cases} q_y^{(I,J-1/2)} C_i^{(I,J-1)} & \text{if } q_y^{(I,J-1/2)} > 0 \\ q_y^{(I,J-1/2)} C_i^{(I,J)} & \text{if } q_y^{(I,J-1/2)} < 0 \end{cases} \tag{24.16}$$

$$q_{y_i}^{\mathrm{A}}\Big|^{(I,J+1/2)} = \begin{cases} q_y^{(I,J+1/2)} C_i^{(I,J)} & \text{if } q_y^{(I,J+1/2)} > 0 \\ q_y^{(I,J+1/2)} C_i^{(I,J+1)} & \text{if } q_y^{(I,J+1/2)} < 0 \end{cases} . \tag{24.17}$$

The dispersive mass fluxes (Eq. 23.20) across the left and right sides of the control volume are calculated by

$$q_{x_i}^{\mathrm{D}}\Big|^{(I-1/2,J)} = -\phi \left[D_{xx} \left(\frac{C_i^{(I,J)} - C_i^{(I-1,J)}}{\Delta x} \right) + D_{xy} \left(\frac{C_i^{(I-1,J+1)} - C_i^{(I-1,J-1)}}{2\Delta x} \right) \right] \tag{24.18}$$

$$q_{x_i}^{\mathrm{D}}\Big|^{(I+1/2,J)} = -\phi \left[D_{xx} \left(\frac{C_i^{(I+1,J)} - C_i^{(I,J)}}{\Delta x} \right) + D_{xy} \left(\frac{C_i^{(I+1,J+1)} - C_i^{(I+1,J-1)}}{2\Delta x} \right) \right] , \tag{24.19}$$

where ϕ is porosity (unitless), D_{xx} and so on are the elements of the dispersion tensor (cm^2 s^{-1}), and Δx is the nodal block spacing along x. The dispersive fluxes across the front and back sides are given by

$$q_{y_i}^{\mathrm{D}}\Big|^{(I,J-1/2)} = -\phi \left[D_{yy} \left(\frac{C_i^{(I,J)} - C_i^{(I,J-1)}}{\Delta y} \right) + D_{yx} \left(\frac{C_i^{(I+1,J-1)} - C_i^{(I-1,J-1)}}{2\Delta y} \right) \right] \tag{24.20}$$

$$q_{y_i}^{\mathrm{D}}\Big|^{(I,J+1/2)} = -\phi \left[D_{yy} \left(\frac{C_i^{(I,J+1)} - C_i^{(I,J)}}{\Delta y} \right) + D_{yx} \left(\frac{C_i^{(I+1,J+1)} - C_i^{(I-1,J+1)}}{2\Delta y} \right) \right] , \tag{24.21}$$

where Δy is the nodal block spacing along y.

24.2.3 Updated Composition

Once the advective and dispersive mass fluxes across each face are known, the rate of mass accumulation of component i, required in substep (iii) of the operator splitting procedure, is given by

$$\begin{aligned} \frac{\partial M_i^{(I,J)}}{\partial t} &= A_x \left[\left(q_{x_i}^{\mathrm{A}} + q_{x_i}^{\mathrm{D}}\right)^{(I-1/2,J)} - \left(q_{x_i}^{\mathrm{A}} + q_{x_i}^{\mathrm{D}}\right)^{(I+1/2,J)} \right] \\ &+ A_y \left[\left(q_{y_i}^{\mathrm{A}} + q_{y_i}^{\mathrm{D}}\right)^{(I,J-1/2)} - \left(q_{y_i}^{\mathrm{A}} + q_{y_i}^{\mathrm{D}}\right)^{(I,J+1/2)} \right] , \end{aligned} \tag{24.22}$$

where A_x and A_y are the cross-sectional areas (cm^2) of the faces of the nodal block normal to x and y. The updated component mass is then computed as

$$M_i^{(I,J)}(t + \Delta t) = M_i^{(I,J)}(t) + \frac{\partial M_i^{(I,J)}}{\partial t} \Delta t , \tag{24.23}$$

where Δt is the length of the time step.

24.3 Example Calculations

We take as an example the fate of benzene (C_6H_6) as it migrates with groundwater flowing through an aquifer. Benzene is a common contaminant because it makes up much of the volatile fraction of gasoline and other petroleum products. It is a suspected carcinogen with an MCL (maximum contamination level) set by the US Environmental Protection Agency of 5 μg kg^{-1}.

Dissolved in groundwater, benzene is gradually degraded by a variety of microorganisms present naturally in the subsurface. Under aerobic conditions, it attenuates by the reaction

$$\underset{\textit{benzene}}{C_6H_6(aq)} + {}^{15}\!/_2\, O_2(aq) \rightarrow 6\, CO_2(aq) + 3\, H_2O\,. \tag{24.24}$$

In aquifers containing significant amounts of natural organic matter, benzene migration is retarded by sorption to the organic surfaces.

Alvarez *et al.* (1991) found that at concentrations less than about 100 mg kg^{-1}, the degradation rate in sediment from a natural aquifer under aerobic conditions can be expressed as

$$r^+ = n_w\, k_+\, [X]\, \frac{m_A}{m_A + K_A}\,, \tag{24.25}$$

where k_+ is the rate constant [mol (g cells)$^{-1}$ s^{-1}], [X] is biomass concentration (g cells kg^{-1} solvent), m_A is molal concentration of the substrate, benzene, and K_A is the half-saturation constant (molal). This rate law is the Monod equation (Eq. 19.15), or for the case of constant biomass, as assumed for this calculation, equivalent to the Michaelis–Menten equation (Eq. 18.18), if [X] is taken as enzyme concentration m_E. They report a rate constant k_+ of 8.3 (g benzene) (g cells)$^{-1}$ day^{-1}, or 1.2×10^{-9} mol (g cells)$^{-1}$ s^{-1}, and a half-saturation constant K_A of 12.2 mg kg^{-1}, or 0.16×10^{-3} molal. Because benzene oxidation is so strongly favored thermodynamically, the reverse reaction is not significant.

We assume the same configuration as the example calculation in the previous chapter (Section 23.5). Groundwater passes along an aquifer 1 km long at a velocity v_x of 100 m yr^{-1}. Setting porosity to 30%, specific discharge $q_x = \phi v_x$ in the aquifer is 30 m^3 m^{-2} yr^{-1}. Initially, the groundwater is clean, but at t = 0 and continuing for 2 years, water recharging the aquifer contains 1 mg kg^{-1} benzene. After this time, the aquifer is flushed with clean water. We set a dispersivity of 30 cm, although the calculation results will reflect a somewhat higher value due to the effects of numerical dispersion, as discussed in the previous chapter (Section 23.4).

In a first case, we account for the attenuation of the benzene by biodegradation, using the rate law above (Eq. 24.25). To do so, we add to the thermodynamic data the reaction,

$$\underset{\textit{benzene}}{C_6H_6(aq)} + 3\, H_2O + {}^{15}\!/_2\, O_2(aq) \rightleftarrows 6\, H^+ + 6\, HCO_3^-\,, \tag{24.26}$$

with a log K at 25 °C of 537.5 and store the result in a dataset "thermo+Benzene.tdat". We take a biomass concentration [X] (or m_E) in our calculation of 1 mg kg^{-1}, or 0.001 (g cells) kg^{-1}.

We assume the initial and recharging waters are dilute and aerobic, with a pH of 5.5 and in equilibrium with CO_2 in the atmosphere. The procedure in X1T is

```
read thermo+Benzene.tdat
decouple Carbon

interval start   = 0 yr, fluid = contaminated
interval elution = 2 yr, fluid = ambient
interval end     = 10 yr

length = 1000 m
Nx = 400
discharge = 30 m3/m2/yr
porosity = 30%
```

```
dispersivity = 30 cm

scope initial
   swap CO2(g) for HCO3-
   P CO2(g) = 10^-3.5 bar
   10 mmolal Na+
   10 mmolal Cl-
   8 mg/kg free O2(aq)
   pH = 5.5
   .001 mg/kg Benzene(aq)
scope contaminated = initial
   1 mg/kg Benzene(aq)
scope ambient = initial

kinetic redox-1 rxn = "Benzene(aq)  + 7.5*O2(aq)  -> 6*CO2(aq)  + 3*H2O", \
   rate_con = 1.2e-9, KA = .00016, mE = .001
dxplot .2
go
```

To compare the results to the nonreactive case, we need only set the rate constant

```
(cont'd)
kinetic redox-1 rate_con = 0.0
go
```

to zero.

Figure 24.2 shows how in the calculation results benzene is transported through the aquifer. The pulse of benzene migrates at the rate of groundwater flow, traversing the aquifer in 10 years. As a result of biodegradation by the natural microbial consortium, however, the benzene concentration decreases markedly with time, compared to the nonreacting case.

In a second calculation, we assume benzene not only biodegrades, but sorbs to organic matter in the aquifer,

$$>C_6H_6 \rightleftarrows \underset{benzene}{C_6H_6(aq)} \quad , \tag{24.27}$$

according to the distribution coefficient model. A retardation factor R_F of two, which might be observed in an aquifer containing about 1 wt% organic matter, translates by Equation 24.6 to a distribution coefficient K'_d of 0.16×10^{-3} mol (g sediment)$^{-1}$. Storing the sorption reaction in dataset "Benzene_Kd.sdat", the X1T commands

```
(cont'd)
kinetic redox-1 rate_con = 1.2e-9
read Benzene_Kd.sdat
go
```

restore the rate constant and set the sorption model.

The calculation results for this case, shown in Figure 24.3, differ from those of the first simulation. The benzene passes more slowly through the aquifer, moving at only half the speed of the groundwater flow, reflecting sorption and retardation. Less benzene degrades by biological activity than in the nonsorbing case, since at any given time only a fraction of the contaminant is available to the microbial consortium in aqueous

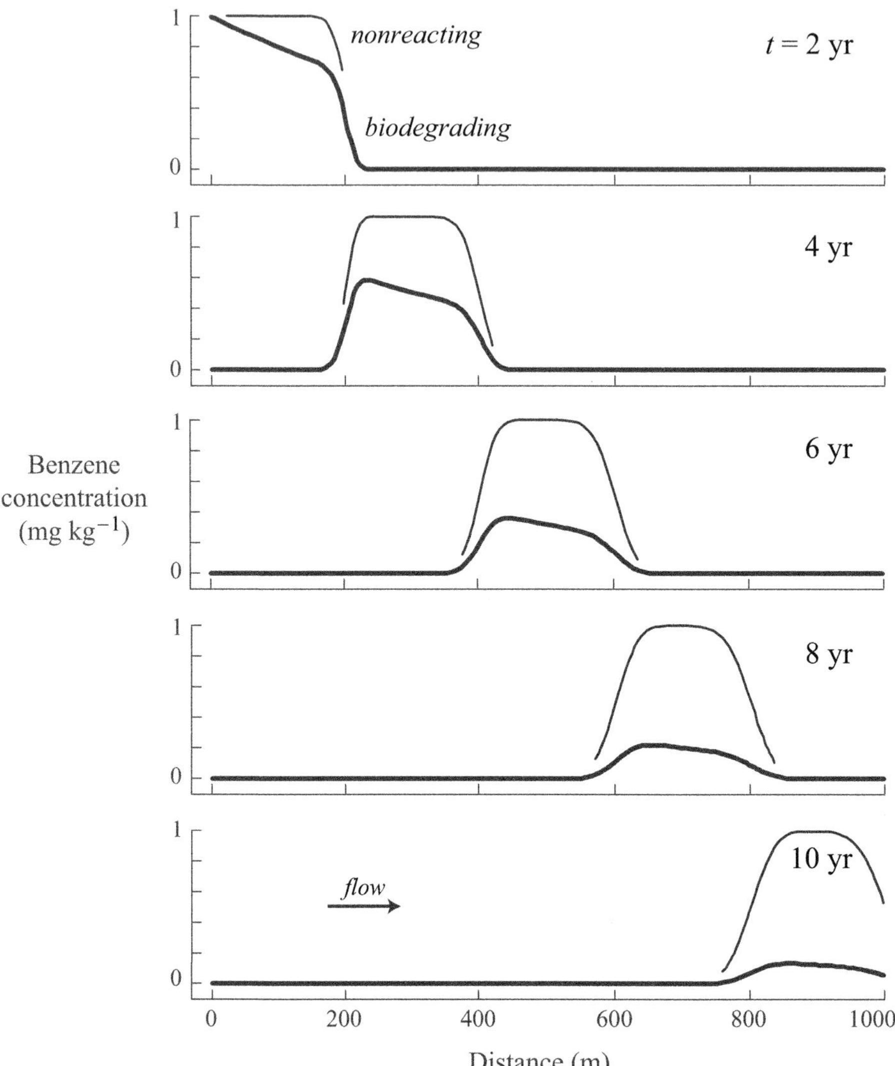

Figure 24.2 Transport of benzene within an aerobic aquifer through which groundwater is flowing at a velocity v_x of 100 m yr^{-1}, calculated accounting for biodegradation, assuming biomass in the aquifer remains constant. Benzene recharges the aquifer at 1 mg kg^{-1} concentration from $t = 0$ to $t = 2$ years, after which time clean water enters the aquifer. Fine lines show transport calculated assuming the species is nonreactive, for comparison.

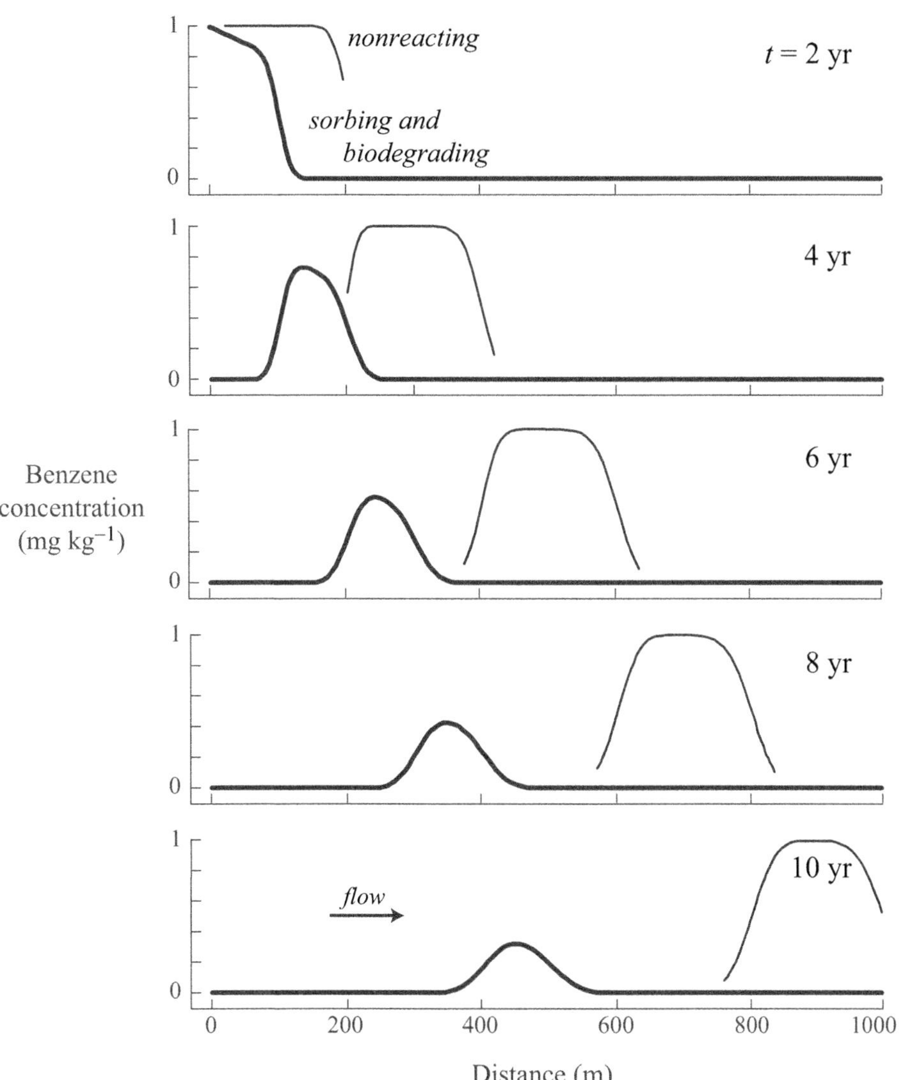

Figure 24.3 Transport of benzene within an aerobic aquifer, as depicted in Figure 24.2, calculated assuming the species not only biodegrades, but sorbs to organic matter in the aquifer. Benzene in the simulation sorbs with a distribution coefficient of 0.16×10^{-3} mol (g sediment)$^{-1}$, equivalent to a retardation factor R_F of 2. Fine lines show nonreactive case.

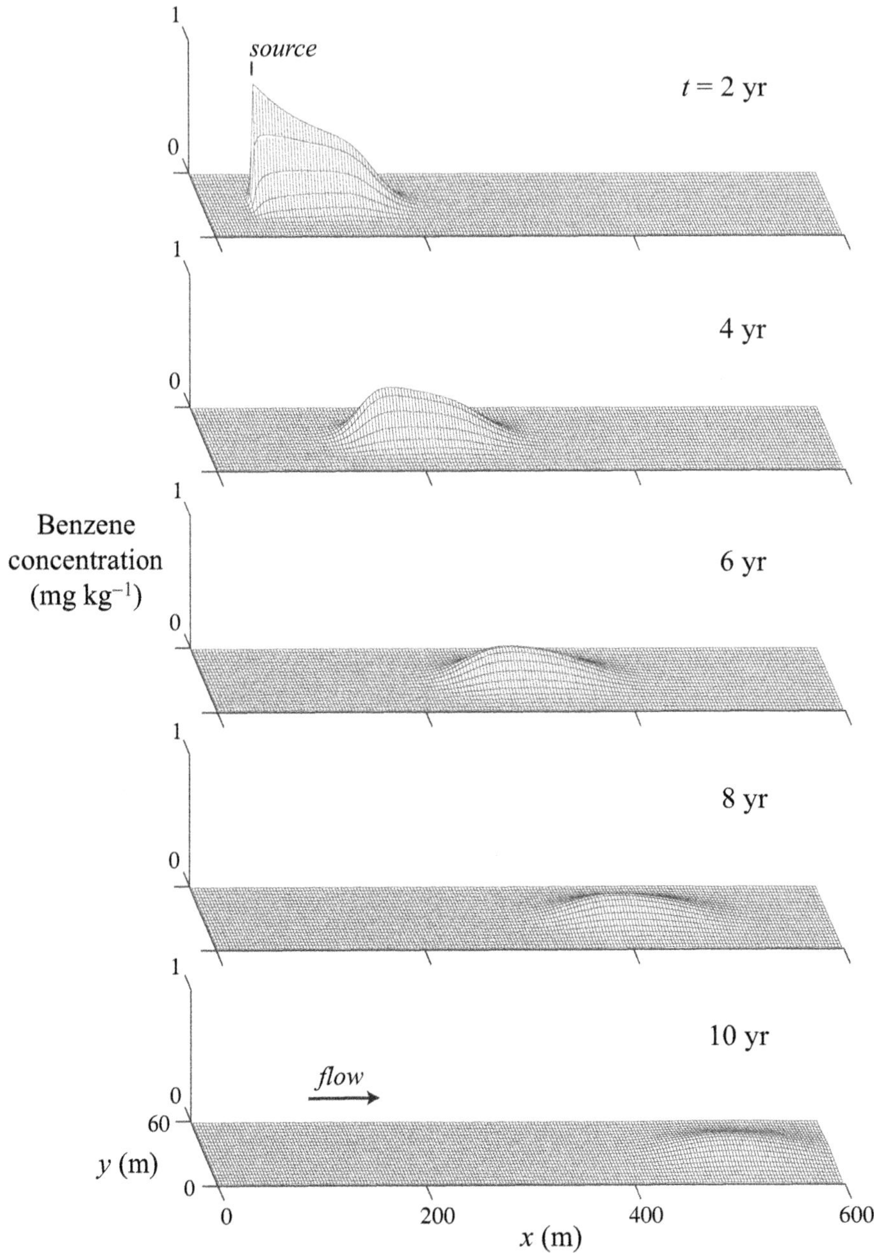

Figure 24.4 Transport of benzene within an aerobic aquifer, modeled in two dimensions. Contaminated water containing 1 mg kg^{-1} benzene leaks into the aquifer over the course of 2 years, at the point indicated. As in the previous model (Fig. 24.3), the benzene is retarded by sorption to organic matter in the aquifer and attenuates due to sorption, biodegradation, and dispersive mixing. Plots were rendered using the MATLAB® software.

form. The latter effect is an observed consequence of the compound's sorption reaction (e.g., Zhang and Bouwer, 1997; Kim *et al.*, 2003).

In a final example, we consider a similar problem in two dimensions. Water containing 1 mg kg^{-1} benzene leaks into an aquifer for a period of 2 years, at a rate of 300 m^3 yr^{-1}. Once in the aquifer, which is 1 m thick, the benzene migrates with the ambient flow, sorbs, and biodegrades. We model flow and reaction over 10 years, within a 600 m $\times$ 60 m area, assuming a dispersivity α_L along the flow of 30 cm, and α_T across flow of 10 cm. All other parameters, including the flow velocity, remain the same as in the previous calculation.

The procedure in X2T is

```
read thermo+Benzene.tdat
read Benzene_Kd.sdat
decouple Carbon

interval start =  0 yr, fluid = ambient
interval end   = 10 yr

length = 600 m; width = 60 m; height = 1 m
Nx = 200; Ny = 20
discharge left = 30 m3/m2/yr
porosity = 30%
dispersivity long = 30 cm, trans = 10 cm

scope initial
   swap CO2(g) for HCO3-
   P CO2(g) = 10^-3.5 bar
   10 mmolal Na+
   10 mmolal Cl-
   8 mg/kg free O2(aq)
   pH = 5.5
   .001 mg/kg Benzene(aq)
scope contaminated = initial
   1 mg/kg Benzene(aq)
scope ambient = initial

well x = 50 m, y = 30 m
well interval leak end = 2 yr, rate = 300 m3/yr, fluid = contaminated

kinetic redox-1 rxn = "Benzene(aq)  + 7.5*O2(aq)  -> 6*CO2(aq)  + 3*H2O", \
   rate_con = 1.2e-9, KA = .00016, mE = .001
go
```

In the calculation results (Fig. 24.4), benzene again is retarded by sorption and attenuates due to sorption, biodegradation, and dispersion along the direction of flow. In this case, it further attenuates due to transverse dispersion, by mixing with clean water flowing beside the plume.

25

Stagnant Zones

To this point in our discussion of mass transport (Chapters 23 and 24) we have considered the effects of groundwater flow through aquifers with uniform properties. The actual subsurface, in contrast, is heterogeneous and complex: areas of high permeability, traversed preferentially by flowing water, are juxtaposed with less permeable zones where transport is much slower. Hydrologists commonly try to account for this complexity by assigning permeability and dispersivity values representative of the overall behavior of the composite medium.

Simple modeling strategies, however, neglect a potentially significant control on mass transport through the subsurface. Specifically, solutes can diffuse from preferential flow paths into areas of slow flow, the *stagnant zones*. The solutes may accumulate over time in the pore space there, perhaps sorbing to surfaces as well, only to later diffuse back into the free-flowing groundwater. Stagnant zones, in other words, are capable of storing and releasing solutes (Bibby, 1981; Tang *et al.*, 1981; Sudicky and Frind, 1982).

The phenomenon is most sharply defined in fractured impermeable rock, where the fractures transport solutes primarily by advection, but molecular diffusion may be the dominant transport process operating within the rock matrix (e.g., March *et al.*, 2018). Contamination in a fractured rock, for example, will over time diffuse from the fractures into the matrix. Later attempts to remediate the site may be frustrated, especially if the acceptable level of contamination is small, by the pollutant diffusing back into the fractures and tainting the water passing through them.

Hydrologists use a class of techniques known as *dual-porosity models* to simulate the effects of stagnant zones on transport through the subsurface. The models work on the assumption that the medium can be separated into two distinct pore systems, each of which has its own hydraulic and transport properties: pore space in an aquifer's free-flowing fraction and the pores in its stagnant zones (e.g., Gerke and van Genuchten, 1993).

25.1 Conceptual Model

To construct a dual-porosity model, we divide each nodal block in the domain between a free-flowing fraction and a collection of stagnant zones. The stagnant zones take up a specified fraction X_{stag} of the node's bulk volume, so the free-flowing portion occupies fraction $1 - X_{\text{stag}}$. Chemical mass—the solutes and solvent—can diffuse from the free-flowing fraction into the stagnant zones, and vice-versa, at a rate that depends on the zones' diffusion coefficient D^*_{stag}, porosity ϕ_{stag}, and retardation coefficient R_{stag}, which is unity in the absence of sorption.

At the onset of the calculation, the concentration of each chemical component throughout the stagnant zones reflects that in the free-flowing fraction, so diffusion is nil. As conditions in the free-flowing fraction change over the course of the simulation, mass may diffuse into and out of the zones. At any point in time, the concentration of each chemical component falls along a diffusion profile extending inward from a zone's margin.

A contaminant introduced to a previously pristine nodal block, for example, will diffuse into the stagnant zones, and hence be depleted from the flowing water. The rate of loss will be most rapid at first, because the concentration gradient at the contact with the stagnant zones is initially steep, and gradually decrease as contaminant accumulates and the gradient lessens. If clean water later invades the nodal block, contaminant built up within the stagnant zones will diffuse in the opposite direction, polluting the flowing water.

Dual-porosity models can further be used to model heat transfer, especially in fractured rock. Stagnant zones in this case act as heat sinks when a nodal block is infiltrated by hot water, and as sources when breached by cold. Applying dual-porosity techniques in this way, we can relax in our modeling the common but many times vulnerable assumption of local thermal equilibrium. The discussion in this chapter is cast in terms of diffusion, but applies equally to heat conduction.

25.2 Governing Equations

To trace a dual-porosity model, we need to determine over the course of the simulation the rate at which chemical components diffuse across the contact between the free-flowing fraction and the stagnant zones, either out of or into the flowing water. To this end, we have to figure the contact area and calculate how each component's diffusion profile within the stagnant zones evolves over time.

25.2.1 Contact Area

The most significant factor controlling the results of a dual-porosity model, in terms of geometry, is the contact area between the free-flowing fraction and the stagnant zones. For a given X_{stag}, small stagnant zones bear more contact area than large zones. A certain volume of baseballs, for example, holds more surface area than the same volume of basketballs, even though the balls occupy the same fraction in each case. Setting smaller stagnant zones in a dual-porosity simulation while holding X_{stag} steady, then, works to increase the zones' effect on the calculation results.

For this reason, the wise modeler is concerned with defining the dimensions of the stagnant zones, along with their volumetric importance. Geometric details such as the precise shapes of the stagnant zones are not only difficult to determine in practice, but unimportant relative to other uncertainties in the calculation (e.g., Gerke and van Genuchten, 1993). As such, a dual-porosity code generally represents the geometry of stagnant zones in nominal terms, using shapes such as spheres, the slabs between fractures, or cubes. In each case, the user specifies a characteristic dimension δ, whether the radius of a sphere or the half-width of slabs or cubes.

Where the nominal geometry is spherical, the contact area A_c follows from the assumed radius δ and volume fraction X_{stag}. A sphere encloses $4\pi\delta^3/3$, so if there are n spheres within a nodal block of bulk volume V_b,

$$X_{\text{stag}} = \frac{n\,4\pi\delta^3}{3V_b} \quad \text{or} \quad n = \frac{3X_{\text{stag}}V_b}{4\pi\delta^3}\,. \tag{25.1}$$

The contact area, then, is

$$A_c = n\,4\pi\delta^2 = \frac{3X_{\text{stag}}V_b}{\delta}\,, \tag{25.2}$$

since $4\pi\delta^2$ is a sphere's surface area. The contact area varies inversely with radius by this relation, becoming quite large where δ is small.

Parallel logic gives the contact area when slabs,

$$A_c = \frac{X_{stag} V_b}{\delta}, \tag{25.3}$$

or cubes,

$$A_c = \frac{3X_{stag} V_b}{\delta}, \tag{25.4}$$

of half-width δ make up the stagnant zones. Again, contact area scales inversely with half-width.

25.2.2 Diffusion Equation

To describe the distribution of chemical mass within the stagnant zones at a nodal block, we account for one-dimensional diffusion of chemical components within the zones. We define the volumetric concentrations c_w, c_i, c_k, and c_m (mol cm^{-3}) for the mobile fraction of each component by analogy to the corresponding concentrations C_w, C_i, C_k, and C_m in the free-flowing fraction (Eqs. 23.14–23.17). We further take the zones to be uniform with respect to D^*_{stag}, ϕ_{stag}, and R_{stag}.

For a spherical nominal geometry, the mass distribution for component i, for example, varies according to

$$R_{stag}\frac{\partial c_i}{\partial t} = D^*_{stag}\frac{1}{r}\frac{\partial}{\partial r}\left(r\frac{\partial c_i}{\partial r}\right), \tag{25.5}$$

where r is the radial direction, ranging from zero at the sphere's center to radius δ at the free-flowing contact. For a slabbed or cubic nominal geometry, the controlling equation is

$$R_{stag}\frac{\partial c_i}{\partial t} = D^*_{stag}\frac{\partial^2 c_i}{\partial r^2}, \tag{25.6}$$

where r is now a linear dimension that varies from zero to half-width δ. Note that invoking a cubic geometry is meaningful only when diffusion affects the outermost reaches of the stagnant zones, since diffusion near a cube's corners cannot be cast in a single dimension.

To solve Equation 25.5 or 25.6, we note that by symmetry there is no concentration gradient at a zone's center, whereas a component's concentration at the contact is fixed to that in the free-flowing zone, as is its initial concentration,

$$\left.\frac{\partial c_i}{\partial r}\right|_{r=0} = 0, \quad c_i\big|_{r=\delta} = C_i, \quad \text{and} \quad c_i\big|^{t=0} = C_i\big|^{t=0}. \tag{25.7}$$

These relations form, respectively, the boundary and initial conditions for solving Equations 25.5 and 25.6.

Diffusion in a dual-porosity simulation may in practice affect only the outermost rinds of the stagnant zones. In such cases, we can define a diffusion depth,

$$\delta_{diff} = \sqrt{4D^*_{stag}T}, \tag{25.8}$$

(e.g., Crank, 1975) that represents the affected thickness of the rinds. Here, T (s) is time span, the time elapsed between the simulation's onset and conclusion. We can now recast the first boundary condition in Equation 25.7 as

$$\left.\frac{\partial c_i}{\partial r}\right|_{r=\delta-\delta_{diff}} = 0, \tag{25.9}$$

by posing it at the base of the rinds, speeding computation.

Knowing the diffusion profile for component i, we can at any point in the simulation calculate the component's transfer rate (mol s^{-1}) into the free-flowing fraction as

$$r_i = -A_c \phi_{stag} D^*_{stag} \left.\frac{\partial c_i}{\partial r}\right|_{r=\delta}, \tag{25.10}$$

from the surface area A_c of the stagnant zones (Eqs. 25.2–25.4) and Fick's law (Eq. 23.22). Whereas we have written Equations 25.5–25.10 for an arbitrary component i, each may equivalently be written for the other components in the simulation by substituting c_w, c_k, and c_m for c_i.

25.2.3 Node-to-Node Transport

Chemical mass migrates from nodal block to nodal block in a dual-porosity model only within the free-flowing fraction of the domain. In coding such a model, we might consider that permeabilities k_x and k_y and coefficients of hydrodynamic dispersion D_{xx}, D_{xy}, and D_{yy} represent properties of the composite medium. In this case, the transport equations in Chapter 23 (Eqs. 23.10–23.13 and 23.20) can be applied directly.

Alternatively, the transport parameters might represent the hydraulic properties of the free-flowing fraction alone, as they do in the GWB software. The mass fluxes predicted by the equations in Chapter 23 need to be reduced by a factor of $1 - X_{stag}$ in this case, since the stagnant zones accommodate neither advection nor hydrodynamic dispersion.

25.3 Numerical Solution

The equation governing the one-dimensional diffusion profiles (Eq. 25.5 or 25.6) within the stagnant zones is readily solved subject to the given boundary and initial conditions (Eqs. 25.7 and 25.9) using finite difference techniques (e.g., Peaceman, 1977), as described in Chapter 23.

To this end, the stagnant zones, or just a diffusive rind around the zones, are discretized into a series of N_{sub} subnodes (Fig. 25.1). As the encompassing reactive transport simulation marches forward from t' to t, the diffusion profiles evolve in response to passing time and changing conditions at the free-flowing contact.

The transfer rates r_w, r_i, r_k, and r_m at each time step are given by the discrete form of Equation 25.10 and its counterparts for c_w, c_k, and c_m. The gain or loss of each chemical component in the free-flowing fraction is the product of the corresponding rate, r_i and so on, and the length $t - t'$ of the current time step. In this way, formulation of a dual-porosity model resembles that of a kinetic model, as described in Chapter 17 (Eqs. 17.10–17.12).

25.4 Example Calculation

As an example of a dual-porosity model, we return to program X1T and the mass transport calculation from Section 23.5. The domain, 1000 m long and 1 m^2 in cross-section, is infiltrated by 30 m^3 yr^{-1} of first contaminated, and then ambient water

```
interval start   =  0 yr, fluid = contaminated
interval elution =  2 yr, fluid = ambient
interval end     = 10 yr
```

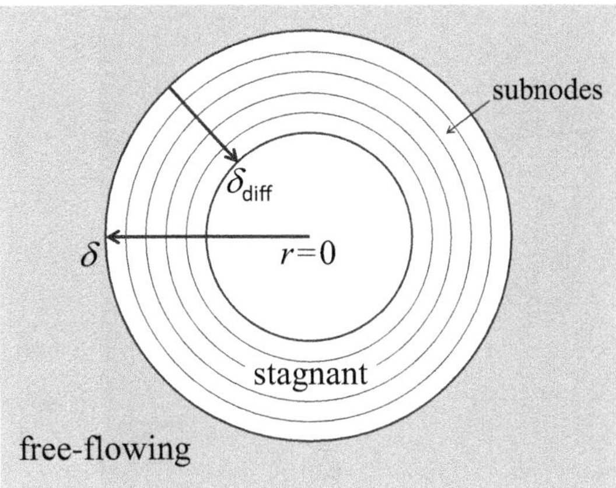

Figure 25.1 Discretization of the diffusive rind around a spherical stagnant zone of radius δ into subnodes, to allow calculation of diffusion profiles using finite differences. Coordinate r is the radial dimension and δ_{diff} is the diffusion depth, given by Equation 25.8.

```
length = 1000 m; width = 1 m; height = 1 m
discharge 30 m3/m2/yr; porosity = 30%
dispersivity = 1 m
Nx = 400

scope initial
   Na+  = 10     mmolal
   Pb++ = 1e-12 mmolal
   balance on Cl-
scope contaminated = initial
   Pb++ = 0.1    mmolal
scope ambient = initial
```

We set spherical stagnant zones of radius $\delta = 100$ cm to comprise 40% of the domain's volume. To solve the diffusion equation, we divide the outermost 20 cm of the zones into 10 subnodes. A small initial step helps the program transition from the first to the second set of boundary conditions

(cont'd)
```
dual_porosity spheres, volfrac = 40%, radius = 100 cm, \
   diff_length = 20 cm, Nsubnode = 10
dx_init = 10^-4
go
```

By default, D^*_{stag} and ϕ_{stag} match the diffusion coefficient and porosity of the free-flowing fraction. We can repeat the calculation with D^*_{stag} set to zero using the commands

(cont'd)
```
dual_porosity diff_con = 0
go
```

To return to the uniform case, we enter

(cont'd)

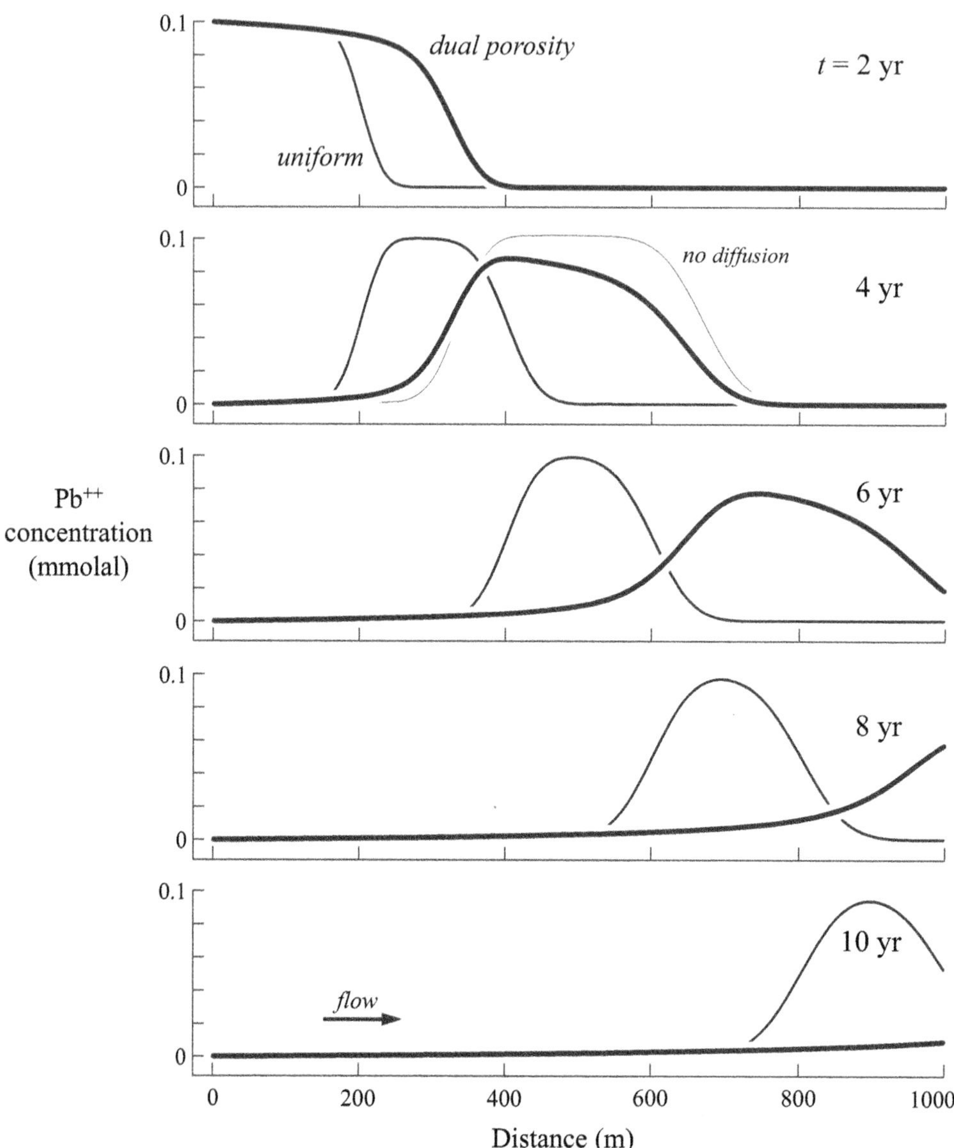

Figure 25.2 Dual-porosity model depicting migration of a pulse of inorganic lead contamination through an aquifer containing 40% stagnant zones (bold lines), at specific time slices in the simulation. Fine lines show the corresponding result assuming a uniform domain and hence not accounting for the effects of stagnant zones on transport. Very fine line at $t = 4$ years shows the dual-porosity model traced with D^*_{stag} set to zero, so that solute diffuses neither into nor out of the stagnant zones.

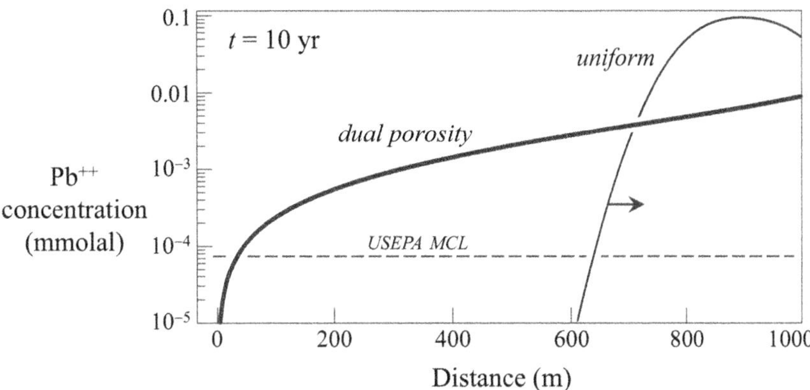

Figure 25.3 Results at the final time slice of the simulations shown in Figure 25.2, with concentration rendered on a logarithmic scale. Concentration at the end of the dual-porosity simulation (bold line) assumes an almost invariant profile along the aquifer, everywhere well in excess of Maximum Contaminant Level set by the U.S. Environmental Protection Agency (dashed line). In the simulation assuming a uniform medium (fine line), in contrast, the contaminant pulse migrates along the aquifer, leaving untainted water upgradient.

```
dual_porosity = off
go
```

to disable the dual-porosity feature and rerun the simulation.

Figure 25.2 shows the calculation results. The contaminant pulse in the dual-porosity simulation migrates along the domain more rapidly than in the uniform case, because flow in the former is funneled through the free-flowing fraction, which makes up 60% of the aquifer. The pulse in the dual-porosity case further assumes a notably asymmetric shape.

The asymmetry arises from diffusion into and out of the stagnant zones, as can be seen by comparing the results with the case where $D^*_{stag} = 0$. Water at the leading edge of the pulse has been in the aquifer longer, and hence has lost more Pb^{++} by diffusion, than water at the trailing edge. As well, the concentration gradient into the stagnant zones is steepest at first encounter with the contaminated water, so by Equation 25.10 flux into the zones is largest at the leading edge. Water at the trailing edge, in contrast, is enriched in Pb^{++} relative to the no-diffusion case, accentuating the asymmetry.

Rendering results at the end of the simulation on a logarithmic scale (Fig. 25.3) reveals an important aspect of dual-porosity modeling. At $t = 10$ years, the contaminant pulse is in the process of passing toward the outlet in the uniform model and has migrated past the domain in the dual-porosity model. Whereas water upstream of the pulse in the uniform case is effectively free of contaminant, the dual-porosity model predicts that water across the domain will hold a significant amount of Pb^{++}.

Contaminant, having been introduced to the stagnant zones as the pulse passed, is now re-entering the free-flowing fraction, tainting ambient water there at levels far in excess of drinking water standards. Worse, whereas flow is sweeping the pulse in the uniform case downgradient, the concentration profile in the dual porosity model is being continually fed by diffusion from the stagnant zones and, as such, assumes an almost invariant position in the aquifer.

PART III

APPLIED REACTION MODELING

26

Hydrothermal Fluids

Hydrothermal fluids, hot groundwaters that circulate within the Earth's crust, play central roles in many geological processes, including the genesis of a broad variety of ore deposits, the chemical alteration of rocks and sediments, and the origin of hot springs and geothermal fields. Many studies have been devoted to modeling how hydrothermal fluids react chemically as they encounter wall rocks, cool, boil, and mix with other fluids. Such modeling proliferated in part because hydrothermal fluids are highly reactive and because the reaction products are commonly well preserved, readily studied, and likely to be of economic interest. Further impetus was provided by the development of reliable modeling software in the 1970s, a period of concern over the availability of strategic and critical minerals and of heightened interest in economic geology and the exploitation of geothermal energy.

As a result, many of the earliest and most imaginative applications of geochemical modeling, beginning with Helgeson's (1970) simulation of ore deposition in hydrothermal veins and the alteration of nearby country rock, have addressed the reaction of hydrothermal fluids. For example, Reed (1977) considered the origin of a precious metal district; Garven and Freeze (1984), Sverjensky (1984, 1987), and Anderson and Garven (1987) studied the role of sedimentary brines in forming Mississippi Valley-type and other ore deposits; Wolery (1978), Janecky and Seyfried (1984), Bowers *et al.* (1985), and Janecky and Shanks (1988) simulated hydrothermal interactions along the midocean ridges; and Drummond and Ohmoto (1985) and Spycher and Reed (1988) modeled how fluid boiling is related to ore deposition.

In this chapter, we develop geochemical models of two hydrothermal processes: the formation of fluorite veins in the Albigeois ore district, and the origin of "black smokers," a name given to hydrothermal vents found along the ocean floor at midocean ridges.

26.1 Origin of a Fluorite Deposit

As a first case study, we borrow from the modeling work of Rowan (1991), who considered the origin of fluorite (CaF_2) veins in the Albigeois district of the southwest Massif Central, France. Production and reserves for the district as a whole total about 7 million metric tons, making it comparable to the more famous deposits of southern Illinois and western Kentucky, USA.

Like other fluorite deposits, the Albigeois ores are notable for their high grade. In veins of the Le Burc deposit, for example, fluorite comprises 90% of the ore volume (Deloule, 1982). Accessory minerals include quartz (SiO_2), siderite ($FeCO_3$), chalcopyrite ($CuFeS_2$), and small amounts of arsenopyrite (AsFeS). The deposits occur in a tectonically complex terrain dominated by metamorphic, plutonic, and volcanic rocks and sediments.

Deloule (1982) studied fluid inclusions from the Montroc and Le Burc deposits. He found that the ore-

Table 26.1 *Composition of ore-forming fluids at the Albigeois district, as determined by analysis of fluid inclusions (Deloule, 1982)*

Component	Concentration (molar)
Na^+	3.19
K^+	0.45
Ca^{++}	0.45
Mg^{++}	$(0.2–1) \times 10^{-2}$
Fe^{++}	$(0.75–3) \times 10^{-2}$
Cu^+	$(0.2–3) \times 10^{-2}$
Cl^-	4.5
HCO_3^-	0.2
SO_4^{--}	0.004–0.2

forming fluid was highly saline at both deposits; Table 26.1 summarizes its chemical composition. Homogenization temperatures in fluorite from the Le Burc veins range from about 110 °C to 150 °C, and most of the measurements fall between 120 °C and 145 °C. Assuming burial at the time of ore deposition to depths as great as 1 km, these values should be corrected upward for the effect of pressure by perhaps 10 °C and no more than 30 °C. We will assume here that the ore was deposited at temperatures between 125 °C and 175 °C.

To model the process of ore formation in the district, we first consider the effects of simply cooling the ore fluid. We begin by developing a model of the fluid at 175 °C, the point at which we assume that it begins to precipitate ore. To constrain the model, we use the data in Table 26.1 and assume that the fluid was in equilibrium with minerals in and near the vein. Specifically, we assume that the fluid's silica and aluminum contents were controlled by equilibrium with the quartz (SiO_2) and muscovite [$KAl_3Si_3O_{10}(OH)_2$] in the wall rocks, and that equilibrium with fluorite set the fluid's fluorine concentration.

We further specify equilibrium with kaolinite [$Al_2Si_2O_5(OH)_4$], which occurs in at least some of the veins as well as in the altered wall rock. Since we know the fluid's potassium content (Table 26.1), assuming equilibrium with kaolinite fixes pH according to the reaction

$$\underset{\textit{kaolinite}}{{}^{3}\!/\!_{2}\ Al_2Si_2O_5(OH)_4} + K^+ \rightleftarrows \underset{\textit{muscovite}}{KAl_3Si_3O_{10}(OH)_2} + {}^{3}\!/\!_{2}\,H_2O + H^+ \,. \tag{26.1}$$

By this reaction, we can expect the modeled fluid to be rather acidic, since it is rich in potassium. We could have chosen to fix pH by equilibrium with the siderite, which also occurs in the veins. It is not clear, however, that the siderite was deposited during the same paragenetic stages as the fluorite. It is difficult on chemical grounds, furthermore, to reconcile coexistence of the calcium-rich ore fluid and siderite with the absence of calcite ($CaCO_3$) in the district. In any event, assuming equilibrium with kaolinite leads to a fluid rich in fluorine and, hence, to an attractive mechanism for forming fluorite ore.

The model calculated in this manner predicts that two minerals, alunite [$KAl_3(OH)_6(SO_4)_2$] and anhydrite ($CaSO_4$), are supersaturated in the fluid at 175 °C, although neither mineral is observed in the district. This result is not surprising, given that the fluid's salinity exceeds the correlation limit for the activity coefficient model (Chapter 8). The observed composition in this case (Table 26.1), furthermore, actually represents the average of fluids from many inclusions and hence a mixture of hydrothermal fluids present over a range of time. As noted in Chapter 6, mixtures of fluids tend to be supersaturated, even if the individual fluids are not.

To avoid starting with a supersaturated initial fluid, we use alunite to constrain the SO_4 content to the limiting case, and anhydrite to similarly set the calcium concentration. The resulting SO_4 concentration will fall near the lower end of the range shown in Table 26.1, and the calcium content will lie slightly above the reported value.

In the program REACT, the procedure to model the initial fluid is

```
T = 175

swap Quartz    for SiO2(aq)
swap Muscovite for Al+++
swap Fluorite  for F-
swap Alunite   for SO4--
swap Kaolinite for H+
swap Anhydrite for Ca++

density = 1.14
TDS = 247500
Na+    =  3.19  molar
K+     =   .45  molar
Mg++   =   .006 molar
Fe++   =  2.e-2 molar
Cu+    =  1.e-2 molar
HCO3-  =  0.2   molar
Cl-    =  4.5   molar

1 free mole Quartz
1 free mole Muscovite
1 free mole Fluorite
1 free mole Alunite
1 free mole Kaolinite
1 free mole Anhydrite
go
```

Since the input constraints are in molar (instead of molal) units, we have specified the dissolved solid content and the fluid density under laboratory conditions, the latter estimated from the correlation of Phillips *et al.* (1981) for NaCl solutions. The resulting fluid is, as expected, acidic, with a predicted pH of 2.9. Neutral pH at 175 °C, for reference, is 5.7.

The fluid is extraordinarily rich in fluorine, as shown in Table 26.2, primarily because of the formation of the complex species AlF_3 and AlF_2^+. The importance of these species results directly from the assumed mineral assemblage,

$$\begin{aligned} AlF_3 + {}^3/_4\, \underset{\textit{muscovite}}{KAl_3Si_3O_{10}(OH)_2} + {}^3/_2\, \underset{\textit{anhydrite}}{CaSO_4} + {}^5/_2\, H_2O \rightleftarrows \\ {}^3/_2\, \underset{\textit{fluorite}}{CaF_2} + {}^5/_4\, \underset{\textit{quartz}}{SiO_2} + {}^3/_4\, \underset{\textit{alunite}}{KAl_3(OH)_6(SO_4)_2} + {}^1/_2\, \underset{\textit{kaolinite}}{Al_2Si_2O_5(OH)_4}\,, \end{aligned} \tag{26.2}$$

and, in the case of AlF_2^+, the activity of K^+,

Table 26.2 *Calculated concentrations of fluorine-bearing species in the Albigeois ore fluid at 175 °C*

Species	Concentration (mmolal)
AlF_3	227.5
AlF_2^+	38.8
AlF^{++}	5.54
HF	1.16
CaF^+	0.44
AlF_4^-	0.0443
F^-	0.0275
total F	767

$$\underset{}{AlF_2^+} + 3/2 \underset{\textit{muscovite}}{KAl_3Si_3O_{10}(OH)_2} + \underset{\textit{anhydrite}}{CaSO_4} + 4\,H_2O \rightleftarrows$$
$$\underset{\textit{fluorite}}{CaF_2} + 1/2 \underset{\textit{quartz}}{SiO_2} + 1/2 \underset{\textit{alunite}}{KAl_3(OH)_6(SO_4)_2} + 2 \underset{\textit{kaolinite}}{Al_2Si_2O_5(OH)_4} + K^+ \tag{26.3}$$

(as can be verified quickly with the program RXN). These complex species make the solution a potent ore-forming fluid. The fluid's acidity and fluorine content might reflect the addition of HF(g) derived from a magmatic source, as Plumlee *et al.* (1995) suggested for the Illinois and Kentucky deposits.

To model the consequences of cooling the fluid, we enter the commands

```
(cont'd)
pickup fluid
T final = 125
```

and type go to trigger the calculation. REACT carries the modeled fluid over a polythermal path, incrementally cooling it from 175 °C to 125 °C.

Figure 26.1 shows the mineralogic results of tracing the reaction path. As the fluid cools by 50 °C, it produces about 1.8 cm^3 of fluorite and 0.02 cm^3 of quartz per kg of solvent. No other minerals form. From a plot of the concentration of fluorine-bearing species (Fig. 26.2), it is clear that the fluorite forms in response to progressive breakdown of the AlF_3 complex with decreasing temperature. The complex sheds fluoride to produce AlF_2^+ according to the reaction

$$AlF_3 + 1/2\,CaCl^+ \rightarrow AlF_2^+ + 1/2 \underset{\textit{fluorite}}{CaF_2} + 1/2\,Cl^-, \tag{26.4}$$

yielding fluorite.

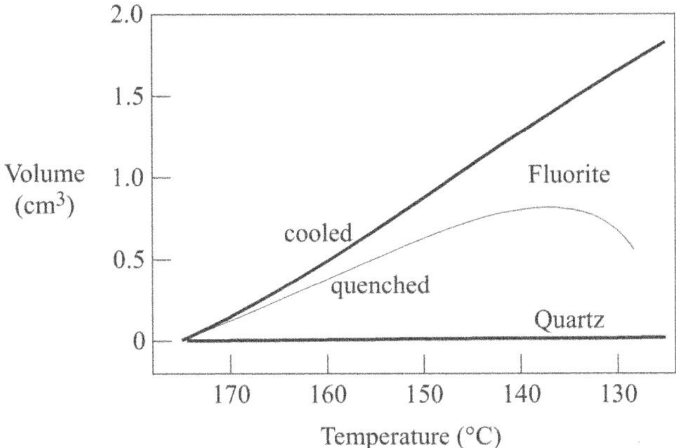

Figure 26.1 Calculated mineralogical consequences of cooling (bold lines) the Albigeois ore fluid from 175 °C to 125 °C, and of quenching it (fine line) with 125 °C water.

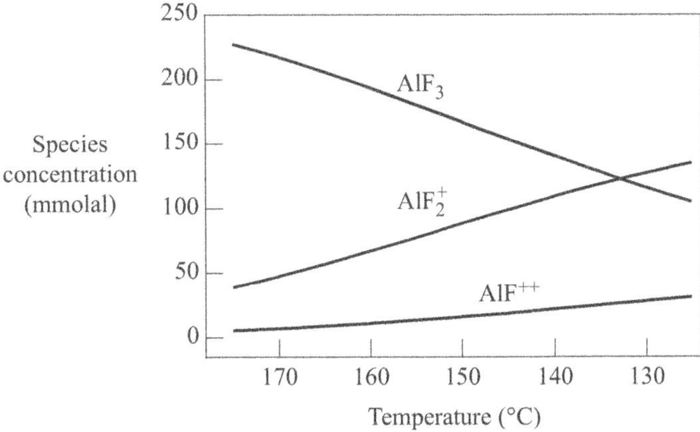

Figure 26.2 Calculated concentrations of predominant fluorine-bearing species in the Albigeois ore fluid as it cools from 175 °C to 125 °C.

As an alternative to simple cooling, we can test the consequences of quenching the ore fluid by incrementally mixing cooler water into it. The commands

```
(cont'd)
react 20 kg H2O
T 175, reactants = 125
```

set up a polythermal path in which temperature is set by the mixing proportions of the fluids, assuming each has a constant heat capacity. Typing `go` triggers the calculation.

As shown in Figure 26.1, quenching is effective at producing fluorite. The process, however, is somewhat less efficient than simple cooling because of the counter-effect of dilution, which limits precipitation and eventually begins to cause the fluorite that formed early in the reaction path to redissolve. Each mechanism of fluorite deposition is quite efficient, nonetheless, producing—over a temperature drop of just 50 °C—about a cm^3 or more mineral per kg of water in the ore fluid.

There are considerable uncertainties in the calculation. The fluid has an ionic strength of about 5 molal, far in excess of the range considered in the correlations for estimating species' activity coefficients, as discussed in Chapter 8. Stability constants for aluminum-bearing species are, in general, rather difficult to determine accurately (e.g., May, 1992). And the model relies on the equilibria assumed between the fluid and minerals in the vein and country rock. Nonetheless, the modeling suggests an attractive explanation for how veins of almost pure fluorite might form and provides an excellent example of the importance of complex species in controlling mineral solubility in hydrothermal fluids.

26.2 Black Smokers

In the spring of 1977, researchers on the submarine ALVIN discovered hot springs on the seafloor of the Pacific Ocean, along the Galapagos spreading center. Later expeditions to the East Pacific Rise and Juan de Fuca spreading center found more springs, some discharging fluids as hot as 350 °C. The hot springs are part of large-scale hydrothermal systems in which seawater descends into the oceanic crust, circulates near magma bodies where it warms and reacts extensively with deep rocks, and then, under the force of its buoyancy, discharges back into the ocean.

Discovery of the hot springs has had an important impact in the geosciences. Geologists today recognize the importance of hydrothermal systems in controlling the thermal structure of the ocean crust and the composition of ocean waters, as well as their role in producing ore deposits. The expeditions, in fact, discovered a massive sulfide deposit along the Galapagos spreading center. The springs also created excitement in the biologic sciences because of the large number of previously unknown species, such as tube worms, discovered near the vents.

Where fluids discharge from hot springs and mix with seawater, they cool quickly and precipitate clouds of fine-grained minerals. The clouds are commonly black with metal sulfides, giving rise to the term "black smokers." Some vents give off clouds of white anhydrite; these are known as "white smokers." Structures composed of chemical precipitates tend to form at the vents, where the hot fluids discharge into the ocean. The structures can extend upward into the ocean for several meters or more, and are composed largely of anhydrite and, in some cases, sulfide minerals.

The chemical processes occurring within a black smoker are certain to be complex because the hot, reducing hydrothermal fluid mixes quickly with cool, oxidizing seawater, allowing the mixture little chance to approach equilibrium. Despite this obstacle, or perhaps because of it, we bravely attempt to construct a chemical model of the mixing process. Table 26.3 shows chemical analyses of fluid from the NGS hot spring, a black smoker along the East Pacific Rise near 21 °N, as well as ambient seawater from the area.

To model the mixing of the hydrothermal fluid with seawater, we begin by equilibrating seawater at 4 °C, "picking up" this fluid as a reactant, and then reacting it into the hot hydrothermal fluid. In REACT, we start by suppressing several minerals:

```
suppress Quartz, Tridymite, Cristobalite, Chalcedony
suppress Hematite
```

According to Mottl and McConachy (1990), amorphous silica (SiO_2) is the only silica polymorph present in the "smoke" at the site. To allow it to form in the calculation, we suppress each of the more stable silica polymorphs. We also suppress hematite (Fe_2O_3) in order to give the iron oxy-hydroxide goethite (FeOOH) a chance to form.

We then equilibrate seawater, using data from Table 26.3,

Table 26.3 *Endmember compositions of a hydrothermal fluid and seawater, East Pacific Rise near 21 °N (Von Damm* et al., *1985; Drever, 1988; Mottl and McConachy, 1990)*

	NGS field	Seawater
	Concentration (mmolal)	
Na^+	529	480
K^+	26.7	10.1
Mg^{++}	~ 0	54.5
Ca^{++}	21.6	10.5
Cl^-	600	559
HCO_3^-	2.0*	2.4
SO_4^{--}	~ 0	29.5
H_2S	6.81	~ 0
SiO_2	20.2	0.17
	Concentration (μmolal)	
Sr^{++}	100.5	90
Ba^{++}	>15	0.20
Al^{+++}	4.1	0.005
Mn^{++}	1 039	<0.001
Fe^{++}	903	<0.001
Cu^+	<0.02	0.007
Zn^{++}	41	0.01
O_2(aq)	~ 0	123
T (°C)	273	4
pH	3.8 (25 °C)	7.8 (25 °C)
	4.2 (273 °C)	8.1 (4 °C)

*Estimated from titration alkalinity.

(cont'd)

```
T = 4
pH = 8.1

Cl-       559.     mmolal
Na+       480.     mmolal
Mg++       54.5    mmolal
SO4--      29.5    mmolal
Ca++       10.5    mmolal
K+         10.1    mmolal
HCO3-       2.4    mmolal
SiO2(aq)     .17   mmolal
Sr++         .09   mmolal
```

```
Ba++       .20   umolal
Zn++       .01   umolal
Al+++      .005  umolal
Cu+        .007  umolal
Fe++       .001  umolal
Mn++       .001  umolal
O2(aq)  123.     free umolal

go

pickup reactant = fluid
reactants times 10
```

and pick it up as a reactant. By multiplying the extent of the reactant system 10-fold, we prescribe mixing of seawater into the hydrothermal fluid in ratios as great as 10:1.

Finally, we define a polythermal path by equilibrating the hot hydrothermal fluid (Table 26.3)

(cont'd)

```
pH = 4.2
swap H2S(aq) for O2(aq)

Cl-      600.    mmolal
Na+      529.    mmolal
K+        26.7   mmolal
Ca++      21.6   mmolal
SiO2(aq)  20.2   mmolal
H2S(aq)    6.81  mmolal
HCO3-      2.    mmolal

Mn++    1039.    umolal
Fe++     903.    umolal
Sr++     100.5   umolal
Zn++      41.    umolal
Ba++      15.    umolal
Al+++      4.1   umolal
Cu+         .02  umolal
Mg++        .01  umolal
SO4--       .01  umolal

dump
T initial 273, reactants = 4
```

and reacting into it the cold seawater. The `dump` command causes the program to remove any minerals present in the initial system, before beginning to trace the reaction path. The final command sets a polythermal path in which temperature depends on the proportion of hot and cold fluids in the mixture, assuming that each has a constant heat capacity. Typing `go` triggers the calculation.

The calculation results (Fig. 26.3) show that anhydrite is the most abundant mineral to form. The mineral forms rapidly during the initial mixing,

$$\mathrm{Ca^{++}} + \mathrm{SO_4^{--}} \rightarrow \underset{anhydrite}{\mathrm{CaSO_4}} \quad , \tag{26.5}$$

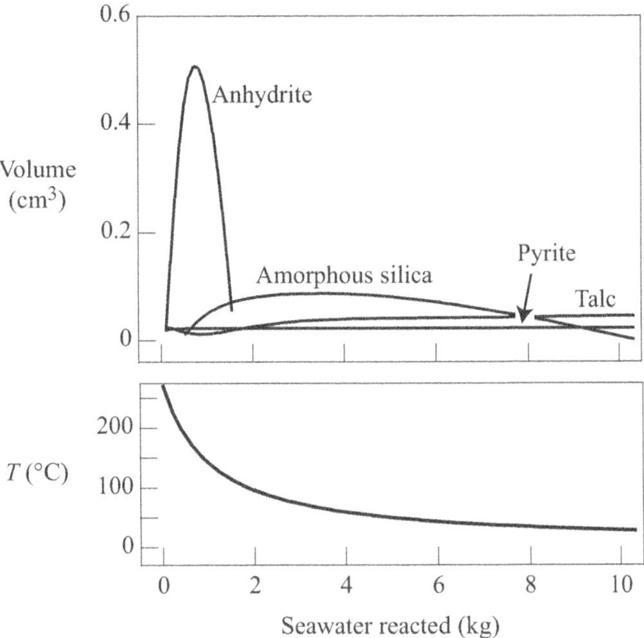

Figure 26.3 Mineralogical results (top) of mixing cold seawater into 1 kg of hot hydrothermal fluid from the NGS field. Minerals present in volumes less than 0.01 cm^3 are not shown. These minerals, in order of decreasing abundance, are: sphalerite, barite, potassium mordenite, calcium clinoptilolite, and covellite. Also shown (bottom) is the temperature of the fluid during mixing.

from reaction of the calcium in the hydrothermal fluid with sulfate in the seawater, eventually forming about a half cm^3 per kg of hydrothermal fluid. At a mixing ratio somewhat less than 1:1, however, anhydrite begins to redissolve, reflecting dilution of the hydrothermal calcium as mixing proceeds. The mineral has completely dissolved by a mixing ratio of about 1.5:1. These results seem in accord with the observation that anhydrite forms in abundance near the vent sites, where the hydrothermal fluid is least diluted.

After anhydrite, the most voluminous minerals to form are amorphous silica, talc [$Mg_3Si_4O_{10}(OH)_2$], and pyrite (FeS_2). Amorphous silica forms because its solubility decreases with cooling faster than it is diluted. Other minerals that form in small quantities over the simulation include sphalerite (ZnS), barite ($BaSO_4$), the zeolites mordenite ($KAlSi_5O_{12} \cdot 3H_2O$) and clinoptilolite ($CaAl_2Si_{10}O_{24} \cdot 8H_2O$), and covellite (CuS). According to the data of Mottl and McConachy (1990), each of these minerals (lumping the talc and zeolites into the category of unidentified aluminosilicates) is observed suspended above the NGS vent. Even though the smoke at this site is described as black, sulfide minerals make up just a small fraction of the mineral volume precipitated over the course of the simulation.

A number of the observed minerals (formulae given in Table 26.4) do not form in the simulation. Wurtzite is metastable with respect to sphalerite, so it cannot be expected to appear in the calculation results. Similarly, the formation of pyrite in the simulation probably precludes the possibility of pyrrhotite precipitating. In the laboratory, and presumably in nature, pyrite forms slowly, allowing less stable iron sulfides to precipitate. Elemental sulfur at the site probably results from incomplete oxidation of H_2S(aq), a process not accounted

Table 26.4 *Minerals in samples taken above black smokers of the East Pacific Rise near 21 °N (Mottl and McConachy, 1990)*

"Smoke" in plume	
Pyrrhotite, $Fe_{1-x}S$	Anhydrite, $CaSO_4$
Sphalerite, ZnS	Barite (trace), $BaSO_4$
Pyrite, FeS_2	Cubanite (trace), $CuFe_2S_3$
Unidentified Fe–S–Si phases	Wurtzite (trace), ZnS
Chalcopyrite, $CuFeS_2$	Covellite (trace), CuS
Amorphous silica, SiO_2	Marcasite (trace), FeS_2
Elemental sulfur, S	Unidentified silicates,
Goethite, etc., FeOOH	aluminosilicates (traces)
Particles dispersed in local seawater	
Anhydrite, $CaSO_4$	Sphalerite, ZnS
Pyrite, FeS_2	Sulfur, S
Gypsum, $CaSO_4 \cdot 2H_2O$	Pyrrhotite, $Fe_{1-x}S$
Chalcopyrite, $CuFeS_2$	

for in the simulation. There are no data in the LLNL database for marcasite or cubanite. Finally, goethite forms after we run the simulation to higher ratios of seawater to hydrothermal fluid than shown in Figure 26.3.

An interesting aspect of the calculation is that when oxidizing seawater mixes into the reduced hydrothermal fluid, the oxygen fugacity decreases (Fig. 26.4). The capacity of seawater to oxidize the large amount of hydrothermal $H_2S(aq)$ is limited by the supply of $O_2(aq)$ in seawater, which is small. Given the reaction

$$H_2S(aq) + 2\,O_2(aq) \rightarrow SO_4^{--} + 2\,H^+ , \tag{26.6}$$

and the data in Table 26.3, more than 100 kg of seawater are needed to oxidize each kg of hydrothermal fluid. The decrease in f_{O_2} shown in Figure 26.4 results from the shift of the sulfide–sulfate buffer with temperature. As long as any significant amount of $H_2S(aq)$ remains, the oxygen fugacity of the mixed fluid tracks the redox buffer as it moves to lower oxygen fugacity during cooling. The ability of the hydrothermal fluid to remain reducing as it mixes with seawater helps explain Mottl and McConachy's (1990) observation that sulfide minerals sampled from the "smoke" show no evidence under the electron microscope of beginning to redissolve.

26.3 Energy Available to Thermophiles

Subsea hydrothermal vents, as mentioned in the previous section, are sites of intense biological activity, relative to the rest of the ocean floor (e.g., Van Dover, 2000; Zierenberg *et al.*, 2000). Life here ranges in complexity from single cells to higher forms such as tube worms. The vent ecosystems are unique in many ways, including the fact that the primary producers create biomass not by photosynthesis, as is familiar in more accessible environments, but by chemosynthesis.

As fluid from the hydrothermal vent mixes with seawater, chemolithotrophic microbes by this process

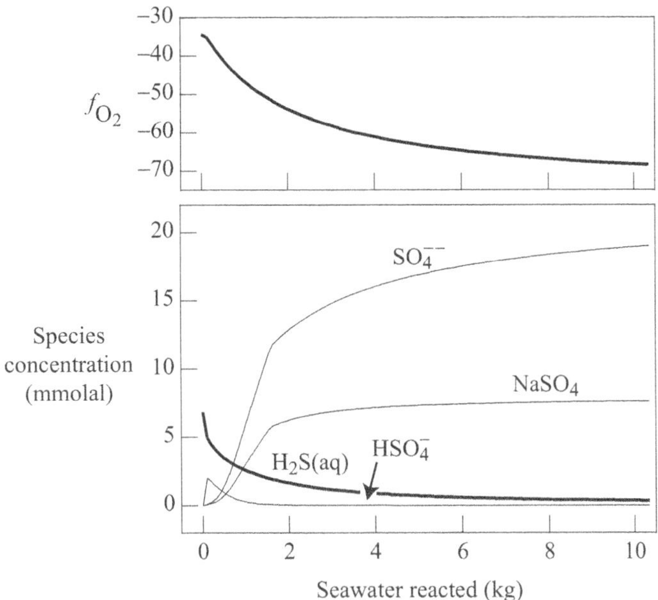

Figure 26.4 Oxygen fugacity (top) and concentrations of the predominant sulfur species (bottom) during the mixing simulation shown in Figure 26.3. Decrease in the $H_2S(aq)$ concentration is mostly in response to dilution with seawater, rather than oxidation.

harvest energy from the chemical disequilibrium among redox reactions, forming the base of the ecosystem's food chain. Microbes can derive energy by converting $CO_2(aq)$ to methane using $H_2(aq)$ from the vent fluid, for example, or using the $O_2(aq)$ in seawater to oxidize $H_2S(aq)$.

We consider in this section the energy available to thermophilic microorganisms as a hydrothermal fluid mixes with seawater, following the work of McCollom and Shock (1997) and Jin and Bethke (2005), for a variety of such metabolisms. To model redox energetics during mixing, we follow the procedure in the previous section, disengaging all redox couples except between $O_2(aq)$ and $H_2(aq)$. We assume the reaction

$$2\ H_2(aq) + O_2(aq) \rightarrow 2\ H_2O \tag{26.7}$$

between these species proceeds spontaneously in the mixture.

We begin in REACT by defining the composition of local seawater. The procedure is as before

```
suppress Quartz, Tridymite, Cristobalite, Chalcedony
suppress Hematite
decouple ALL
couple H2(aq)
T = 4

pH = 8.1
Cl-      559.    mmolal
Na+      480.    mmolal
Mg++      54.5   mmolal
SO4--     29.5   mmolal
Ca++      10.5   mmolal
```

```
K+         10.1    mmolal
HCO3-       2.4    mmolal
SiO2(aq)     .17   mmolal
Sr++         .09   mmolal
Ba++         .20   umolal
Zn++         .01   umolal
Al+++        .005  umolal
Cu+          .007  umolal
Fe++         .001  umolal
Mn++         .001  umolal
O2(aq)    123.     free umolal
```

with the addition of commands to set the redox coupling. Following McCollom and Shock (1997) and Jin and Bethke (2005), we set small concentrations of acetate, methane, and sulfide

(cont'd)

```
CH3COO-    3.      umolal
CH4(aq)     .002   umolal
HS-         .001   umolal
```

The commands

(cont'd)

```
go
pickup reactant = fluid
reactants times 50
```

cause the program to equilibrate the seawater and pick it up as a reactant.

In a similar fashion, we constrain the composition of fluid from the hydrothermal vent

(cont'd)

```
swap H2S(aq) for HS-
swap H2(aq) for O2(aq)

pH = 4.2
Cl-       600.    mmolal
Na+       529.    mmolal
K+         26.7   mmolal
Ca++       21.6   mmolal
SiO2(aq)   20.2   mmolal
HCO3-       2.    mmolal
Mn++     1039.    umolal
Fe++      903.    umolal
Sr++      100.5   umolal
Zn++       41.    umolal
Ba++       15.    umolal
Al+++       4.1   umolal
Cu+          .02  umolal
Mg++         .01  umolal
SO4--        .01  umolal
H2(aq)      1.7   mmolal

H2S(aq)     6.81  mmolal
```

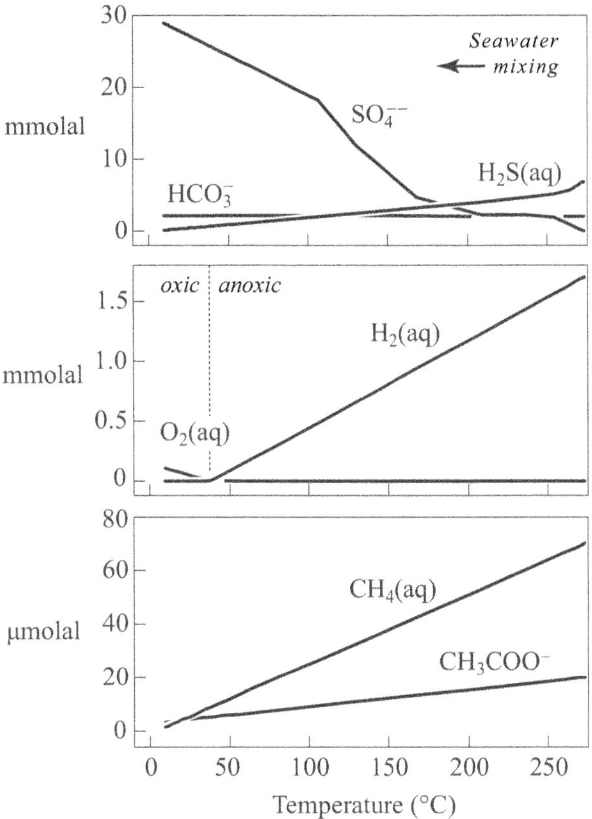

Figure 26.5 Concentrations of components (sulfate, sulfide, carbonate, methane, and acetate) and species (O_2 and H_2) that make up redox couples, plotted against temperature, during a model of the mixing of fluid from a hot subsea hydrothermal vent with cold seawater. Model assumes redox couples remain in chemical disequilibrium, except between $O_2(aq)$ and $H_2(aq)$. As the mixture cools past about 38 °C, the last of the dihydrogen from the vent fluid is consumed by reaction with dioxygen in the seawater. At this point the anoxic mixture becomes oxic as dioxygen begins to accumulate.

```
CH4(aq)     .07  mmolal
CH3COO-     .02  mmolal
```

The commands

```
(cont'd)
dump
T initial 273, reactants = 4
dx_init = 10^-4
step_increase = 2
dxplot = 0
go
```

tell the model to equilibrate the hot vent fluid and gradually add cold seawater into it. Figure 26.5 shows the calculation results.

During the mixing, concentrations of the $H_2S(aq)$, $CH_4(aq)$, and CH_3COO^- components decrease as the vent fluid is diluted, and the SO_4^{--} concentration increases toward that of seawater. $H_2(aq)$ in the vent fluid attenuates not only by dilution, but by reaction with $O_2(aq)$. As the mixture reaches a temperature of about 38 °C, the last of the $H_2(aq)$ in the anoxic fluid is oxidized in this way. Beyond this point, $O_2(aq)$ accumulates, leaving the mixture oxic. The temperature at which the transition from anoxic to oxic conditions occurs reflects the assumed supply of $H_2(aq)$ from the vent fluid, relative to $O_2(aq)$ from the seawater. If the vent produced a fluid poorer in dihydrogen, for example, or ambient seawater was more oxic, the transition would occur at a higher temperature.

As discussed in detail in Section 7.4, the energy liberated by a redox reaction depends on the redox potential of the electron donating half-cell reaction, relative to the electron accepting reaction. In the calculation results, we can trace the redox potentials of half-cell reactions among sulfur species,

$$HS^- + 4\,H_2O \rightleftarrows SO_4^{--} + 9\,H^+ + 8\,e^- , \tag{26.8}$$

and carbon species,

$$CH_3COO^- + 4\,H_2O \rightleftarrows 2\,HCO_3^- + 9\,H^+ + 8\,e^- \tag{26.9}$$

$$CH_4(aq) + 3\,H_2O \rightleftarrows HCO_3^- + 9\,H^+ + 8\,e^- \tag{26.10}$$

$$2\,CH_4(aq) + 2\,H_2O \rightleftarrows CH_3COO^- + 9\,H^+ + 8\,e^- , \tag{26.11}$$

as well as the hydrolysis reactions,

$$H_2O \rightleftarrows 1/2\,O_2(aq) + 2\,H^+ + 2\,e^- \tag{26.12}$$

and

$$H_2(aq) \rightleftarrows 2\,H^+ + 2\,e^- , \tag{26.13}$$

as shown in Figure 26.6. We could extend this list by including in the calculation both oxidized and reduced form of metals such as iron. The redox potentials for the last two reactions listed, for electron acceptance by dioxygen and donation by dihydrogen, are equivalent in the model results, since we have set $O_2(aq)$ and $H_2(aq)$ to be in equilibrium.

Taking the half-cell reactions in pairs, one electron donating and the other accepting, we can write a number of metabolisms by which microbes might derive energy during the mixing process, as shown in Table 26.5. Several hydrogentrophic metabolisms are viable where the fluid mixture is anoxic: sulfate reduction,

$$SO_4^{--} + 4\,H_2(aq) + 2\,H^+ \rightarrow H_2S(aq) + 4\,H_2O , \tag{26.14}$$

acetogenesis,

$$2\,CO_2(aq) + 4\,H_2(aq) \rightarrow CH_3COO^- + H^+ + 2\,H_2O , \tag{26.15}$$

and methanogenesis,

$$CO_2(aq) + 4\,H_2(aq) \rightarrow CH_4(aq) + 2\,H_2O . \tag{26.16}$$

Other possible anaerobic metabolisms are acetotrophic sulfate reduction,

$$SO_4^{--} + CH_3COO^- + 3\,H^+ \rightarrow H_2S(aq) + 2\,CO_2(aq) + 2\,H_2O , \tag{26.17}$$

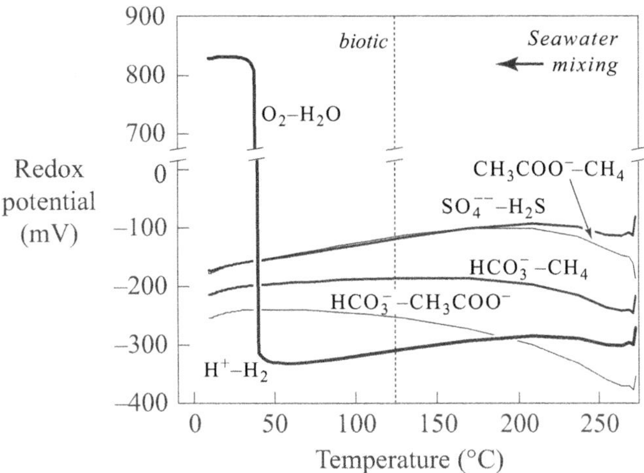

Figure 26.6 Redox potentials (mV) of various half-cell reactions during mixing of fluid from a subsea hydrothermal vent with seawater, as a function of the temperature of the mixture. Since the model is calculated assuming O_2(aq) and H_2(aq) remain in equilibrium, the potential for electron acceptance by dioxygen is the same as that for donation by dihydrogen. Dotted line shows currently recognized upper temperature limit (121 °C) for microbial life in hydrothermal systems. A redox reaction is favored thermodynamically when the redox potential for the electron donating half-cell reaction falls below that of the accepting half-reaction.

Table 26.5 *Electron donating and accepting redox couples and limiting reactant species for various anaerobic and aerobic microbial metabolisms favored in fluid from a subsea hydrothermal vent, as it mixes with seawater*

Metabolism	Donating couple	Accepting couple	Limiting reactant
Anaerobic ($T \gtrsim 38$ °C)			
H_2-trophic sulfate reduction	H_2–H^+	SO_4–H_2S	H_2(aq)
H_2-trophic acetogenesis	H_2–H^+	CO_2–CH_3COO^-	H_2(aq)
H_2-trophic methanogenesis	H_2–H^+	CO_2–CH_4	H_2(aq)
Ac-trophic sulfate reduction	CH_3COO^-–CO_2	SO_4–H_2S	CH_3COO^-
Ac-clastic methanogenesis	CH_3COO^-–CO_2	CH_3COO^-–CH_4	CH_3COO^-
Aerobic ($T \lesssim 38$ °C)			
Sulfide oxidation	H_2S–SO_4	O_2–H_2O	O_2 or H_2S
Methanotrophy	CH_4–HCO_3	O_2–H_2O	CH_4(aq)
Acetotrophy	CH_3COO^-–HCO_3	O_2–H_2O	CH_3COO^-

and acetoclastic methanogenesis,

$$2\,CH_3COO^- + 2\,H^+ \rightarrow 2\,CH_4(aq) + 2\,CO_2(aq)\,. \tag{26.18}$$

In each case, we write the metabolic reaction for the transfer of eight electrons, so the resulting energies can be compared on an equal basis.

Where the fluid is oxic, several aerobic metabolisms can proceed, including sulfide oxidation,

$$H_2S(aq) + 2\,O_2(aq) \rightarrow SO_4^{--} + 2\,H^+\,, \tag{26.19}$$

methanotrophy,

$$CH_4(aq) + 2\,O_2(aq) \rightarrow HCO_3^- + H^+ + H_2O\,, \tag{26.20}$$

and homoacetotrophy,

$$CH_3COO^- + 2\,O_2(aq) \rightarrow 2\,HCO_3^- + H^+\,. \tag{26.21}$$

Reflecting the higher pH of the oxic relative to the anoxic fluid, we have written these reactions in terms of HCO_3^-, rather than $CO_2(aq)$. Again, each reaction is balanced on an eight-electron basis.

The thermodynamic drive for each reaction is its negative free-energy change, $-\Delta G_r$. From Equation 7.15, this quantity is given as

$$\begin{aligned} -\Delta G_r &= nF\ \Delta \text{Eh} \\ &= nF\ (\text{Eh}_{\text{acc}} - \text{Eh}_{\text{don}})\ , \end{aligned} \tag{26.22}$$

directly from the difference in redox potentials between the electron accepting and donating half-reactions. Here ΔG_r is in J mol^{-1}, or V C mol^{-1}; n is the number of electrons transferred, eight in our case; F is the Faraday constant, 96 485 C mol^{-1}; and Eh_{acc} and Eh_{don} are the redox potentials (V) of the accepting and donating reactions.

Figure 26.7 shows how the thermodynamic drive for each metabolism varies as the fluids mix, as determined from the redox potentials in Figure 26.6. Microbial life is known to exist near the subsea vents to temperatures as high as 121 °C (Kashefi and Lovley, 2003) and it seems unlikely that it will be observed beyond 150 °C. We need consider, therefore, only the latter part of the mixing, where temperature falls below such limits.

Each of the hydrogentrophic metabolisms is favored in the anoxic fluid, but cannot proceed once the mixture has become oxic, in the absence of dihydrogen. Of these, sulfate reduction has the largest thermodynamic drive. The other two anaerobic metabolisms, acetotrophic sulfate reduction and acetoclastic methanogenesis, do not depend on the presence of $H_2(aq)$ and are favored throughout the mixing. The organisms that utilize these metabolisms, however, are strict anaerobes (e.g., Konhauser, 2007) and so can live only where the mixture is anoxic, regardless of the energy available to them.

The aerobic metabolisms—sulfide oxidation, methanotrophy, and acetotrophy – cannot proceed in the anoxic fluid, which is devoid of dioxygen, but are strongly favored in the oxic fluid. The thermodynamic drive for the aerobic processes under oxic conditions is about 800 kJ mol^{-1}, much larger than the drives of less than about 150 kJ mol^{-1} observed under anoxic conditions for the anaerobic metabolisms.

The thermodynamic drives cited are the energy released instantaneously by the metabolic reaction, at the moment reaction commences. The drives tell whether the reaction can proceed, and whether it can supply enough energy for a cell to conserve energy by synthesizing ATP, as discussed in Section 7.4. The values, however, do not describe how much energy microbes can extract from a fluid, and hence how much microbial growth the fluid can sustain.

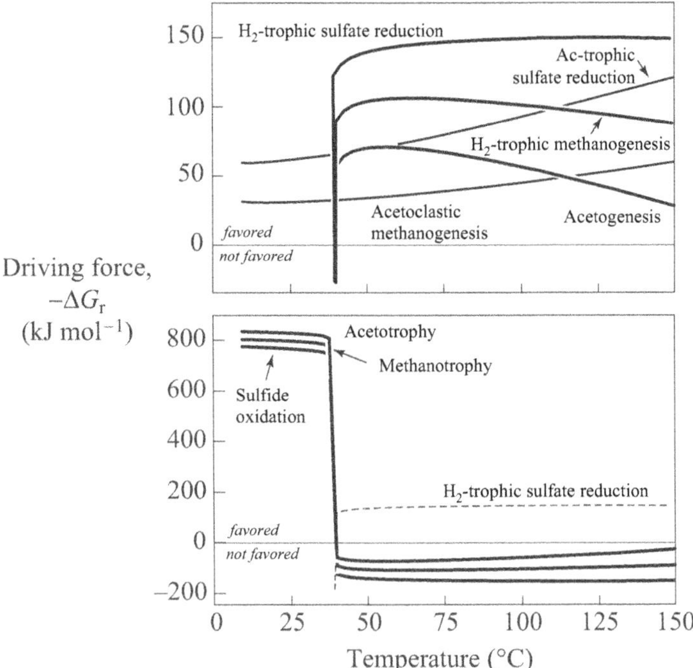

Figure 26.7 Thermodynamic driving forces for various anaerobic (top) and aerobic (bottom) microbial metabolisms during mixing of a subsea hydrothermal fluid with seawater, as a function of temperature. Since the driving force is the negative free-energy change of reaction, metabolisms with positive drives are favored thermodynamically; those with negative drives cannot proceed. The drive for sulfide oxidation is the mirror image of that for hydrogentrophic sulfate reduction, since in the calculation O_2(aq) and H_2(aq) are in equilibrium.

The net amount of energy available from a fluid is its *energy yield*, given in units of energy per mass fluid, or J kg^{-1}. Since the thermodynamic drive is given by Equation 26.22 from a difference in electrical potential (ΔEh), we can properly consider it a voltage. Thinking this way, the energy yield accounts not only for the reaction's voltage, but its "capacitance," the number of electrons in the fluid that can be transferred.

The simplest way to calculate the energy yield is to multiply the thermodynamic drive by the mass of the limiting reactant, the reactant that will be first exhausted from the fluid as the reaction proceeds. Figure 26.8 shows energy yields calculated in this way, for the various metabolisms considered. An energy yield calculated in this manner is approximate, since the reaction's thermodynamic drive does not remain constant, but diminishes as its reactants are consumed. This effect, however, is commonly minor, because the driving force varies with the logarithm of reactant concentration. Once 90% of a limiting reactant has been consumed, assuming a reaction coefficient of one, the driving force at 25 °C has been reduced by just 5.7 kJ mol^{-1}; at 99% consumption, the reduction is only 11.4 kJ mol^{-1}.

Taking sulfide oxidation (Reaction 26.19) as an example, when the fluid mixture reaches 25 °C , there are about 5 mmol of H_2S(aq) and 0.6 mmol of O_2(aq) in the unreacted fluid, per kg of vent water. The O_2(aq) will be consumed first, after about 0.3 mmol of reaction turnover, since its reaction coefficient is two; it is the limiting reactant. The thermodynamic drive for this reaction at this temperature is about 770 kJ mol^{-1}. The energy yield, then, is $(0.3 \times 10^{-3}$ mol kg$^{-1}) \times (770 \times 10^3$ J mol$^{-1})$, or about 230 J kg^{-1} vent water

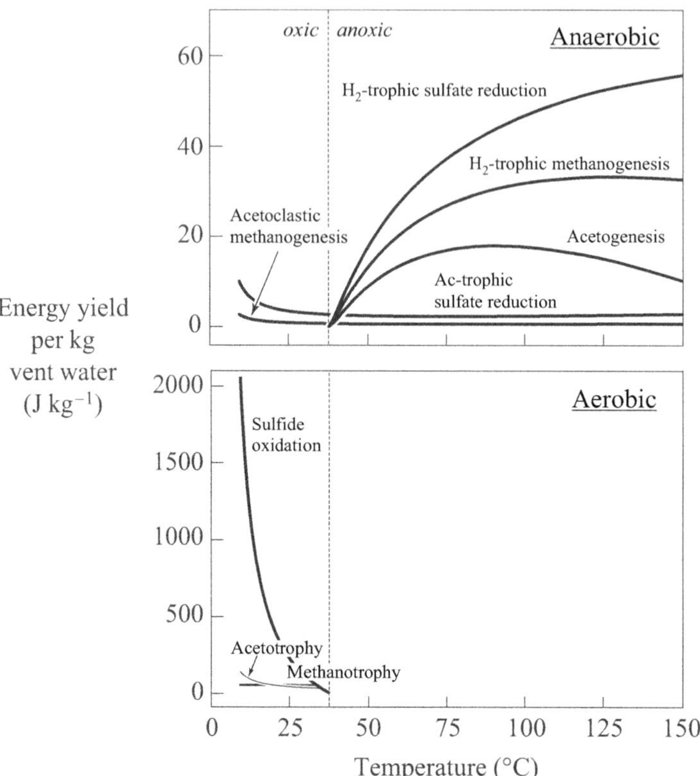

Figure 26.8 Energy yields for various anaerobic (top) and aerobic (bottom) metabolisms during mixing of a subsea hydrothermal fluid with seawater, expressed as a function of temperature, per kg of hydrothermal water. Energy yields for acetoclastic methanogenesis and acetotrophic sulfate reduction under oxic conditions are hypothetical, since microbes from these functional groups are strict anaerobes and cannot live in the presence of dioxygen.

(Fig. 26.8). In reality, of course, this entire yield would not necessarily be available at this point in the mixing. If some of the O_2(aq) had been consumed earlier, or was taken up by reaction with other reduced species, less of it, and hence less energy would be available for sulfide oxidation.

In the calculation results (Fig. 26.8), we see that while the fluid remains anoxic, modest amounts of energy are available, most notably to hydrogentrophic sulfate reducers and hydrogentrophic methanogens. When the mixture turns oxic, however, much larger amounts of energy become available, primarily to the sulfide oxidizers. The total amount of energy that can be supplied to sulfide oxidizers during the mixing process dwarfs that available to all the other metabolisms combined.

The energetics depicted in this way are in accord with the microbial ecology observed at deep sea hydrothermal systems (e.g., Kelley *et al.*, 2002; Huber *et al.*, 2003; Schrenk *et al.*, 2003). Sediments and black smoker walls invaded by hydrothermal fluids there contain sparse microbial populations of mostly thermophilic methanogens and sulfate reducers. Abundant populations of mesophilic aerobes dominated by sulfide reducers, in contrast, are found in the open ocean where hydrothermal fluids mix freely with seawater.

27
Geothermometry

Geothermometry is the use of a fluid's (or, although not discussed here, a rock's) chemical composition to estimate the temperature at which it equilibrated in the subsurface. The specialty is important, for example, in exploring for and exploiting geothermal fields, characterizing deep groundwater flow systems, and understanding the genesis of ore deposits.

Several chemical geothermometers are in widespread use. The silica geothermometer (Fournier and Rowe, 1966) works because the solubilities of the various silica minerals (e.g., quartz and chalcedony, SiO_2) increase monotonically with temperature. The concentration of dissolved silica, therefore, defines a unique equilibrium temperature for each silica mineral. The Na–K (White, 1970) and Na–K–Ca (Fournier and Truesdell, 1973) geothermometers take advantage of the fact that the equilibrium points of cation exchange reactions among various minerals (principally, the feldspars) vary with temperature.

In applying these methods, it is necessary to make a number of assumptions or corrections (e.g., Fournier, 1977). First, the minerals with which the fluid reacted must be known. Applying the silica geothermometer assuming equilibrium with quartz, for example, would not give the correct result if the fluid's silica content is controlled by reaction with chalcedony. Second, the fluid must have attained equilibrium with these minerals. Many studies have suggested that equilibrium is commonly approached in geothermal systems, especially for ancient waters at high temperature, but this may not be the case in young sedimentary basins like the Gulf of Mexico basin (Land and Macpherson, 1992). Third, the fluid's composition must not have been altered by separation of a gas phase, mineral precipitation, or mixing with other fluids. Finally, corrections may be needed to account for the influence of certain dissolved components, including CO_2 and Mg^{++}, which affect the equilibrium composition (Pačes, 1975; Fournier and Potter, 1979; Giggenbach, 1988).

Using geochemical modeling, we can apply chemical geothermometry in a more generalized manner. By utilizing the entire chemical analysis rather than just a portion of it, we avoid some of the restricting assumptions mentioned in the preceding paragraph (see Michard *et al.*, 1981; Michard and Roekens, 1983; and especially Reed and Spycher, 1984). Having constructed a theoretical model of the fluid in question, we can calculate the saturation state of each mineral in the database, noting the temperature at which each is in equilibrium with the fluid. Hence, we need make no a-priori assumption about which minerals control the fluid's composition in the subsurface.

Given sufficient field data, we can use modeling techniques to restore flashed gases or precipitated minerals to the fluid, minimizing another potential source of error. In the final section of this chapter, for example, we use production data from wet-steam wells to reconstitute geothermal fluids as they existed before gas separation. Finally, since the geochemical model is a relatively complete description of the fluid's chemistry, we avoid the necessity of the various corrections that might have had to be applied had we used a simpler calculation method. A disadvantage of applying geochemical modeling, however, is that the technique requires a reasonably complete and accurate chemical analysis of the fluid in question, which is not always available.

In this chapter, we explore how we can use chemical analyses and pH determinations made at room temperature to deduce details about the origins of natural fluids. These same techniques are useful in interpreting laboratory experiments performed at high temperature, since analyses made at room temperature need to be projected to give pH, oxidation state, gas pressures, saturation indices, and so on under experimental conditions.

27.1 Principles of Geothermometry

The most direct way to demonstrate the principles of geothermometry is to construct a synthetic example on the computer. We start by "sampling" a hypothetical geothermal water at 250 °C and letting it cool to room temperature as a closed system. We assume a water that is initially in equilibrium with albite ($NaAlSi_3O_8$), muscovite [$KAl_3Si_3O_{10}(OH)_2$], quartz (SiO_2), potassium feldspar ($KAlSi_3O_8$; "maximum microcline" in the LLNL database), and calcite ($CaCO_3$).

In REACT, the commands

```
T = 250
swap Albite for Na+
swap "Maximum Microcline" for K+
swap Muscovite for Al+++
swap Quartz for SiO2(aq)
swap Calcite for HCO3-

pH = 5
Ca++ =  .05 molal
Cl-  = 3.   molal

1 free cm3 Albite
1 free cm3 "Maximum Microcline"
1 free cm3 Muscovite
1 free cm3 Quartz
1 free cm3 Calcite
```

define the initial system. To describe sampling and cooling of the fluid, we enter the commands

```
(cont'd)
dump
precip = off
T final = 25
go
```

which set up a polythermal reaction path.

With the `dump` command, we cause the program to discard the minerals present in the initial system before beginning the reaction path. In this way, we simulate the separation of the fluid from reservoir minerals as it flows into the wellbore. The `precip = off` command prevents the program from allowing minerals to precipitate as the fluid cools. In practice, samples are acidified immediately after they have been sampled and their pH determined. Preservation by this procedure helps to prevent solutes from precipitating, which would alter the fluid's composition before it is analyzed.

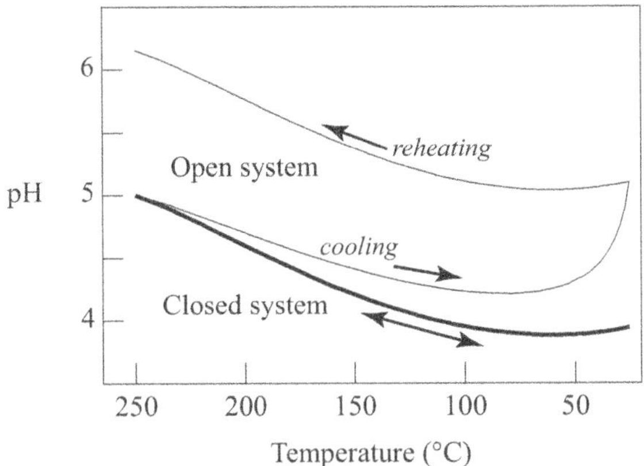

Figure 27.1 Variation in pH in a computer simulation of sampling, cooling, and then reheating a hypothetical geothermal fluid. Bold line shows path followed when system is held closed; fine lines show variations in pH when fluid is allowed to degas CO_2 as it cools.

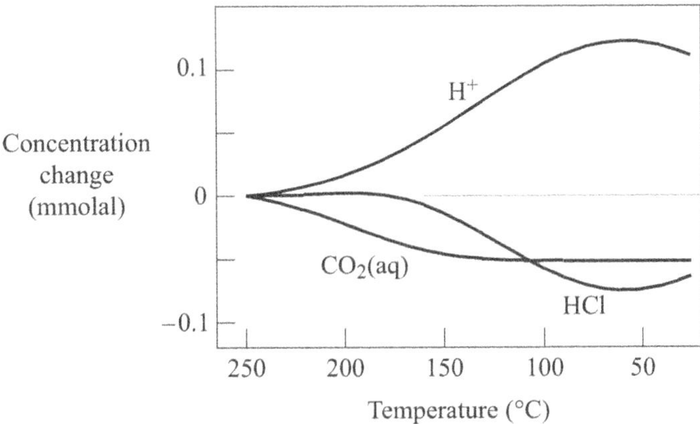

Figure 27.2 Changes in concentration of aqueous species H^+, CO_2(aq), and HCl with temperature during cooling of a geothermal fluid as a closed system. A positive value indicates an increase in concentration relative to 250 °C; a negative value represents a decrease.

Since we have provided initial and final temperatures but have not specified any reactants, the program traces a polythermal path for a closed system (see Chapter 15). The fluid's pH (Fig. 27.1) changes with temperature from its initial value of 5 at 250 °C to less than 4 at 25 °C. The change is entirely due to variation in the stabilities of the aqueous species in solution. As shown in Figure 27.2, the H^+ concentration increases in response to the dissociation of the HCl ion pair,

$$\mathrm{HCl} \rightarrow \mathrm{H^+} + \mathrm{Cl^-}, \tag{27.1}$$

and the breakdown of CO_2(aq),

$$\mathrm{CO_2(aq)} + \mathrm{H_2O} \rightarrow \mathrm{H^+} + \mathrm{HCO_3^-}, \tag{27.2}$$

to form HCO_3^-.

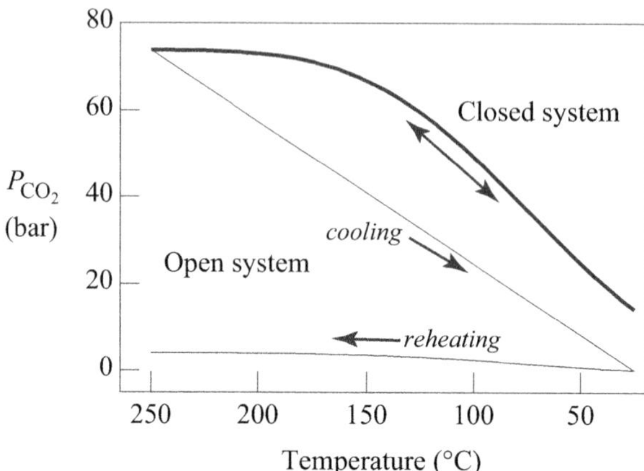

Figure 27.3 Variation in the partial pressure of CO_2 in a computer simulation of sampling, cooling, and reheating a hypothetical geothermal fluid. Bold line shows path followed when system is held closed. Fine lines show effects of an open system in which fluid is allowed to degas CO_2 as it cools.

The partial pressure of CO_2 decreases sharply during cooling (Fig. 27.3), as would be expected, since gas solubility increases as temperature decreases. In the calculation, the decrease in CO_2 pressure results almost entirely from variation in the equilibrium constant for the reaction

$$CO_2(g) \rightleftarrows CO_2(aq)\,. \tag{27.3}$$

The log K for this reaction increases from -2.12 at 250 °C to -1.45 at 25 °C. The final CO_2 pressure is about 15 bar, considerably in excess of atmospheric P_{CO_2}. We would certainly need to take extraordinary measures to prevent the fluid from effervescing, if we were actually performing this experiment instead of simulating it.

Now that we have simulated sampling the fluid and letting it cool, let us predict the fluid's original temperature (which we already know to be 250 °C). The REACT commands

```
(cont'd)
pickup
T final = 300
go
```

cause the program to "pick up" the results at the end of the previous path as the starting point for the current calculation. In other words, the cooled fluid will constitute the new initial system. We specify that the program heats the fluid to 300 °C and triggers the calculation.

The values predicted for pH and CO_2 pressure retrace the paths followed during the cooling calculation (Figs. 27.1 and 27.3). Since the system is closed to mass transfer, its equilibrium state depends only on temperature. Figure 27.4 shows the saturation indices calculated for various minerals. There are two salient points to consider in this plot. First, each of the minerals (albite, muscovite, quartz, potassium feldspar, and calcite) present in the formation when the fluid was sampled is in equilibrium (i.e., log $Q/K = 0$) at 250 °C. The minerals are in equilibrium together at no other temperature. Second, minerals in the database that were not present in the formation appear undersaturated at the equilibrium temperature. These two criteria allow

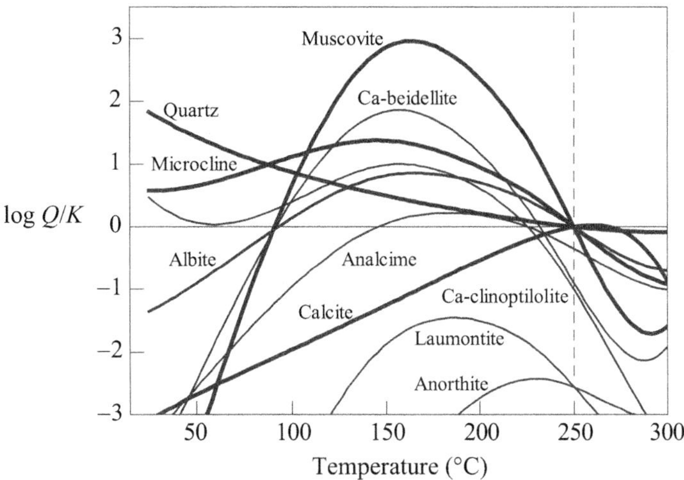

Figure 27.4 Mineral saturation indices (log Q/K) over the course of simulating the reheating of a hypothetical geothermal fluid. Bold lines show indices for minerals assumed to be present in the initial formation; fine lines show values for other minerals. Dashed line marks sampling temperature (250 °C).

us to uniquely identify the fluid's original temperature and hence form the basis of our generalized chemical geothermometer.

As a second experiment, let us simulate the sampling of the same fluid as an open system. This time, we allow it to effervesce CO_2 as we bring it to the surface and let it cool. We start as before, but include a `slide` command to vary CO_2 fugacity from its initial value to one, corresponding to the confining pressure of the atmosphere. The procedure (starting anew in REACT) is

(Enter initial system as before, at the beginning of this section)

```
dump
precip = off
T final = 25
slide f CO2(g) to 1
go
```

In the open system, CO_2 pressure (Fig. 27.3) varies linearly over the reaction path from 75 bar to a value of about 1 bar, as we prescribed, tracing a path of lower pressure than predicted for the closed system. To drive down the gas pressure, the program allows CO_2 to escape from the fluid into the external gas buffer (see Chapter 15). The pH (Fig. 27.1) follows a path of higher values than in a closed system, reflecting the transfer of CO_2, an acid gas, into the buffer.

Now we apply our geothermometer by simulating the reheating of the fluid:

(cont'd)

```
pickup
T final = 300
go
```

The system was not closed during cooling, so neither the pH nor CO_2 pressure (Figs. 27.2 and 27.3) returns to its original value at 250 °C.

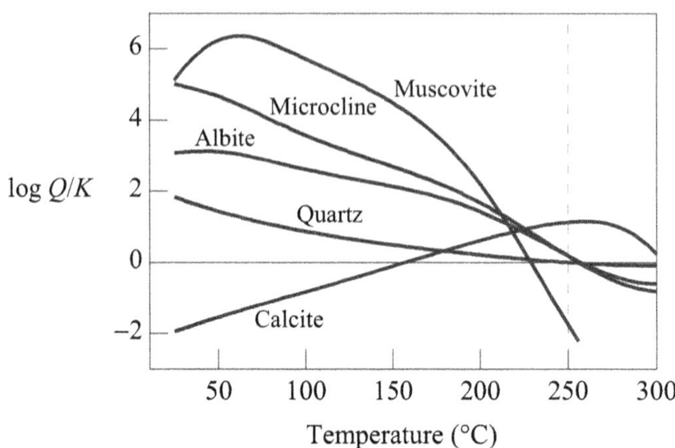

Figure 27.5 Mineral saturation indices (log Q/K) over the course of simulating the reheating of a hypothetical geothermal fluid that degassed CO_2 during sampling. Dashed line marks original formation temperature (250 °C). Although each of the minerals shown was present in the formation, the saturation profiles do not clearly identify the formation temperature because of the CO_2 loss.

The predicted saturation states of the formation minerals (Fig. 27.5), furthermore, no longer identify a unique formation temperature. Whereas the temperatures suggested by albite, quartz, and potassium feldspar are quite close to the 250 °C formation temperature, those predicted by assuming that the fluid was in equilibrium with muscovite and calcite are too low, respectively, by margins of about 25 °C and 100 °C. To avoid error of this sort, we would need to determine the amount of gas lost from the sample and reintroduce it to the equilibrium system before calculating saturation indices.

27.2 Hot Spring at Hveravik, Iceland

To see how we might apply geochemical modeling to the geothermometry of natural waters, we consider the effluent of a hot spring at Gjögur, Hveravik, Iceland. The hot spring is part of the surface expression of Iceland's well-known geothermal resources, which are developed within basaltic rocks in or near active volcanic belts. Arnórsson *et al.* (1983), part of a group noted for its high-quality analyses of geothermal waters, provide the water's chemical composition (Table 27.1).

The spring yields a Na–Ca–Cl water at about 72 °C with a pH (measured at 11 °C) of 7.1. The water appears to be oxidized, since there is abundant sulfate but no detected dissolved sulfide. Details of the origin of the spring water are unknown. Does it circulate deeply, reaching high temperatures only to cool near the discharge point? Is it a hot saline water from depth that has mixed with local groundwater near the spring (and, hence, is likely to be out of equilibrium with most minerals)? Did it degas significantly as it discharged? Or is it a relatively shallow groundwater that has reached equilibrium with its host rock near its discharge temperature?

In REACT, the commands

```
swap CO2(aq) for  HCO3-
pH =  7.10
SiO2(aq) =   49.0   mg/kg
Na+      =  715.7   mg/kg
```

Table 27.1 *Chemical composition of water emanating from a hot spring at Gjögur, Hveravik, Iceland (Arnórsson* et al., *1983)*

SiO_2 (mg kg^{-1})	49
Na^+	715.5
K^+	17.3
Ca^{++}	759.4
Mg^{++}	3.68
Fe^{++}	0.018
Al^{+++}	0.01
CO_2(aq)	13.3
SO_4^{--}	297.6
H_2S(aq)	<0.01
Cl^-	2 460
F^-	0.82
Dissolved solids	4 366
Sampling temperature (°C)	72
pH at 11 °C	7.10

```
K+        =   17.3   mg/kg
Ca++      =  759.4   mg/kg
Mg++      =    3.68  mg/kg
Fe++      =    .018  mg/kg
Al+++     =    0.01  mg/kg
CO2(aq)   =   13.3   mg/kg
SO4--     =  297.6   mg/kg
Cl-       = 2460.    mg/kg
F-        =     .82  mg/kg

precip = off
T initial = 11, final = 150
go
```

set up the geothermometry calculation, as discussed in the previous section.

Figure 27.6 shows the saturation states of the aluminosilicate minerals in the LLNL database, plotted against temperature. Many minerals are supersaturated at low and high temperature, but a clear equilibrium point appears at about 80 °C, slightly warmer than the discharge temperature of 72 °C. At this point, the fluid's composition seems to be controlled by calcium and potassium clinoptilolite (zeolites; e.g., $CaAl_2Si_{10}O_{24} \cdot 8H_2O$) and calcium and magnesium saponite [smectite clays; e.g., $Ca_{.165}Mg_3Al_{.33}Si_{3.67}O_{10}(OH)_2$] minerals.

From a plot of the saturation states of the silica polymorphs (Fig. 27.7), the fluid's equilibrium temperature with quartz is about 100 °C. Quartz, however, is commonly supersaturated in geothermal waters below about 150 °C and so can give erroneously high equilibrium temperatures when applied in geothermometry (Fournier, 1977). Chalcedony is in equilibrium with the fluid at about 76 °C, a temperature consistent with that suggested by the aluminosilicate minerals.

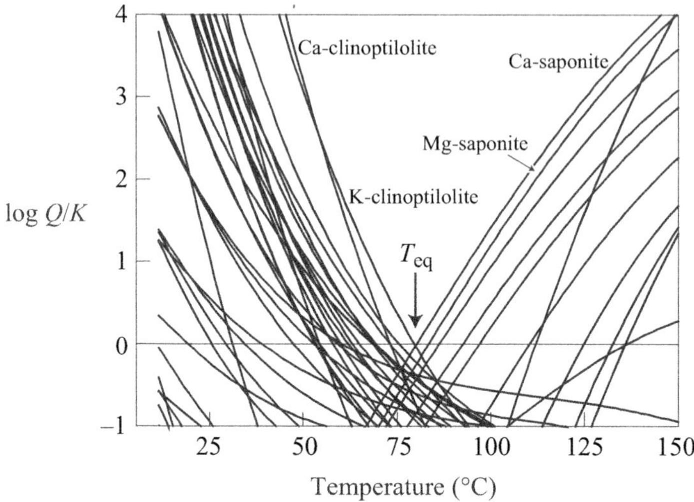

Figure 27.6 Calculated saturation indices (log Q/K) of aluminum-bearing minerals plotted versus temperature for a hot spring water from Gjögur, Hveravik, Iceland. Lines for most of the minerals are not labeled, due to space limitations. Sampling temperature is 72 °C and predicted equilibrium temperature (arrow) is about 80 °C. Clinoptilolite (zeolite) minerals are the most supersaturated minerals below this temperature and saponite (smectite clay) minerals are the most supersaturated above it.

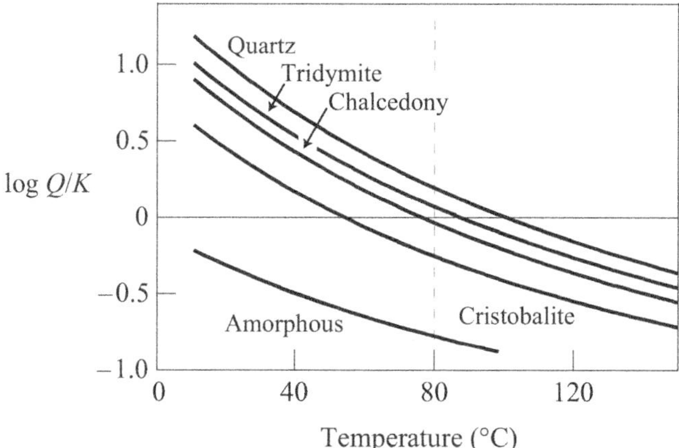

Figure 27.7 Calculated saturation indices (log Q/K) of silica minerals for Gjögur hot spring water. Chalcedony is approximately in equilibrium at 80 °C, but quartz is supersaturated at this temperature.

According to the calculation, the partial pressure of CO_2 varies in the range 10^{-2} bar to 10^{-3} bar, somewhat higher than the atmospheric pressure of this gas, which is about $10^{-3.5}$ bar. The CO_2 pressure, however, is much lower than the total atmospheric pressure, suggesting that the fluid has not effervesced CO_2. In light of these results, we might reasonably argue that the fluid is a fresh water that has circulated to relatively shallow depths into the geothermal area, obtaining a moderate solute content but remaining oxidized. The fluid probably circulated through fractures in the basalt, reacting with zeolites, smectite clays, and chalcedony lining the fracture surfaces. The fluid last attained equilibrium with these minerals at about 75 °C to 80 °C, a temperature just slightly higher than the 72 °C discharge temperature at the hot spring.

27.3 Geothermal Fields in Iceland

In a final application, following Reed and Spycher (1984), we consider fluids produced from wet-steam wells (i.e., wells that produce both vapor and liquid phases) at three geothermal fields in Iceland. Arnórsson *et al.* (1983) again supply the analytical data, given in Table 27.2. The calculations in this case are more complicated than those for the spring water considered in the previous section because, before applying our geothermometer, we must recombine the vapor and liquid phases sampled at the wellhead to find the composition of the original fluid.

To find the mass ratio of vapor to liquid produced, we note the discharge enthalpy H_{tot} and sampling pressure P_s from Table 27.2. From the steam tables (Keenan *et al.*, 1969), we find the sampling temperature T_s corresponding to the boiling point at P_s, and the enthalpies H_{liq} and H_{vap} of liquid water and steam at this temperature. The mass fraction X_{vap} of vapor produced by the well is given (e.g., Henley, 1984) by energy balance,

$$X_{vap} = \frac{H_{tot} - H_{liq}}{H_{vap} - H_{liq}} . \tag{27.4}$$

The resulting values for the three wells are

	H_{tot}	P_s	T_s	H_{liq}	H_{vap}	X_{vap}
Reykjanes #8	275	20	213.1	217.8	668.7	12.7%
Hveragerdi #4	183	6.8	164.4	165.9	659.9	3.5%
Namafjall #8	261	9.8	179.6	181.9	663.5	16.4%

Here, enthalpy is given in kcal kg^{-1}, temperature in °C, and pressure in atm. Note that the sampling temperatures are considerably lower than subsurface temperatures because of the energy used to produce the vapor phase.

Next, we need to calculate the amount of each component in the vapor phase. At room temperature, the vapor separates into a condensate that is mostly water and a gas phase that is mostly CO_2. Table 27.2 provides the composition of each. The mole number of each component (H_2O, CO_2, and H_2S) in the condensate, expressed per kg H_2O in the liquid, is derived by multiplying the concentration (g kg^{-1}) by the vapor fraction X_{vap} and dividing by the component's mole weight.

The mole numbers per kg liquid for the gases (H_2O, CO_2, H_2S, H_2, and CH_4) that separated from the condensate are obtained by multiplying the volume fraction of each by (1) the volume of gas per kg condensate, (2) the mass fraction X_{vap} of the vapor phase produced, and (3) the number of moles per liter of gas. The latter value can be calculated from the ideal gas law $PV = nRT_K$; at 20 °C and 1 atm pressure,

Table 27.2 *Chemical compositions of water and steam discharged from wet-steam geothermal wells in Iceland (Arnórsson* et al., *1983)*

	Reykjanes Well #8	Hveragerdi Well #4	Namafjall Well #8
Fluid			
pH; °C	6.38; 20	8.82; 20	8.20; 22
SiO_2 (mg kg^{-1})	631.1	281.0	446.3
Na^+	11 150	153.3	154.8
K^+	1 720	13.4	24.0
Ca^{++}	705	1.73	4.52
Mg^{++}	1.44	0.002	0.085
Fe^{++}	0.329	0.008	0.019
Al^{+++}	0.07	0.14	0.10
CO_2(aq)	63.1	74.2	88.2
SO_4^{--}	28.4	43.7	48.7
H_2S(aq)	2.21	19.2	132.6
Cl^-	22 835	109.5	16.6
F^-	0.21	1.82	0.43
Dissolved solids	39 124	765	902
Condensate			
CO_2 (mg kg^{-1})	584	627	172
H_2S	65.6	84.5	277
Gas with condensate			
CO_2 (vol. %)	96.2	84.5	36.8
H_2S	2.9	3.0	17.0
H_2	0.2	2.8	37.4
CH_4	0.1	0.3	2.9
N_2	0.6	9.4	5.9
l gas (kg condensate)$^{-1}$; °C	2.63; 20	1.06; 20	6.25; 20
Sampling pressure (bars abs.)	20	6.8	9.8
Discharge enthalpy (kcal kg^{-1})	275	183	261

there are 0.0416 moles per liter of gas. The final values for each well,

	Reykjanes #8	Hveragerdi #4	Namafjall #8
H_2O	7.045	1.941	9.099
CO_2	0.01505	1.80×10^{-3}	0.01633
H_2S	6.47×10^{-4}	1.33×10^{-4}	8.58×10^{-3}
H_2	2.78×10^{-5}	4.32×10^{-5}	0.01595
CH_4	1.39×10^{-5}	4.63×10^{-6}	3.59×10^{-3}

are sums of the mole numbers for the condensate and gas, expressed in moles per kg of H_2O in the liquid phase.

To run REACT for the Reykjanes #8 well, we start by defining the initial system

```
TDS = 39124
swap H2S(aq) for O2(aq)
swap CO2(aq) for HCO3-

pH = 6.38
SiO2(aq) =   631.1   mg/kg
Na+      = 11150     mg/kg
K+       =  1720     mg/kg
Ca++     =  1705     mg/kg
Mg++     =     1.44  mg/kg
Fe++     =      .329 mg/kg
Al+++    =      .07  mg/kg
CO2(aq)  =    63.1   mg/kg
SO4--    =    28.4   mg/kg
H2S(aq)  =     2.21  mg/kg
Cl-      = 22835     mg/kg
F-       =     0.21  mg/kg
```

Here, we use the ratio of sulfate to sulfide to constrain oxidation state in the fluid.

To invoke our geothermometer, we need to recombine the vapor and fluid phases and then heat the mixture to determine saturation indices as functions of temperature. We could do this in two steps, first titrating the vapor phase into the liquid and then picking up the results as the starting point for a polythermal path. We will employ a small trick, however, to accomplish these steps in a single reaction path. The trick is to add the vapor phase quickly during the first part of the reaction path but use the `cutoff` option to prevent mass transfer over the remainder of the path.

The commands to set the mass transfer are

```
(cont'd)
react 70.045   moles H2O    cutoff = 7.045
react   .1505  moles CO2(g) cutoff =  .01505
react  6.47e-3 moles H2S(g) cutoff = 6.47e-4
react  2.78e-4 moles H2(g)  cutoff = 2.78e-5
react  1.39e-4 moles CH4(g) cutoff = 1.39e-5
```

The cutoff values are the mole numbers calculated above, and the corresponding reaction rates are the mole numbers augmented by a factor of ten.

We type the commands

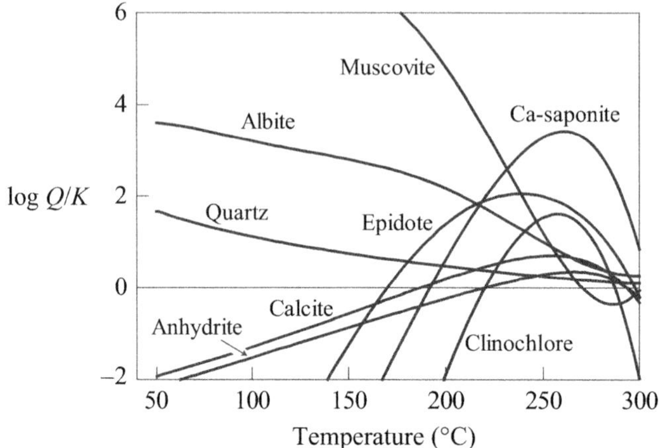

Figure 27.8 Calculated saturation indices ($\log Q/K$) of various minerals for the Reykjanes #8 wet-steam geothermal well. The well produces from seven intervals at temperatures varying from 274 °C to 292 °C. The calculated saturation indices suggest an equilibrium temperature between about 285 °C and slightly above 300 °C.

(cont'd)

```
T initial = 20, final = 300
precip = off
go
```

to set up the polythermal path and trigger the calculation.

In the calculation results (Fig. 27.8), a number of minerals converge to equilibrium with the fluid in the range of about 285 °C to somewhat above 300 °C (slightly beyond the calculation's high-temperature limit). The well, for comparison, produces fluid from seven zones at temperatures ranging from 274 °C to 292 °C (Arnórsson *et al.*, 1983). Several minerals [i.e, epidote, $Ca_2FeAl_2Si_3O_{12}OH$, and calcium saponite, $Ca_{0.165}Mg_3Al_{0.33}Si_{3.67}O_{10}(OH)_2$] are supersaturated at temperatures less than about 300 °C. This result might reasonably be interpreted to reflect a fluid that equilibrated at a temperature somewhat higher than observed in the well. Alternatively, the supersaturation may be due to the mixing in the wellbore of fluids from the well's various producing zones. As discussed in Chapter 6, fluid mixing tends to leave minerals supersaturated.

For the Hveragerdi #4 well, we follow the same procedure, using the data in Table 27.2 and the calculations already shown. In this case, the model predicts that a number of minerals in the LLNL database are supersaturated near the inflow temperature of 181 °C. Close examination reveals that each of the supersaturated minerals contains either Mg^{++}, Ca^{++}, or Fe^{++}, components that are characteristically depleted in geothermal fluids. The Mg^{++} concentration in this fluid, for example, is just 2 μg kg^{-1}.

The minor amounts of Mg^{++}, Ca^{++}, and Fe^{++} in the fluid might easily be accounted for by contamination by a small volume of a shallow groundwater or even by dissolution of concrete and steel in the wellbore. If we discount the results for minerals containing these components, we arrive at a well-defined equilibrium temperature slightly above 200 °C (Fig. 27.9). Again, the equilibrium temperature is slightly higher than the temperature measured in the wellbore.

The analysis for Namafjall well #8 is similar to that of the Hveragerdi well in that a number of Ca, Mg, and Fe-bearing minerals appear supersaturated over the temperature range of interest. Again, this result probably reflects mixing or contamination. The equilibrium temperatures for quartz, albite, potassium feldspar,

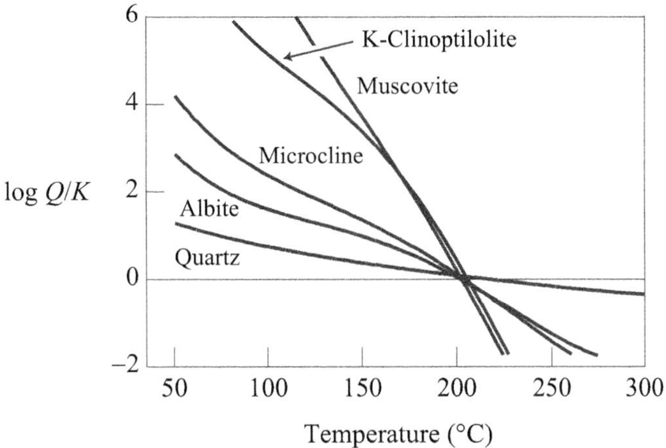

Figure 27.9 Calculated saturation indices ($\log Q/K$) of various minerals for the Hveragerdi #4 wet-steam geothermal well. The inflow temperature is 181 °C. The calculated saturation indices suggest an equilibrium temperature near 200 °C.

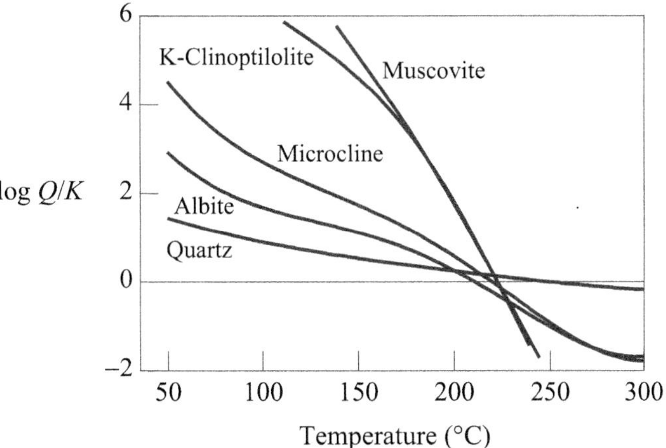

Figure 27.10 Calculated saturation indices ($\log Q/K$) of various minerals for the Namafjall #8 wet-steam geothermal well. The inflow temperature is 246 °C. The calculated saturation indices suggest an equilibrium temperature in the range 205 °C to 250 °C.

potassium clinoptilolite, and muscovite (Fig. 27.10) bracket a relatively broad temperature range of 205 °C to 250 °C, which can be compared to the well's inflow temperature of 246 °C. In this case, the equilibrium temperature is notably less well defined than in the previous example, perhaps reflecting the mixing of a significant amount of shallow groundwater into the geothermal fluid.

28
Evaporation

The process of evaporation, including transpiration (evaporation from plants), returns to the atmosphere more than half of the water reaching the Earth's land surface; thus, it plays an important role in controlling the chemistry of surface water and groundwater, especially in relatively arid climates. Geochemists study the evaporation process to understand the evolution of water in desert playas and lakes as well as the origins of evaporite deposits. They also investigate environmental aspects of evaporation (e.g., Appelo and Postma, 2005), such as its effects on the chemistry of rainfall and, in areas where crops are irrigated, the quality of groundwater and runoff.

To model the chemical effects of evaporation, we construct a reaction path in which H_2O is removed from a solution, thereby progressively concentrating the solutes. We also must account in the model for the exchange of gases such as CO_2 and O_2 between fluid and atmosphere. In this chapter we construct simulations of this sort, modeling the chemical evolution of water from saline alkaline lakes and the reactions that occur as seawater evaporates to desiccation.

28.1 Springs and Saline Lakes of the Sierra Nevada

We choose as a first example the evaporation of spring water from the Sierra Nevada mountains of California and Nevada, USA, as modeled by Garrels and Mackenzie (1967). Their hand calculation, the first reaction path traced in geochemistry (see Chapter 1), provided the inspiration for Helgeson's (1968 and later) development of computerized methods for reaction modeling.

Garrels and Mackenzie wanted to test whether simple evaporation of groundwater discharging from the mountains, which is the product of the reaction of rainwater and CO_2 with igneous rocks, could produce the water compositions found in the saline alkaline lakes of the adjacent California desert. They began with the mean of analyses of perennial springs from the Sierra Nevada (Table 28.1). The springs are Na–Ca–HCO_3 waters, rich in dissolved silica.

Using REACT to distribute species

```
26.4  mg/kg SiO2(aq)
10.4  mg/kg Ca++
 1.7  mg/kg Mg++
 5.95 mg/kg Na+
 1.57 mg/kg K+
54.6  mg/kg HCO3-
 2.38 mg/kg SO4--
 1.06 mg/kg Cl-
pH = 6.8
```

Table 28.1 *Mean composition of spring water from the Sierra Nevada, California and Nevada, USA (Garrels and Mackenzie, 1967)*

	mg kg^{-1}	mmolal
$SiO_2(aq)$	24.6	0.410
Ca^{++}	10.4	0.260
Mg^{++}	1.70	0.070
Na^+	5.95	0.259
K^+	1.57	0.040
HCO_3^-	54.6	0.895
SO_4^{--}	2.38	0.025
Cl^-	1.06	0.030
pH	6.8*	

*Median value.

```
balance on HCO3-
go
```

we calculate a CO_2 pressure for the water of $10^{-2.1}$ bar, typical of soil waters but somewhat higher than the atmospheric partial pressure of $10^{-3.5}$ bar. In their model, Garrels and Mackenzie assumed that the evaporating water remained in equilibrium with the CO_2 in the atmosphere. To prepare the reaction path, therefore, they computed the effects of letting the spring water exsolve CO_2 until it reached atmospheric fugacity.

They made several assumptions about which minerals could precipitate from the fluid. The alkaline lakes tend to be supersaturated with respect to each of the silica polymorphs (quartz, tridymite, and so on) except amorphous silica, so they suppressed each of the other silica minerals. They assumed that dolomite [$CaMg(CO_3)_2$], a highly ordered mineral known not to precipitate at 25 °C except from saline brines, would not form. Finally, they took the clay mineral sepiolite [$Mg_4Si_6O_{15}(OH)_2 \cdot 6H_2O$] in preference to minerals such as talc [$Mg_3Si_4O_{10}(OH)_2$] as the magnesium silicate likely to precipitate during evaporation.

The procedure to suppress these minerals and adjust the fluid's CO_2 fugacity is

(cont'd)
```
suppress Quartz, Tridymite, Cristobalite, Chalcedony
suppress Dolomite, Dolomite-ord, Dolomite-dis
suppress Talc, Tremolite, Antigorite

slide f CO2(g) to 10^-3.5
go
```

The resulting fluid, which provides the starting point for modeling the effects of evaporation, has a pH of 8.2.

To model the process of evaporation, we take the fluid resulting from the previous calculation and, while holding the CO_2 fugacity constant, remove almost all (999.9 g of the original kg) of its water. The procedure is

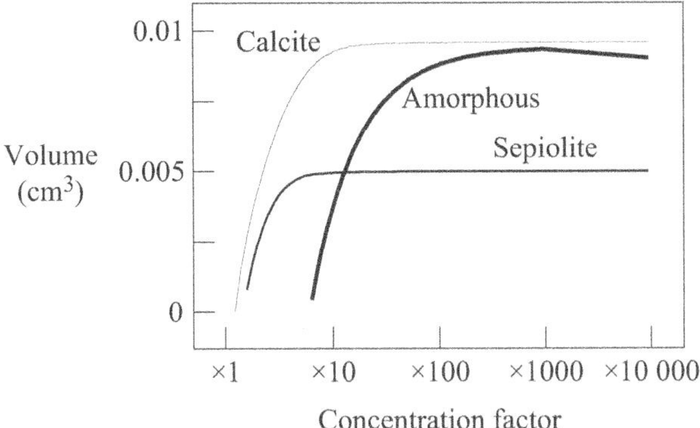

Figure 28.1 Volumes of minerals (amorphous silica, calcite, and sepiolite) precipitated during a reaction model simulating at 25 °C the evaporation of Sierra Nevada spring water in equilibrium with atmospheric CO_2, plotted against the concentration factor. For example, a concentration factor of ×100 means that of the original 1 kg of water, 10 g remain.

(cont'd)

```
pickup fluid
react -999.9 g H2O
fix f CO2(g)
delxi  = .001
dxplot = .001
go
```

The `delxi` and `dxplot` commands set a small reaction step to provide increased detail near the end of the reaction path.

In the calculation results (Fig. 28.1), amorphous silica, calcite ($CaCO_3$), and sepiolite precipitate as water is removed from the system. The fluid's pH and ionic strength increase with evaporation as the water evolves toward an Na–CO_3 brine (Fig. 28.2). The concentrations of the components Na^+, K^+, Cl^-, and SO_4^{--} rise monotonically (Fig. 28.2), since they are not consumed by mineral precipitation. The HCO_3^- and SiO_2(aq) concentrations increase sharply but less regularly, since they are taken up in forming the minerals. The components Ca^{++} and Mg^{++} are largely consumed by the precipitation of calcite and sepiolite. Their concentrations, after a small initial rise, decrease with evaporation.

Two principal factors drive reaction in the evaporating fluid. First, the loss of solvent concentrates the species in solution, causing the saturation states of many minerals to increase. The precipitation of amorphous silica, for example,

$$SiO_2(aq) \rightarrow \underset{amorphous}{SiO_2} \quad , \tag{28.1}$$

results almost entirely from the increase in SiO_2(aq) concentration as water evaporates.

A second and critical factor is the loss of CO_2 from the fluid to the atmosphere. Evaporation concentrates CO_2(aq), driving CO_2(g) to exsolve,

$$CO_2(aq) \rightarrow CO_2(g)\,. \tag{28.2}$$

About 0.5 mmol of CO_2, or about 60% of the fluid's carbonate content at the onset of the calculation, is lost in this way.

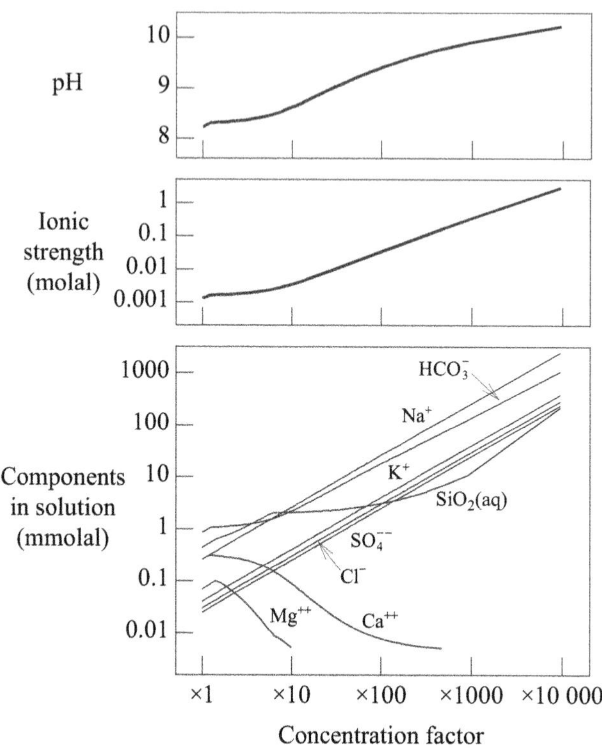

Figure 28.2 Calculated effects of evaporation at 25 °C on the chemistry of Sierra Nevada spring water. Top figures show how pH and ionic strength vary over the reaction path in Figure 28.1; bottom figure shows variation in the fluid's bulk composition.

The escape of CO_2, an acid gas, affects the fluid's pH. By mass action, the loss of $CO_2(aq)$ from the system causes readjustment among the carbonate species in solution,

$$HCO_3^- + H^+ \rightarrow H_2O + CO_2(aq) \quad (28.3)$$

$$CO_3^{--} + 2\,H^+ \rightarrow H_2O + CO_2(aq)\,. \quad (28.4)$$

The reactions consume H^+, driving the fluid toward alkaline pH, as shown in Figure 28.2. This effect explains the alkalinity of saline alkaline lakes.

Calcite and sepiolite precipitate in large part because of the effects of the escaping CO_2. The corresponding reactions are

$$Ca^{++} + 2\,HCO_3^- \rightarrow \underset{\textit{calcite}}{CaCO_3} + H_2O + CO_2(g) \quad (28.5)$$

and

$$4\,Mg^{++} + 6\,SiO_2(aq) + 8\,HCO_3^- + 3\,H_2O \rightarrow \underset{\textit{sepiolite}}{Mg_4Si_6O_{15}(OH)_2 \cdot 6H_2O} + 8\,CO_2(g)\,. \quad (28.6)$$

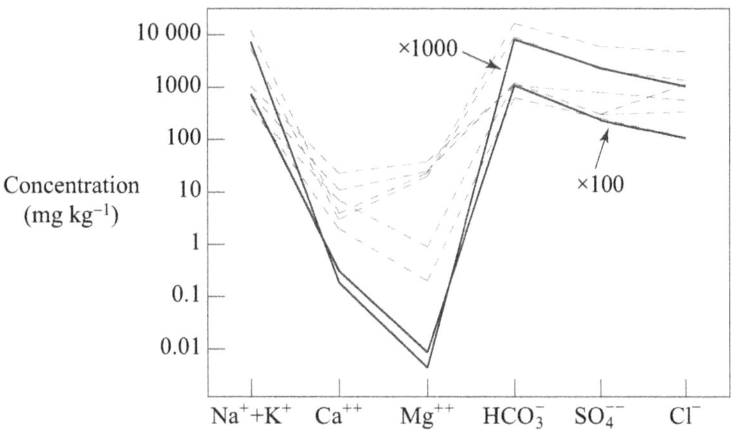

Figure 28.3 Calculated composition of evaporated spring water (from the reaction path shown in Figures 28.1 and 28.2) concentrated 100-fold and 1000-fold (solid lines) compared with the compositions of six saline alkaline lakes (dashed lines), as compiled by Garrels and Mackenzie (1967).

Figure 28.3 compares the calculated composition of the evaporated water, concentrated 100-fold and 1000-fold, with analyses of waters from six saline alkaline lakes (compiled by Garrels and Mackenzie, 1967). The field for the modeled water overlaps that for the analyzed waters, except that Ca^{++} and Mg^{++} are more depleted in the model than in the lake waters. This discrepancy might be explained if in nature the calcite and sepiolite begin to precipitate but remain supersaturated in the fluid.

In the reaction path we calculated (Fig. 28.2), the precipitation of calcite consumes nearly all of the calcium originally in solution so that no further calcium-bearing minerals form. Calcite precipitation, on the other hand, does not deplete the dissolved carbonate, because the original fluid was considerably richer in carbonate than in calcium ($M_{HCO_3^-} > M_{Ca^{++}}$). The carbonate concentration, in fact, increases during evaporation.

If the fluid had been initially richer in calcium than carbonate ($M_{Ca^{++}} > M_{HCO_3^-}$), as noted by Hardie and Eugster (1970), it would have followed a distinct reaction path. In such a case, calcite precipitation would deplete the fluid in carbonate, allowing the calcium concentration to increase until gypsum ($CaSO_4 \cdot 2H_2O$) saturates and forms. The point at which the calcium and carbonate are present at equal initial concentration ($M_{Ca^{++}} = M_{HCO_3^-}$) is known as a *chemical divide*.

According to Hardie and Eugster's (1970) model and its later variants (see discussions in Eugster and Jones, 1979; Drever, 1988, pp. 232–250; and Jankowski and Jacobson, 1989), a natural water, as it evaporates, encounters a series of *chemical divides* that controls the sequence of minerals that precipitate. The reaction pathway specific to the evaporation of a water of any initial composition can be traced in detail using a reaction model like the one applied in this section to Sierra spring water.

28.2 Chemical Evolution of Mono Lake

In a second example, we consider the changing chemistry of Mono Lake, a saline alkaline lake that occupies a closed desert basin in California, USA, and why gaylussite [$CaNa_2(CO_3)_2 \cdot 5H_2O$], a rare hydrated carbonate mineral, has begun to form there. The lake has shrunk dramatically since 1941, when the Los Angeles Department of Water and Power began to divert tributary streams to supply water for southern California.

As it shrank, the lake became more saline. Salinity has almost doubled from 50 000 mg kg^{-1} in 1940 to about 90 000 mg kg^{-1} in recent years. The change in the lake's chemistry threatens to damage a unique ecosystem that supports large flocks of migratory waterfowl.

In 1988, Bischoff *et al.* (1991) discovered gaylussite crystals actively forming in the lake. The crystals were found growing on hard surfaces, especially in the lake's deeper sections. Gaylussite had also formed earlier, because pseudomorphs after gaylussite were observed. The earlier gaylussite has been replaced by aragonite, leaving the porous skeletal pseudomorphs.

Bischoff *et al.* (1991) attributed the occurrence of gaylussite to the lake's increase in Na^{+} content and pH since diversion began. Table 28.2 shows chemical analyses of lake water sampled at ten points in time from 1956 to 1988. We can use SPECE8 to calculate for each sample the saturation state of gaylussite. The procedure for the 1956 sample, for example, is

```
Ca++    =     4.3 mg/kg
Mg++    =    38   mg/kg
Na+     = 22540   mg/kg
K+      =  1124   mg/kg
SO4--   = 12000   mg/kg
Cl-     = 13850   mg/kg
B(OH)3  =  1720   mg/kg
HCO3-   = 17600   mg/kg

TDS = 67686
pH  =  9.49
go
```

Figure 28.4 shows the trend in gaylussite saturation plotted against time and against the salinity of the lake water.

In the calculation results, gaylussite appears increasingly saturated in the lake water as salinity increases irregularly over time. The calculations suggest that between about 1975 and 1980, as salinities reached about 75 000 to 80 000 mg kg^{-1}, the mineral became supersaturated at summer temperatures.

The solubility of gaylussite, however, varies strongly with temperature. Unlike calcite and aragonite, gaylussite grows less soluble (or more stable) as temperature decreases. We can recalculate mineral solubility under winter conditions by setting temperature to 0 °C

```
(cont'd)
T = 0
go
```

These calculations (Fig. 28.5) suggest that, to the extent that the analyses in Table 28.2 are representative of the wintertime lake chemistry, gaylussite has been supersaturated in Mono Lake during the winter months since sampling began.

The saturation state of aragonite (Fig. 28.5), on the other hand, is affected little by temperature. Aragonite remains supersaturated by a factor of about ten (one log unit) over the gamut of analyses. The supersaturation probably arises from the effect of orthophosphate, present at concentrations of about 100 mg kg^{-1} in Mono Lake water; orthophosphate is observed in the laboratory (Bischoff *et al.*, 1993) to inhibit the precipitation of calcite and aragonite.

We can use our results to predict the conditions favorable for the transformation of gaylussite to aragonite. The porous nature of the pseudomorphs and the small amounts of calcium available in the lake water

Table 28.2 *Chemical analyses (mg kg^{-1}) of surface water from Mono Lake, 1956 to 1988 (James Bischoff, personal communication*)*

	1956	1957	1974	1978	1979
Ca^{++}	4.3	4.2	3.5	4.5	4.6
Mg^{++}	38	37	30	34	42
Na^{+}	22 540	22 990	31 100	21 500	37 200
K^{+}	1 124	1 140	1 500	1 170	1 580
SO_4^{--}	12 000	7 810	11 000	7 380	12 074
Cl^{-}	13 850	14 390	18 000	13 500	20 100
$B(OH)_3$	1 720	2 000	1 850	—	2 760
Alkalinity†	21 854	21 600	30 424	21 617	34 818
HCO_3^{-}‡	17 600	15 900	22 700	17 000	25 000
TDS	67 686	59 500	80 370	59 312	92 540
pH	9.49	9.7	9.66	9.6	9.68

	1982	1983	1984	1985	1988
Ca^{++}	4.4	4.3	3.1	3.6	3.3
Mg^{++}	37	36	30	34	37
Na^{+}	34 000	29 300	26 600	28 900	32 000
K^{+}	1 980	1 600	1 240	1 300	1 500
SO_4^{--}	11 100	10 700	9 590	10 190	11 200
Cl^{-}	20 400	18 860	17 673	20 370	19 700
$B(OH)_3$	2 460	2 230	1 720	1 720	2 120
Alkalinity†	34 700	31 240	27 240	29 170	31 900
HCO_3^{-}‡	25 300	22 800	18 600	19 600	20 500
TDS	92 200	83 800	74 700	80 600	85 600
pH	9.66	9.68	9.90	9.93	10.03

*From published (see Bischoff *et al.*, 1991) and unpublished sources.

†As $CaCO_3$.

‡Calculated from alkalinity, using method described in Section 16.1.

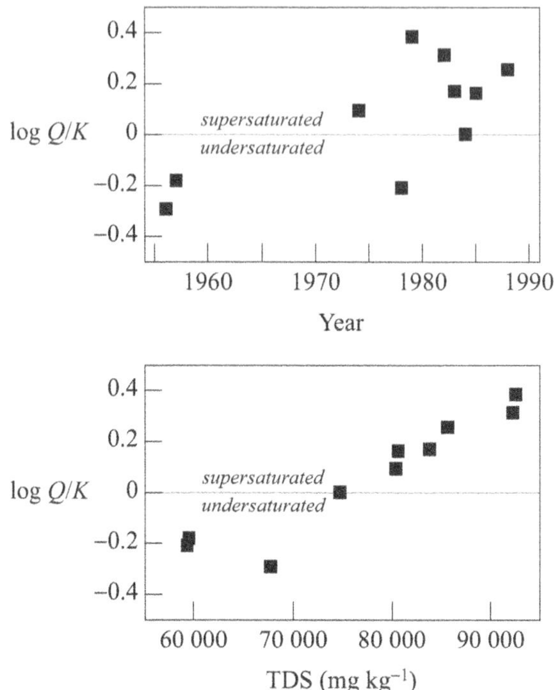

Figure 28.4 Saturation indices for gaylussite in Mono Lake water at 25 °C for various years between 1956 and 1988, calculated from analyses in Table 28.2 and plotted against time (top diagram) and salinity (bottom).

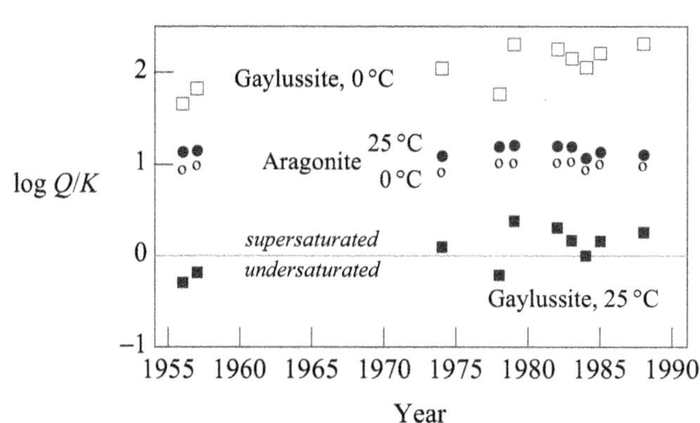

Figure 28.5 Saturation indices of gaylussite (□, ■) and aragonite (○, ●) calculated for Mono Lake water in various years from analyses in Table 28.2. Open symbols (□, ○) represent values calculated for 0 °C and solid symbols (■, ●) show those for 25 °C.

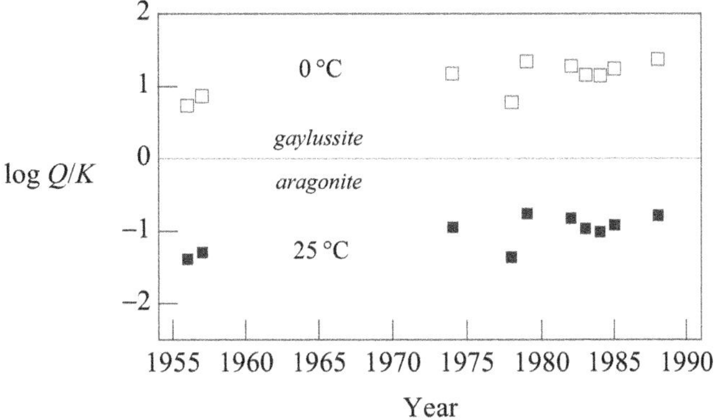

Figure 28.6 The state (log Q/K) of Reaction 28.7 between gaylussite and aragonite at 0 °C (□) and 25 °C (■), showing which mineral is favored to form at the expense of the other.

(Table 28.2) suggest that the replacement occurs by the incongruent dissolution of gaylussite, according to the reaction

$$\underset{\textit{gaylussite}}{CaNa_2(CO_3)_2 \cdot 5H_2O} \rightarrow \underset{\textit{aragonite}}{CaCO_3} + 2\,Na^+ + CO_3^{--} + 5\,H_2O \qquad (28.7)$$

(Bischoff *et al.*, 1991). Figure 28.6 compares the activity product for this reaction to its equilibrium constant, showing which mineral is favored to form at the expense of the other. According to the calculations, gaylussite is prone to transform into aragonite during the summer (even though gaylussite is supersaturated in the lake water then), but is not likely to be replaced during the winter months.

28.3 Evaporation of Seawater

Since the experimental studies of van't Hoff at the turn of the century, geochemists have sought a quantitative basis for describing the chemical evolution of seawater and other complex natural waters, including the minerals that precipitate from them, as they evaporate. The interest has stemmed in large part from a desire to understand the origins of ancient deposits of evaporite minerals, a goal that remains mostly unfulfilled (Hardie, 1991).

The results of the early experimental studies, although of great significance, were limited by the complexity of a chemical system that can be portrayed on paper within a phase diagram. There was little possibility, furthermore, of calculating a useful reaction model of the evaporation process using conventional correlations to compute activity coefficients for the aqueous species (see Chapter 8), given the inherent inaccuracy of the correlations at high ionic strength.

In a series of papers, Harvie and Weare (1980), Harvie *et al.* (1980), and Eugster *et al.* (1980) attacked this problem by presenting a virial method for computing activity coefficients in complex solutions (see Chapter 8) and applying it to construct a reaction model of seawater evaporation. Their calculations provided the first quantitative description of this process that accounted for all of the abundant components in seawater.

To reproduce their results, we trace the reaction path taken by seawater at 25 °C as it evaporates to desiccation. Our calculations follow those of Harvie *et al.* (1980) and Eugster *et al.* (1980), except that we employ the more recent Harvie–Møller–Weare activity model (Harvie *et al.*, 1984), which accounts for bicarbonate. We include an HCO_3^- component in our calculations, assuming that the fluid as it evaporates remains in equilibrium with the CO_2 in the atmosphere.

In a first calculation, we specify that the fluid maintains equilibrium with whatever minerals precipitate. Minerals that form, therefore, can redissolve into the brine as evaporation proceeds. In REACT, we set the Harvie–Møller–Weare model and specify that our initial system contains seawater

```
read thermo_hmw.tdat

swap CO2(g) for H+
log P CO2(g) = -3.5 bar

TDS = 35080
Cl-    = 19350 mg/kg
Ca++   =   411 mg/kg
Mg++   =  1290 mg/kg
Na+    = 10760 mg/kg
K+     =   399 mg/kg
SO4--  =  2710 mg/kg
HCO3-  =   142 mg/kg
```

just as we did in Chapter 6. We then set a reaction path in which we fix the CO_2 fugacity and remove solvent from the system

```
(cont'd)
fix fugacity of CO2(g)
react -996 grams of H2O

delxi = .001
dxplot = 0
dump
go
```

The `dump` command serves to eliminate the small mineral masses that precipitate when, at the onset of the calculation, the program brings seawater to its theoretical equilibrium state (see Chapter 6). The `delxi` and `dxplot` commands serve to set a small reaction step, assuring that the results are rendered in sufficient detail.

By removing 996 g of H_2O, we eliminate all of the 1 kg of solvent initially present in the system; the remaining 4 g are consumed by the precipitation of hydrated minerals. In fact, just slightly less than 996 g of water can be removed before the system is completely desiccated. The program continues until less than 1 μg of solvent remains and then abandons its efforts to trace the path. At this point, it gives a warning message, which can be ignored.

Figure 28.7 shows the minerals that precipitate over the reaction path (Table 28.3 lists their compositions), and Figure 28.8 shows how fluid chemistry in the calculation varies. Initially, dolomite and gypsum precipitate. When the fluid is concentrated about 10-fold, the decreasing water activity causes the gypsum to dehydrate,

$$\underset{gypsum}{CaSO_4 \cdot 2H_2O} \rightarrow \underset{anhydrite}{CaSO_4} + 2\,H_2O\,, \tag{28.8}$$

forming anhydrite.

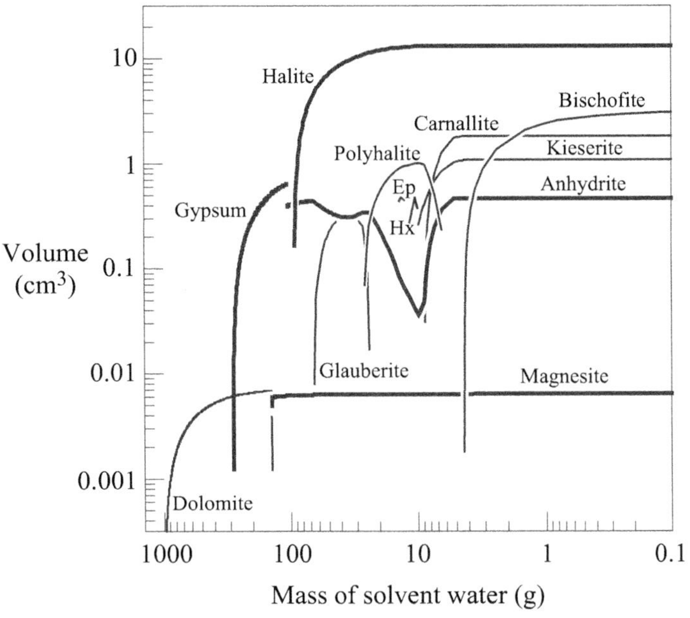

Figure 28.7 Volumes of minerals precipitated during a reaction model simulating the evaporation of seawater as an equilibrium system at 25 °C, calculated using the Harvie–Møller–Weare activity model. Abbreviations: Ep = Epsomite, Hx = Hexahydrite.

Table 28.3 *Minerals formed during the simulated evaporation of seawater*

Anhydrite	$CaSO_4$
Bischofite	$MgCl_2 \cdot 6H_2O$
Bloedite	$Na_2Mg(SO_4)_2 \cdot 4H_2O$
Carnallite	$KMgCl_3 \cdot 6H_2O$
Dolomite	$CaMg(CO_3)_2$
Epsomite	$MgSO_4 \cdot 7H_2O$
Glauberite	$Na_2Ca(SO_4)_2$
Gypsum	$CaSO_4 \cdot 2H_2O$
Halite	NaCl
Hexahydrite	$MgSO_4 \cdot 6H_2O$
Kainite	$KMgClSO_4 \cdot 3H_2O$
Kieserite	$MgSO_4 \cdot H_2O$
Magnesite	$MgCO_3$
Polyhalite	$K_2MgCa_2(SO_4)_4 \cdot 2H_2O$

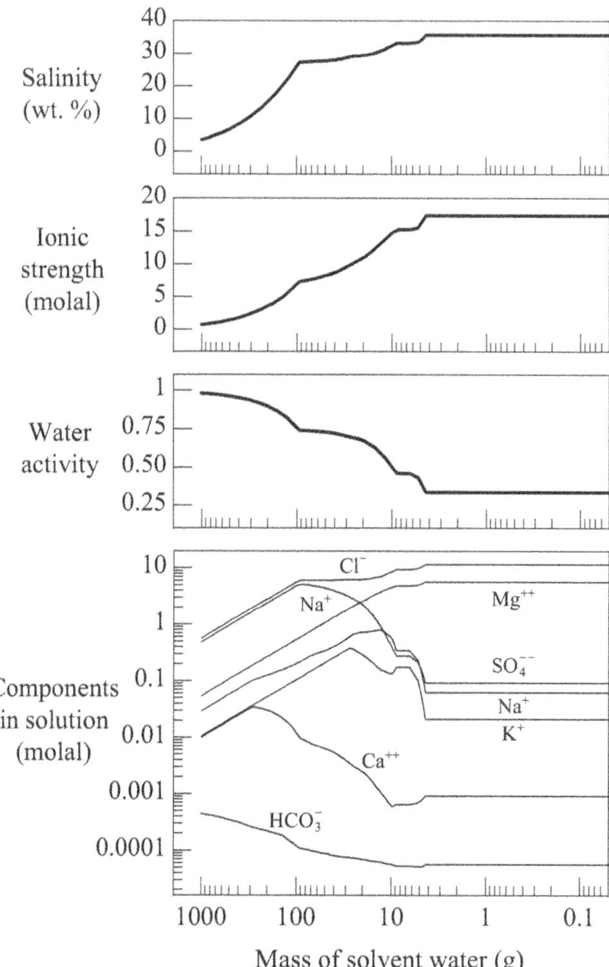

Figure 28.8 Evolution of fluid chemistry during the simulated evaporation of seawater as an equilibrium system at 25 °C, calculated using the Harvie–Møller–Weare activity model. Upper figures show variation in salinity, ionic strength (I), and water activity (a_w) over the reaction path in Figure 28.7; bottom figure shows how the fluid's bulk composition varies.

Shortly afterwards, halite becomes saturated and begins to precipitate,

$$\mathrm{Na^+ + Cl^- \rightarrow \underset{\textit{halite}}{NaCl}} \ . \tag{28.9}$$

Halite forms in the calculation in far greater volume than any other mineral, reflecting the fact that seawater is dominantly an NaCl solution. The precipitation reaction represents a chemical divide, as discussed in Section 28.1. Since Na^+ is less concentrated (on a molal basis) in seawater than Cl^- ($M_{Na^+} < M_{Cl^-}$), it becomes depleted in solution. As a result (Fig. 28.8), seawater evolves with evaporation from a dominantly NaCl solution into an $MgCl_2$ bittern.

With further evaporation, a small amount of glauberite precipitates,

$$\mathrm{\underset{\textit{anhydrite}}{CaSO_4} + 2\,Na^+ + SO_4^{--} \rightarrow \underset{\textit{glauberite}}{Na_2Ca(SO_4)_2}} \ , \tag{28.10}$$

at the expense of anhydrite. The increasing activities of K^+, Mg^{++}, and SO_4^{--} in solution, however, soon drive the glauberite and some of the anhydrite to form polyhalite according to the reaction

$$\begin{aligned} &\mathrm{\underset{\textit{glauberite}}{Na_2Ca(SO_4)_2} + \underset{\textit{anhydrite}}{CaSO_4} + 2\,K^+ + Mg^{++} + SO_4^{--} +} \\ &\qquad \mathrm{2\,H_2O \rightarrow \underset{\textit{polyhalite}}{K_2MgCa_2(SO_4)_4 \cdot 2H_2O} + 2\,Na^+} \ . \end{aligned} \tag{28.11}$$

In accord with these predictions (Harvie *et al.*, 1980), pseudomorphs of glauberite after gypsum and anhydrite are observed in marine evaporites, and polyhalite, in turn, is known to replace glauberite.

As it evolves toward a dominantly $MgCl_2$ solution, the fluid becomes supersaturated with respect to kieserite and carnallite. The reaction to form these minerals,

$$\begin{aligned} &\mathrm{\underset{\textit{polyhalite}}{K_2MgCa_2(SO_4)_4 \cdot 2H_2O} + Mg^{++} + 6\,Cl^- + 12\,H_2O \rightarrow} \\ &\qquad \mathrm{2\,\underset{\textit{kieserite}}{MgSO_4 \cdot H_2O} + 2\,\underset{\textit{carnallite}}{KMgCl_3 \cdot 6H_2O} + 2\,\underset{\textit{anhydrite}}{CaSO_4}} \ , \end{aligned} \tag{28.12}$$

consumes the polyhalite. Finally, when about 4.5 g of solvent remain, bischofite forms,

$$\mathrm{Mg^{++} + 2\,Cl^- + 6\,H_2O \rightarrow \underset{\textit{bischofite}}{MgCl_2 \cdot 6H_2O}} \ , \tag{28.13}$$

in response to the high Mg^{++} and Cl^- activities in the residual fluid.

With the precipitation of bischofite, the system reaches an invariant point at which the mineral assemblage (magnesite, anhydrite, kieserite, carnallite, bischofite, and halite) fully constrains the fluid composition. Further evaporation causes more of these phases (principally bischofite) to form, but the fluid chemistry no longer changes, as can be seen in Figure 28.8.

In a second calculation, we model the reaction path taken when the minerals, once precipitated, cannot redissolve into the fluid. In this model, the solutes in seawater fractionate into the minerals as they precipitate, irreversibly altering the fluid composition. As discussed in Chapter 2, we set up such a model using the "flow-through" configuration. The procedure is

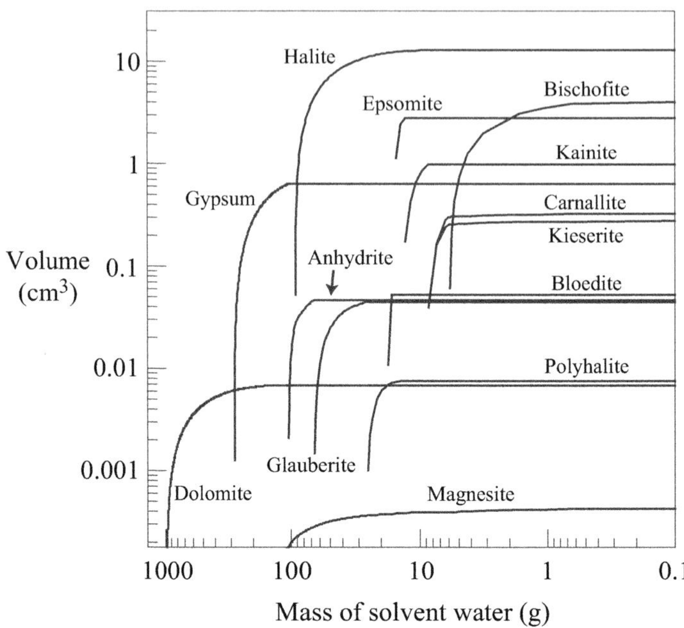

Figure 28.9 Volumes of minerals precipitated during a reaction model simulating the evaporation of seawater as a fractionating system (the "flow-through" configuration) at 25 °C, calculated using the Harvie–Møller–Weare activity model.

(cont'd)

```
flow-through
go
```

The results of the fractionation model (Fig. 28.9) differ from the equilibrium model in two principal ways. First, the mineral masses can only increase in the fractionation model, since they are protected from resorption into the fluid. Therefore, the lines in Figure 28.9 do not assume negative slopes. Second, in the equilibrium calculation the phase rule limits the number of minerals present at any point along the reaction path. In the fractionation calculation, on the other hand, no limit to the number of minerals present exists, since the minerals do not necessarily maintain equilibrium with the fluid. Therefore, the fractionation calculation ends with thirteen minerals in the system, whereas the equilibrium calculation reaches an invariant point at which only six minerals are present.

The fractionation calculation is notable in that it predicts the formation of two minerals (bloedite and kainite) that did not precipitate in the equilibrium model. Also, hexahydrite, which appeared briefly in the equilibrium model, does not form in the fractionation model. The two classes of models, therefore, represent qualitatively distinct pathways by which an evaporating water can evolve.

29

Sediment Diagenesis

Diagenesis is the set of processes by which sediments evolve after they are deposited and begin to be buried. Diagenesis includes physical effects such as compaction and the deformation of grains in the sediment (or sedimentary rock), as well as chemical reactions such as the dissolution of grains and the precipitation of minerals to form cements in the sediment's pore space. The chemical aspects of diagenesis are of special interest here.

Formerly, geologists considered chemical diagenesis to be a process by which the minerals and pore fluid in a sediment reacted with each other in response to changes in temperature, pressure, and stress. As early as the 1960s and especially since the 1970s, however, geologists have recognized that many diagenetic reactions occur in systems open to groundwater flow and mass transfer. The reactions proceed in response to a supply of reactants introduced into the sediments by flowing groundwater, which also serves to remove reaction products.

Hay (1963, 1966), in studies of the origin of diagenetic zeolite, was perhaps the first to emphasize the effects of mass transport on sediment diagenesis. He showed that sediments open to groundwater flow followed reaction pathways different from those observed in sediments through which flow was restricted. Sibley and Blatt (1976) used cathodoluminescence microscopy to observe the Tuscarora orthoquartzite of the Appalachian basin. The almost nonporous Tuscarora had previously been taken as a classic example of pressure welding, but the microscopy demonstrated that the rock is not especially well compacted but, instead, tightly cemented. The rock consists of as much as 40% quartz (SiO_2) cement that was apparently deposited by advecting groundwater.

By the end of the decade, Hayes (1979) and Surdam and Boles (1979) argued forcefully that the extent to which diagenesis has altered sediments in sedimentary basins can be explained only by recognition of the role of groundwater flow in transporting dissolved mass. This view has become largely accepted among geoscientists, although it is clear that the scale of groundwater flow might range from the regional (e.g., Bethke and Marshak, 1990) to circulation cells perhaps as small as tens of meters (e.g., Bjorlykke and Egeberg, 1993; Aplin and Warren, 1994). Since the possible reactions occurring in open geochemical systems are numerous and complex, the study of diagenesis has become a fertile field for applying reaction modeling (e.g., Bethke *et al.*, 1988; Baccar and Fritz, 1993).

In this chapter we consider how reaction modeling applied to open systems might be used to study the nature of diagenetic alteration. We develop examples in which modeling of this type can aid in interpreting the diagenetic reactions observed to have occurred in sedimentary rocks.

Table 29.1 *Analyses of formation water sampled at two oil wells producing from the Latrobe group, Gippsland basin (Harrison, 1990)*

	"Fresh" water (Barracouta A-3)	Saline water (Kingfish A-19)
Ca^{++} (mg l^{-1})	32	220
Mg^{++}	9	1 000
Na^{+}	2 943	11 000
SO_4^{--}	1 461	900
HCO_3^{-}	1 135	198
Cl^{-}	2 953	19 000
Dissolved solids	8 530	32 320
pH (measured)	7.0	5.6
pH (corrected to 60 °C)	6.95	5.55

29.1 Dolomite Cement in the Gippsland Basin

As a first example, we consider the diagenesis of clastic sandstones in the Gippsland basin, southeastern Australia, basing our model on the work of Harrison (1990). The Gippsland basin is the major offshore petroleum province in Australia. Oil production is from the Latrobe group, a fluvial to shallow marine sequence of Late Cretaceous to early Eocene age that partly fills a Mesozoic rift valley.

In the fluvial sandstones, the distribution of diagenetic cements in large part controls reservoir quality and the capacity for petroleum production. These sandstones are composed of quartz and potassium feldspar ($KAlSi_3O_8$) grains and detrital illite [which we will represent by muscovite, $KAl_3Si_3O_{10}(OH)_2$], kaolinite [$Al_2Si_2O_5(OH)_4$], and lithic fragments. Where cementation is minor, reservoir properties are excellent. Porosity can exceed 25%, and permeabilities greater than 2 darcy (2×10^{-8} cm^2) have been noted.

In more diagenetically altered areas, however, cements including dolomite [$CaMg(CO_3)_2$], clay minerals (principally kaolinite), and quartz nearly destroy porosity and permeability. In these facies, potassium feldspar grains are strongly leached, and pyrite (FeS_2) is corroded. Dolomite cement occupies up to 40% of the rock's volume, and quartz cement takes up an average of several percent. Understanding the processes that control the distribution of cements, therefore, is of considerable practical importance in petroleum exploration.

Harrison (1990) proposed that the diagenetic alteration observed in the Latrobe group resulted from the mixing within the formation of two types of groundwaters. Table 29.1 shows analyses of waters sampled from two oil wells, which she took to be representative of the two water types as they exist in the producing areas of the basin.

The first water is considerably less saline than seawater and hence is termed a "fresh" water, although it is far too saline to be potable. This water is apparently derived from meteoric water that recharges the Latrobe group where it outcrops onshore. The water flows basinward through an aquifer that extends 60 km offshore and 2 km subsea. A second, more saline water exists in deeper strata and farther offshore. This water is very similar in composition to seawater, although slightly depleted in Ca^{++} and SO_4^{--}, as can be seen by comparing the analysis to Table 6.2. On the basis of isotopic evidence, the diagenetic alteration probably occurred at temperatures of 60 °C or less.

To test whether the mixing hypothesis might explain the diagenetic alteration observed, we begin by equilibrating the fresh water, assuming equilibrium with the potassium feldspar ("maximum microcline" in the database), quartz, and muscovite (a proxy for illite) in the formation. In REACT, we enter the commands

```
swap "Maximum Microcline" for Al+++
swap Quartz for SiO2(aq)
swap Muscovite for K+

T = 60
TDS = 8530

pH = 6.95
Ca++   =   32 mg/l
Mg++   =    9 mg/l
Na+    = 2943 mg/l
HCO3-  = 1135 mg/l
SO4--  = 1461 mg/l
Cl-    = 2953 mg/l

1 free cm3 Muscovite
1 free cm3 Quartz
1 free cm3 "Maximum Microcline"

go
```

causing the program to iterate to a description of the fluid's equilibrium state.

To model fluid mixing, we will use the fresh water as a reactant, titrating it into a system containing the saline water and formation minerals. To do so, we "pick up" the fluid from the previous step to use as a reactant:

(cont'd)
```
pickup reactants = fluid
reactants times 100
```

The latter command multiplies the amount of fluid (1 kg) to be used as a reactant by 100. Hence, we will model mixing in ratios from zero to as high as 100 parts fresh to one part saline water.

To prepare the initial system, we use the analysis in Table 29.1 for the saline water, which we assume to be in equilibrium with potassium feldspar, quartz, muscovite, and dolomite ("dolomite-ord" is the most stable variety in the database). The commands

(cont'd)
```
swap "Maximum Microcline" for Al+++
swap Quartz for SiO2(aq)
swap Muscovite for K+
swap Dolomite-ord for HCO3-

T = 60
TDS = 32320
```

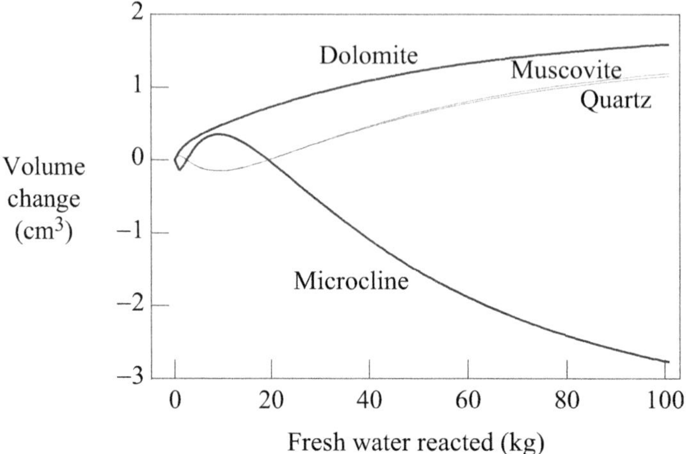

Figure 29.1 Mineralogical consequences of mixing the two fluids shown in Table 29.1 at 60 °C in the presence of microcline, muscovite, quartz, and dolomite. Results shown as the volume change for each mineral (precipitation is positive, dissolution negative), expressed per kg of pore water.

```
pH = 5.55
Ca++   =   220 mg/l
Mg++   =  1000 mg/l
Na+    = 11000 mg/l
SO4-- =    900 mg/l
Cl-    = 19000 mg/l

10 free cm3 Muscovite
10 free cm3 Quartz
10 free cm3 "Maximum Microcline"
10 free cm3 Dolomite-ord
```

set the initial system, with the minerals present in excess amounts. Typing go triggers the reaction path. Figure 29.1 shows the calculation results.

The mixing calculation is interesting in that it demonstrates a common ion effect by which dolomite precipitation drives feldspar alteration. In the model, dolomite forms because the saline water is rich in Ca^{++} and Mg^{++}, whereas the fresh water contains abundant HCO_3^-. As the fluids mix, dolomite precipitates according to the reaction

$$Ca^{++} + Mg^{++} + 2\,HCO_3^- \rightarrow \underset{\text{dolomite}}{CaMg(CO_3)_2} + 2\,H^+ , \tag{29.1}$$

producing H^+. Many of the hydrogen ions produced are consumed by driving the reaction of potassium feldspar,

$$3\ \underset{\text{microcline}}{KAlSi_3O_8} + 2\,H^+ \rightarrow \underset{\text{muscovite}}{KAl_3Si_3O_{10}(OH)_2} + 6\ \underset{\text{quartz}}{SiO_2} + 2\,K^+ , \tag{29.2}$$

to produce muscovite (illite) and quartz. Hence, the diagenetic reactions for carbonate and silicate minerals in the formation are closely linked.

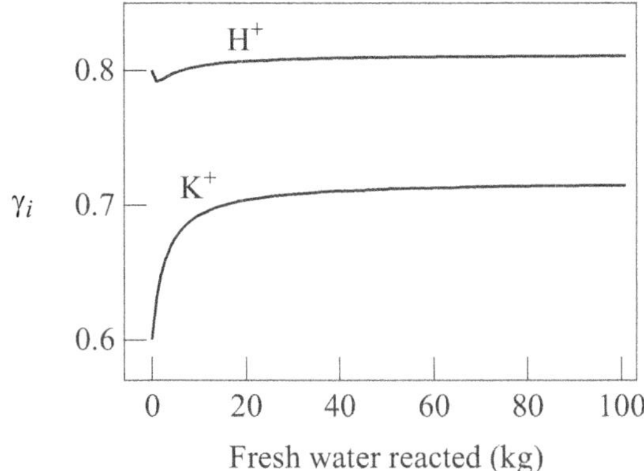

Figure 29.2 Activity coefficients γ_i for the aqueous species H^+ and K^+ over the course of the mixing reaction shown in Figure 29.1.

Strangely, Reaction 29.2 proceeds backward in the early part of the calculation (Fig. 29.1), producing a small amount of potassium feldspar at the expense of muscovite and quartz. This result, quite difficult to explain from the perspective of mass transfer, is an activity coefficient effect. As seen in Figure 29.2, the activity coefficient for K^+ increases rapidly as the fluid is diluted over the initial segment of the reaction path, whereas that for H^+ remains nearly constant. (The activity coefficients differ because the $\mathring{a}$ parameter in the Debye–Hückel model is 3 Å for K^+ and 9 Å for H^+.) As a result, a_{K^+} increases more quickly than a_{H^+}, temporarily driving Reaction 29.2 from right to left.

The model shown is quite simple and, although certainly useful from a conceptual point of view, might be expanded to better describe petrographic observations. The reaction path shown does not predict that kaolinite forms, because we assumed rather arbitrarily that the fresh water begins in equilibrium with potassium feldspar and muscovite. If we choose a lower value for the initial activity ratio a_{K^+}/a_{H^+} (or select a less evolved meteoric water than shown in Table 29.1), the reaction eventually produces kaolinite, once the available microcline is consumed. We could also account for the oxidation of pyrite as it reacts with dissolved oxygen carried by the fresh water. Pyrite oxidation produces hydrogen ions, which might further drive the reactions to produce clay minerals (Harrison, 1990), but the fact that the formation fluid is depleted in sulfate relative to seawater (Table 29.1) suggests that sulfide oxidation plays a minor role in the overall diagenetic reaction.

29.2 Lyons Sandstone, Denver Basin

As a second example, we consider the origin of anhydrite ($CaSO_4$) and dolomite cements of the Permian Lyons sandstone in the Denver basin, which lies in Colorado and Wyoming, USA, to the east of the Front Range of the Rocky Mountains. The Lyons is locally familiar as a red building stone that outcrops in hogbacks (known as the Flatirons) along the Front Range. This red facies is a quartz sand in which the sand grains are coated with iron oxides and clays and cemented by quartz overgrowths and at least two generations of calcite ($CaCO_3$; Hubert, 1960). The facies, where buried shallowly, provides a source of potable water.

Petroleum reservoirs, however, occur in a gray facies of the Lyons found in the deep basin (Levandowski *et al.*, 1973). This facies contains no ferric oxides or calcite. Many grains in the facies are coated with bitumen, the remnants of oil that migrated through the rock, and the rock is cemented with anhydrite and dolomite. The anhydrite and dolomite cements occupy as much as 25% and 15%, respectively, of the rock's pre-cement pore volume. The origin of these cements is of special interest because of their relationship to the distribution of petroleum reservoirs in the basin.

Anhydrite and dolomite cements are known to occur together in sediments that, shortly after burial in a sabkha environment, were invaded by evaporated seawater (e.g., Butler, 1969). The Denver basin contained evaporite sub-basins in the late Paleozoic (Martin, 1965), but textural and isotopic evidence argues that the cements are unlikely to have formed in a sabkha. Cements in the gray facies overlie bitumen, so they must have formed after basin strata were buried deeply enough to generate oil. The oxygen isotopic composition of the dolomite (Levandowski *et al.*, 1973), for which $\delta^{18}O_{SMOW}$ values are as low as +8.8 ‰, argues that the cement precipitated after the formation was buried from a fluid containing varying amounts of meteoric water. The dolomite would be composed of isotopically heavier oxygen if it had formed at surface temperatures from evaporated seawater.

The cements also might have formed if H_2S had migrated into the formation and oxidized to make anhydrite. The sulfur isotopic composition of the anhydrite, however, does not allow such an explanation. Values for $\delta^{34}S_{CDT}$, which vary from +9.6 to +12.5 ‰ (Lee and Bethke, 1994), span the worldwide range for Permian evaporite minerals, but are much heavier than values associated with H_2S in sedimentary basins. Furthermore, the cements could not have precipitated from the original pore fluid of the Lyons, because the solubility of anhydrite (as well as gypsum, $CaSO_4 \cdot 2H_2O$) in aqueous solution is much too low to account for the amounts of cement observed. For these reasons, the cements of the gray facies must have formed after the Lyons was buried, in a system open to groundwater flow.

Lee and Bethke (1994) suggested that the gray Lyons facies formed as an alteration product of the red facies in a groundwater flow regime set up by uplift of the Front Range, which began to rise in the early Tertiary and reached its peak in Eocene time (McCoy, 1953). Groundwater in the basin today flows from west to east (Belitz and Bredehoeft, 1988) in response to the elevation of the Front Range. Past flow was more rapid than in the present day because erosion has reduced the elevation of the basin's western margin. Paleohydrologic models calculated for the basin (Lee and Bethke, 1994) suggest that in the Eocene groundwater flowed eastward through the Lyons at an estimated discharge of about 1 m yr^{-1}.

Flow in the Pennsylvanian Fountain formation, a sandstone aquifer that underlies the Lyons and is separated from it by an aquitard complex, was more restricted because the formation grades into less permeable dolomites and evaporites in the deep basin. Groundwater in the Fountain recharged along the Front Range and flowed eastward at an estimated discharge of about 0.1 m yr^{-1}. Where Fountain groundwater encountered less permeable sediments along the basin axis, it discharged upward and mixed by dispersion into the Lyons formation.

According to Lee and Bethke's (1994) interpretation, the gray facies formed in this zone of dispersive mixing when saline Fountain groundwater reacted with Lyons sediments and groundwater. To simulate the mixing reaction, we start by developing chemical models of the two groundwaters, assuming that the mixing occurred at 100 °C.

To set the initial composition of the Lyons fluid, we use an analysis of modern Lyons groundwater sampled at 51 °C (McConaghy *et al.*, 1964), which we correct to the temperature of the simulation by heating it in the presence of calcite and quartz. In REACT, the commands

```
T = 51
swap Calcite for H+
swap Quartz for SiO2(aq)
```

Table 29.2 *Predicted compositions of Lyons groundwater and Fountain brine, before mixing (Lee and Bethke, 1994)*

	Lyons	Fountain
Na^+ (mg kg^{-1})	108	56 400
Ca^{++}	16	516
K^+	4.6	—
Mg^{++}	0.5	3 450
SiO_2(aq)	49	23
HCO_3^-	419	28 780
Cl^-	9	87 000
SO_4^{--}	36	14 200
pH (100 °C)	6.7	4.6

```
Na+     = 108    mg/kg
Ca++    =  40    mg/kg
K+      =    4.6 mg/kg
Mg++    =   1    mg/kg
Cl-     =   9    mg/kg
SO4--   =  36    mg/kg
HCO3-   = 340    mg/kg
balance on HCO3-

1 free cm3 Calcite
1 free cm3 Quartz
```

describe an initial system containing a kilogram of groundwater and excess amounts of calcite and quartz. The commands

```
(cont'd)
T final = 100
go
```

cause the program to equilibrate the fluid and heat it to the temperature of interest. The resulting fluid is a predominantly sodium-bicarbonate solution (Table 29.2).

Since we have no direct information about the chemistry of the Fountain fluid, we assume that its composition reflects reaction with minerals in the evaporite strata that lie beneath the Lyons. We take this fluid to be a 3 molal NaCl solution that has equilibrated with dolomite, anhydrite, magnesite ($MgCO_3$), and quartz. The choice of NaCl concentration reflects the upper correlation limit of the B-dot (modified Debye–Hückel) equations (see Chapter 8). To set pH, we assume a CO_2 partial pressure of 50 bar, which we will show leads to a reasonable interpretation of the isotopic composition of the dolomite cement.

In fact, the choice of CO_2 partial pressure has little effect on the mineralogical results of the mixing calculation. In the model, the critical property of the Fountain fluid is that it is undersaturated with respect to calcite, so that calcite dissolves when the fluid mixes into the Lyons. Because we assume equilibrium with dolomite and magnesite, the saturation index ($\log Q/K$) of calcite is fixed by the reaction

$$\underset{\textit{calite}}{CaCO_3} \rightleftarrows \underset{\textit{dolomite}}{CaMg(CO_3)_2} - \underset{\textit{magnesite}}{MgCO_3} \tag{29.3}$$

to a value of −1.3, and hence, is independent of pH and CO_2 pressure.

To calculate the composition of the Fountain brine, we start anew in REACT, enter the commands

```
T = 100
swap CO2(g) for H+
swap Magnesite for Mg++
swap Anhydrite for SO4--
swap Dolomite-ord for Ca++
swap Quartz for SiO2(aq)

Na+ = 3 molal
Cl- = 3 molal

P CO2(g) = 50 bar
1 free mole Magnesite
1 free mole Anhydrite
1 free mole Dolomite-ord
1 free mole Quartz

balance on HCO3-
```

and type go. In contrast to the Lyons groundwater, the Fountain brine (Table 29.2) is a sodium chloride water.

To model the mixing of these fluids in contact with quartz and calcite of the red Lyons formation, we follow three steps. First, we calculate the composition of Lyons groundwater, as before, and save it to a file. Second, we compute the composition of the Fountain brine. Finally, we "pick up" the Fountain brine as a reactant, multiply its mass by 15 (giving 15 kg of solvent plus the solute mass), and titrate it into a system containing 1 kg of Lyons groundwater and excess amounts of quartz and calcite. Here, the factor 15 is largely arbitrary; we choose it to give a calculation endpoint with a high ratio of brine to Lyons groundwater. The REACT procedure, starting anew, is:

— Step 1 —

(constrain composition of Lyons groundwater, as before)
```
T initial = 51, final = 100
go

pickup fluid
save Lyons_100.rea
```

— Step 2 —

```
reset
```
(constrain composition of Fountain brine, as before)
```
T = 100
go
```

— Step 3 —

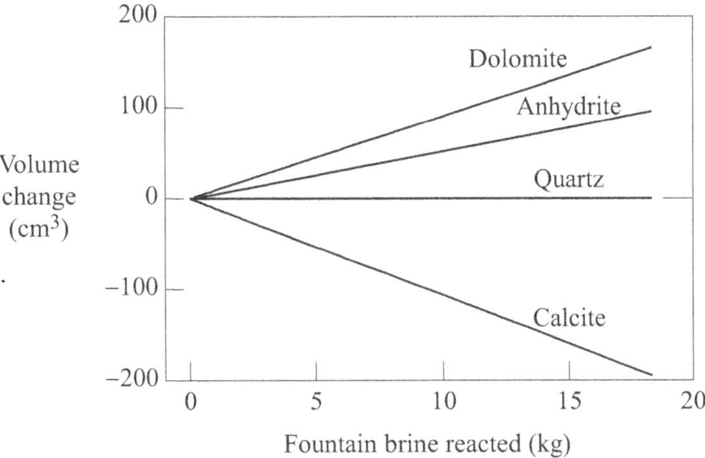

Figure 29.3 Mineralogical consequences of mixing Fountain brine into the Lyons formation. Vertical axis shows changes in mineral volume, expressed per kg of Lyons groundwater; positive changes indicate precipitation and negative, dissolution.

```
pickup reactants = fluid
reactants times 15

read Lyons_100.rea
swap Quartz for SiO2(aq)
swap Calcite for HCO3-
100 free mol Quartz
100 free mol Calcite
balance on Cl-
go
```

In the resulting reaction (Fig. 29.3), calcite progressively dissolves as the Fountain brine mixes into the Lyons formation. The Ca^{++} and HCO_3^- added to solution drive precipitation of anhydrite and dolomite by a common ion effect. The overall reaction,

$$\underset{\textit{calcite}}{5\ CaCO_3} + 2\ SO_4^{--} + 5/2\ Mg^{++} \rightarrow \underset{\textit{anhydrite}}{2\ \ CaSO_4} + \underset{\textit{dolomite}}{5/2\ CaMg(CO_3)_2} + 1/2\ Ca^{++}\,, \tag{29.4}$$

explains the origin of the cements in the gray facies as well as the facies' lack of calcite.

We can predict the oxygen and carbon isotopic compositions of the dolomite produced by this reaction path, using the techniques described in Chapter 22. Figure 29.4 shows the compositions of calcite and dolomite cements in the Lyons, as determined by Levandowski *et al.* (1973). The calcite and dolomite show broad ranges in oxygen isotopic content. The dolomite, however, spans a much narrower range in carbon isotopic composition than does the calcite.

To set up the calculation, we specify initial isotopic compositions for the fluid and calcite. We choose a value of −13 ‰ for $\delta^{18}O_{SMOW}$ of the Lyons fluid, reflecting Tertiary rainfall in the region, and set the calcite composition to +11 ‰, the mean of the measured values (Fig. 29.4). We further set $\delta^{13}C_{PDB}$ for the fluid to

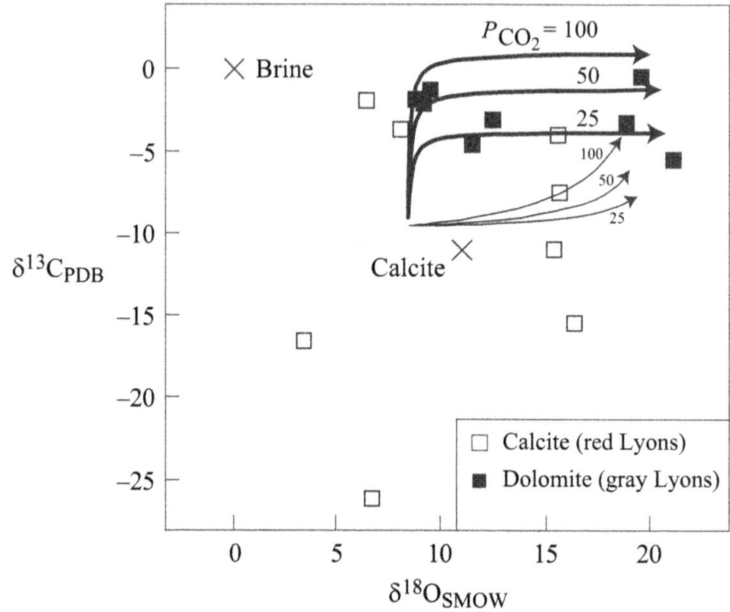

Figure 29.4 Oxygen and carbon stable isotopic compositions of calcite (□) and dolomite (■) cements from Lyons sandstone (Levandowski *et al.*, 1973), and assumed initial compositions (×) of calcite and brine. Bold arrows trace isotopic trends predicted for dolomite cements produced by the mixing reaction shown in Figure 29.3, assuming various CO_2 partial pressures (25 bar, 50 bar, and 100 bar) for the Fountain brine. Fine arrows, for comparison, show isotopic trends predicted in calculations which assume (improperly) that fluid and minerals maintain isotopic equilibrium over the course of the simulation. Figure after Lee and Bethke (1996).

−12 ‰. We do not specify an initial carbon composition for the calcite, so the model sets this value to −11 ‰, in isotopic equilibrium with the fluid. Again, this value is near the mean of the measurements.

We then set the isotopic compositions of each oxygen and carbon-bearing component in the reactant, the Fountain brine, to $\delta^{18}O$ and $\delta^{13}C$ values of zero, as might be expected in a sedimentary brine. Finally, we segregate each mineral in the calculation from isotopic exchange, as discussed in Chapter 22. The procedure for small water/rock ratios is

(cont'd)

```
oxygen initial = -13, Calcite = +11
carbon initial = -12

carbon HCO3- = 0
oxygen H2O = 0, SiO2(aq) = 0, HCO3- = 0, SO4-- = 0
segregate Calcite, Quartz, Dolomite-ord, Anhydrite

reactants times 1
go
```

We then repeat the calculation to carry the model to high ratios with the commands

(cont'd)
```
reactants times 50
go
```

Figure 29.4 shows the predicted isotopic trends for the dolomite produced by the reaction path, calculated assuming several values for the CO_2 partial pressure in the Fountain brine. These results suggest that the cement's carbon isotopic composition reflects the composition of a CO_2-rich brine more closely than that of the precursor calcite cement, explaining the narrow range observed in $\delta^{13}C$. The spread in oxygen composition results from mixing of fresh Lyons water and Fountain brine in varying proportions.

As noted by Lee and Bethke (1996), if we had calculated this model without holding minerals segregated from isotopic exchange, we would have predicted broadly different isotopic trends that are not in accord with the observed data. To verify this point, we enter the command

(cont'd)
```
unsegregate ALL
```

and type `go`. The resulting isotopic trends for the equilibrium case are shown in Figure 29.4.

As we have demonstrated, reaction modeling provides an explanation of the observed diagenetic mineralogy as well as the isotopic compositions of the cements. The model also helps explain the association of the gray facies with oil reservoirs. Fractures that developed along the basin axis when the Front Range was uplifted provided pathways for oil to migrate by buoyancy upward into the Lyons. Continued uplift set up regional groundwater flow that drove brine from the Fountain formation upward along the same pathways as the oil. The petroleum or brine, or both, reduced iron oxides in the Lyons, changing its color from red to gray. As brine mixed into the Lyons, it dissolved the existing calcite cement and, by a common ion effect, precipitated anhydrite and dolomite on top of bitumen left behind by migrating oil.

Oil in the Cretaceous Dakota sandstone, a shallower aquifer than the Lyons, has migrated laterally as far as 150 km into present-day reservoirs (Clayton and Swetland, 1980). In contrast, oil has yet to be found in the Lyons outside the deep strata where it was generated. The formation of anhydrite and dolomite cements may have served to seal the oil into reservoirs, preventing it from migrating farther.

30
Kinetics of Water–Rock Interaction

In calculating most of the reaction paths in this book, we have measured reaction progress with respect to the dimensionless variable ξ. We showed in Chapter 17, however, that by incorporating kinetic rate laws into a reaction model, we can trace reaction paths describing mineral precipitation and dissolution using time as the reaction coordinate.

In this chapter we construct a variety of kinetic reaction paths to explore how this class of model behaves. Our calculations in each case are based on kinetic rate laws determined by laboratory experiment. In considering the calculation results, therefore, it is important to keep in mind the uncertainties entailed in applying laboratory measurements to model reaction processes in nature, as discussed in detail in Section 17.2.

In the next chapter (Chapter 31) we show how calculations of this type can be integrated into mass transport models to produce models of weathering in soils and sediments open to groundwater flow. In later chapters, we consider redox kinetics in geochemical systems in which a mineral surface or enzyme acts as a catalyst (Chapter 32), and those in which the reactions are catalyzed by microbial populations (Chapter 37).

30.1 Approach to Equilibrium and Steady State

In Chapter 17 we considered how quickly quartz dissolves into water at 100 °C, using a kinetic rate law determined by Rimstidt and Barnes (1980). In this section we take up the reaction of silica (SiO_2) minerals in more detail, this time working at 25 °C. We use kinetic data for quartz and cristobalite from the same study, as shown in Table 30.1.

Each silica mineral dissolves and precipitates in our calculations according to the rate law

$$r_{SiO_2} = A_S\, k_+ \left(1 - \frac{Q}{K}\right) \tag{30.1}$$

(Eq. 17.23), as discussed in Chapter 17. Here, r_{SiO_2} is the reaction rate (mol s^{-1}; positive for dissolution), A_S and k_+ are the mineral's surface area (cm^2) and rate constant (mol cm^{-2} s^{-1}), and Q and K are the activity product and equilibrium constant for the dissolution reaction. The reaction for quartz, for example, is

$$\underset{quartz}{SiO_2} \rightleftarrows SiO_2(aq)\,. \tag{30.2}$$

According to Knauss and Wolery (1988), this rate law is valid for neutral to acidic solutions; a distinct rate law applies in alkaline fluids, reflecting the dominance of a second reaction mechanism under conditions of high pH.

The procedure in REACT is similar to that used in the earlier calculation (Section 17.4)

Table 30.1 *Rate constants* k_+ *(mol cm*$^{-2}$ *s*$^{-1}$*) for the reaction of silica minerals with water at various temperatures, as determined by Rimstidt and Barnes (1980)*

T(°C)	Quartz	α-Cristobalite	Amorphous silica
25	4.20×10^{-18}	1.71×10^{-17}	7.32×10^{-17}
70	2.30×10^{-16}	6.47×10^{-16}	2.19×10^{-15}
100	1.88×10^{-15}	4.48×10^{-15}	1.33×10^{-14}
150	3.09×10^{-14}	6.12×10^{-14}	1.49×10^{-13}
200	2.67×10^{-13}	4.81×10^{-13}	9.81×10^{-13}
250	1.46×10^{-12}	2.55×10^{-12}	4.43×10^{-12}
300	5.71×10^{-12}	1.01×10^{-11}	1.51×10^{-11}

```
time end = 1 year
1 mg/kg SiO2(aq)

react 5000 grams Quartz
kinetic Quartz  rate_con = 4.2e-18, surface = 1000
go
```

except that we work at 25 °C and, since reaction proceeds more slowly at low temperature, set a longer time span. By including 5 kg of quartz sand in the calculation, we imply that the system's porosity is about 35%, since the density of quartz is 2.65 g cm^{-3}. At the specified silica concentration of 1 mg kg^{-1}, the initial fluid is undersaturated with respect to quartz, so we can expect quartz to dissolve over the reaction path.

In the calculation results (Fig. 30.1), the silica concentration gradually increases from the initial value, asymptotically approaching the equilibrium value of 6 mg kg^{-1} after about half a year of reaction. We repeat the calculation, this time starting with a supersaturated fluid

```
(cont'd)
12 mg/kg SiO2(aq)
suppress Tridymite, Chalcedony
go
```

To keep our discussion simple for the moment, we suppress the silica polymorphs tridymite and chalcedony. In the calculation results (Fig. 30.1), the silica concentration gradually decreases from its initial value and, as in the previous calculation, approaches equilibrium with quartz after about half a year.

We could have anticipated the results in Figure 30.1 from the form of the rate law (Eq. 30.1). If we let $m_{\mathrm{SiO_2}}$ represent the molality of SiO_2(aq) and m_{eq} represent this value at equilibrium, we can rewrite the rate law as

$$\frac{dm_{\mathrm{SiO_2}}}{dt} = -\frac{A_S\, k_+}{n_w\, m_{\mathrm{eq}}} \left(m_{\mathrm{SiO_2}} - m_{\mathrm{eq}}\right) . \tag{30.3}$$

Here, we have assumed that the activity coefficient $\gamma_{\mathrm{SiO_2}}$ does not vary with silica concentration. As before, n_w is the mass of solvent water in the system.

Since we can take each variable except $m_{\mathrm{SiO_2}}$ to be constant, Equation 30.3 has the form of an ordinary differential equation in time. We can use standard techniques to solve the equation for $m_{\mathrm{SiO_2}}(t)$. The solution

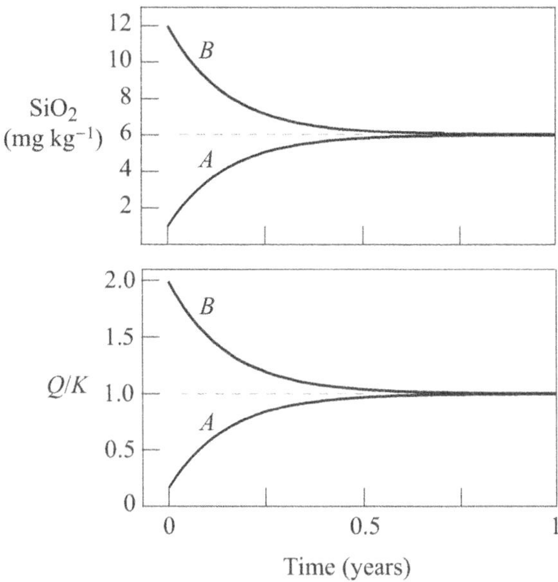

Figure 30.1 Reaction of quartz with water at 25 °C, showing approach to equilibrium (dashed lines) with time. Top diagram shows variation in $SiO_2(aq)$ concentration and bottom plot shows change in quartz saturation. In calculation *A*, the fluid is initially undersaturated with respect to quartz; in *B* it is supersaturated.

corresponding to the initial condition $m_{SiO_2} = m_o$ at $t = 0$ is

$$m_{SiO_2} = \left(m_o - m_{eq}\right) e^{-(A_S k_+/n_w m_{eq})t} + m_{eq} \,. \tag{30.4}$$

The first term on the right side of the solution represents the extent to which the silica concentration deviates from equilibrium. Since the term appears as a negative exponential function in time, its value decays to zero (as can be seen in Figure 30.1) at a rate that depends on the surface area and rate constant. As t becomes large, the first term disappears, leaving only the equilibrium concentration m_{eq}.

The calculation results in Figure 30.1 suggest that groundwaters in quartz sand aquifers should approach equilibrium with quartz in less than a year, a short period compared to typical residence times in groundwater flow regimes. At low temperature, however, groundwaters in nature seldom appear to be in equilibrium with quartz, often appearing supersaturated. As discussed in Section 17.2, this discrepancy between calculation and observation might be accounted for if the surfaces of quartz grains in real aquifers were coated in a way that inhibited reaction. Alternatively, the discrepancy may arise from the effects of reactions with minerals other than quartz that consume and produce silica. If these reactions proceed rapidly compared to the dissolution and precipitation of quartz, they can control the fluid's silica content.

To see how a second kinetic reaction might affect the fluid's silica concentration, we add 250 g of cristobalite to the system. The mass ratio of quartz to cristobalite, then, is twenty to one. Taking the fluid to be in equilibrium with quartz initially, the procedure in REACT is

```
time end = 1 year
swap Quartz for SiO2(aq)
5000 free grams Quartz
```

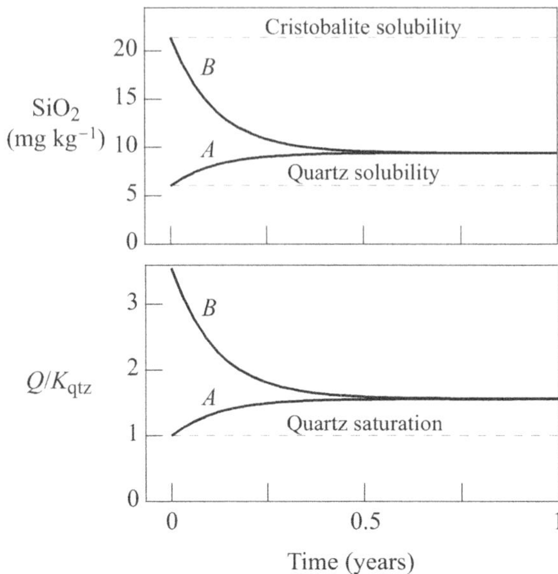

Figure 30.2 Kinetic reaction of quartz and cristobalite with water at 25 °C. In calculation *A* the fluid is originally in equilibrium with quartz, in *B* with cristobalite. The top diagram shows how the $SiO_2(aq)$ concentration varies with time, and the bottom plot shows the change in quartz saturation. The reaction paths approach a steady state in which the fluid is supersaturated with respect to quartz and undersaturated with respect to cristobalite.

```
react 250 grams of Cristobalite
kinetic Cristobalite  rate_con = 1.7e-17  surface = 5000
kinetic Quartz        rate_con = 4.2e-18  surface = 1000

suppress Tridymite, Chalcedony
go
```

Here, we take a specific surface area for cristobalite five times greater than the value assumed for quartz. To calculate a second case in which the fluid starts the reaction path in equilibrium with cristobalite, we enter the commands

(cont'd)
```
swap Cristobalite for SiO2(aq)
250 free grams Cristobalite

react 0 grams Cristobalite
react 5000 grams Quartz
go
```

Figure 30.2 shows the results of the two calculations.

As in the previous example (Fig. 30.1), silica concentration in the calculations asymptotically approaches a single value, regardless of the initial concentration. The final silica concentration, however, does not represent a thermodynamic equilibrium, since (as we can see in Fig. 30.2) it is supersaturated with respect to quartz and undersaturated with respect to cristobalite.

Instead, this concentration marks a *steady state* (or *dynamic equilibrium*, a term that persists despite being an oxymoron) at which the rate of cristobalite dissolution matches the rate at which quartz precipitates. The steady-state concentration will decay toward equilibrium with continued reaction only to the extent that the surface area of cristobalite decreases as the mineral dissolves. Because only a few tens of milligrams of cristobalite dissolve each year (as we can quickly compute using Equation 30.1), the steady state will persist for a long time, decaying only gradually over tens of thousands of years.

We can calculate the value of the steady-state silica concentration m_{ss} directly from the rate law (Eq. 30.1). Noting that the steady state is marked by the condition $r_{qtz} = -r_{cri}$, we can write

$$m_{ss} = \frac{1}{\gamma_{SiO_2}} \times \frac{(A_S\, k_+)_{qtz} + (A_S\, k_+)_{cri}}{(A_S\, k_+/K)_{qtz} + (A_S\, k_+/K)_{cri}} . \tag{30.5}$$

To evaluate this equation, we use the values of the rate constants k_+ and surface areas A_S (the latter given as the product of specific surface area and mineral mass) for the two minerals and the equilibrium constants K for quartz (1.00×10^{-4}) and cristobalite (3.56×10^{-4}), and take γ_{SiO_2} to be one. The resulting steady-state concentration is 1.57×10^{-4} molal, or 9.4 mg kg^{-1}, which agrees with the simulation results in Figure 30.2.

It is interesting to note that adding only a small amount of cristobalite to the system gives rise to a significant departure from equilibrium with the predominant mineral, quartz. The cristobalite plays a role in the calculation disproportionate to its abundance because the assumed values for its surface area and rate constant are considerably larger than those for quartz. In nature, therefore, we might expect highly reactive minerals to have significant effects on fluid chemistry, even when they are present in small quantities.

It is further interesting to observe that the behavior of a system approaching a thermodynamic equilibrium differs little from one approaching a steady state. According to the kinetic interpretation of equilibrium, as discussed in Chapter 17, a mineral is saturated in a fluid when it precipitates and dissolves at equal rates. At a steady state, similarly, the net rate at which a component is consumed by the precipitation reactions of two or more minerals balances with the net rate at which it is produced by the minerals' dissolution reactions. Thermodynamic equilibrium viewed from the perspective of kinetic theory, therefore, is a special case of the steady state.

In experimental studies, it is common practice to attempt to "bracket" a measured solubility by reacting a sample with undersaturated as well as supersaturated solutions. As is shown in Figure 30.2, however, this technique might equally well identify a steady-state condition as an equilibrium state.

30.2 Quartz Deposition in a Fracture

In a second application, we consider the rate at which quartz might precipitate in an open fracture as a hydrothermal water flows along it, gradually cooling from 300 °C to surface temperature. We assume that the fracture has an aperture δ and is lined with quartz. We further assume that the fluid is initially in equilibrium with quartz and that temperature varies linearly along the fracture. The latter condition imposes an important constraint on the calculation since at high rates of discharge the fluid can in reality control temperature along the fracture, causing temperature to deviate broadly from the assumed linear gradient. To investigate such conditions, we would want to construct a more sophisticated model in which we specifically account for advective heat transfer.

In our calculation, we need not be concerned with the dimensions of the fracture or the velocity of the fluid. Instead, we need specify only the length of time Δt it takes the fluid to travel from high-temperature conditions at depth to cool surface conditions.

To model the problem, we take a packet of water in contact with the fracture walls over a polythermal reaction path. The fact that the packet moves relative to the walls is of no concern, since the fracture surface area exposed to the packet is approximately constant. Since the system contains 1 kg of water, we can show from geometry that the surface area A_S (in cm^2) of the fracture lining is

$$A_S = \frac{2000\ \psi}{\rho \delta} \,. \tag{30.6}$$

Here, ψ is the surface roughness (surface area per unit area in cross section) of the fracture walls, ρ is fluid density in g cm^{-3}, and the aperture δ is taken in cm. For our purposes, it is sufficiently accurate to choose a value of 2 for ψ and set ρ to 1 g cm^{-3}. In a fracture with an aperture of 10 cm, for example, each kg of water is exposed to 400 cm^2 of quartz surface.

We use the Arrhenius equation,

$$k_+ = A\, e^{-E_A/RT_K} \tag{30.7}$$

(Eq. 17.3), in our calculations to set the rate constant k_+ as a function of temperature along the fracture. A pre-exponential factor A of about 2.35×10^{-5} mol cm^{-2} s^{-1} and an activation energy E_A of 72 800 J mol^{-1} fit the kinetic data for quartz in Table 30.1.

To assign the amount of quartz in the system, we arbitrarily specify a specific surface area of 1 cm^2 g^{-1}. Then, we need only set the quartz mass to a value in grams equal to the desired surface area in cm^2. Finally, we set for each run the amount of time Δt it takes the packet of water to flow along the fracture.

To model the effects of flow through a 10-cm-wide fracture, assuming a time span of 1 year, for example, the procedure in REACT is

```
time end = 1 year
T initial = 300, final = 25

swap Quartz for SiO2(aq)
400 free grams Quartz
kinetic Quartz  surface = 1
kinetic Quartz  pre-exp = 2.35e-5, act_eng = 72800

suppress Tridymite Chalcedony Cristobalite Amrph^silica
delxi = .001
go
```

By suppressing the other silica polymorphs, we prevent them from forming in the calculation. We will show, however, that except at large Δt these minerals tend to become supersaturated in the low-temperature end of the fracture. In reality, therefore, these minerals would be likely to form within the fracture under such conditions.

Figure 30.3 shows how SiO_2 concentration in the fluid varies along the fracture, calculated assuming different traversal times Δt. For slow flow rates (values of Δt of about 1000 years or longer), the fluid has enough reaction time to remain near equilibrium with quartz. When flow is more rapid, however, the fluid maintains much of its silica content, quickly becoming supersaturated with respect to quartz. Farther along the fracture, the fluid also becomes supersaturated with respect to cristobalite, and at traversal times of less than about 100 years, with respect to amorphous silica.

It is interesting to examine Figure 30.3 in light of our discussion in Chapter 27 of the silica geothermometer. If we wish to derive an estimate for the fluid's original temperature of 300 °C from its silica content under

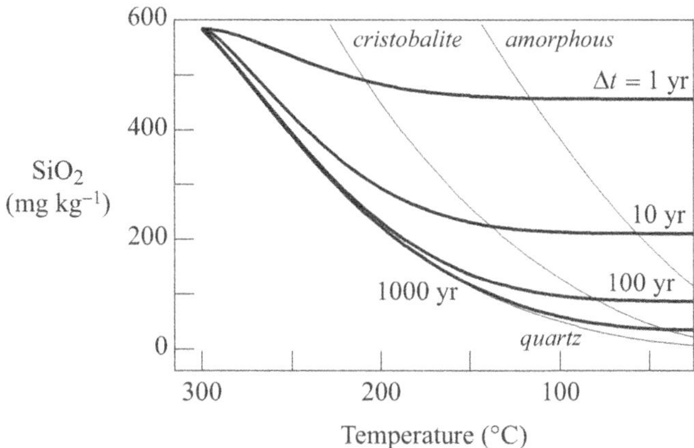

Figure 30.3 Silica concentration (bold lines) in a fluid packet that cools from 300 °C as it flows along a quartz-lined fracture of 10 cm aperture, calculated assuming differing traversal times Δt. Fine lines show solubilities of the silica polymorphs quartz, cristobalite, and amorphous silica.

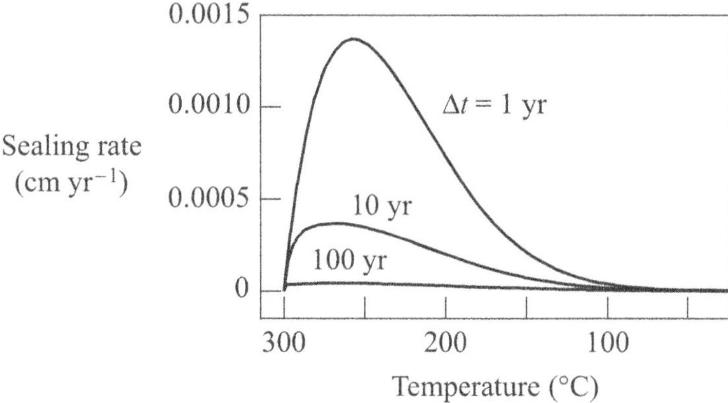

Figure 30.4 Fracture sealing rates (cm yr^{-1}) for the quartz precipitation calculations shown in Figure 30.3.

surface conditions, then our calculations suggest that under the modeled conditions the fluid must traverse the fracture (e.g., a fracture through the cap rock in a geothermal field, assuming that temperature in the cap rock varies linearly with depth) in less than 1 year.

We can use the calculated reaction rates (Fig. 30.3) to compute how rapidly quartz precipitation seals the fracture. The sealing rate, the negative rate at which fracture aperture changes, can be expressed as

$$-\frac{d\delta}{dt} = \frac{r_{\text{qtz}}\, M_V\, \rho\, \delta}{1000}\,. \tag{30.8}$$

Here, r_{qtz} is the rate of quartz precipitation (mol s^{-1} from a kg of solvent water) and M_V is the mineral's molar volume (22.7 $\text{cm}^3\ \text{mol}^{-1}$). Figure 30.4 shows the resulting sealing rates calculated for several traversal times. For a Δt of 1 year, for example, we expect the fracture to become occluded near its high-temperature end over a time scale of about 10 000 years.

30.3 Silica Transport in an Aquifer

Next we look at how temperature gradients along an aquifer might affect the silica content of flowing groundwater. We consider a symmetrical aquifer that descends from a recharge area at the surface to a depth of about 2 km and then ascends to a discharge area. Temperature in the calculation varies linearly from 20 °C at the surface to 80 °C at the aquifer's maximum depth.

To set up the calculation, we take a quartz sand of the same porosity as in the calculations in Section 30.1 and assume that the quartz reacts according to the same rate law (Eq. 30.1). We let the rate constant vary with temperature according to the Arrhenius equation (Eq. 30.7), using the values for the preexponential factor and activation energy given in Section 30.2. As in the previous section, we need only be concerned with the time available for water to react as it flows through the aquifer. We need not specify, therefore, either the aquifer length or the flow velocity.

To model reaction within the descending leg of the flow path, along which water warms with depth, the procedure in REACT is

```
time end = 1 year
T initial = 20, final = 80
swap Quartz for SiO2(aq)
5000 free grams Quartz

kinetic Quartz  surface = 1000
kinetic Quartz  pre-exp = 2.35e-5  act_eng = 72800

suppress Tridymite, Chalcedony
go
```

Then, to model the ascending limb, we start with the final composition of the descending fluid and let it cool. The corresponding commands are

```
(cont'd)
pickup
T final 20
react 5000 grams Quartz
kinetic Quartz  surface = 1000
kinetic Quartz  pre-exp = 2.35e-5  act_eng = 72800
go
```

Here we take a reaction time Δt (the time it takes water to descend to maximum depth or to ascend back to the surface) of 1 year. We then repeat the calculation assuming other reaction times.

In the calculation results (Fig. 30.5), silica concentration increases as the fluid flows downward, reflecting the increase in quartz solubility with depth, and then decreases as fluid moves back toward the surface. In calculations in which the reaction time exceeds about 10 years, the fluid remains close to equilibrium with quartz. Given shorter reaction times, however, the fluid does not react quickly enough to respond to the changing quartz solubility along the flow path. Along the descending limb, the fluid appears undersaturated with silica, and it becomes supersaturated along the upflowing leg of the flow path. If we do not suppress the other silica polymorphs in runs assuming short reaction times, tridymite (the most stable) tends to precipitate as the fluid cools.

The results in Figure 30.5 are interesting because they suggest that gradual temperature variations along quartz aquifers are unlikely to produce silica concentrations that deviate significantly from equilibrium, except

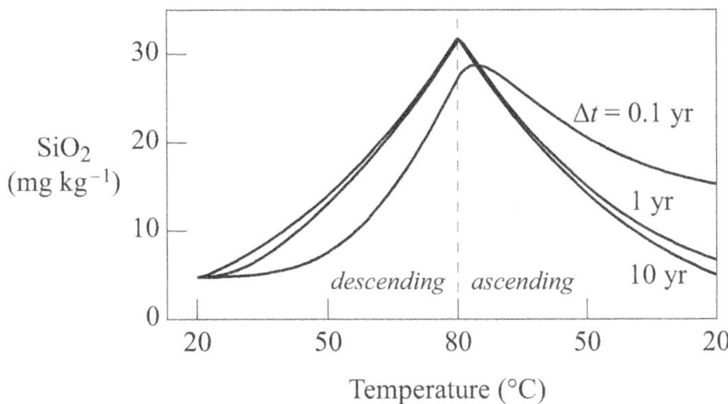

Figure 30.5 Calculated silica concentration in a fluid packet flowing through a quartz sand aquifer. The fluid descends from the surface ($T = 20\,°C$) to a depth of about 2 km (80 °C) and then returns to the surface (20 °C). Results are shown for time spans Δt (representing half of the time the fluid takes to migrate through the aquifer) of 0.1, 1, and 10 years. In the latter calculation, the fluid remains near equilibrium with quartz.

for fluids that quickly traverse the flow system. The effects of other reactions that consume and produce silica, as discussed in Section 30.1, would appear more likely to cause disequilibrium between groundwater and quartz than would flow along temperature gradients.

30.4 Ostwald's Step Rule

Ostwald's step rule holds that a thermodynamically unstable mineral reacts over time to form a sequence of progressively more stable minerals (e.g., Morse and Casey, 1988; Steefel and Van Cappellen, 1990; Nordeng and Sibley, 1994). The step rule is observed to operate, especially at low temperature, in a number of mineralogic systems, including the carbonates, silica polymorphs, iron and manganese oxides, iron sulfides, phosphates, clay minerals, and zeolites.

Various theories, ranging from qualitative interpretations to those rooted in irreversible thermodynamics and geochemical kinetics, have been put forward to explain the step rule. A kinetic interpretation of the phenomenon, as proposed by Morse and Casey (1988), may provide the most insight. According to this interpretation, Ostwald's sequence results from the interplay of the differing reactivities of the various phases in the sequence, as represented by A_S and k_+ in Equation 30.1, and the thermodynamic drive for their dissolution and precipitation of each phase, represented by the $(1 - Q/K)$ term.

To investigate the kinetic explanation for the step rule, we model the reaction of three silica polymorphs—quartz, cristobalite, and amorphous silica—over time. We consider a system that initially contains 100 cm^3 of amorphous silica, the least stable of the polymorphs, in contact with 1 kg of solvent water, and assume that the fluid is initially in equilibrium with this phase. We include in the system small amounts of cristobalite and quartz, thereby avoiding the question of how best to model nucleation. In reality, nucleation, crystal growth, or both of these factors might control the nature of the reaction; we will consider only the effect of crystal growth in our simple calculation.

Each mineral in the calculation dissolves and precipitates according to the kinetic rate law (Eq. 30.1) used in the previous examples and the rate constants listed in Table 30.1. We take the same specific surface areas for quartz and cristobalite as we did in our calculations in Section 30.1, and assume a value of 20 000 cm^2 g^{-1}

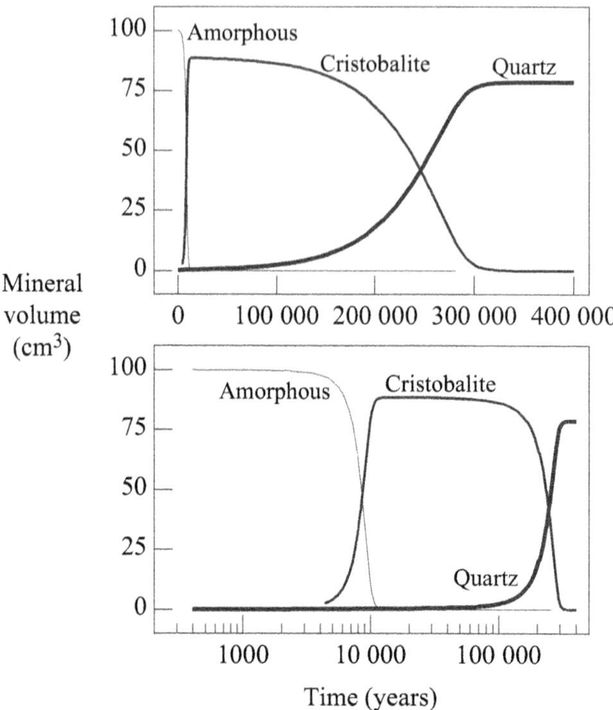

Figure 30.6 Variation in mineral volumes over a kinetic reaction path designed to illustrate Ostwald's step sequence. The calculation traces the reaction at 25 °C among the minerals amorphous silica (fine line), cristobalite (medium line), and quartz (bold line). The top diagram shows results plotted against time on a linear scale; the time scale on the bottom diagram is logarithmic. The decrease in total volume with time reflects the differing molar volumes of the three minerals.

for the amorphous silica, consistent with measurements of Leamnson *et al.* (1969). The procedure in REACT is

```
time end = 400000 years
swap Amrph^silica for SiO2(aq)
100 free cm3 Amrph^silica

react 0.1 cm3 of Cristobalite
react 0.1 cm3 of Quartz

kinetic Amrph^silica  rate_con = 7.3e-17  surface = 20000
kinetic Cristobalite  rate_con = 1.7e-17  surface = 5000
kinetic Quartz        rate_con = 4.2e-18  surface = 1000

suppress Tridymite, Chalcedony
delxi  = .001
dxplot = .001
go
```

In the calculation results (Fig. 30.6), the initial segment of the path is marked by the disappearance of the amorphous silica as it reacts to form cristobalite. The amorphous silica is almost completely consumed after about 10 000 years of reaction. The mineral's mass approaches zero asymptotically because (as can be seen in Equation 30.1) as its surface area A_S decreases, the dissolution rate slows proportionately. During the initial period, only a small amount of quartz forms.

Once the amorphous silica has nearly disappeared, the cristobalite that formed early in the calculation begins to redissolve to form quartz. The cristobalite dissolves, however, much more slowly than it formed, reflecting the slow rate of quartz precipitation. After about 300 000 years of reaction, nearly all of the cristobalite has been transformed into quartz, the most stable silica polymorph, and the reaction has virtually ceased.

The step sequence in the calculation results arises from the fact that the values assumed for the specific surface areas A_{sp} ($cm^2\ g^{-1}$) and rate constants k_+ (mol $cm^{-2}\ s^{-1}$), and hence the product of these two terms, decrease among the minerals with increasing thermodynamic stability:

Mineral	log K	A_{sp}	$k_+ \times 10^{18}$	$A_{sp}\,k_+ \times 10^{15}$
Amorphous silica	−2.71	20 000	73	1460
Cristobalite	−3.45	5 000	17	85
Quartz	−4.00	1 000	4.2	4.2

The product $A_{sp}\,k_+$ for amorphous silica is about 17 times greater than it is for cristobalite, and this value in turn exceeds the value for quartz about 20-fold. The least stable minerals, therefore, are the most reactive.

In the first segment of the calculation, the high reactivity of the amorphous silica assures that the fluid remains near equilibrium with this mineral, as shown in Figure 30.7. Only after the mineral has almost disappeared does the silica concentration begin to decrease. Since the surface area and rate constant for cristobalite are considerably greater than those of quartz, the fluid remains near equilibrium with cristobalite until it in turn nearly disappears. Finally, after several hundred thousand years of reaction, the fluid approaches saturation with quartz and hence thermodynamic equilibrium.

30.5 Dissolution of Albite

In a final application of kinetic reaction modeling, we consider how sodium feldspar (albite, $NaAlSi_3O_8$) might dissolve into a subsurface fluid at 70 °C. We consider a Na–Ca–Cl fluid initially in equilibrium with kaolinite [$Al_2Si_2O_5(OH)_4$], quartz, muscovite [$KAl_3Si_3O_{10}(OH)_2$, a proxy for illite], and calcite ($CaCO_3$), and in contact with a small amount of albite. Feldspar cannot be in equilibrium with quartz and kaolinite, since the minerals will react to form a mica or a mica-like clay mineral such as illite. Hence, the initial fluid is necessarily undersaturated with respect to the albite.

We assume that albite and quartz react with the fluid according to kinetic rate laws. We take a rate law for albite,

$$r_{alb} = A_S\, k_+ \left(1 - \frac{Q}{K}\right), \tag{30.9}$$

of the form discussed in Chapter 17 (Eq. 17.5). According to Knauss and Wolery (1986), this form is valid at 70 °C over a range in pH of 2.9 to 8.9. The corresponding rate constant k_+ at this temperature is 7.9×10^{-16} mol $cm^{-2}\ s^{-1}$. The law for quartz (Eq. 30.1) is the same one used in the previous examples in this chapter. The rate constant for quartz at 70 °C (Table 30.1) is 2.3×10^{-16} mol $cm^{-2}\ s^{-1}$. For both minerals we assume

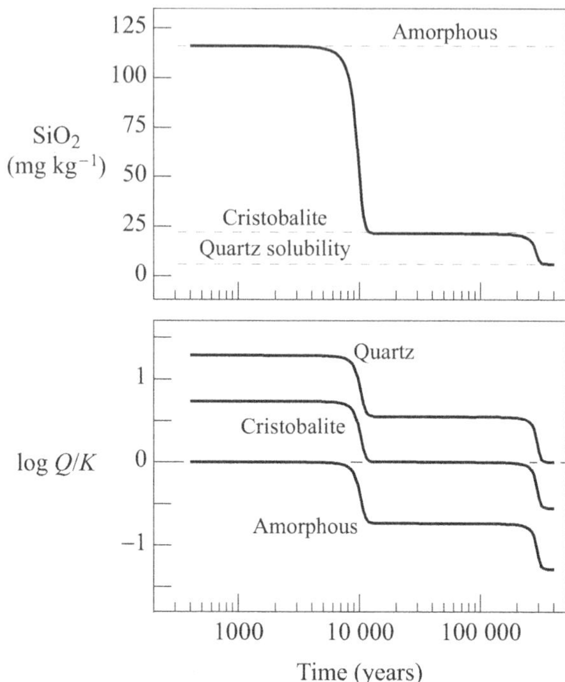

Figure 30.7 Variation in silica concentration (top) and saturation indices (log Q/K) of the silica polymorphs (bottom) over the course of the reaction path shown in Figure 30.6. The dashed lines in the top diagram show SiO_2(aq) concentrations in equilibrium with quartz, cristobalite, and amorphous silica.

a specific surface area of 1000 $cm^2\ g^{-1}$, which is reasonable for sand-sized grains. All other minerals in the system remain in equilibrium with the fluid over the reaction path.

The procedure in REACT is

```
time end = 1500 years
T = 70

pH = 5.7
Na+   = .3  molal
Ca++  = .05 molal
Cl-   = .4  molal

swap Kaolinite for Al+++
swap Quartz    for SiO2(aq)
swap Muscovite for K+
swap Calcite   for HCO3-
 10 free grams Kaolinite
100 free grams Quartz
 10 free grams Muscovite
  1 free gram  Calcite

react 25 grams Albite
```

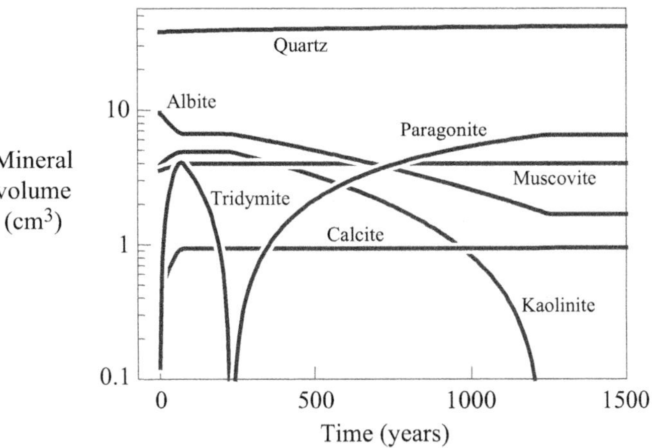

Figure 30.8 Mineralogical results of a reaction path in which albite dissolves and quartz precipitates at 70 °C according to kinetic rate laws.

```
kinetic Albite  rate_con = 7.9e-16  surface = 1000
kinetic Quartz  rate_con = 2.3e-16  surface = 1000

delxi = .001
go
```

The predicted reaction path (Fig. 30.8) is interesting because the albite twice achieves saturation with the fluid. In the initial part of the calculation, the albite dissolves,

$$\underset{\text{albite}}{NaAlSi_3O_8} + 0.4\,CO_2(aq) + 0.1\,HCO_3^- + 1/2\,Ca^{++} + 0.1\,K^+ + 0.9\,H_2O \rightarrow Na^+ + 0.4\underset{\text{kaolinite}}{Al_2Si_2O_5(OH)_4} + 0.1\underset{\text{muscovite}}{KAl_3Si_3O_{10}(OH)_2} + 1/2\underset{\text{calcite}}{CaCO_3} + 2\,SiO_2(aq)\,, \qquad (30.10)$$

to produce kaolinite, muscovite, and calcite and to add silica to the solution. Some of the silica precipitates to form quartz,

$$1/2\,SiO_2(aq) \rightarrow 1/2\underset{\text{quartz}}{SiO_2}\,, \qquad (30.11)$$

but the reaction rate is not sufficient to consume all of the silica produced by Reaction 30.10. The remaining silica accumulates in solution, quickly causing the silica polymorph tridymite to become saturated,

$$3/2\,SiO_2(aq) \rightarrow 3/2\underset{\text{tridymite}}{SiO_2}\,, \qquad (30.12)$$

and precipitate.

When tridymite has formed, the $SiO_2(aq)$ activity and hence the saturation state of quartz is fixed at a constant value (Fig. 30.9). Albite continues to dissolve for about 100 years until, according to the reaction

$$\underset{\text{albite}}{NaAlSi_3O_8} + 1/2\,H_2O + H^+ \rightleftarrows Na^+ + 1/2\underset{\text{kaolinite}}{Al_2Si_2O_5(OH)_4} + 2\underset{\text{tridymite}}{SiO_2}\,, \qquad (30.13)$$

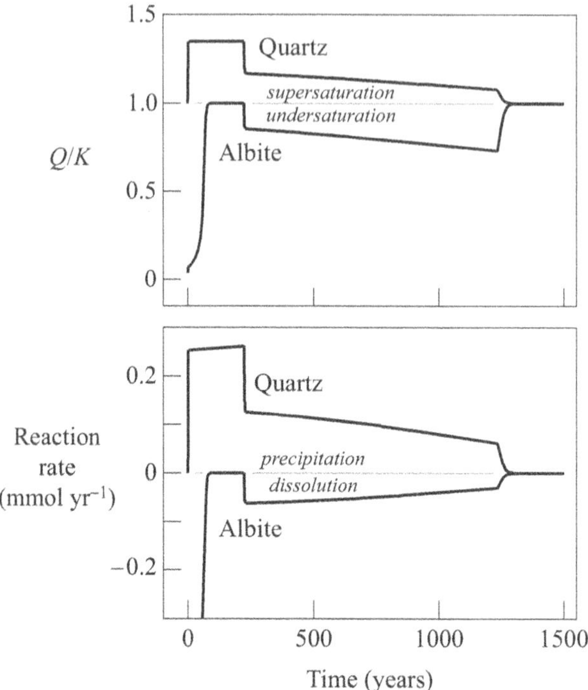

Figure 30.9 Variation in quartz and albite saturation (top) and the kinetic reaction rates for these minerals (bottom) over the course of the reaction path shown in Figure 30.8.

it becomes saturated in the fluid. Note that although albite and kaolinite cannot achieve equilibrium in the presence of quartz, they can coexist in a metastable equilibrium with tridymite.

From this point until about 225 years of reaction time, when the last tridymite disappears, the only reaction occurring in the system,

$$\underset{tridymite}{SiO_2} \rightarrow \underset{quartz}{SiO_2}, \tag{30.14}$$

is the conversion of tridymite to quartz. Once the conversion is complete, the silica activity decreases, causing the albite to become undersaturated once again (Fig. 30.9). It begins to dissolve,

$$\underset{albite}{NaAlSi_3O_8} + \underset{kaolinite}{Al_2Si_2O_5(OH)_4} \rightarrow \underset{paragonite}{NaAl_3Si_3O_{10}(OH)_2} + 2\ \underset{quartz}{SiO_2} + H_2O, \tag{30.15}$$

consuming kaolinite in the process while producing paragonite and quartz (Fig. 30.8). In this segment of the reaction path, the rate of quartz precipitation is sufficient to consume the silica produced as the albite dissolves. The reaction continues until all of the kaolinite in the system is consumed, after which point the albite and quartz quickly reach equilibrium with the fluid.

31

Weathering

Weathering is the interaction of the atmosphere on Earth with the exposed crust. It is an especially complex phenomenon, comprising the cumulative effect of a number of physical, chemical, and biological processes operating at and near the Earth's surface. Weathering plays a central role in the formation of soils, the origin of ores such as alumina, and the regulation of the CO_2 content of the atmosphere and hence climate change.

In this chapter, we build on applications in the previous chapter (Chapter 30), where we considered the kinetics of mineral dissolution and precipitation. Here, we construct simple reactive transport models of the chemical weathering of minerals, as it might occur in shallow aquifers and soils.

31.1 Rainwater Infiltration in an Aquifer

In a first reactive transport model (Bethke, 1997), we consider the reaction of silica as rainwater infiltrates an aquifer containing quartz (SiO_2) as the only mineral. Initially, groundwater is in equilibrium with the aquifer, giving an SiO_2(aq) concentration of 6 mg kg^{-1}. The rainwater contains only 1 mg kg^{-1} SiO_2(aq), so as it enters the aquifer, quartz there begins to dissolve,

$$\underset{quartz}{SiO_2} \rightarrow SiO_2(aq)\,. \tag{31.1}$$

As in the previous chapter (Section 30.1; Eq. 30.1), we use the rate law,

$$r = A_S\, k_+ \left(1 - \frac{Q}{K}\right), \tag{31.2}$$

to describe the dissolution reaction. Here, r is the reaction rate (mol s^{-1}), A_S and k_+ are the mineral's surface area (cm^2) and rate constant (mol cm^{-2} s^{-1}), and Q and K are the activity product and equilibrium constant for the dissolution reaction. We use a value for k_+ at 25 °C of 4.2×10^{-18} mol cm^{-2} s^{-1} from Rimstidt and Barnes (1980), and calculate A_S assuming a specific surface area A_{sp} for quartz of 1000 cm^2 g^{-1}, as given for sand grains by Leamnson *et al.* (1969).

We consider a portion of the aquifer extending 100 m from the recharge point, dividing this segment into 200 nodal blocks. The aquifer contains 70% quartz grains by volume, leaving a porosity of 30%. Groundwater in the aquifer is replaced with rainwater 100 times over the course of the simulation, which spans 100 years. Since the domain is 100 m long and the pore volume displaced once per year, flow velocity v_x in the model is 100 m yr^{-1}.

The procedure in X1T is

```
length = 100 m
```

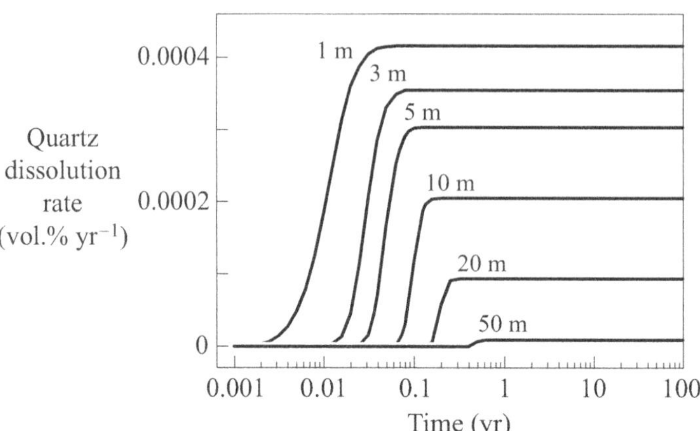

Figure 31.1 Rate at which quartz sand dissolves at 25 °C in an aquifer invaded by rainwater, plotted against time at various approximate distances from the recharge point. Beginning at $t = 0$, rainwater recharges the aquifer and flows along it at a velocity v_x of 100 m yr^{-1}. Reaction is most rapid near the recharge point and decreases along the direction of flow as SiO_2(aq) accumulates in the groundwater. Reaction rate and groundwater composition adjust to the infiltration once the rainwater arrives at a point along the aquifer, quickly approaching a near-steady state.

```
Nx = 200
interval start = 0 yr, fluid = rainwater
interval end   = 1 yr
discharge = 100 pore_volumes
dispersivity = 10 cm

scope initial
   swap Quartz for SiO2(aq)
   Quartz = 70 vol%
scope rainwater
   SiO2(aq)  = 1 mg/kg

kinetic Quartz rate_con = 4.2e-18, surface = 1000

dxplot log = 0.1
go
```

In order to observe details of how the model responds to the onset of flow, we have set in the final commands an output interval on a logarithmic scale, and a small initial time step.

As rainwater enters the aquifer and reaction begins, the dissolution rate, initially zero, quickly adjusts to an almost constant value (Fig. 31.1), as does silica concentration. The simulation end result is not a steady state, strictly speaking, because quartz is gradually dissolving, decreasing its surface area and causing the dissolution rate to diminish slowly with time. Quartz is only sparingly soluble, however, so it will take many thousands of years to dissolve to any significant extent; over shorter time intervals, its consumption is hardly perceptible. Lichtner (1988) introduced the term *stationary state* to refer to such a quasi-steady state.

We can see this point by evaluating the characteristic time required for chemical reaction to change the concentration of dissolved silica, and that required to affect the mass of quartz grains. These intervals are the *relaxation times* (s) for the groundwater and aquifer minerals (Lasaga and Rye, 1993; Lichtner, 1996). In

Chapter 24 (Eq. 24.11), we determined the former quantity as

$$\tau_{SiO_2(aq)} = \frac{C_{eq}}{(A_S/V)\,k_+}, \tag{31.3}$$

where C_{eq} is the equilibrium silica concentration, 0.1×10^{-6} mol cm^{-3}, and A_S/V is quartz surface area per volume of fluid (cm^2 cm^{-3}). Expressing this ratio in terms of the specific surface area A_{sp}, the relaxation time becomes

$$\tau_{SiO_2(aq)} = \frac{\phi}{1-\phi} \times \frac{C_{eq}}{\rho_{qtz}\,A_{sp}\,k_+}, \tag{31.4}$$

where ϕ is porosity and ρ_{qtz} is quartz density, 2.65 g cm^{-3}. By parallel logic, the relaxation time for minerals in the aquifer is given by

$$\tau_{qtz} = \frac{1}{A_{sp}\,M_W\,k_+}, \tag{31.5}$$

where M_W is the mole weight of quartz, 60.08 g mol^{-1}.

In our example, the relaxation time for the groundwater is 2.7×10^6 s, according to Equation 31.4, or about 0.1 yr. The relaxation time for the aquifer (Eq. 31.5), in contrast, is 4.0×10^{12} s, or 130 000 yr. The infiltrating fluid adjusts to reaction with the aquifer on a time scale six orders of magnitude shorter than the scale on which the aquifer adjusts to the fluid. Whereas we could observe the chemical evolution of the rainwater at any point along the aquifer within less than a month of its arrival, the aquifer grains dissolve at a rate imperceptible on the human time scale.

We can repeat the simulation for a range of flow velocities, by adjusting the time span of the simulation. For a simulation spanning 1 year, for example, we type the X1T commands

```
(cont'd)
interval end = 1 yr
go
```

This time span corresponds to a flow velocity of 10 000 m yr^{-1}, whereas in a 1000 year simulation, velocity is 10 m yr^{-1}.

Figure 31.2 shows the results calculated for velocities spanning this range. The simulations correspond to Damköhler numbers Da (see Chapter 24) much less than one for rapid flow, to much greater than one when flow is slow. At low Da, transport overwhelms reaction and the fluid across the domain remains fully undersaturated with respect to quartz. Since quartz saturation changes little with position, the dissolution rate remains nearly constant. At high Da, in contrast, reaction controls transport and the infiltrating rainwater reacts toward equilibrium before migrating far from the inlet. In this case, the reaction rate is nearly zero across most of the domain.

Only for the intermediate cases—those with velocities in the range of about 100 m yr^{-1} to 1000 m yr^{-1}—does silica concentration and reaction rate vary greatly across the main part of the domain. Significantly, only these cases benefit from the extra effort of calculating a reactive transport model. For more rapid flows, the same result is given by a lumped parameter simulation, or box model, as we could construct in REACT. And for slower flow, a local equilibrium model suffices.

Perhaps the most notable observation about these results is how poorly they reflect field observations. Even in clean orthoquartzite aquifers undergoing rapid recharge, groundwater below about 60 °C is not observed to approach equilibrium with quartz, or even necessarily follow a clear trend along the direction of flow. Instead, silica concentration is highly variable and commonly in excess of quartz saturation, rather than below it.

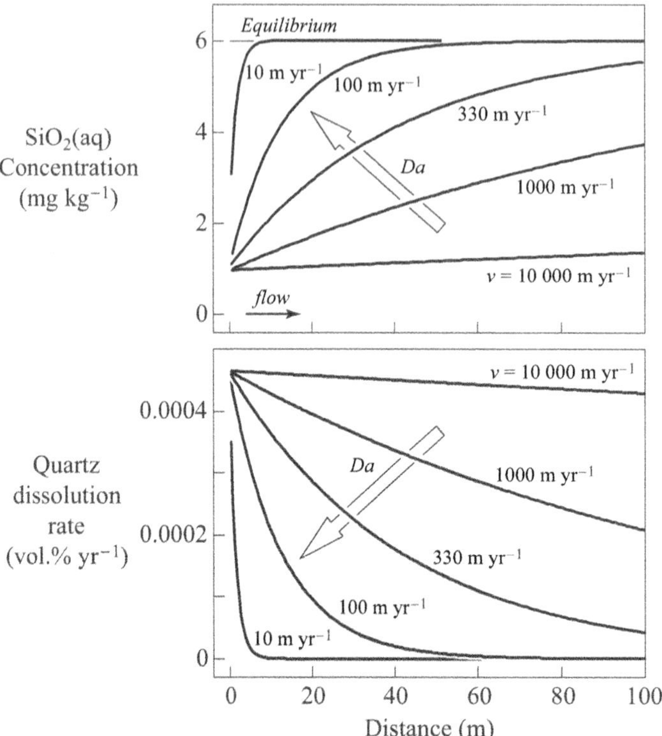

Figure 31.2 Concentration of dissolved silica and the quartz dissolution rate along a quartz sand aquifer being recharged at left by rainwater, for the scenario considered in Figure 31.1. Results were calculated assuming a range of flow velocities; rapid flow corresponds to a Damköhler number Da less than one, whereas Da is greater than one for slow flow.

A number of factors contribute to the disparity between the predictions of kinetic theory and conditions observed in the field, as discussed in Section 17.2. In this case, we might infer the dissolution and precipitation of minerals such as opal CT (cristobalite and tridymite, SiO_2), smectite and other clay minerals, and zeolites help control silica concentration. The minerals may be of minor significance in the aquifer volumetrically, but their high rate constants and specific surface areas allow them to react rapidly.

31.2 Weathering in a Soil

As a second example, we construct a simple model of how minerals might dissolve and precipitate as rainwater percolates through a soil (Bethke, 1997). The soil, 1 m thick, is composed initially of 50% quartz by volume, 5% potassium feldspar ($KSiAl_3O_8$), and 5% albite (sodium feldspar, $NaSiAl_3O_8$). The remaining 40% of the soil's volume is taken up by soil gas (15% of the bulk) and water (25%).

Rainwater recharges the top of the profile, and reacted water drains from the bottom. We take discharge through the soil to be 4 $m^3\ m^{-2}\ yr^{-1}$ and assume the dispersivity α_L (see Chapter 23) is 1 cm. The rainwater is dilute and in equilibrium with CO_2 partial pressure in the atmosphere, $10^{-3.5}$ bar. Within the soil, however, the soil gas is taken to contain additional CO_2 as a result of the decay of organic matter, and root respiration.

The pore fluid is assumed to maintain equilibrium with the soil gas, and CO_2 pressure within the soil is held constant over the simulation, at 10^{-2} bar.

Minerals in the soil can dissolve or, if they become supersaturated, precipitate according to the kinetic rate law in the previous section (Eq. 31.2). We take a rate constant of 4.2×10^{-18} mol cm^{-2} s^{-1} for quartz, as before, and of 30×10^{-18} mol cm^{-2} s^{-1} for potassium feldspar and 100×10^{-18} mol cm^{-2} s^{-1} for albite, from Blum and Stillings (1995). We assume a specific surface area of 1000 cm^2 g^{-1}, typical of sand-sized grains (Leamnson *et al.*, 1969), for each of the minerals.

Furthermore, we allow minerals kaolinite [$Al_2Si_2O_5(OH)_4$], gibbsite [$Al(OH)_3$], and tridymite (SiO_2, a proxy here for opal CT) to form in the simulation, according to the same rate law. We assign rate constants of 10×10^{-18} mol cm^{-2} s^{-1} for kaolinite and 50×10^{-18} mol cm^{-2} s^{-1} for gibbsite, from Nagy (1995), and of 6.3×10^{-18} mol cm^{-2} s^{-1} for tridymite, from Rimstidt and Barnes (1980). We further assume specific surface areas of 10^5 cm^2 g^{-1} for kaolinite (Carroll and Walther, 1990), 4000 cm^2 g^{-1} for gibbsite (Nagy and Lasaga, 1992), and, as above, 1000 cm^2 g^{-1} for tridymite.

These minerals cannot form directly, according to the kinetic law, since they are not present initially and hence have no surface area. We must provide a means for the minerals to nucleate in the simulation, most plausibly by growing on other surfaces in the soil. Lacking guidance in specifying the surface area available in a soil for such heterogeneous nucleation, we rather arbitrarily specify that each mineral can nucleate and grow on an area of 4000 cm^2 (cm fluid)$^{-3}$. For comparison, the internal surface areas of soils commonly fall in the range 20 000–300 000 cm^2 (cm bulk)$^{-3}$ (White, 1995). The nucleation area set in this way will exceed the actual surface areas of these minerals over the course of the simulation and hence control the rates at which the minerals form.

The procedure in X1T is

```
interval start =  0 years, fluid = rainwater
interval end   = 10 years
length = 1 m
Nx = 100
porosity = 25%
discharge = 4 m3/m2/yr
dispersivity = 1 cm

scope rainwater
   swap CO2(g) for HCO3-
   P CO2(g) = 10^-3.5 bar
   pH = 5
   Na+       =  .5 mg/kg
   Cl-       = 1.3 mg/kg
   K+        =  .2 mg/kg
   Al+++     = 1   ug/kg
   SiO2(aq)  =  .2 mg/kg
scope initial = rainwater
   P CO2(g) = 10^-2 bar

fix fugacity of CO2(g)
react  5 vol% K-feldspar
react  5 vol% Albite
react 50 vol% Quartz
```

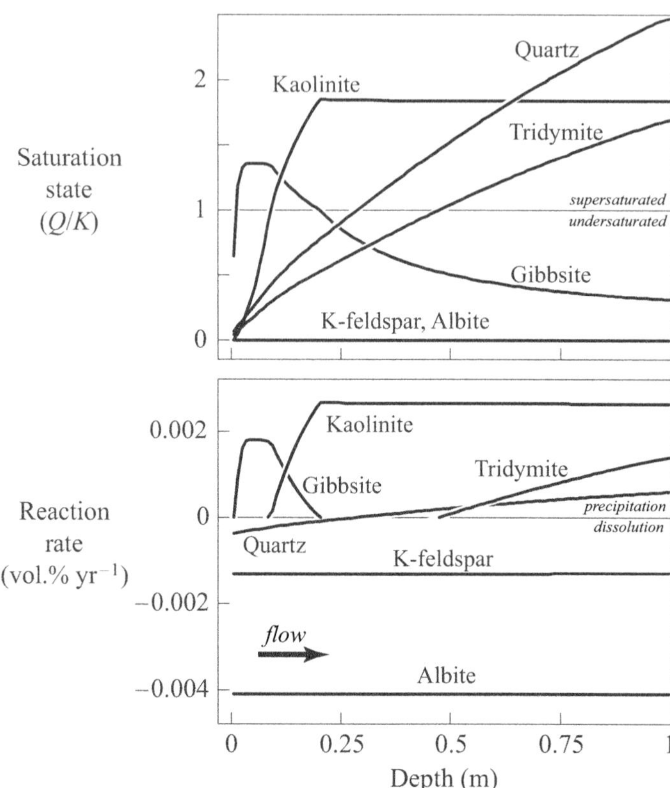

Figure 31.3 Saturation states (top) and reaction rates (bottom) for minerals in a simulation of weathering in a soil profile, at the calculation's stationary state. Rainwater in the simulation recharges the top of the profile (left side of plots) at 4 m yr^{-1}, and reacted fluid drains from the bottom (right).

```
kinetic K-feldspar  rate_con =  30e-18  surface = 1000
kinetic Albite      rate_con = 100e-18  surface = 1000
kinetic Quartz      rate_con = 4.2e-18  surface = 1000
kinetic Kaolinite   rate_con =  10e-18  surface =  1e5  nucleus = 4000
kinetic Gibbsite    rate_con =  50e-18  surface = 4000  nucleus = 4000
kinetic Tridymite   rate_con = 6.3e-18  surface = 1000  nucleus = 4000

precip = off
go
```

As in the previous simulation, the calculation results assume a stationary state (Fig. 31.3) at which the rates of mineral dissolution and precipitation at any point in the profile are nearly invariant in time.

In the simulation, CO_2 in the soil gas reacts with the feldspars, leading to the alkali leaching and separation of silica from alumina observed to result from soil weathering. Near the top of the profile, the reaction produces gibbsite and adds Na^+, K^+, and $SiO_2(aq)$ to the migrating pore fluid, according to the reactions

$$\underset{\textit{albite}}{NaAlSi_3O_8} + CO_2(aq) + 2\,H_2O \rightarrow \underset{\textit{gibbsite}}{Al(OH)_3} + Na^+ + 3\,SiO_2(aq) + HCO_3^- \qquad (31.6)$$

and

$$\underset{K\text{-feldspar}}{KAlSi_3O_8} + CO_2(aq) + 2\,H_2O \rightarrow \underset{gibbsite}{Al(OH)_3} + K^+ + 3\,SiO_2(aq) + HCO_3^- . \qquad (31.7)$$

At the same time, quartz at the top of the profile dissolves congruently,

$$\underset{quartz}{SiO_2} \rightarrow SiO_2(aq) , \qquad (31.8)$$

adding additional silica to solution.

As a result of these reactions, silica activity in the fluid increases with depth. Below about 10 cm, as a result, kaolinite becomes saturated and precipitates, and below about 25 cm the mineral replaces gibbsite as the sink for alumina. The reactions here are

$$\underset{albite}{NaAlSi_3O_8} + CO_2(aq) + 3/2\,H_2O \rightarrow 1/2\,\underset{kaolinite}{Al_2Si_2O_5(OH)_4} + Na^+ + 2\,SiO_2(aq) + HCO_3^- \qquad (31.9)$$

and

$$\underset{K\text{-feldspar}}{KAlSi_3O_8} + CO_2(aq) + 3/2\,H_2O \rightarrow 1/2\,\underset{kaolinite}{Al_2Si_2O_5(OH)_4} + K^+ + 2\,SiO_2(aq) + HCO_3^- . \qquad (31.10)$$

At about this depth, furthermore, quartz becomes supersaturated and starts to precipitate,

$$SiO_2(aq) \rightarrow \underset{quartz}{SiO_2} , \qquad (31.11)$$

rather than dissolve.

The rate of quartz precipitation, however, is insufficient to consume all the excess silica released by the conversion of feldspar to kaolinite, so silica continues to accumulate in the migrating fluid. At a depth of about 50 cm, tridymite, a proxy in the calculation for opal CT, becomes saturated and begins to form,

$$SiO_2(aq) \rightarrow \underset{tridymite}{SiO_2} . \qquad (31.12)$$

Near the bottom of the profile, more tridymite forms than quartz, reflecting the former mineral's larger rate constant and specific surface area.

The simulation predicts salient features of mineral weathering in a soil—alkali leaching and the separation of alumina and silica—but just as notable is its limited scope. Mineral surfaces in a real soil are likely to be occluded by other minerals, organic matter, and biofilm, but we have no basis for projecting specific surface areas measured in the laboratory to this setting. Treatment of the heterogeneous nucleation of secondary minerals, as noted, is of necessity ad hoc. We have taken no account of seasonal variation in temperature and infiltration. And, despite soils being the most biologically active zones on Earth, we have reduced biological activity, including microbial fermentation and respiration, root activity, and transpiration, to a single value of CO_2 partial pressure. So far, a full model of weathering within a soil eludes us.

32

Oxidation and Reduction

In the previous two chapters (Chapters 30 and 31), we showed how kinetic laws describing the rates at which minerals dissolve and precipitate can be integrated into reaction path and reactive transport simulations. The purpose of this chapter is to consider how we can trace the reaction paths that arise when redox reactions proceed according to kinetic rate laws.

We take two cases in which mineral surfaces catalyze oxidation or reduction, and one in which a consortium of microbes, modeled as if it were a simple enzyme, promotes a redox reaction. In Chapter 37, we treat the question of modeling the interaction of microbial populations with geochemical systems in a more general way.

32.1 Uranyl Reduction by Ferrous Iron

The reduction of uranyl (U^{VI}, or UO_2^{++}) is a reaction important to geochemists because it transforms oxidized uranium, an anthropogenic and naturally occurring contaminant that is highly mobile, to insoluble reduced forms. Liger *et al.* (1999) studied the role ferric oxides play at pH 6–7.5 and 25 °C in catalyzing the reduction of uranyl by ferrous iron. The study provides a good example of how surface complexation modeling can be used to develop a kinetic description of heterogeneous catalysis.

Liger *et al.* (1999) found the reaction

$$UO_2^{++} + 2\,Fe^{++} + 2\,H_2O \rightarrow U(OH)_4 + 2\,Fe^{+++} \tag{32.1}$$

proceeds in the presence of hematite (Fe_2O_3) nanoparticles, a proxy for particulate matter in natural waters, according to a rate law

$$r = n_w\, k_+\, m_{>FeO\cdot UO_2OH}\, m_{>FeO\cdot FeOH} \left(1 - \frac{Q}{K}\right), \tag{32.2}$$

where r is reaction rate (mol s^{-1}), n_w is water mass (kg), k_+ is the rate constant (molal^{-1} s^{-1}), and $m_{>FeO\cdot UO_2OH}$ and $m_{>FeO\cdot FeOH}$ are the molal concentrations of two surface complexes, one of uranyl and the other of ferrous iron. The law is first order with respect to the surface complexes, and the rate constant k_+ is 6.7 molal^{-1} s^{-1}. The Fe^{+++} ions produced probably precipitate on the hematite surface, or as a distinct oxide or oxy-hydroxide phase.

The rate law is based on a surface complexation model Liger *et al.* (1999) developed for the hematite nanoparticles (see Chapter 10, "Surface Complexation"). The >FeOH surface sites react by protonation and deprotonation to form $>FeOH_2^+$ and $>FeO^-$, by complexation with ferrous iron to form $>FeOFe^+$ and >FeOFeOH, and to make a complex $>FeOUO_2OH$ with uranyl. Table 32.1 shows the reactions and

Table 32.1 *Surface complexation reactions considered by Liger* et al. *(1999) in describing the kinetics of the catalytic oxidation of uranyl by ferrous iron*

Reaction	log K
(1) $>FeOH_2^+ \rightleftarrows >FeOH + H^+$	−8.08
(2) $>FeO^- + H^+ \rightleftarrows >FeOH$	8.82
(3) $>FeOFe^+ + H^+ \rightleftarrows >FeOH + Fe^{++}$	1.15
(4) $>FeOFeOH + 2\,H^+ \rightleftarrows >FeOH + Fe^{++} + H_2O$	10.05
(5) $>FeOUO_2OH + 2\,H^+ \rightleftarrows >FeOH + UO_2^{++} + H_2O$	4.65

corresponding log K values. The nanoparticles are taken to have a specific surface area of 109 $m^2\ g^{-1}$, and a site density of 0.06 per Fe_2O_3.

To see how we can use the surface complexation model to trace the kinetics of this reaction, we simulate an experiment conducted at pH 7.5 (Liger *et al.*, 1999, their figure 6). They started with a solution containing 100 mmolar $NaNO_3$, 0.16 mmolar $FeSO_4$, and 0.53 g l^{-1} of hematite nanoparticles. At t = 0, they added enough uranyl to give an initial concentration of 5×10^{-7} molar, almost all of which sorbed to the nanoparticles. They then observed how the mass of sorbed uranyl, which they recovered by $NaHCO_3$ extraction, varied with time.

To run the simulation, we save the surface complexation model to a dataset "FeOH_UO2.sdat", decouple the relevant redox reactions, set the system's initial composition, and define the rate law. The procedure in REACT is

```
time end 80 hours
read FeOH_UO2.sdat
decouple Iron, Uranium, Nitrogen

100 mmolal Na+
100 mmolal NO3-
.16 mmolal Fe++
.16 mmolal SO4--
5e-4 mmolal UO2++
sorbate include
1e-3 umolal U++++
swap Hematite for Fe+++
.53 free gram Hematite
pH = 7.5
fix pH
balance on NO3-

kinetic redox-1 rxn = "UO2++  + 2 H2O  + 2 Fe++  -> U(OH)4  + 2 Fe+++", \
  rate_con = 6.7, mpower(>FeOFeOH) = 1, mpower(>FeOUO2OH) = 1

delxi = .001
dxplot = 0
go
```

Figure 32.1 shows the experimental and calculation results.

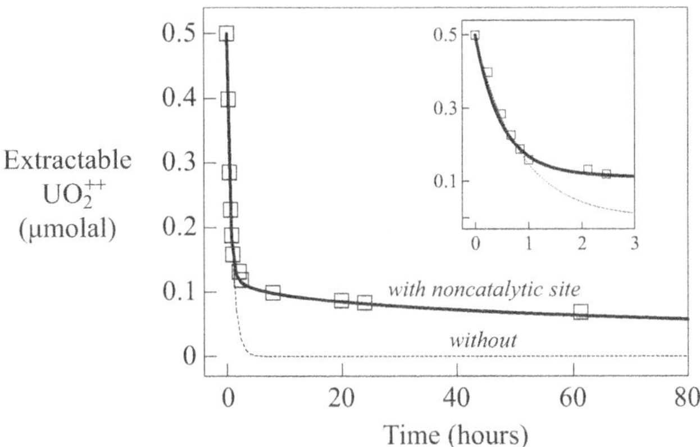

Figure 32.1 Results (symbols) and simulations (lines) of an experiment at 25 °C by Liger *et al.* (1999; their figure 6) in which uranyl was oxidized by ferrous iron in the presence of nanoparticulate hematite, which served as a catalyst. Vertical axis is amount of $NaHCO_3$-extractable uranyl, which includes uranyl present in solution as well as that sorbed to the nanoparticles; in the experiment, nearly all the uranyl was sorbed. Broken line shows results of a simulation assuming uranyl forms a single surface complex, $>FeOUO_2OH$, which is catalytically active; solid line shows simulation in which a noncatalytic site of this stoichiometry is also present. Inset is an expanded view of the first few hours of reaction.

Reaction in the simulation follows the trend observed in the experiment for about the first hour. After an hour, however, the rate law (Eq. 32.2) predicts the reaction will proceed to completion, consuming the uranyl in about 3 hours. In the experiment, in contrast, reaction proceeds more slowly and a considerable amount of uranyl remains unreacted after more than 60 hours.

This discrepancy might be explained if after about an hour the reaction approached equilibrium and slowed due to a diminishing thermodynamic drive. If the Fe^{+++} produced did not precipitate on the hematite surface, and did not form either hematite or goethite (FeOOH), it would accumulate in solution and weaken the drive for uranyl reduction. As the saturation index for hematite reached about 1.7, or about 1.25 for goethite, reaction would cease.

Liger *et al.* (1999) suggested another explanation that, in fact, better explains the pattern of their results. Whereas sorbed uranyl commonly shares an edge of its polyhedron with the hematite octahedron, they proposed that some of the uranyl shares only an apex. If the apical site were not catalytic, reaction would slow as the uranyl was depleted at the normal, but not the apical sites. For such a scenario to explain the data, the site would need to be less abundant than the normal sites, but sorb uranyl more strongly.

To simulate this effect, we extend the surface complexation model to include a noncatalytic site, denoted $>$(s)FeOH, at which uranyl sorbs strongly, forming a species $>$(s)$FeOUO_2OH$, according to the reaction in line 5 of Table 32.1. We can figure the site's density as about 3.5×10^{-5} per Fe_2O_3 from the amount of uranyl remaining at the end of the experiment and the mass of the hematite. The stability of the apical complex controls the convexity of the trend in uranyl concentration with time. Adjusting the stability, we find a log K for the complexation reaction of about −0.7. Finally, once these changes are made, we find we can better fit the data at early time using an adjusted rate constant k_+ of 11 $\text{molal}^{-1}\ \text{s}^{-1}$.

Saving the revised surface complexation model in dataset "FeOH_UO2s.sdat", the procedure is

```
(cont'd)
read FeOH_UO2s.sdat
kinetic redox-1 rate_con = 11
go
```

and the simulation results are shown in Figure 32.1.

32.2 Autocatalytic Oxidation of Manganese

An autocatalytic reaction is one promoted by its own reaction products. A good example in geochemistry is the oxidation and precipitation of dissolved Mn^{II} by $O_2(aq)$. The reaction is slow in solution, but is catalyzed by the precipitated surface and so proceeds increasingly rapidly as the oxidation product accumulates. Morgan (1967) studied in the laboratory the kinetics of this reaction at 25 °C and pH $\geq$ 9. He added an Mn^{II} solution to a stirred beaker containing a pH buffer, through which a mixture of oxygen and nitrogen gas was being bubbled.

The $O_2(aq)$ reacted with the Mn^{II}, forming a colloidal precipitate of oxidized manganese, likely composed of a mixture of Mn^{III} and Mn^{IV}. We can write this reaction in a simple form as

$$Mn^{++} + 1/4\, O_2(aq) + 5/2\, H_2O \rightarrow Mn(OH)_3(s) + 2\, H^+ \,, \tag{32.3}$$

where $Mn(OH)_3(s)$ represents the colloidal product. As pH increases above 9, the Mn^{++} ions in the reaction, in fact, convert progressively to various hydroxy species, such as $Mn_2(OH)_3^+$. Brewer (1975) proposed the rate law

$$r^+ = n_w\, k_+\, m_{Mn^{II}}\, m_{Mn(OH)(s)}\, m_{O_2(aq)}\, m^2_{OH^-} \,, \tag{32.4}$$

to describe Morgan's (1967) experiments. Here, $m_{Mn^{II}}$ is the sum of the concentrations of the Mn^{II} species in solution, and $m_{Mn(OH)(s)}$ is the catalyst mass per kg of solvent water. By this equation, the rate of the reaction increases as it proceeds, in response to accumulation of the catalyst.

As we start to model the reaction, we note the LLNL thermodynamic dataset "thermo.tdat" does not contain a redox couple linking Mn^{III} and Mn^{II}, as required for our purposes, but this is easily remedied by adding to the database the coupling reaction

$$Mn^{+++} + 1/2\, H_2O \rightleftarrows 1/4\, O_2(aq) + H^+ + Mn^{++} \,, \tag{32.5}$$

with a log K of 4.08 at 25 °C. We then rebalance the reactions for the Mn^{III} minerals in the database in terms of Mn^{+++} and save the result as "thermo+MnIII.tdat". The Mn^{III} minerals are: bixbyite (Mn_2O_3), hausmannite (Mn_3O_4), manganite (MnOOH), and $Mn(OH)_3$(c); we will use only the latter.

A second complication to modeling the reaction is that it is necessary to allow oxidation at the start of the experiment, so the catalyst can begin to form. The nature of the initial reaction is not known, but it may be promoted by small amounts of colloids or particles, perhaps $MnCO_3$, present at the onset. A simple strategy, in light of our lack of knowledge, is to set in the initial system a small amount of $Mn(OH)_3(s)$ to represent the system's initial capacity to catalyze the reaction.

We take the results of a series of experiments conducted by Morgan (1967, his figure 23) at pH 9, 9.3, and 9.5. He used an initial Mn^{II} concentration of 4.5×10^{-4} molar, a carbonate concentration of 1.6×10^{-3} molar, and an oxygen partial pressure of 1 atm. We can figure an approximate value for the rate constant k_+ from oxidation rate at the end of the experiment, when the mass of catalyst is known from the depletion of Mn^{II}, then estimate the initial catalyst mass from that value and the oxidation rate at the onset of reaction. Running the simulation, we can refine the two numbers until prediction matches observation.

The procedure in REACT to model the experiment at pH 9.5 is

```
read thermo+MnIII.tdat
time end 180 minutes
decouple Manganese
suppress ALL
unsuppress Mn(OH)3(c)

pH 9.5
swap O2(g) for O2(aq)
P O2(g) = 1 atm

.45 mmolal Mn++
swap Mn(OH)3(c) for Mn+++
2 free mg Mn(OH)3(c)
.45 mmolal Cl-

1.6 mmolal HCO3-
1.6 mmolal Na+
balance on Na+

kinetic redox-1 \
  rxn = "Mn++ + 1/4 O2(aq) + 5/2 H2O -> Mn(OH)3(c) + 2 H+", \
  rate_con = 3e12, \
  rate_law = 'Wmass * rate_con * molality("Mn(OH)3(c)") * \
    totmolal("Mn++") * molality("O2(aq)") * molality("OH-")^2'

fix pH
go
```

Here, we take the initial catalyst mass as 2 mg, and set a rate constant of $3{\times}10^{12}$ molal^{-4} s^{-1}, which gives good results. As shown in Figure 32.2, the reaction starts slowly, increases in rate as the catalyst accumulates, then decreases to zero as the supply of reduced manganese is depleted.

To see the autocatalytic nature of the reaction, we can compare the simulation to one made assuming a constant amount of catalyst. Taking $m_{\mathrm{Mn(OH)(s)}}$ to be 4.5×10^{-4} molal, its value when oxidation in the previous run is complete, the procedure is

```
(cont'd)
kinetic redox-1 \
  rate_law = 'Wmass * rate_con * 4.5e-4 * \
    totmolal("Mn++") * molality("O2(aq)") * molality("OH-")^2'
go
```

In the result for simple catalysis, shown in Figure 32.2, reaction is most rapid at the onset of the experiment and decreases with time, in contrast to the autocatalytic simulation.

We can model the experiments at pH 9.3 and 9.0 in a similar way. Returning to the original rate law, the procedure

```
(cont'd)
kinetic redox-1 \
```

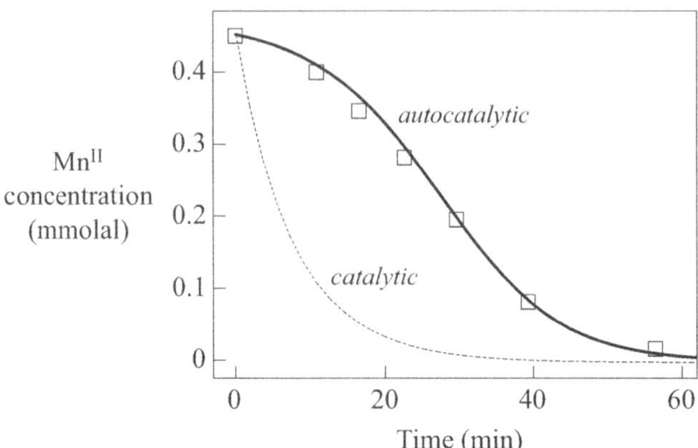

Figure 32.2 Variation in the concentration of Mn^{II} versus time in a simulation of the autocatalytic oxidation of manganese at pH 9.5 and 25 °C. Squares are results of a laboratory experiment by Morgan (1967). Broken line shows the result of a simulation of catalytic oxidation, assuming the catalyzing surface has a constant area.

```
  rate_law = 'Wmass * rate_con * molality("Mn(OH)3(c)") * \
    totmolal("Mn++") * molality("O2(aq)") * molality("OH-")^2'

pH 9.3
kinetic redox-1 rate_con = 4.8e12
go

pH 9.0
kinetic redox-1 rate_con = 4e12
go
```

repeats the simulation at the pH values of interest. We find good fits using the same value as before for initial catalyst mass, and somewhat differing rate constants of 4.8×10^{12} molal^{-4} s^{-1} for the pH 9.3 experiment, and 4×10^{12} molal^{-4} s^{-1} at pH 9.

The oxidation rate should, in principle, be described by a law using a rate constant independent of pH, as long as a single reaction mechanism is involved. The rate law (32.4) is unusual in that the rate varies with the concentration of the Mn^{II} component, rather than an individual species. If we hypothesize that the catalytic activity is promoted by a surface complex >MnOMnOH, a slightly different form of the rate law may be appropriate. Since the surface complex would form according to

$$>\text{MnOH} + \text{Mn}^{++} + 2\,\text{OH}^- \rightleftarrows\ >\text{MnOMnOH} + \text{H}_2\text{O}\,, \tag{32.6}$$

we might expect to find a rate law,

$$r^+ = n_w\, k_+\, m_{\text{Mn}^{++}}\, m_{\text{Mn(OH)(s)}}\, m_{\text{O}_2\text{(aq)}}\, m^2_{\text{OH}^-}\,, \tag{32.7}$$

that is first order in the concentration of the Mn^{++} free species, and second order with respect to OH^-. We can repeat our simulations

(cont'd)

```
kinetic redox-1 \
  rate_con = 1e13, \
```

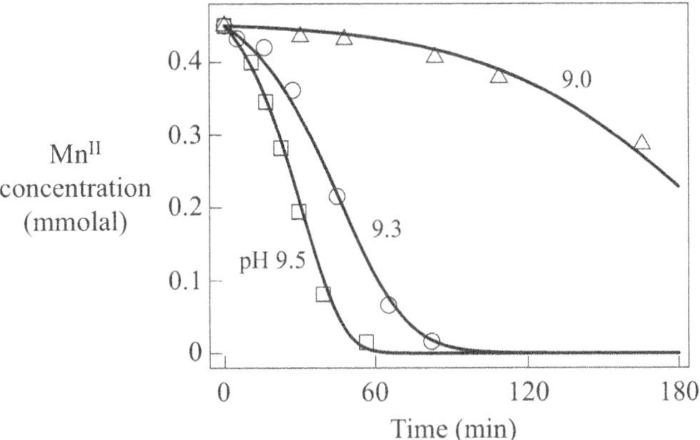

Figure 32.3 Concentration of Mn^{II} versus time in simulations of the autocatalytic oxidation of manganese at pH 9.0, 9.3, and 9.5, at 25 °C, compared to results of laboratory experiments (symbols) by Morgan (1967). Simulations made assuming rate law of a form carrying $m_{Mn^{++}}$, rather than $m_{Mn^{II}}$. Rate constant in the simulations is taken to be 10^{13} molal^{-4} s^{-1}, and the initial catalyst mass is 0.6 mg (pH 9.0), 5 mg (9.3), and 6 mg (9.5).

```
  rate_law = 'Wmass * rate_con * molality("Mn(OH)3(c)") * \
    molality("Mn++") * molality("O2(aq)") * molality("OH-")^2'

pH 9.5
6 free mg Mn(OH)3(c)
go

pH 9.3
5 free mg Mn(OH)3(c)
go

pH 9.0
0.6 free mg Mn(OH)3(c)
go
```

In this case, we find we can match the experimental results using a single value for the rate constant, but an initial catalyst mass that depends on pH, as shown in Figure 32.3. Here, we take the rate constant to be 10^{13} molal^{-4} s^{-1}, and use initial catalyst masses of 6 mg at pH 9.5, 5 mg at pH 9.3, and 0.6 mg at pH 9. This form of the rate law seems more satisfactory than the previous form, since we might reasonably expect the amount of catalyst, perhaps an $MnCO_3$ colloid, initially present to be greatest in the experiments conducted at highest pH, but the rate constant to be invariant.

32.3 Microbial Degradation of Phenol

Bekins *et al.* (1998) followed the concentration of phenol, $C_6H_5OH(aq)$, in a laboratory experiment as it was degraded over time within a microcosm containing a consortium of methanogens. The degradation can be represented by a reaction

$$\underset{phenol}{C_6H_5OH(aq)} + 13/2\ H_2O \rightarrow 7/2\ \underset{methane}{CH_4(aq)} + 5/2\ HCO_3^- + 5/2\ H^+ , \tag{32.8}$$

in which phenol dismutates to methane and bicarbonate.

The experiment contained sufficient nutrients and phenol was present at sufficiently low levels, about 40 mg kg^{-1} initially, that the substrate was the rate-limiting reactant. Methanogenic consortia are slow growing when observed on the time scale of the experiment, which lasted about 40 days. The biomass concentration, in fact, was not observed to change significantly. The reaction, furthermore, remained far from equilibrium, so the reverse reaction rate was negligible, compared to the forward reaction.

We can, therefore, reasonably expect the degradation to behave as an enzymatically catalyzed reaction in which phenol is the substrate and the microbial consortium serves as the enzyme. As discussed in Chapter 18, reaction rate in this case can be represented as

$$r^+ = n_w\, r_{max} \frac{m_{C_6H_5OH(aq)}}{m_{C_6H_5OH(aq)} + K_A}, \tag{32.9}$$

using the Michaelis–Menten equation (Eq. 18.18).

Since the enzyme concentration was not observed separately from the rate constant, we carry the product $k_+\ m_E$ in this equation as r_{max}, the maximum reaction rate. Bekins *et al.* (1998) fitted their results using values of 1.4 mg kg^{-1} day^{-1} (or 1.7×10^{-10} molal s^{-1}) for r_{max}, and 1.7 mg kg^{-1} (1.8×10^{-5} molal) for K_A, the half-saturation constant. In a field application lasting many years, of course, the assumption that enzyme concentration remains constant might not be valid, and we might need to apply a model accounting for growth of the methanogenic consortium, as discussed in Chapter 37.

The default thermodynamic dataset does not contain an entry for phenol, so to trace the reaction we need to add the reaction

$$\underset{phenol}{C_6H_5OH(aq)} + 7\,O_2(aq) + 3\,H_2O \rightleftarrows 6\,H^+ + 6\,HCO_3^- , \tag{32.10}$$

which has a log K at 25 °C of 503.4, to the database as a redox coupling reaction. We save the modified version of "thermo.tdat" as dataset "thermo+Phenol.tdat".

The REACT procedure

```
read thermo+Phenol.tdat
time end 42 days
decouple Phenol(aq), CH4(aq)

swap CO2(g) for HCO3-
Na+          = 10    mmolal
Cl-          = 10    mmolal
P CO2(g)     = 10^-3.5 bar
Phenol(aq)   = 38.7 mg/kg
CH4(aq)      = .01  mg/kg
pH = 7

kinetic redox-1 rxn = "Phenol(aq) + 13/2 H2O -> \
    7/2 CH4(aq) + 5/2 HCO3- + 5/2 H+" \
  rate_con = 1.7e-10, enzyme = on, mE = 1, KA = 1.8e-5
go
```

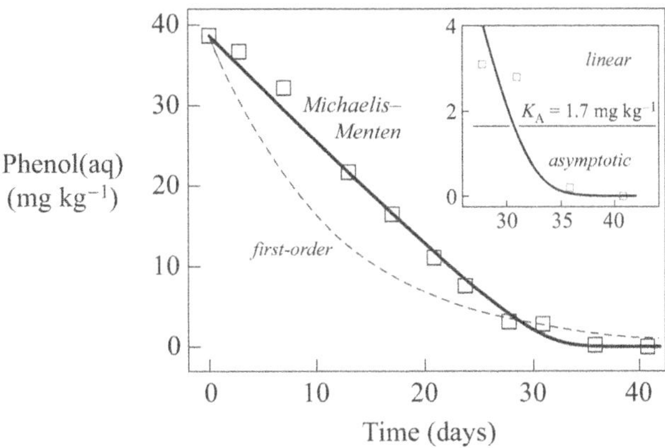

Figure 32.4 Degradation of phenol by a consortium of methanogens, as observed in a laboratory experiment by Bekins *et al.* (1998; symbols), and modeled using the Michaelis–Menten equation (solid line). Inset shows detail of transition from linear or zero-order trend at concentrations greater than K_A, to asymptotic, first-order kinetics below this level. Broken line is result of assuming a first-order rather than Michaelis–Menten law.

traces progress with time of the degradation reaction, according to the Michaelis–Menten equation. Note that we have set enzyme concentration to one, so the rate constant in the model serves to carry the value of r_{max}.

In the calculation results, shown in Figure 32.4, phenol concentration decreases with time at a constant rate for about the first 30 days of reaction. Over this interval, the concentration is greater than the value of K_A, the half-saturation constant, so the ratio $m/(m + K_A)$ in Equation 32.9 remains approximately constant, giving a zero-order reaction rate. Past this point, however, concentration falls below K_A and the reaction rate becomes first order. Now, phenol concentration does not decrease linearly, but asymptotically approaches zero.

We might ask what would happen if instead of taking degradation to be enzymatically catalyzed, we instead represent it as a first-order decay reaction, as is common practice in environmental hydrology. The steps

(cont'd)
```
kinetic redox-1 enzyme = off, rate_con = 1e-6, mpower(Phenol(aq)) = 1
go
```

set and evaluate a first-order law, using a value for the rate constant that reflects the decay rate over the full course of the experiment. As we can see in the calculation results (Fig. 32.4), a first-order law cannot explain the pattern of the results, since reaction in the experiment's initial leg was limited by the size of the microbial consortium, rather than the supply of phenol.

33

Waste Injection Wells

Increasingly since the 1930s, various industries around the world that generate large volumes of liquid byproducts have disposed of their wastes by injecting them into the subsurface of sedimentary basins. In the United States, according to a 1985 survey by Brower *et al.* (1989), 411 "Class I" wells were licensed to inject hazardous and nonhazardous waste into deep strata, and 48 more were proposed or under construction. Legal restrictions on the practice vary geographically, as does the suitability of geologic conditions. Nonetheless, the practice of deep-well injection had increased over time, partly in response to environmental laws that emphasize protection of surface water and shallow groundwater. More restrictive regulations introduced in the late 1980s and 1990s have begun to cause a decrease in the number of operating Class I wells.

Some injected wastes are persistent health hazards that need to be isolated from the biosphere indefinitely. For this reason, and because of the environmental and operational problems posed by loss of permeability or formation caving, well operators seek to avoid deterioration of the formation accepting the wastes and its confining layers. When wastes are injected, they are commonly far from chemical equilibrium with the minerals in the formation and, therefore, can be expected to react extensively with them (Boulding, 1990). The potential for subsurface damage by chemical reaction, nonetheless, has seldom been considered in the design of injection wells.

According to Brower *et al.* (1989; Fig. 33.1), nine wells at seven industrial sites throughout the state of Illinois were in use in the late 1980s for injecting industrial wastes into deeply buried formations; these wells accepted about 300 million gallons of liquid wastes per year. In this chapter, we look at difficulties stemming from reaction between waste water and rocks of the host formation at several of these wells and consider how geochemical modeling might be used to help predict deterioration and prevent blowouts.

33.1 Caustic Waste Injected in Dolomite

Velsicol Chemical Corporation maintained two injection wells at its plant near Marshall, Illinois, to dispose of caustic wastes from pesticide production as well as contaminated surface runoff. In September 1965, the company began to inject the wastes into Devonian dolomites of the Grand Tower formation at a depth of about 2600 feet. The wells accepted about 6 million gallons of waste monthly.

The waste contained about 3.5% dissolved solids, 1.7% chlorides, 0.4% sodium hydroxide, and tens to hundreds of parts per million of chlorinated hydrocarbons and chlordane; its pH was generally greater than 13 (Brower *et al.*, 1989). At the time of drilling, analysis of formation samples indicated that the injection zone was composed of nearly pure dolomite [$CaMg(CO_3)_2$]. The carbonate formation was thought to be safe for accepting an alkaline waste water because carbonates are considered stable at high pH.

With time, however, the company encountered problems, including caving of the formation into the

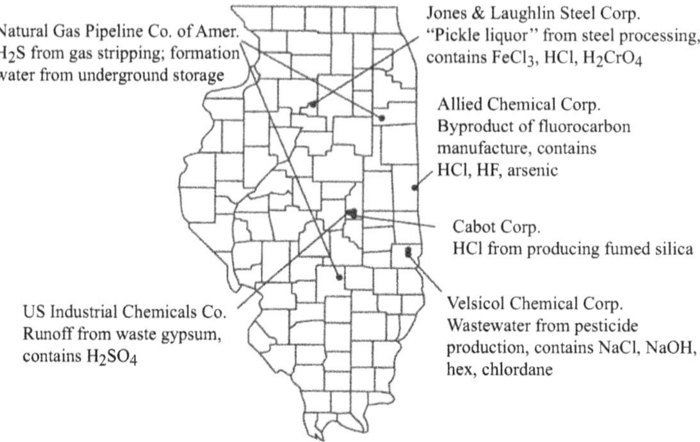

Figure 33.1 Locations of "Class I" injection wells in Illinois, from Brower *et al.* (1989).

wellbore and the loss of permeability in zones that had accepted fluid. In June 1987, a number of sidewall cores were taken from the formation (Mehnert *et al.*, 1990). Mineralogic analysis by X-ray diffraction showed that significant amounts of calcite ($CaCO_3$) and brucite [$Mg(OH)_2$], as well as some amorphous matter, had formed from the original dolomite. In some samples, the dolomite was completely consumed and the rock was found to be composed entirely of a mixture of brucite and calcite.

The plant eventually closed for environmental reasons, including surface contamination unrelated to the injection wells, causing a loss of jobs in an economically depressed area. Could geochemical modeling techniques have predicted the wells' deterioration? Using REACT, we trace the irreversible reaction of dolomite into the NaOH–NaCl waste. To calculate the waste's initial state and then titrate dolomite into it, we enter the commands

```
T = 35
pH = 13
balance on Na+
Cl-   =  .5  molal
Ca++  = 1    mg/kg
HCO3- = 1    mg/kg
Mg++  =  .01 ug/kg

react 40 grams Dolomite
go
```

The fluid contains arbitrarily small amounts of Ca^{++}, Mg^{++}, and HCO_3^-, as is necessary in order for the program to be able to recognize dolomite. The initial magnesium content is set small to assure that the hydroxide mineral brucite is not supersaturated in the alkaline fluid.

Figure 33.2 shows the mineralogic results of the calculation. Dolomite dissolves, since it is quite undersaturated in the waste fluid. The dissolution adds calcium, magnesium, and carbonate to solution. Calcite and brucite precipitate from these components, as observations from the wells indicated. The fluid reaches equilibrium with dolomite after about 11.6 cm^3 of dolomite have dissolved per kg water. About 11 cm^3 of calcite and brucite form during the reaction. Since calculation predicts a net decrease in mineral volume, damage to

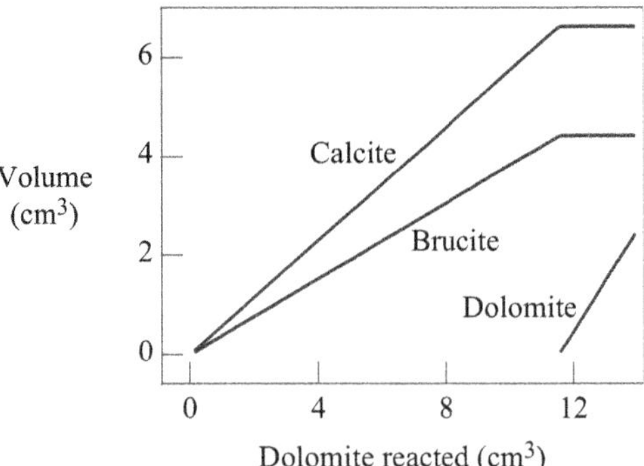

Figure 33.2 Mineralogic results of reacting at 35 °C alkaline waste water from the Velsicol plant with dolomite.

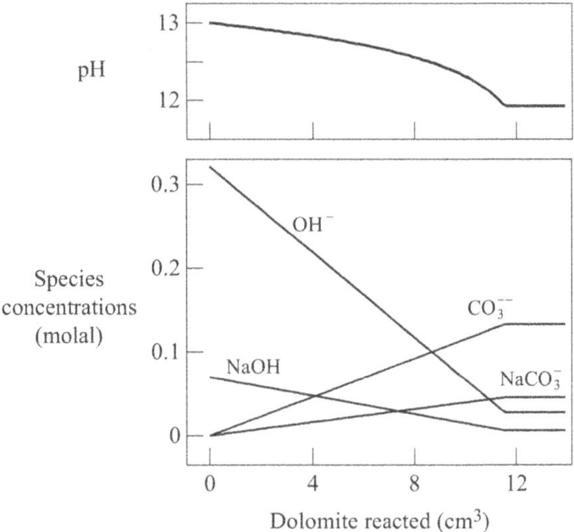

Figure 33.3 Variation in pH and species concentrations during reaction at 35 °C of alkaline waste water with dolomite.

the dolomite formation is evidently not due to loss of porosity. Instead, the reaction products likely formed a fine-grained, poorly coherent material capable of clogging pore openings and slumping into the wellbore.

At the point of dolomite saturation, where reaction ceases, pH has decreased from 13 to about 11.9 (Fig. 33.3). The overall reaction between the fluid and dolomite,

$$\underset{dolomite}{CaMg(CO_3)_2} + 2\,OH^- \rightarrow \underset{calcite}{CaCO_3} + \underset{brucite}{Mg(OH)_2} + CO_3^{--}\,, \tag{33.1}$$

is given by the slopes-of-the-lines method, as discussed in Section 14.2. For simplicity, we have lumped the ion pairs NaOH and $NaCO_3^-$ with the ions OH^- and CO_3^{--} in writing the reaction.

An interesting further experiment is to test the effects of letting the waste fluid equilibrate with the atmosphere before it is injected. Because the waste is so alkaline, its calculated CO_2 partial pressure is $10^{-13.5}$ bar, about ten orders of magnitude less than the atmospheric value of $10^{-3.5}$ bar. To simulate the effects of leaving the waste in a lagoon long enough to equilibrate with atmospheric CO_2, we trace two reaction steps, which, for simplicity, we assume to occur at the same temperature.

(cont'd)
```
remove reactant Dolomite
slide log f CO2(g) to -3.5
go

pickup fluid
react 1 mg Dolomite
go
```

The first step adjusts the fluid's CO_2 fugacity to the atmospheric value, reducing pH to less than 10 by the reaction

$$CO_2(g) + H_2O \rightarrow CO_3^{--} + 2\,H^+ \,. \tag{33.2}$$

After this step, dolomite appears only slightly undersaturated in the calculation results. In the second step, which simulates reaction of the formation into the equilibrated fluid, only about 2×10^{-5} cm^3 of dolomite dissolve per kilogram of waste water. This result suggests that the plant's waste stream could have been neutralized inexpensively by aeration.

33.2 Gas Blowouts

According to the Illinois Environmental Protection Agency (IEPA), a series of gas blowouts has occurred at two waste injection wells in the state (Brower *et al.*, 1989). In each case, well operators were injecting concentrated hydrochloric acid into a dolomite bed. At its plant near Tuscola, the Cabot Corporation injects acid waste from the production of fumed silica into the Cambrian Eminence and Potosi Formations below 5000 ft (1500 m) depth. Allied Chemical Corporation injects acid into the Potosi formation below about 3600 ft (1100 m). The acid, which is contaminated with arsenic, is a byproduct of the manufacture of refrigerant gas. Since some of the blowouts have caused damage such as fish kills, there is environmental interest as well as operational concern in preventing such accidents.

The blowouts seem to have occurred at times when especially acidic wastes were being injected. The acid apparently reacted with carbonate in the formations to produce a CO_2 gas cap at high pressure. In some cases, the injected waste was more than 31 wt.% HCl. As a temporary measure, the plants are now limited by the IEPA to injecting wastes containing no more than 6 wt.% HCl.

We can use reaction modeling techniques to test the conditions under which dolomite will react with hydrochloric acid to produce gas in the injection zones. Equivalent values of wt.% and molality for HCl are:

Wt.% HCl	molal HCl
2	0.57
4	1.2
6	1.8
10	3.1
20	6.9
30	11.9

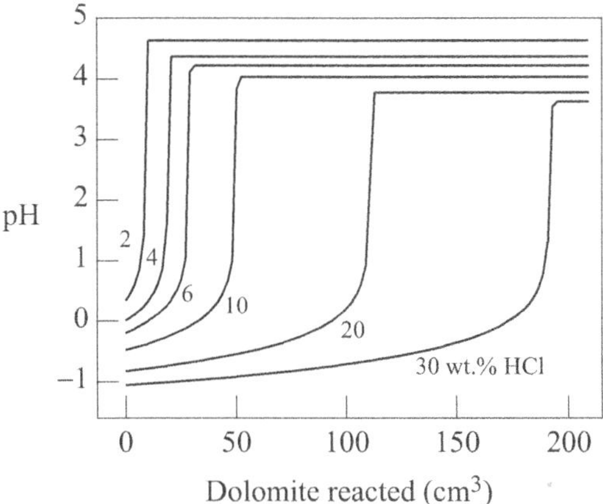

Figure 33.4 Calculated variation in pH as HCl solutions of differing initial concentrations react at 50 °C with dolomite.

To configure a model of the reaction of dolomite with 30 wt.% hydrochloric acid, we start REACT and enter the commands

```
T = 50
H+    = 11.9  molal
Cl-   = 11.9  molal
Na+   =   .01 molal
Ca++  =  1    mg/kg
HCO3- =  1    mg/kg
Mg++  =  1    mg/kg

react 600 grams Dolomite
go
```

We then repeat the calculation for solutions of differing HCl contents, according to the chart above.

Reacting dolomite into the waste water increases the pH as well as the partial pressure of CO_2. The predicted reaction is

$$\underset{dolomite}{CaMg(CO_3)_2} + 4\,H^+ \rightarrow 2\,CO_2(aq) + Ca^{++} + Mg^{++} + 2\,H_2O\,. \tag{33.3}$$

Here, we lump the ion pairs $CaCl^+$ and $MgCl^+$ with the free Ca^{++} and Mg^{++} ions. When the CO_2 pressure exceeds the confining pressure in the formation, CO_2 exsolves as a free gas,

$$CO_2(aq) \rightarrow CO_2(g)\,, \tag{33.4}$$

producing a gas cap in the subsurface.

Figure 33.4 shows how pH changes as hydrochloric acid in differing concentrations reacts at 50 °C with dolomite, and Figure 33.5 shows how CO_2 partial pressure varies. Solutions of greater HCl contents dissolve more dolomite and, hence, produce more CO_2. At 30 wt.%, for example, each kilogram of waste water consumes almost 200 cm^3 of dolomite. At an injection rate of 25 000 kg day^{-1} at this concentration, about 200 000 m^3 of dolomite would dissolve each year.

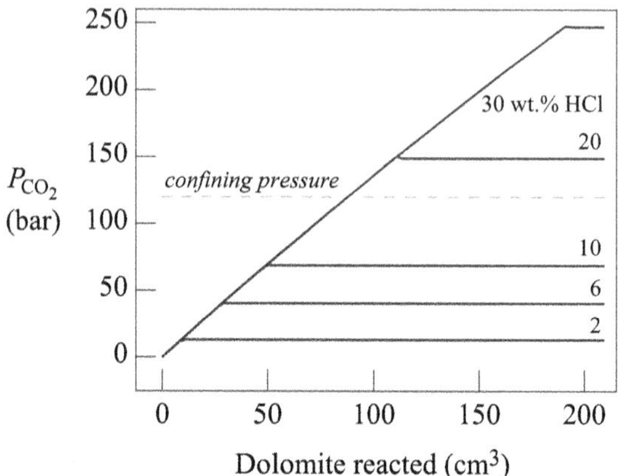

Figure 33.5 Calculated variation in CO_2 partial pressure as HCl solutions of differing initial concentrations react at 50 °C with dolomite.

The most acidic solutions, as expected, produce the greatest CO_2 pressures. For the 30 wt.% fluid, the partial pressure of CO_2 escaping from the fluid would approach 250 atm. Assuming a confining pressure of about 120 atm at the Allied well, solutions containing more than 15 wt.% HCl are likely to exsolve CO_2. The calculations indicate, on the other hand, that even the most acidic waste can be injected without fear of a gas blowout if it is first diluted by an equal amount of water.

34

Petroleum Reservoirs

In efforts to increase and extend production from oil and gas fields, as well as to keep wells operational, petroleum engineers pump a wide variety of fluids into the subsurface. Fluids are injected into petroleum reservoirs for a number of purposes, including:

- Waterflooding, where an available fresh or saline water is injected into the reservoir to displace oil toward producing wells.
- Improved Oil Recovery (IOR), where a range of more exotic fluids such as steam (hot water), caustic solutions, carbon dioxide, foams, polymers, surfactants, and so on are injected to improve recovery beyond what might be obtained by waterflooding alone.
- Near-well treatments, in which chemicals are injected into producing and sometimes injector wells, where they are intended to react with the reservoir rock. Well stimulation techniques such as acidization, for example, are intended to increase the formation's permeability. Alternatively, producing wells may receive "squeeze treatments" in which a mineral scale inhibitor is injected into the formation. In this case, the treatment is designed so that the inhibitor sorbs onto mineral surfaces, where it can gradually desorb into the formation water during production.
- Pressure management, where fluid is injected into oil fields in order to maintain adequate fluid pressure in reservoir rocks. Calcium carbonate may precipitate as mineral scale, for example, if pressure is allowed to deteriorate, especially in fields where formation fluids are rich in Ca^{++} and HCO_3^-, and CO_2 pressure is high.

In each of these procedures, the injected fluid can be expected to be far from equilibrium with sediments and formation waters. As such, it is likely to react extensively once it enters the formation, causing some minerals to dissolve and others to precipitate. Hutcheon (1984) appropriately refers to this process as "artificial diagenesis," drawing an analogy with the role of groundwater flow in the diagenesis of natural sediments (see Chapter 29). Further reaction is likely if the injected fluid breaks through to producing wells and mixes there with formation waters.

There is considerable potential, therefore, for mineral scale, such as barium sulfate (see the next section), to form during these procedures. The scale may be deposited in the formation, in the wellbore, or in production tubing. Scale that forms in the formation near wells, known as "formation damage," can dramatically lower permeability and throttle production. When it forms in the wellbore and production tubing, mineral scale is costly to remove and may lead to safety problems if it blocks release valves.

In this chapter, in an attempt to devise methods for helping to foresee such unfavorable consequences, we construct models of the chemical interactions between injected fluids and the sediments and formation waters in petroleum reservoirs. We consider two cases: the effects of using seawater as a waterflood, taking oil fields

Table 34.1 *Compositions of formation fluids from three North Sea oil fields (Edward Warren, personal communication) and seawater*

	Miller	Forties	Amethyst	Seawater
Na^+ (mg kg^{-1})	27 250	27 340	51 900	10 760
K^+	1 730	346	1 100	399
Mg^{++}	110	469	2 640	1 290
Ca^{++}	995	2 615	16 320	411
Sr^{++}	105	534	1 000	8
Ba^{++}	995	235	10	0.01
Cl^-	45 150	48 753	121 550	19 350
HCO_3^-	1 980	462	85	142
SO_4^{--}	0	10	0	2 710
T (°C)	121	96	88	20
pH (20 °C)	7	7	6.7	8.1
TDS (mg kg^{-1})	78 300	80 800	194 600	35 100

of the North Sea as an example, and the potential consequences of using alkali flooding (i.e., the injection of a strong caustic solution) in order to increase oil production from a clastic reservoir.

34.1 Sulfate Scaling in North Sea Oil Fields

A common problem in offshore petroleum production is that sulfate scale may form when seawater is injected into the formation during waterflooding operations. The scale forms when seawater, which is rich in sulfate but relatively poor in Ca^{++} and nearly depleted in Sr^{++} and Ba^{++}, mixes with formation fluids, many of which contain bivalent cations in relative abundance but little sulfate. The mixing causes minerals such as gypsum ($CaSO_4 \cdot 2H_2O$), anhydrite ($CaSO_4$), celestite ($SrSO_4$), and barite ($BaSO_4$, an almost insoluble salt) to become saturated and precipitate as scale.

Sulfate scaling poses a special problem in oil fields of the North Sea (e.g., Todd and Yuan, 1990, 1992; Yuan *et al.*, 1994), where formation fluids are notably rich in barium and strontium. The scale can reduce permeability in the formation, clog the wellbore and production tubing, and cause safety equipment (such as pressure release valves) to malfunction. To try to prevent scale from forming, reservoir engineers use chemical inhibitors such as phosphonate (a family of organic phosphorus compounds) in "squeeze treatments," as described in the introduction to this chapter.

Table 34.1 shows the compositions of formation waters from three North Sea oil fields, and the composition of seawater (from Drever, 1988). The origin of the scaling problem is clear. Seawater contains more than 2500 mg kg^{-1} of sulfate but little strontium and almost no barium. The formation waters, however, are depleted in sulfate, but they contain strontium and barium in concentrations up to about 1000 mg kg^{-1} and significant amounts of calcium. A mixture of seawater and formation fluid, therefore, will contain high concentrations of both sulfate and the cations, and hence, will probably be supersaturated with respect to sulfate minerals.

To model the results of the mixing process, we calculate the effects of titrating seawater into the formation waters. Two aspects of the modeling results are of interest. First, the volume of mineral scale produced during mixing provides a measure of the potential severity of a scaling problem. Second, in simulations in which scale is prevented from forming (simulating the case of a completely successful inhibition treatment), the saturation states of the sulfate minerals give information about the thermodynamic driving force for precipitation. This information is of value because it provides a measure of the difficulty of inhibiting scale formation (Sorbie *et al.*, 1994). In general, to be effective against scaling, the inhibitor must be present at greater concentration where minerals are highly supersaturated than where they are less supersaturated (e.g., He *et al.*, 1994).

To start the simulation, we equilibrate seawater (as we did in Chapter 6), using REACT to carry it to formation temperature, and then "pick up" the resulting fluid as a reactant in the mixing calculation. The procedure (taking the temperature of 121 °C reported for the Miller field) is

```
pH     = 8.1
Na+    = 10760    mg/kg
K+     =   399    mg/kg
Mg++   =  1290    mg/kg
Ca++   =   411    mg/kg
Sr++   =     8    mg/kg
Ba++   =       .01 mg/kg
Cl-    = 19350    mg/kg
HCO3-  =   142    mg/kg
SO4--  =  2710    mg/kg
TDS = 35100
T initial = 20, final = 121
go

pickup reactants = fluid
```

We then equilibrate the formation fluid, using data from Table 34.1. Since pH measurements from saline solutions are not reliable, we assume that pH in the reservoir is controlled by equilibrium with the most saturated carbonate mineral, which turns out to be witherite ($BaCO_3$) or, for the Amethyst field, strontianite ($SrCO_3$). Using the Miller analysis, the procedure for completing the calculation is

(cont'd)

```
T      = 121
Na+    = 27250 mg/kg
K+     =  1730 mg/kg
Mg++   =   110 mg/kg
Ca++   =   995 mg/kg
Sr++   =   105 mg/kg
Ba++   =   995 mg/kg
Cl-    = 45150 mg/kg
HCO3-  =  1980 mg/kg
SO4--  =    10 ug/kg
swap Witherite for H+
1 free cm3 Witherite
TDS = 78300

reactants times 10
```

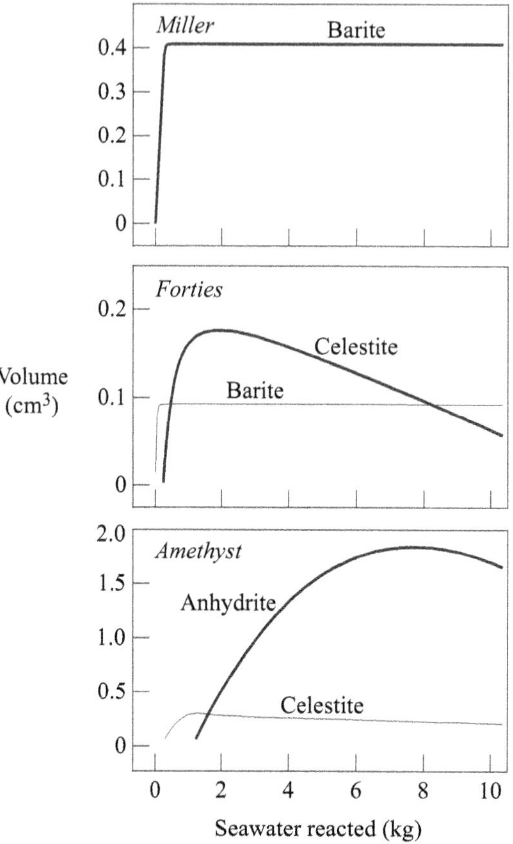

Figure 34.1 Volumes of minerals precipitated during a reaction model simulating the mixing at reservoir temperature of seawater into formation fluids from the Miller, Forties, and Amethyst oil fields in the North Sea. The reservoir temperatures and compositions of the formation fluids are given in Table 34.1. The initial extent of the system in each case is 1 kg of solvent water. Not shown for the Amethyst results are small volumes of strontianite, barite, and dolomite that form during mixing.

```
delxi = .001
dump
go
```

To model the case in which scale is prevented from forming, we repeat the calculation

```
(cont'd)
precip = off
go
```

with mineral precipitation disabled. Figure 34.1 shows the volumes of minerals precipitated during the mixing reactions for fluids from the three fields; Figure 34.2 shows the saturation states of sulfate minerals during mixing in the absence of precipitation.

For the Miller fluid, barite precipitates,

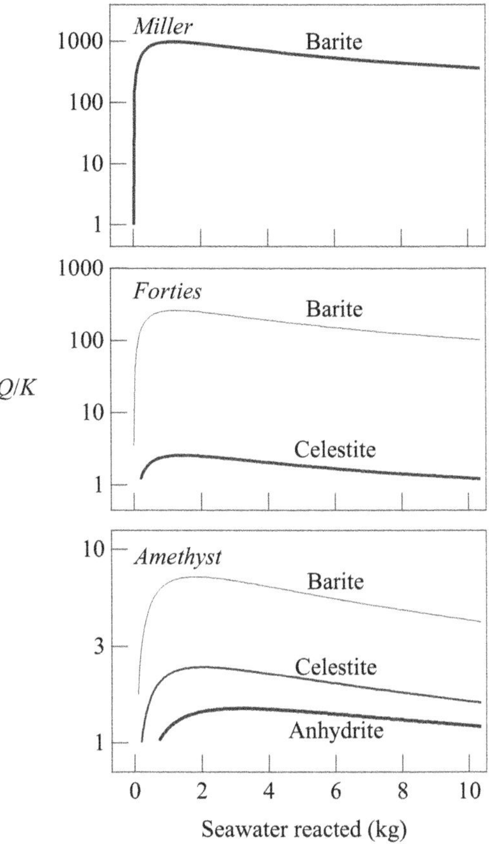

Figure 34.2 Saturation states (Q/K) of supersaturated sulfate minerals over the courses of simulations in which seawater mixes at reservoir temperature with formation fluids from three North Sea oil fields. Reaction paths are the same as shown in Figure 34.1, except that minerals are not allowed to precipitate.

$$Ba^{++} + SO_4^{--} \rightarrow \underset{barite}{BaSO_4} , \tag{34.1}$$

early in the mixing reaction. Because of the mineral's low solubility, virtually all of the seawater sulfate added to the system is consumed by the precipitation reaction until the barium is depleted from the fluid. Barite is the only mineral to form in the simulation, reaching a maximum volume of 0.4 cm^3 (per kg of solvent water in the formation fluid) after reaction with just a small amount of seawater.

The Forties fluid contains less barium but is richer in strontium. Barite forms initially (according to Reaction 34.1), but in a lesser amount than from the Miller fluid. Celestite forms,

$$Sr^{++} + SO_4^{--} \rightarrow \underset{celestite}{SrSO_4} , \tag{34.2}$$

shortly thereafter. In reality, the precipitate would occur as a barium–strontium solid solution rather than as separate phases. Later in the mixing reaction, as seawater comes to dominate the mixed fluid, the celestite starts to redissolve, reflecting the small amount of strontium in seawater. Less than about 0.3 cm^3 of scale forms during the mixing reaction.

The Amethyst fluid is richer in strontium and calcium than the other fluids, but nearly depleted in barium. Celestite becomes saturated first, and more of this mineral forms from this fluid than from the Forties fluid. Anhydrite becomes saturated later in the mixing process and precipitates,

$$Ca^{++} + SO_4^{--} \rightarrow \underset{anhydrite}{CaSO_4} \quad . \tag{34.3}$$

More than 2 cm^3 of scale form during the simulation, making this fluid potentially more damaging than the Miller or Forties formation waters by a factor of five or greater.

In the three simulations, the sulfate minerals form at mixing ratios related to their solubilities. Barite, the least soluble, forms early, when small amounts of seawater are added. The more soluble celestite forms only after the addition of somewhat larger quantities of seawater. Anhydrite, the most soluble of the minerals, forms from the Amethyst fluid at still higher ratios of seawater to formation fluid.

Comparing the volumes of scale produced in simulations in which minerals are allowed to form (Fig. 34.1) with the minerals' saturation states when precipitation is disabled (Fig. 34.2), it is clear that no direct relationship exists between these values. In the Forties and Amethyst simulations, the saturation state of barite is greater than those of the other minerals that form, although the other minerals form in greater volumes. Whereas the Amethyst fluid is capable of producing about five times as much scale as fluids from the Miller and Forties fields, furthermore, the saturation states (Q/K) predicted for it are 1.5 to 2 orders of magnitude lower than for the other fluids. This result indicates that although the Amethyst fluid is the most potentially damaging of the three, it is also the fluid for which scale formation might be most readily inhibited by chemical treatment.

We could, of course, attempt more sophisticated simulations of scale formation. Since the fluid mixture is quite concentrated early in the mixing, we might use a virial model to calculate activity coefficients (see Chapter 8). The Harvie–Møller–Weare (1984) activity model is limited to 25 °C and does not consider barium or strontium, but Yuan and Todd (1991) suggested a similar model for the Na–Ca–Ba–Sr–Mg–SO_4–Cl system in which the virial coefficients can be extrapolated to typical reservoir temperatures in the North Sea.

Given a specific application, we might also include precipitation kinetics in our calculations, as described in Chapter 17. Wat *et al.* (1992) present a brief study of the kinetics of barite formation, including the effects of scale inhibitors on precipitation rates. For a variety of reasons (see Section 17.2), however, it remains difficult to construct reliable models of the kinetics of scale precipitation.

34.2 Alkali Flooding

Some of the most radical changes to the geochemistry of a petroleum reservoir are induced by the highly reactive fluids injected in well stimulation and IOR procedures. Stimulation (e.g., acidization) is generally a near-well treatment designed to improve the productivity of a formation, sometimes by reversing previous formation damage. The fluids used in IOR may react with the formation water, with the mineral assemblage in the formation, and with the crude oil itself. Alkali flooding is an example, considered in this section, of an IOR procedure employing an extremely reactive fluid.

The purpose of alkali flooding (Jennings *et al.*, 1974) is to introduce alkali into a reservoir where it can react with organic acids in the oil to produce organic salts, which act as surfactants. The surfactants (or "petroleum soaps") generated reduce the surface tension between the oil and water and this in turn reduces the level of capillary trapping of the oil. Thus, more oil is recovered because less of it remains trapped in the formation's pore spaces.

This type of flood can be successful only if, as the fluid moves through the reservoir, a sufficient amount of the alkali remains in solution to react with the oil. Reaction of the flood with minerals and fluid in the reservoir, however, can consume the flood's alkali content. Worse, the reactions may precipitate minerals in the formation's pore space, decreasing permeability near the wellbore where free flow is most critical. A special problem for this type of flood is the reaction of clay minerals to form zeolites (Sydansk, 1982).

The effectiveness of alkali flooding, and, in fact, most reservoir treatments, varies widely from formation to formation in a manner that is often difficult to predict. Quantitative techniques have been applied to model the migration and consumption of alkali as it moves through a reservoir (e.g., Bunge and Radke, 1982; Zabala *et al.*, 1982; Dria *et al.*, 1988). There have been fewer attempts, however, to predict the specific chemical reactions that might occur in a reservoir or the effects of the initial mineralogy of the reservoir and the composition of the flood on those reactions (Bethke *et al.*, 1992).

To consider how such predictions might be made, we model how three types of alkali floods might affect a hypothetical sandstone reservoir. The floods, which are marketed commercially for this purpose, are NaOH, Na_2CO_3, and Na_2SiO_3. We take each flood at 0.5 N strength and assume that reaction occurs at a temperature of 70 °C.

The reservoir rock in our model is composed of quartz grains, carbonate cement, and clay minerals in the following proportions, by volume:

Quartz	SiO_2	85%
Calcite	$CaCO_3$	6%
Dolomite	$CaMg(CO_3)_2$	4%
Muscovite (illite)	$KAl_3Si_3O_{10}(OH)_2$	3%
Kaolinite	$Al_2Si_2O_5(OH)_4$	2%

The initial formation fluid is a solution at pH 5 of 1 molal NaCl and 0.2 molal $CaCl_2$ which is in chemical equilibrium with the minerals in the reservoir. The extent of the system is 1 kg solvent water, which at 15% porosity corresponds to a 1-cm-thick slice of the formation extending about 20 cm from the wellbore.

Using the "flush" configuration (Chapter 2), we continuously displace the pore fluid with the flooding solution. In this way, we replace the pore fluid in the system a total of ten times over the course of the simulated flood, which lasts 20 days. Because of the short interval selected for the flood, we assume that the pore fluid does not remain in equilibrium with quartz or framework silicates such as feldspar.

We set quartz dissolution and precipitation according to a kinetic rate law (Knauss and Wolery, 1988; see Chapter 17)

$$r_{\text{qtz}} = A_S\, k_+\, a_{\text{H}^+}^{-1/2} \left(1 - \frac{Q}{K}\right), \tag{34.4}$$

valid at 70 °C for pH values greater than about 6, which are quickly reached in the simulation; the corresponding rate constant k_+ is 1.6×10^{-18} mol cm^{-2} s^{-1}. As in our calculations in Chapter 30, we assume a specific surface area for quartz of 1000 cm^2 g^{-1}. Feldspars are suppressed so they will not precipitate if they become supersaturated, since these minerals presumably do not have time to nucleate and grow. We assume, however, that the carbonate and clay minerals in the simulation maintain equilibrium with the fluid as it moves through the system. In a more sophisticated simulation, we might also set kinetic rate laws for these minerals.

For the NaOH flood, the complete procedure in REACT is

```
T = 70

pH = 5
Na+  = 1   molal
Ca++ = 0.2 molal
Cl-  = 1   molal

swap Dolomite-ord for Mg++
swap Muscovite for K+
swap Kaolinite for Al+++
swap Quartz for SiO2(aq)
swap Calcite for HCO3-

 365 free cm3 Calcite
 235 free cm3 Dolomite-ord
 180 free cm3 Muscovite
 120 free cm3 Kaolinite
5150 free cm3 Quartz

react 1    kg H2O
react 0.5  moles Na+
react 0.5  moles OH-

suppress Albite, "Albite high", "Albite low"
suppress "Maximum Microcline", K-feldspar, "Sanidine high"
kinetic Quartz  rate_con = 1.8e-18,  apower(H+) = -1/2,  surface = 1000

reactants times 10
time end = 20 days
flush
go
```

To run the Na_2CO_3 and Na_2SiO_3 simulations, we need only alter the composition of the reactant fluid:

(cont'd)

```
remove reactant OH-
react 0.25 moles CO3--
go
```

and

(cont'd)

```
remove reactant CO3--
remove reactant Na+
react 0.25 moles Na2O
react 0.25 moles SiO2(aq)
go
```

Figures 34.3–34.4 show the results of the simulations.

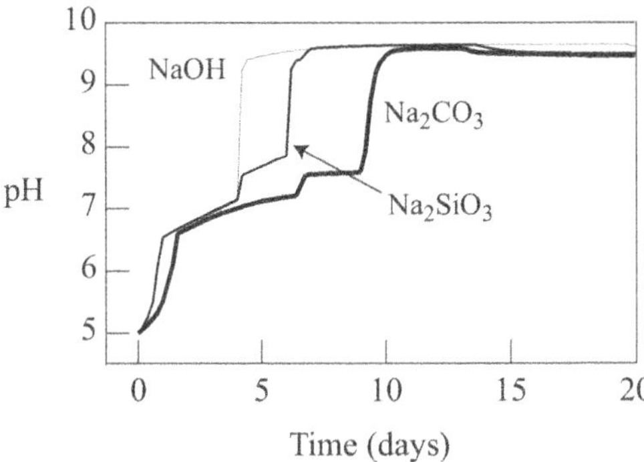

Figure 34.3 Variation in pH during simulated alkali floods of a clastic petroleum reservoir at 70 °C, using 0.5 N NaOH, Na_2CO_3, and Na_2SiO_3 solutions. Pore fluid is displaced by unreacted flooding solution at a rate of one-half of the system's pore volume per day.

As the alkaline solution enters the system, pH increases in each of the simulations, eventually reaching a value of about 9.5. In the early portions of the simulations, however, some of the alkali is consumed in the conversion of kaolinite to paragonite [$NaAl_3Si_3O_{10}(OH)_2$]. In the NaOH flood, for example, the reaction

$$\underset{}{2\,NaOH} + 3\,\underset{kaolinite}{Al_2Si_2O_5(OH)_4} \rightarrow 2\,\underset{paragonite}{NaAl_3Si_3O_{10}(OH)_2} + 5\,H_2O \tag{34.5}$$

consumes most of the sodium hydroxide and helps maintain a low pH until the kaolinite has been exhausted.

Later in the simulations, the zeolite analcime ($NaAlSi_2O_6 \cdot H_2O$) begins to form, largely at the expense of micas, which serve as proxies for clay minerals. In the NaOH flood, the overall reaction (expressed per formula unit of analcime) is

$$\begin{aligned} &1.75\,NaOH + \underset{dolomite}{CaMg(CO_3)_2} + 0.9\,\underset{quartz}{SiO_2} + 0.33\,\underset{muscovite}{KAl_3Si_3O_{10}(OH)_2} + 0.36\,\underset{paragonite}{NaAl_3Si_3O_{10}(OH)_2} \\ &+0.6\,H_2O \rightarrow \underset{calcite}{CaCO_3} + \underset{analcime}{NaAlSi_2O_6 \cdot H_2O} + 0.75\,\underset{dawsonite}{NaAlCO_3(OH)_2} + \\ &0.33\,\underset{phlogopite}{KAlMg_3Si_3O_{10}(OH)_2} + 0.4\,Na^+ + 0.15\,HCO_3^- + 0.1\,CO_3^{--}\,. \end{aligned} \tag{34.6}$$

In part because of the open crystal structure and resulting low density characteristic of zeolite minerals, analcime is the most voluminous reaction product in the simulations.

Table 34.2 summarizes the simulation results. In each case, the flood dissolves clay minerals and quartz. The simulations, however, predict the production of significant volumes of analcime, in accord with the observation that zeolites are prone to form during alkali floods. The volume of analcime produced in each case is sufficient to offset the volumes of the dissolved minerals and leads to a net decrease in porosity. Of the simulations, the Na_2SiO_3 flood leads to the production of the most analcime and to the greatest loss in porosity.

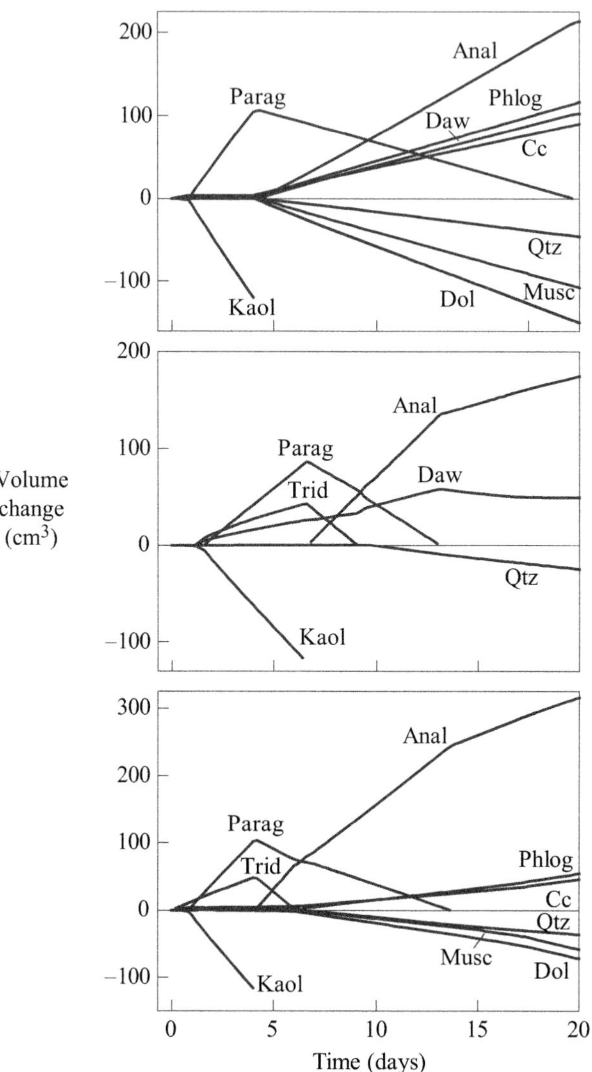

Figure 34.4 Changes in the volumes of minerals in the reservoir rock during the simulated alkali floods (Fig. 34.3) of a clastic petroleum reservoir using NaOH, Na_2CO_3, and Na_2SiO_3 solutions. Minerals that react in small volumes are omitted from the plots. Abbreviations: Anal = analcime, Cc = calcite, Daw = dawsonite, Dol = dolomite, Kaol = kaolinite, Musc = muscovite, Parag = paragonite, Phlog = phlogopite, Qtz = quartz, Trid = tridymite.

Table 34.2 *Comparison of the alkali flood simulations**

	NaOH	Na_2CO_3	Na_2SiO_3
Net change in mineral volume (cm^3)			
zeolite	+213	+175	+316
carbonate	+42	+53	−27
quartz	−46	−25	−36
clay (& mica)	−111	−126	−118
	+98	+77	+135
Change in pore volume (Δ%)	−9	−7	−13
Alkali consumed (%)	78	53	65

*Extent of system is 1 kg solvent water.

In the simulations, a significant fraction (about 50% to 80%) of the alkali present in solution is consumed by reactions near the wellbore with the reservoir minerals (as shown in Reaction 34.6 for the NaOH flood), mostly by the production of analcime, paragonite, and dawsonite [$NaAlCO_3(OH)_2$]. In the clastic reservoir considered, therefore, alkali floods might be expected to cause formation damage (mostly due to the precipitation of zeolites) and to be less effective at increasing oil mobility than in a reservoir where they do not react extensively with the formation.

35
Acid Drainage

Acid drainage is a persistent environmental problem in many mineralized areas. The problem is especially pronounced in areas that host or have hosted mining activity (e.g., Lind and Hem, 1993), but it also occurs naturally in unmined areas. The acid drainage results from weathering of sulfide minerals that oxidize to produce hydrogen ions and contribute dissolved metals to solution (e.g., Blowes *et al.*, 2005).

These acidic waters are toxic to plant and animal life, including fish and aquatic insects. Streams affected by acid drainage may be rendered nearly lifeless, their streambeds coated with unsightly yellow and red precipitates of oxy-hydroxide minerals. In some cases, the heavy metals in acid drainage threaten water supplies and irrigation projects.

Where acid drainage is well developed and extensive, the costs of remediation can be high. In the Summitville, Colorado district (USA), for example, efforts to limit the contamination of fertile irrigated farmlands in the nearby San Luis Valley and protect aquatic life in the Alamosa River will cost an estimated \$100 million or more (Plumlee, 1994a).

Not all mine drainage, however, is acidic or rich in dissolved metals (e.g., Ficklin *et al.*, 1992; Mayo *et al.*, 1992; Plumlee *et al.*, 1992). Drainage from mining districts in the Colorado Mineral Belt ranges in pH from 1.7 to greater than 8 and contains total metal concentrations ranging from as low as about 0.1 mg kg^{-1} to more than 1000 mg kg^{-1}. The primary controls on drainage pH and metal content seem to be (1) the exposure of sulfide minerals to weathering, (2) the availability of atmospheric oxygen, and (3) the ability of nonsulfide minerals to buffer acidity.

In this chapter we construct geochemical models to consider how the availability of oxygen and the buffering of host rocks affect the pH and composition of acid drainage. We then look at processes that can attenuate the dissolved metal content of drainage waters.

35.1 Role of Atmospheric Oxygen

Acid drainage results from the reaction of sulfide minerals with oxygen in the presence of water. As we show in this section, water in the absence of a supply of oxygen gas becomes saturated with respect to a sulfide mineral after only a small amount of the mineral has dissolved. The dissolution reaction in this case (when oxygen gas is not available) causes little change in the water's pH or composition. In a separate effect, it is likely that atmospheric oxygen further promotes acid drainage because of its role in the metabolism of bacteria that catalyze both the dissolution of sulfide minerals and the oxidation of dissolved iron (Nordstrom, 1982).

For these reasons, there is a clear connection between the chemistry of mine drainage and the availability of oxygen. Plumlee *et al.* (1992) found that the most acidic, metal-rich drainage waters in Colorado tend to develop in mine dumps, which are highly permeable and open to air circulation, and in mineral districts like Summitville (see King, 1995), where abandoned workings and extensive fracturing give atmospheric oxygen access to the ores. Drainage waters that are depleted in dissolved oxygen, on the other hand, tend to be less acidic and have lower heavy metal concentrations.

To investigate how the presence of atmospheric oxygen affects the reaction of pyrite (FeS_2) with oxidizing groundwater, we construct a simple model in REACT. First we take a hypothetical groundwater at 25 °C that has equilibrated with atmospheric oxygen but is no longer in contact with it. We suppress hematite (Fe_2O_3), which does not form directly at low temperature, and goethite (FeOOH); each of these minerals is more stable thermodynamically than the ferric precipitate observed to form in acid drainage. To set the initial fluid, we type

```
swap O2(g) for O2(aq)
f O2(g) = 0.2
pH = 6.8

Ca++   = 10 mg/kg
Mg++   =  2 mg/kg
Na+    =  6 mg/kg
K+     =  2 mg/kg
HCO3-  = 75 mg/kg
SO4--  =  2 mg/kg
Cl-    =  1 mg/kg
Fe++   =  1 ug/kg
balance on HCO3-

suppress Hematite, Goethite
go
```

The resulting fluid contains about 8 mg kg^{-1} (0.25 mmol) of dissolved oxygen.

We then add pyrite to the system

```
(cont'd)
react 1 cm3 Pyrite
go
```

letting it react to equilibrium with the fluid. The reaction proceeds until the O_2(aq) has been consumed, dissolving a small amount of pyrite (about 0.08 mmol) according to

$$\underset{pyrite}{FeS_2} + 7/2\, O_2(aq) + 2\, HCO_3^- \rightarrow Fe^{++} + 2\, SO_4^{--} + 2\, CO_2(aq) + H_2O\,. \tag{35.1}$$

The fluid's pH in the model changes slightly (Fig. 35.1), decreasing from 6.8 to about 6.6.

To see how contact with atmospheric oxygen might affect the reaction, we repeat the calculation, assuming this time that oxygen fugacity is fixed at its atmospheric level

```
(cont'd)
fix f O2(g)
go
```

In this case, the reaction proceeds without exhausting the oxygen supply, which in the calculation is limitless, driving pH to a value of about 1.7 (Fig. 35.1). We could, in fact, continue to dissolve pyrite into the fluid indefinitely, thereby reaching even lower pH values.

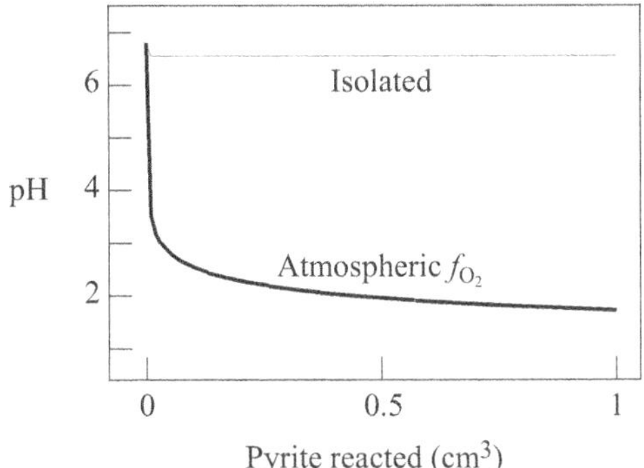

Figure 35.1 Calculated variation in pH during reaction of pyrite with a hypothetical groundwater at 25 °C, assuming that the fluid is isolated from (fine line) and in contact with (bold line) atmospheric oxygen.

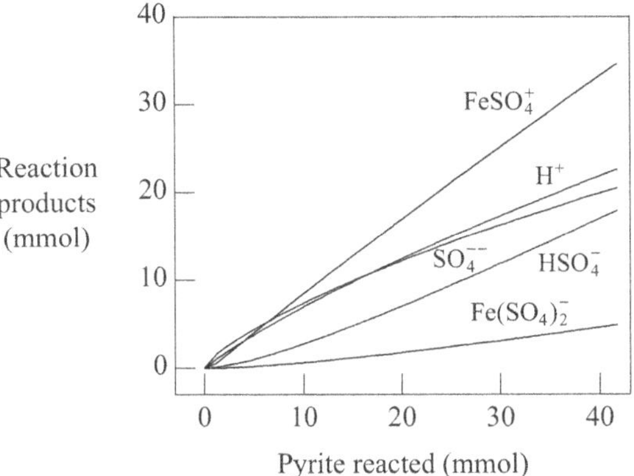

Figure 35.2 Masses of species produced by reacting pyrite with a hypothetical groundwater that is held in equilibrium with atmospheric oxygen, according to the reaction path calculation shown in Figure 35.1.

Initially, pyrite oxidation in the model proceeds according to the reaction

$$\underset{pyrite}{FeS_2} + {}^{15}/_4\, O_2(g) + {}^{1}/_2\, H_2O \rightarrow FeSO_4^+ + SO_4^{--} + H^+ , \tag{35.2}$$

as shown in Figure 35.2, producing H^+ and thus driving the fluid acidic. According to the model, the pyrite dissolution produces ferric iron in an ion pair with sulfate. As the pH decreases, HSO_4^- comes to dominate SO_4^{--} and a second reaction,

$$\underset{pyrite}{FeS_2} + {}^{15}/_4\, O_2(g) + {}^{1}/_2\, H_2O \rightarrow FeSO_4^+ + HSO_4^- , \tag{35.3}$$

becomes important. This reaction produces no free hydrogen ions and hence does not contribute to the fluid's acidity.

Our calculated reaction path may reasonably well represent the overall reaction of pyrite as it oxidizes, but it does little to illustrate the steps that make up the reaction process. Reaction 35.2, for example, involves the transfer of 16 electrons to oxygen, the electron acceptor in the reaction, from the iron and sulfur in each FeS_2 molecule. Elementary reactions (those that proceed on a molecular level), however, seldom transfer more than one or two electrons. Reaction 35.2, therefore, would of necessity represent a composite of elementary reactions.

In nature, at least two aqueous species, $O_2(aq)$ and Fe^{+++}, can serve as electron acceptors during the pyrite oxidation (Moses *et al.*, 1987). In the case of Fe^{+++}, the oxidation reaction proceeds as

$$\underset{pyrite}{FeS_2} + 14\,Fe^{+++} + 8\,H_2O \rightarrow 15\,Fe^{++} + 2\,SO_4^{--} + 16\,H^+ . \quad (35.4)$$

This reaction, while still not an elementary reaction, more closely describes how the oxidation proceeds on a molecular level. Even where Reaction 35.4 operates, however, our model may still represent the overall process occurring in nature, since O_2 is needed to produce the Fe^{+++} that drives the reaction forward and is, therefore, the ultimate oxidant in the system.

Even neglecting the question of the precise steps that make up the overall reaction, our calculations are a considerable simplification of reality. The implicit assumption that iron in the fluid maintains redox equilibrium with the dissolved oxygen, as described in Chapter 7, is especially vulnerable. In reality, the ferrous iron added to solution by the dissolving pyrite must react with dissolved oxygen to produce ferric species, a process that may proceed slowly. To construct a more realistic model, we could treat the dissolution in two steps by disenabling the Fe^{++}/Fe^{+++} redox couple. In the first step we would let pyrite dissolve, and in the second, let the ferrous species oxidize.

35.2 Buffering by Wall Rocks

The most important control on the chemistry of drainage from mineralized areas (once we assume access of oxygen to the sulfide minerals) is the nature of the nonsulfide minerals available to react with the drainage before it discharges to the surface (e.g., Sherlock *et al.*, 1995). These minerals include gangue minerals in the ore, the minerals making up the country rock, and the minerals found in mine dumps. The drainage chemistry of areas in which these minerals have the ability to neutralize acid differs sharply from that of areas in which they do not.

In the Colorado Mineral Belt (USA), for example, deposits hosted by argillically altered wall rocks (as in Summitville) tend to produce highly acidic and metal-rich drainage, because the wall rocks have a negligible capacity for buffering acid (Plumlee, 1994b). In contrast, ores that contain carbonate minerals or are found in carbonate terrains, as well as those with propylitized wall rocks, can be predicted to produce drainages of near-neutral pH. These waters may be rich in zinc but are generally not highly enriched in other metals.

In the historic silver mining districts of the Wasatch Range (Utah, USA), Mayo *et al.* (1992) found that few springs in the area discharge acid drainage. Acid drainage occurs only where groundwater flows through aquifers found in rocks nearly devoid of carbonate minerals. Where carbonate minerals are abundant in the country rock, the discharge invariably has a near-neutral pH.

To model the effect of carbonate minerals on drainage chemistry, we continue our calculations from the previous section (in which we reacted pyrite with a hypothetical groundwater in contact with atmospheric oxygen). This time, we include calcite ($CaCO_3$) in the initial system

```
(cont'd)
swap Calcite for HCO3-
10 free cm3 Calcite
```

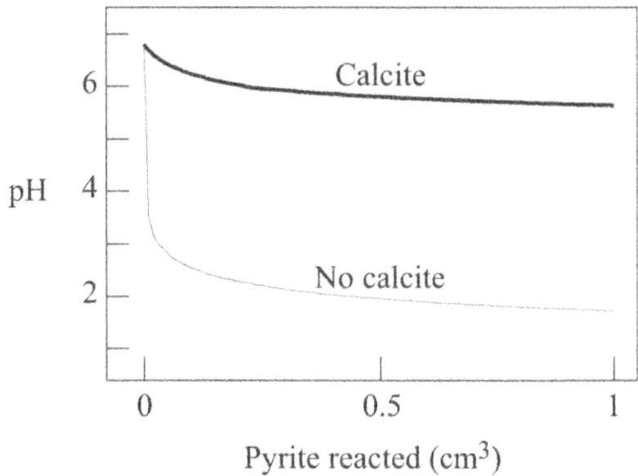

Figure 35.3 Variation in pH as pyrite reacts at 25 °C with a groundwater held in equilibrium with atmospheric O_2, calculated assuming that the reaction occurs in the absence of buffering minerals (fine line, from Fig. 35.1) and in the presence of calcite (bold line).

```
balance on Ca++
go
```

In contrast to the previous calculation, the fluid maintains a near-neutral pH (Fig. 35.3), reflecting the acid-buffering capacity of the calcite.

As the pyrite dissolves by oxidation, calcite is consumed and ferric hydroxide precipitates (Fig. 35.4) according to the reaction

$$\underset{\textit{pyrite}}{FeS_2} + 2\,\underset{\textit{calcite}}{CaCO_3} + 3/2\,H_2O + 15/4\,O_2(g) \rightarrow \underset{\textit{ferric hydroxide}}{Fe(OH)_3} + 2\,Ca^{++} + 2\,SO_4^{--} + 2\,CO_2(aq)\,. \tag{35.5}$$

Ca^{++} and SO_4^{--} accumulate in the fluid, eventually causing gypsum ($CaSO_4 \cdot 2H_2O$) to saturate and precipitate. At this point, the overall reaction becomes

$$\underset{\textit{pyrite}}{FeS_2} + 2\,\underset{\textit{calcite}}{CaCO_3} + 11/2\,H_2O + 15/4\,O_2(g) \rightarrow \underset{\textit{ferric hydroxide}}{Fe(OH)_3} + 2\,\underset{\textit{gypsum}}{CaSO_4 \cdot 2H_2O} + 2\,CO_2(aq)\,. \tag{35.6}$$

In the calculation, reaction of 1 cm^3 of pyrite consumes about 3.4 cm^3 of calcite, demonstrating that considerable quantities of buffering minerals may be required in mineralized areas to neutralize drainage waters.

The accumulation of $CO_2(aq)$ over the reaction path (according to Reactions 35.5–35.6) raises the fluid's CO_2 fugacity to a value greater than atmospheric pressure. In nature, $CO_2(g)$ would begin to effervesce,

$$CO_2(aq) \rightarrow CO_2(g)\,, \tag{35.7}$$

at about this point in the reaction, causing pH to increase to values somewhat greater than predicted by the reaction path, which did not account for degassing. (As a variation of the calculation, we could account for

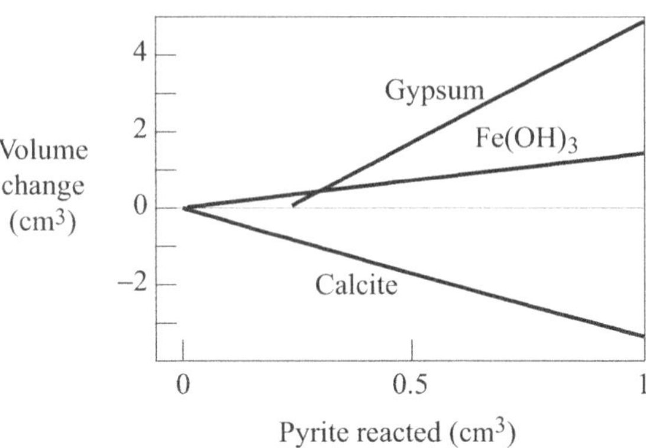

Figure 35.4 Volumes (cm^3) of minerals consumed (negative values) and precipitated (positive values) as pyrite reacts at 25 °C with a groundwater in equilibrium with atmospheric O_2 in the presence of calcite.

the degassing by reacting enough pyrite to bring f_{CO_2} to one, "picking up" the calculation results, fixing the fugacity, and then reacting the remaining pyrite.)

In the two calculations (one including, the other excluding calcite), the resulting fluids differ considerably in composition. After reaction of 1 cm^3 of pyrite at atmospheric oxygen fugacity, the compositions are

	No calcite	Calcite
pH	1.7	5.6
Fe (mg kg^{-1})	2300	0.03
SO_4	8000	1800
Ca	9.9	1100
HCO_3	74	6000

Whereas pyrite oxidation in the absence of calcite produces H–Fe–SO_4 drainage, the reaction in the presence of calcite yields a Ca–HCO_3–SO_4 drainage.

35.3 Fate of Dissolved Metals

The metal concentrations in acid drainage can be alarmingly high, but in many cases they are attenuated to much lower levels by natural processes (and sometimes by remediation schemes) near the drainage discharge area. Attenuation results from a variety of processes. The drainage, of course, may be diluted by base flow and by mixing with other surface waters. More significantly, drainage waters become less acidic after discharge as they react with various minerals, mix with other waters (especially those rich in HCO_3^-), and in some cases lose CO_2 to the atmosphere.

As pH rises, the metal content of drainage water tends to decrease. Some metals precipitate directly from solution to form oxide, hydroxide, and oxy-hydroxide phases. Iron and aluminum are notable in this regard. They initially form colloidal and suspended phases known as hydrous ferric oxide (HFO, $FeOOH \cdot nH_2O$) and hydrous aluminum oxide (HAO, $AlOOH \cdot nH_2O$), both of which are highly soluble under acidic conditions but nearly insoluble at near-neutral pH.

The concentrations of other metals attenuate when the metals sorb onto the surfaces of precipitating minerals (see Chapters 10–11). Hydrous ferric oxide, the behavior of which is well studied (Dzombak and Morel, 1990), has a large specific surface area and is capable of sorbing metals from solution in considerable amounts, especially at moderate to high pH; HAO may behave similarly (e.g., Karamalidis and Dzombak, 2010). The process by which HFO and HAO form and then adsorb metals from solution, known as *coprecipitation*, represents an important control on the mobility of heavy metals in acid drainages (e.g., Chapman *et al.*, 1983; Johnson, 1986; Davis *et al.*, 1991; Smith *et al.*, 1992).

To see how this process works, we construct a model in which reaction of a hypothetical drainage water with calcite leads to the precipitation of ferric hydroxide [$Fe(OH)_3$, which we use to represent HFO] and the sorption of dissolved species onto this phase. We assume that the precipitate remains suspended in solution with its surface in equilibrium with the changing fluid chemistry, using the surface complexation model described in Chapter 10. In our model, we envisage the precipitate eventually settling to the streambed and hence removing the sorbed metals from the drainage.

We do not concern ourselves with the precipitate that lines sediments in the streambed, since it formed earlier while in contact with the drainage, and hence would not be expected to continue to sorb from solution. Smith *et al.* (1992), for example, found that in an acid drainage from Colorado (USA), sorption on the suspended solids, rather than the sediments along the streambed, controls the dissolved metal concentrations.

As the first step in the coprecipitation process, ferric hydroxide precipitates either from the effect of the changing pH on the solubility of ferric iron,

$$Fe^{+++} + 3\,H_2O \rightarrow \underset{\textit{ferric hydroxide}}{Fe(OH)_3} + 3\,H^+\,, \tag{35.8}$$

or by the oxidation of ferrous species,

$$Fe^{++} + {}^1\!/\!{}_4\,O_2(aq) + {}^5\!/\!{}_2\,H_2O \rightarrow \underset{\textit{ferric hydroxide}}{Fe(OH)_3} + 2\,H^+\,, \tag{35.9}$$

in solution. For simplicity, we assume in our calculations that the dissolved iron has already oxidized, so that Reaction 35.8 is responsible for forming the sorbing phase.

In REACT, we prepare the calculation by disenabling the redox couple between trivalent and pentavalent arsenic (arsenite and arsenate, respectively). As well, we disenable the couples for ferric iron and cupric copper, since we will not consider either ferrous or cupric species. We load dataset "FeOH+.sdat", which contains the reactions from the Dzombak and Morel (1990) surface complexation model, including those for which binding constants have only been estimated. The procedure is

```
decouple AsO4---
decouple Cu++
decouple Fe+++
read FeOH+.sdat
```

We then define an initial fluid representing the unreacted drainage. We set the fluid's iron content by assuming equilibrium with jarosite [$NaFe_3(SO_4)_2(OH)_6$] and prescribe a high content of dissolved arsenite, arsenate, cupric copper, lead, and zinc. Finally, we suppress hematite and goethite (which are more stable than ferric hydroxide) two ferrite minerals (e.g., $ZnFe_2O_4$), which we consider unlikely to form, and the $PbCO_3$ ion pair, which is erroneously stable in the thermodynamic database.

```
(cont'd)
swap Jarosite-Na for Fe+++
1 free ug Jarosite-Na
```

```
pH = 3
Na+      =    10 mg/kg
Ca++     =    10 mg/kg
Cl-      =    20 mg/kg
HCO3-    =   100 mg/kg
balance on SO4--

As(OH)4- =   200 ug/kg
AsO4---  =  1000 ug/kg
Cu++     =   500 ug/kg
Pb++     =   200 ug/kg
Zn++     = 18000 ug/kg

suppress Hematite, Goethite, Ferrite-Zn, Ferrite-Cu, PbCO3
```

To calculate the effect of reacting calcite with the drainage, we enter the command

(cont'd)
```
react 4 mmol Calcite
delxi = .002
```

and type go to trigger the calculation.

In the calculation, calcite dissolves into the drainage according to

$$\underset{calcite}{CaCO_3} + 2\,H^+ \rightarrow Ca^{++} + CO_2(aq) + H_2O \tag{35.10}$$

and

$$\underset{calcite}{CaCO_3} + H^+ \rightarrow Ca^{++} + HCO_3^- , \tag{35.11}$$

consuming H^+ and causing pH to increase. The shift in pH drives the ferric iron in solution to precipitate as ferric hydroxide, according to Reaction 35.8. A total of about 0.89 mmol of precipitate forms over the reaction path.

The distribution of metals between solution and the ferric hydroxide surface varies strongly with pH (Fig. 35.5). As discussed in Sections 10.6 and 15.3, pH exerts an important control over the sorption of metal ions for two reasons. First, the electrical charge on the sorbing surface tends to decrease as pH increases, lessening the electrical repulsion between surface and ions. More importantly, because hydrogen ions are involved in the sorption reactions, pH affects ion sorption by mass action. The sorption of bivalent cations such as Cu^{++},

$$>(s)FeOH_2^+ + Cu^{++} \rightarrow >(s)FeOCu^+ + 2\,H^+ , \tag{35.12}$$

is favored as pH increases, and there is a similar effect for Pb^{++} and Zn^{++}. Sorption of arsenite is also favored by increasing pH, according to the reaction

$$>(w)FeOH_2^+ + As(OH)_3 \rightarrow >(w)FeH_2AsO_3 + H_2O + H^+ . \tag{35.13}$$

As a result, the metal ions are progressively partitioned onto the ferric hydroxide surface as pH increases.

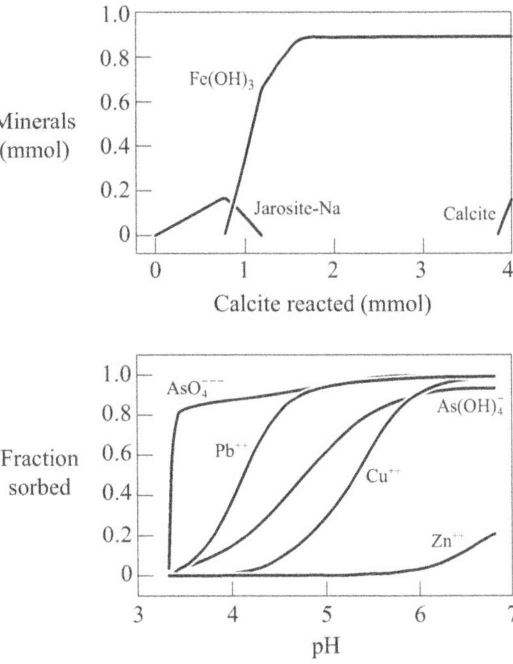

Figure 35.5 Minerals formed during reaction at 25 °C of a hypothetical acid drainage water with calcite (top), and fractions of the amounts of arsenite, arsenate, copper, lead, and zinc present initially in solution that sorb onto ferric hydroxide over the course of the reaction path (bottom). Bottom figure is plotted against pH, which increases as the water reacts with calcite.

Arsenate sorbs onto the ferric hydroxide surface over the pH range of the calculation because of its ability to bond tightly with the surface's weakly sorbing sites. Initially, the arsenate sorbs,

$$>\text{(w)FeOH}_2^+ + \text{H}_2\text{AsO}_4^- \rightarrow > \text{(w)FeH}_2\text{AsO}_4 + \text{H}_2\text{O}\,, \tag{35.14}$$

by complexing with the protonated weak sites. As pH increases, the surface species $>\text{(w)FeHAsO}_4^-$ and $>\text{(w)FeOHAsO}_4^{---}$ come to predominate, and the arsenate is partitioned even more strongly onto the sorbing surface.

It is useful to compare the capacity for each metal to be sorbed (the amount of each that could sorb if it occupied every surface site) with the metal concentrations in solution. To calculate the capacities, we take into account the amount of ferric precipitate formed in the calculation (0.89 mmol), the number of moles of strongly and weakly binding surface sites per mole of precipitate (0.005 and 0.2, respectively, according to the surface complexation model), and the site types that accept each metal [As(OH)_4^- and AsO_4^{---} sorb on weak sites only, whereas Pb^{++}, Cu^{++}, and Zn^{++} sorb on both strong and weak].

Comparing the initial metal concentrations to the resulting capacities for sorption,

Component	Maximum sorbed ($mg\ kg^{-1}$)	In solution ($mg\ kg^{-1}$)
$As(OH)_4^-$	25.4	0.2
AsO_4^{---}	24.7	1.0
Cu^{++}	11.6	0.5
Pb^{++}	37.8	0.2
Zn^{++}	11.9	18.0

we see that the precipitate forms in sufficient quantity to sorb the arsenic initially present in solution as well as the lead and copper, but only a fraction of the zinc. An initial consideration in evaluating the ability of coprecipitation to attenuate metal concentrations in a drainage water, therefore, is whether the sorbing mineral forms in a quantity sufficient to sorb the water's metal content.

36

Contamination and Remediation

Groundwater remediation is the often expensive process of restoring an aquifer after it has been contaminated, or at least limiting the ability of contaminants there to spread. In this chapter, we consider the widespread problem of the contamination of groundwater flows with heavy metals. We use reactive transport modeling to look at the reactions that occur as contaminated water enters a pristine aquifer, and those accompanying remediation efforts.

The method or methods employed to remediate an aquifer vary, depending on the type, degree, and extent of contamination. Where pollution is shallow and dispersed over a small area, the sediments can sometimes be dug up and transported to a landfill designed especially to isolate the contaminants. *Permeable reaction barriers* can be installed to intercept a contaminant plume and strip pollutants from it, if the plume is shallow and narrowly focused.

Metal contaminants can in some cases be immobilized in situ by oxidation or reduction, or precipitated by reaction with sulfide. Sulfate-reducing bacteria are sometimes stimulated to produce sulfide, or a sulfur-bearing compound such as calcium polysulfide can be injected into the subsurface as a reductant and sulfide source. In certain cases where the contamination poses little immediate threat, it can safely be left to attenuate naturally (e.g., Brady *et al.*, 1998), a procedure known as *monitored natural attenuation*.

Remediation more commonly proceeds by a *pump-and-treat* scheme in which contaminated water is drawn from the aquifer and treated at the surface to remove contaminant metals, before being discharged or reinjected into the aquifer. A pump-and-treat remedy can be prolonged, projected not uncommonly to proceed over the course of many decades before the contaminants might be largely flushed from the aquifer.

Any remedial scheme involves considerable expense to design, license, implement, operate, and monitor. As such, there is much emphasis currently on understanding in a quantitative sense the reactions operating during a remedy, in order to optimize its design and better predict its efficacy and duration.

36.1 Contamination With Inorganic Lead

We construct in this section a model of how inorganic lead reacts as it infiltrates and contaminates an aquifer, and then as the aquifer is flushed with fresh water during pump-and-treat remediation (Bethke, 1997; Bethke and Brady, 2000). We assume groundwater in the aquifer contacts hydrous ferric oxide [$Fe(OH)_3$, for simplicity] which sorbs Pb^{++} ions according to the surface complexation model of Dzombak and Morel (1990), as discussed in Chapter 10.

We employ the LLNL thermodynamic data for aqueous species, as before, omitting the $PbCO_3$ ion pair, which in the dataset is erroneously stable by several orders of magnitude. The reactions comprising the surface complexation model, including those for which equilibrium constants have only been estimated, are stored in dataset "FeOH+.sdat".

We consider a 100-m length of an aquifer with a porosity of 30% and a nominal dispersivity of 10 cm; the dispersivity reflected in the calculation results will be somewhat larger than this value, due to the effects of numerical dispersion. The domain is divided into 100 nodal blocks, each 1 m long. We assume local equilibrium, so time enters into the calculation only as a measure of the cumulative volume of fluid that has passed through the aquifer. Specifying the aquifer's pore volume be replaced 30 times over the course of the simulation, and setting the time span to 30 years, each year in the simulation corresponds to a single replacement of the aquifer's pore fluid.

Initially, the aquifer contains a dilute Ca–HCO_3 groundwater, including a negligible amount of Pb^{++} as well as an equal quantity of Br^-, which serves in the calculation as a nonreactive tracer. At the onset of the simulation, water containing 1 mmolal Pb^{++} and Br^- passes into the aquifer until half its pore volume has been displaced. At this point, the composition of water entering the aquifer changes to that of the initial fluid, uncontaminated water nearly devoid of lead and bromide. The simulation continues until water in the aquifer has been replaced 30 times.

We assume Pb^{++} during imbibition of the contaminated fluid is retarded by a factor of two ($R_F = 2$; see Section 24.1). The retardation prescribes equal molal concentrations of dissolved and sorbed metal ions at any point in the domain. Accounting for the concentrations of the two forms of sorbed lead, >(w)$FeOPb^+$ and >(s)$FeOPb^+$, after reaction with the contaminated fluid, this retardation is expected when the sorbing mineral $Fe(OH)_3$ makes up 0.0085 volume percent of the aquifer.

The procedure in X1T is

```
read FeOH+.sdat
decouple Fe+++
suppress PbCO3

interval start   =   0 yr, fluid = contaminated
interval elution = 1/2 yr, fluid = ambient
interval end     = 30 yr

length = 100 m
Nx = 100
discharge = 30 pore_volumes
porosity = 30%
dispersivity = 10 cm

scope initial
   swap Fe(OH)3(ppd) for Fe+++
   Fe(OH)3(ppd) = .0085 volume%
   pH = 6
   Ca++  =    2.0 mmolal
   balance on HCO3-
   Pb++  = 1e-12 mmolal
   Br-   = 1e-12 mmolal
scope contaminated = initial
   Pb++  =      1 mmolal
   Br-   =      1 mmolal
scope ambient = initial
```

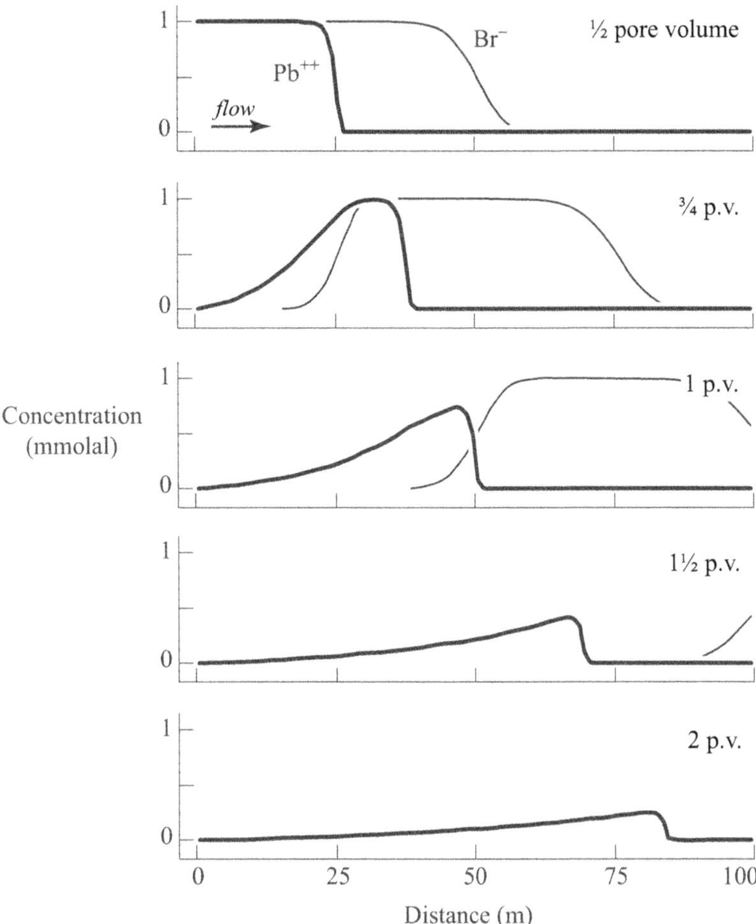

Figure 36.1 Simulation of the contamination at 25 °C of an aquifer with inorganic lead. The 100-m-long section of aquifer contains a small amount of $Fe(OH)_3$, to which Pb^{++} sorbs. Aquifer is initially uncontaminated, but at $t = 0$ water containing 1 mmolal Pb^{++} and 1 mmolal Br^-, which serves as a nonreactive tracer, passes into the left side. Pb^{++} is taken to sorb according to surface complexation theory, and the amount of $Fe(OH)_3$ is chosen so that migration of the metal is retarded by a factor of two relative to the groundwater flow. After half the groundwater has been displaced by the contaminated water (1/2 p.v.), clean water is flushed through the aquifer.

```
dxplot = 1/120
precip = off
go
```

In the calculation results (Fig. 36.1), Pb^{++} passing into the aquifer sorbs strongly to the ferric surface,

$$\begin{aligned} >\text{(w)FeOH} + \text{Pb}^{++} &\rightarrow >\text{(w)FeOPb}^{+} + \text{H}^{+} \\ >\text{(s)FeOH} + \text{Pb}^{++} &\rightarrow >\text{(s)FeOPb}^{+} + \text{H}^{+}, \end{aligned} \tag{36.1}$$

occupying most of the weak sites and almost all the strong sites. Due to the strong sorption, the Pb^{++} forms a

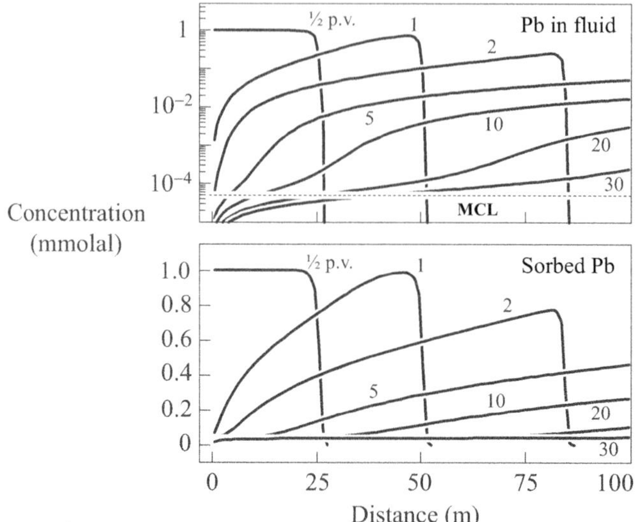

Figure 36.2 Remediation of the aquifer shown in Figure 36.1, as the simulation continues. After water contaminated with Pb^{++} displaces half of the aquifer's pore volume, clean water is flushed through the aquifer until a total of 30 pore volumes have been replaced. Flushing attenuates Pb^{++} concentration in the groundwater (top), so that it gradually approaches drinking water standards (MCL, or Maximum Contaminant Level), and slowly displaces most of the sorbed metal from the $Fe(OH)_3$ surface, primarily from the weak surface sites.

sharp reaction front that passes along the aquifer at half the speed of the groundwater flow, which is reflected by the advance of the nonreacting Br^-. The sorbing sites behind the front are almost fully complexed with the Pb^{++}, which is present in solution at the inlet concentration. Ahead of the front, the lead has been stripped from the flow, leaving the aquifer and groundwater there uncontaminated.

When half the aquifer's pore volume has been displaced, the inlet fluid in the simulation changes to uncomtaminated water. At this point, the contamination has progressed across one-quarter of the aquifer, reflecting the retardation factor of two. Clean water pushes the remaining contaminated water across the reaction front, causing the residual Pb^{++} to sorb. Contamination in the groundwater attenuates in this way, as lead is taken up by the sorbing surface.

As clean water passes into the aquifer, some of the lead, primarily that sorbed to weak sites, begins to desorb,

$$>(w)FeOPb^+ + H^+ \rightarrow >(w)FeOH + Pb^{++}\,, \tag{36.2}$$

as shown in Figure 36.2. The metal is carried downstream with the flow, where it sorbs to clean aquifer sediments as it crosses the reaction front (Reactions 36.1). In this manner, the extent of the contaminated aquifer sediments gradually increases.

The reaction front reaches the end of the aquifer in the simulation after somewhat less than three pore volumes have been displaced. At this point, the contaminant begins to pass out of the domain. Remediation of the aquifer, however, proceeds slowly, as shown in Figure 36.2. Even after 30 pore volumes have been flushed, some lead remains sorbed to weak sites in the aquifer sediments, and the Pb^{++} concentration in the groundwater remains above the limits set as drinking water standards (the MCL, or Maximum Contaminant Level, set by the USEPA is 0.015 mg kg^{-1}, or about 7×10^{-5} mmolal) across half the aquifer.

A notable aspect of the calculation results is their asymmetry: metal sorbs to the aquifer in a pattern that differs sharply from the way it desorbs. When the aquifer is invaded by contaminated water, the Pb^{++} complexes strongly with the ferric surface along a sharply defined reaction front. The front, separating clean from polluted water, migrates along the aquifer at one-half the velocity of the groundwater flow. Contaminating any portion of the aquifer, and the groundwater within it, then, requires that the water be replaced with polluted water just twice.

During elution by clean water, in contrast, a well-defined desorption front fails to develop. Instead, the metal desorbs gradually across a broad area. To clear enough lead from the aquifer to meet drinking water standards, the pore fluid in the simulation must be replaced dozens of times, much more often than was required initially to contaminate the aquifer.

These results differ sharply from the behavior predicted by the distribution coefficient (K_d) approach. This approach, despite being broadly acknowledged as too simplistic to describe the behavior of heavy metals, is nonetheless the sorption model most commonly applied in studying aquifer remediation.

To compare the approaches, we repeat the simulation using the reaction K_d method (Section 9.1) instead of surface complexation theory. By Equation 24.6, the distribution coefficient K_d' corresponding to a retardation factor of two has a value of 2.4×10^{-4} mol g^{-1}. Saving this value in dataset "Pb_Kd.sdat", we enter the X1T commands

```
(cont'd)
surface_data OFF
read Pb_Kd.sdat
go
```

to rerun the simulation.

In the calculation results (Fig. 36.3), we see that the Pb^{++} contamination migrates along the aquifer as a pulse, at half the speed of the groundwater flow. Unlike the previous simulation, in which the contaminant during flushing decreased in concentration as it sorbed to an increasingly large portion of the aquifer, metal concentration attenuates by dispersive mixing only. Behind the pulse, also in contrast to the previous simulation, the aqueous and sorbed concentrations of Pb^{++} fall rapidly to zero, leaving the aquifer and groundwater within it clean.

These results are misleading in two ways (Bethke and Brady, 2000; Brady and Bethke, 2000). First, they suggest that the lead contamination, once introduced to the aquifer, is highly mobile. In fact, once the source has been eliminated, the contaminant can migrate only slowly through the aquifer, because most of the metal sorbs to the aquifer and desorbs only slowly, limiting mobility. Second, the K_d results predict that flushing only a few pore volumes of clean water through the aquifer can displace the contamination, suggesting pump-and-treat remediation will be quick and effective. Models constructed with the surface complexation model, in contrast, depict pump-and-treat as a considerably slower and less effective remedy.

36.2 Groundwater Chromatography

In a second example of contaminant transport in the subsurface (Bethke, 1997), we consider the phenomenon known as "groundwater chromatography." When two or more species in a groundwater flow sorb at different strengths, the species may be subject to chromatographic effects. Such effects include segregation of the various species into individual bands in the sorbate and development of species concentrations in the flow greater than present originally in the unreacted fluid (Valocchi *et al.*, 1981).

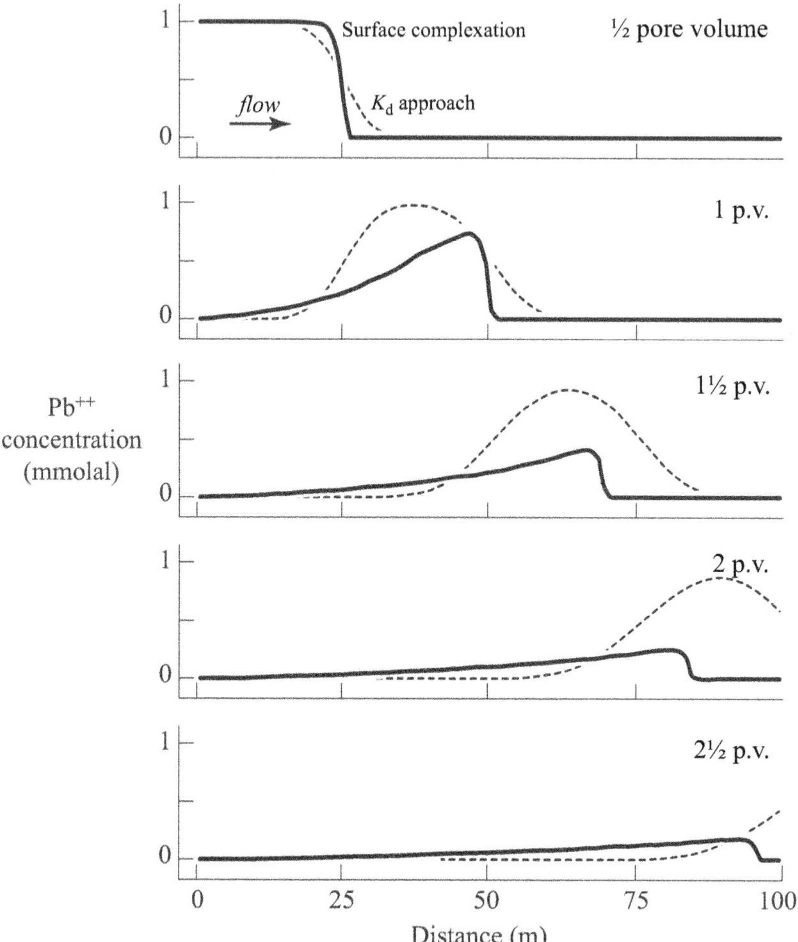

Figure 36.3 Comparison of the simulation results from Figure 36.1 (solid lines), which were calculated using a surface complexation model, with a parallel simulation in which sorption is figured by the reaction K_d approach (dashed lines). In each case, the retardation factor R_F for Pb^{++} transport is two.

The simulation is similar to the one in the previous section, except the inlet fluid contains three metal ions, Hg^{++}, Pb^{++}, and Zn^{++}. The ions sorb to weak sites according to

$$\begin{aligned} >\!(w)FeOH + Hg^{++} &\rightarrow >\!(w)FeOHg^{+} + H^{+} \\ >\!(w)FeOH + Pb^{++} &\rightarrow >\!(w)FeOPb^{+} + H^{+} \\ >\!(w)FeOH + Zn^{++} &\rightarrow >\!(w)FeOZn^{+} + H^{+}, \end{aligned} \tag{36.3}$$

and by parallel reactions for the strong sites, giving $>\!(s)FeOHg^{+}$, $>\!(s)FeOPb^{+}$, and $>\!(s)FeOZn^{+}$. From the corresponding binding constants,

	$\log K(w)$	$\log K(s)$
Hg^{++}	6.45	7.76
Pb^{++}	0.30	4.65
Zn^{++}	−1.99	0.99

(the constant for Pb^{++} at weak sites is estimated), we see that Hg^{++} binds more strongly to hydrous ferric oxide than Pb^{++}, which binds more strongly than Zn^{++}.

The procedure in X1T is

```
read FeOH+.sdat
decouple Fe+++
suppress PbCO3

interval start = 0 yr, fluid = contaminated
interval end = 1 yr

length = 100 m
Nx = 100
discharge = 1 pore_volume
porosity = 30%
dispersivity = 10 cm

scope initial
   swap Fe(OH)3(ppd) for Fe+++
   Fe(OH)3(ppd) = .005 volume%
   pH = 6
   Ca++  =    2.0 mmolal
   balance on HCO3-
   Hg++  = 1e-12 mmolal
   Pb++  = 1e-12 mmolal
   Zn++  = 1e-12 mmolal
   Br-   = 1e-12 mmolal
scope contaminated = initial
   Hg++  =      1 mmolal
   Pb++  =      1 mmolal
   Zn++  =      1 mmolal
   Br-   =      1 mmolal

precip = off
go
```

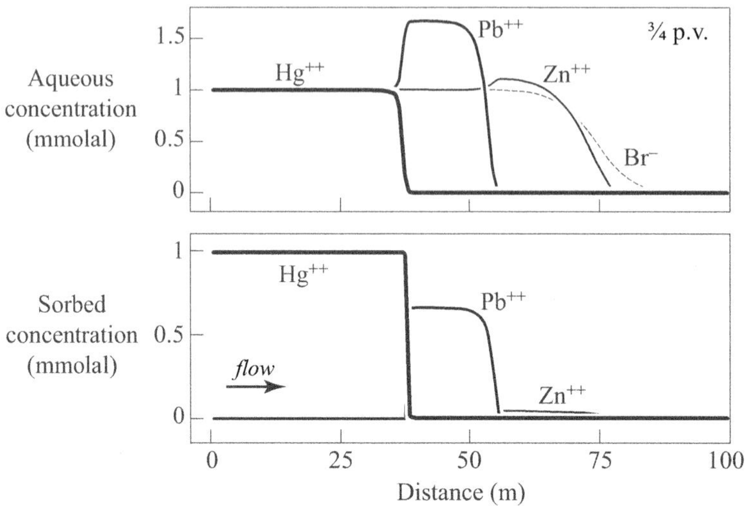

Figure 36.4 Chromatographic separation of metal contaminants in a groundwater flow at 25 °C, due to differential sorption. According to the surface complexation model used, Hg^{++} in the simulation sorbs more strongly to the ferric surface in the aquifer than Pb^{++}, which sorbs more strongly than Zn^{++}. Plot at top shows concentrations of the metal ions in groundwater, and bottom plot shows sorbed metal concentrations.

Figure 36.4 shows the calculation results.

As the contaminated fluid in the simulation passes into the aquifer, the Hg^{++} ions sorb to the ferric surface tightly enough to exclude the other metals. Once the Hg^{++} is exhausted from the fluid, the Pb^{++} begins to sorb, and the Zn^{++} sorbs where the Pb^{++} is depleted, leaving each metal bound within its own chromatographic band. Where Hg^{++} in the migrating fluid passes from the first to second band, it displaces Pb^{++},

$$>\text{(w)FeOPb}^{+} + \text{Hg}^{++} \rightarrow >\text{(w)FeOHg}^{+} + \text{Pb}^{++}\,, \tag{36.4}$$

from the sorbate, since both metals compete for the same sites.

At this point, there are two sources of dissolved Pb^{++}: the metal carried in solution from the inlet and that desorbed at this interface, due to competition with Hg^{++}. The Pb^{++} concentration downstream from this reaction, for this reason, rises well above that of the unreacted fluid. At the interface between the second and third bands, similarly, Pb^{++} displaces Zn^{++},

$$>\text{(w)FeOZn}^{+} + \text{Pb}^{++} \rightarrow >\text{(w)FeOPb}^{+} + \text{Zn}^{++}\,, \tag{36.5}$$

allowing the zinc to build up to levels above those at the inlet. As these reactions proceed, the interfaces between the bands—and hence the bands themselves—migrate slowly downstream, as do the pulses of highest Pb^{++} and Zn^{++} concentration.

37
Microbial Communities

Geochemists increasingly find a need to better understand the distribution of microbial life within the geosphere, and the interaction of the communities of microbes there with the fluids and minerals they contact. How do geochemical conditions determine where microbial communities develop, and what groups of microbes they contain? And how do those communities affect the geochemistry of their environments?

In many cases, microbial life in nature develops into zones within which communities are dominated by one or a few functional groups, such as aerobes, sulfate reducers, or methanogens. Distinct zoning is characteristic, for example, of microbial mats (Konhauser, 2007), hot springs (Fouke *et al.*, 2003), marine sediments and freshwater muds (Berner, 1980), contaminated aquifers (Bekins *et al.*, 1999), and pristine groundwater flows (Chapelle and Lovley, 1992). Communities develop as well in laboratory experiments, when microbes are cultivated in pure or mixed culture.

In this chapter, we consider how to construct quantitative models of the dynamics of microbial communities, building on our discussion of microbial kinetics in Chapter 19. In our modeling, we take care to account for how the ambient geochemistry controls microbial growth, and the effect of the growth on geochemical conditions.

37.1 Arsenate Reduction by *B. arsenicoselenatis*

Blum *et al.* (1998) isolated a bacterial strain *Bacillus arsenicoselenatis* from muds of Mono Lake, a hypersaline alkaline lake in northern California (see Section 28.2). Under anaerobic conditions in saline water, over an optimum pH range of 8.5–10, the strain can respire using As^{V}, or arsenate, as the electron acceptor, reducing it to As^{III}, arsenite.

In a batch experiment, Blum *et al.* (1998; their figure 4) were able to grow the strain at 20 °C on arsenate in saline water, using lactate as the electron donor. In this section, we develop a kinetic description of arsenate reduction in the experiment. Following Jin and Bethke (2003), we pay attention to thermodynamic as well as kinetic controls on the reaction rate.

In the experiment (Fig. 37.1), the strain oxidized the lactate to carbonate plus acetate, according to the reaction

$$\underset{\textit{lactate}}{CH_3CH(OH)COO^-} + 2\,HAsO_4^{--} + 2\,H_2O \rightarrow CO_3^{--} + \underset{\textit{acetate}}{CH_3COO^-} + 2\,As(OH)_4^- . \qquad (37.1)$$

Capturing energy liberated by the reaction, the microbial population grew almost 40-fold over the course of the experiment, which lasted about 3 days. Reaction proceeded increasingly rapidly as the microbial

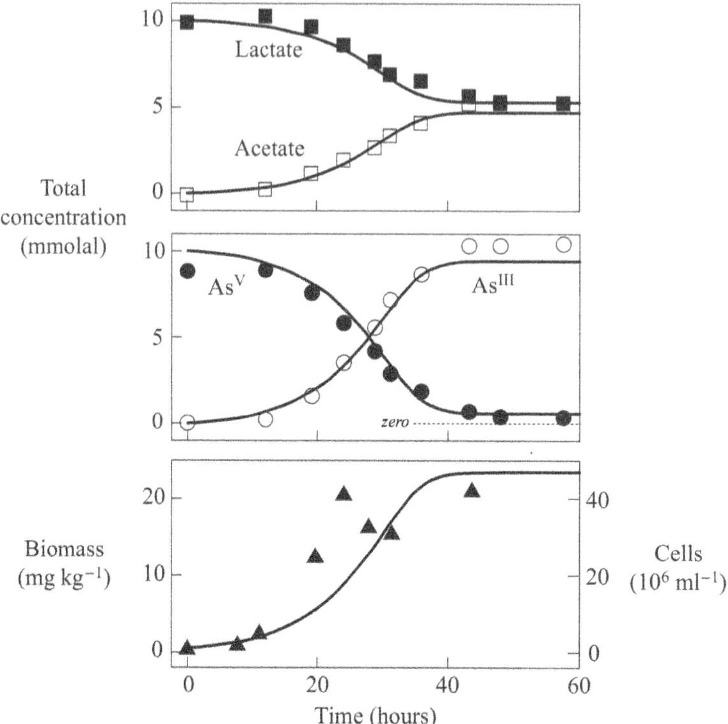

Figure 37.1 Results of a batch experiment (symbols) by Blum *et al.* (1998) in which *Bacillus arsenicoselenatis* grows on lactate, using arsenate (As^{V}) as an electron acceptor. Solid lines show results of integrating a kinetic rate model describing microbial respiration and growth.

population, and hence the system's catalytic ability increased. After about a day, however, the reaction started to slow, and then it stopped altogether, even though all the lactate and arsenate had not been consumed.

The growth medium was a saline fluid buffered to pH 9.8, containing as background electrolyte 80 g l^{-1} NaCl and 2.5 g l^{-1} Na_2CO_3, as well as various nutrients. As the experiment progressed, the authors sampled and analyzed the fluid for its lactate, acetate, arsenate, and arsenite content, and measured its cell density. To convert cell densities to biomass, we assume here that microbes on average weigh 0.5×10^{-12} g, as suggested by their dimensions in photomicrographs.[1]

Following the discussion in Chapter 19, the rate r (mol s^{-1}) at which the strain catalyzes the progress of Reaction 37.1 can be expressed as

$$r = n_w \, k_+ \, [X] \, F_D F_A F_T \, , \tag{37.2}$$

where n_w is solvent mass (kg), k_+ is the rate constant (mol mg^{-1} s^{-1}), $[X]$ is biomass (mg kg^{-1} of solvent), F_D and F_A are unitless kinetic factors representing the electron donating and accepting reactions, and F_T is the thermodynamic potential factor, also unitless. The latter is given by

$$F_T = 1 - \exp\left(\frac{\Delta G_r + m \, \Delta G_P}{\chi \, R T_K}\right) , \tag{37.3}$$

[1] This cell mass differs somewhat from that assumed by Jin and Bethke (2003), leading to slightly different values of kinetic parameters, when expressed per mg of biomass.

where ΔG_r is the free-energy change of Reaction 37.1, the energy available in the fluid; m is the number of ATPs produced per reaction turnover; ΔG_P is the energy (kJ mol^{-1}) needed to synthesize ATP in the cell; χ is the reaction's average stoichiometric number; R is the gas constant (8.3143 J mol^{-1} K^{-1}); and T_K is absolute temperature (K).

From the donating and accepting half-reactions, we assume in the absence of other information that the kinetic factors take the following forms:

$$F_D = \frac{m_{Lac}}{m_{Lac} + K_D\, m_{Ac} m_{CO_3^{--}}} \tag{37.4}$$

and

$$F_A = \frac{m_{HAsO_4^{--}}}{m_{HAsO_4^{--}} + K_A\, m_{As(OH)_4^-}}, \tag{37.5}$$

where m_{Lac} and m_{Ac} are the molal concentrations of lactate and acetate ions, the kinetic constant K_D has units of $molal^{-1}$, and K_A is unitless. We will assume these simple Monod-like forms are sufficiently accurate for our needs, but as discussed in Chapter 19 the nature of the kinetic factors can be demonstrated rigorously only through careful experimentation.

We can estimate n_{ATP} from energetics, assuming reaction ceases when the energy available in the fluid balances that needed to synthesize ATP, or independently from details of the lactate oxidation and arsenate reduction pathways (Jin and Bethke, 2003). Either way, the value is about 2.5. We can take ΔG_{ATP} under cellular conditions to be about 50 kJ mol^{-1}. Assuming the translocation of protons across the cellular membrane is the rate-limiting step, the average stoichiometric number χ is four, since one proton is translocated for each electron passing through the transport chain, and Reaction 37.1 transfers four electrons. The thermodynamic potential factor is now given by

$$F_T = 1 - \exp\left(\frac{\Delta G_r + 125\ \text{kJ mol}^{-1}}{4\, R T_K}\right), \tag{37.6}$$

from Equation 37.3.

We use Equation 19.11,

$$\frac{d[X]}{dt} = \frac{Yr}{n_w} - D\,[X], \tag{37.7}$$

to account for the accumulation of biomass due to microbial growth. The growth yield Y in this equation is given by the amount of biomass created in the experiment per mole of lactate consumed, about 5000 mg mol^{-1}, and we take the decay constant D to be zero over the brief experiment. Finally, the cell density of about 10^6 ml^{-1} observed at the onset of the experiment equates to 0.5 mg kg^{-1} of initial biomass. The only unknown values needed to integrate the rate law are the rate constant k_+ and the kinetic constants K_A and K_D.

Before running the model, we need to include lactate ion in the thermodynamic database. To do so, we add a redox couple,

$$\underset{lactate}{CH_3CH(OH)COO^-} + 3\,O_2(aq) \rightleftarrows 2\,H^+ + 3\,HCO_3^-, \tag{37.8}$$

with a log K at 20 °C of 231.4 and store the result in a dataset "thermo+Lactate.tdat". To model the experiment in REACT, we load the reaction dataset; decouple redox reactions for lactate, acetate, and arsenate; set the background electrolyte; and set initial concentrations of lactate and arsenate, and arbitrarily small quantities of the metabolic products acetate and arsenite

```
read thermo+Lactate.tdat
time end = 60 hours
T = 20
decouple Lactate, CH3COO-, AsO4---

swap CO3-- for HCO3-
Na+      =  1448     mmolal
Cl-      =  1400     mmolal
CO3--    =    24     mmolal
pH       =     9.8

Lactate =     10     mmolal
AsO4--- =     10     mmolal
CH3COO- =       .001 mmolal
As(OH)4- =      .001 mmolal
```

We then set a pH buffer and define the microbial rate law

```
(cont'd)
fix pH
kinetic microbe-Ba \
  rxn = "Lactate + 2*HAsO4-- + 2*H2O -> CH3COO- + CO3-- + 2*As(OH)4-", \
  biomass = .5, growth_yield = 5000, \
  ATP_energy = -50, ATP_number = 2.5, order1 = 1/4, \
  mpower(Lactate)    = 1, mpowerD(Lactate) = 1, \
  mpowerD(CH3COO-)  = 1, mpowerD(CO3--)    = 1, \
  mpower(HAsO4--)    = 1, mpowerA(HAsO4--) = 1, \
  mpowerA(As(OH)4-) = 1
```

To complete the calculation, we need to find values for the rate constant k_+ and kinetic constants K_D and K_A.

We set the kinetic constants to zero initially, forcing the kinetic factors F_D and F_A to one, their likely values at the onset of reaction. Adjusting k_+, we find a rate constant of 7×10^{-9} mol mg^{-1} s^{-1}

```
(cont'd)
kinetic microbe-Ba KD = 0, KA = 0
kinetic microbe-Ba rate_con = 7e-9
go
```

reproduces well the results of first stages of reaction. We then adjust K_D and K_A to obtain a slightly improved fit to data from later in the experiment

```
(cont'd)
kinetic microbe-Ba KD = 10, KA = .1
go
```

settling on values of 10 molal^{-1} for K_D and 0.1 (no units) for K_A. Figures 37.1–37.2 show the modeling results. The results, in fact, are not especially sensitive to the choice of K_D and K_A because, as shown in Figure 37.2, for appropriate choices of these variables the kinetic factors do not deviate from unity until after the thermodynamic potential factor has started to decrease sharply.

The principal controls on the microbial reaction rate in our example, then, are biomass and thermodynamic drive (Fig. 37.2). Initially, in the presence of abundant lactate and arsenate, the rate is controlled by the size of the microbial population available to catalyze lactate oxidation. As the population increases, so does reaction

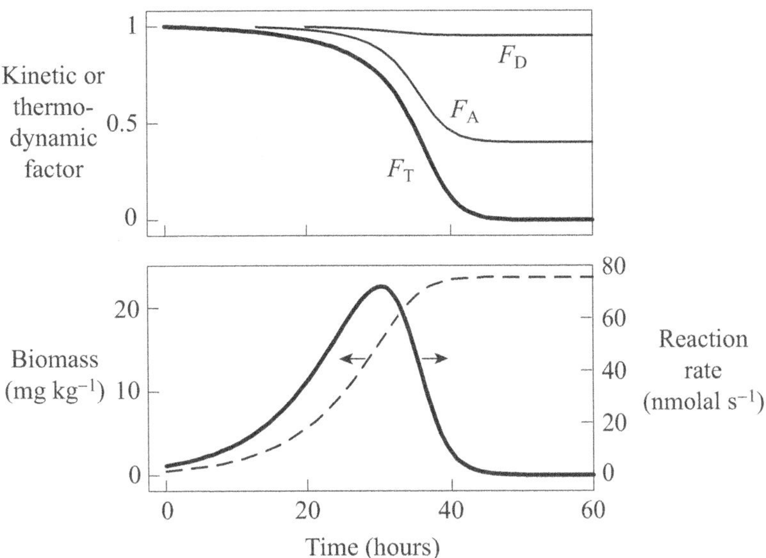

Figure 37.2 Factors controlling reaction rate (expressed per kg water, as r/n_w) in the simulation of bacterial arsenic reduction, including kinetic factors F_D and F_A, thermodynamic potential factor F_T, and biomass concentration $[X]$. Biomass concentration determines the rate early in the simulation, but later the thermodynamic drive exerts the dominant control.

rate. Later, as reactants are consumed and products accumulate, the reaction approaches the point at which the energy liberated by its progress is balanced by that needed in the cell to synthesize ATP. Reaction rate is governed now by the energy available to drive forward the cellular metabolism, this energy represented by the thermodynamic potential factor F_T; over the course of the experiment, the kinetic factors F_D and F_A play minor roles.

Such a pattern is not typical of microbial metabolisms that exploit more energetically favorable reactions. Aerobic oxidation of organic molecules, for example, generally liberates sufficiently large amounts of energy that, in the presence of detectable organic species and dioxygen, the reaction is favored strongly. As the reaction progresses, its rate is controlled by the kinetic factors, rather than the thermodynamic potential factor, which remains close to unity.

37.2 Zoning in an Aquifer

Pristine groundwater flows commonly pass through zones within which the activity of a single functional group of microorganisms appears to dominate microbial activity. Groundwater rich in dissolved iron, for example, seems to develop in anaerobic aquifers where iron-reducing bacteria are more active than sulfate reducers (Chapelle and Lovley, 1992). Methane accumulates where favored acceptors such as dioxygen, ferric iron, and sulfate are absent; under these conditions, hydrogentrophic methanogens can reduce bicarbonate to methane, and acetoclastic methanogens can cleave acetate into methane and bicarbonate (Roden and Wetzel, 2003). And where bacterial sulfate reduction dominates microbial activity, the concentrations of natural contaminants such as heavy metals and arsenic may be held low (Kirk *et al.*, 2004).

We consider here how reactive transport modeling might be used to describe the development of such zoning. Following Bethke *et al.* (2008), we take a clastic aquifer, 200 km long and with a porosity of 30%, through which groundwater flows at a discharge of 10 m^3 m^{-2} yr^{-1}. The groundwater contains 1 mmolal Ca^{++}, 2 mmolal HCO_3^-, 40 μmolal SO_4^{--}, and negligibly small initial amounts of acetate (CH_3COO^-), sulfide (HS^-), and methane (CH_4); its pH is 7.5. The simulation runs for 100 000 years, long enough for water to be replaced five times, and for microbial populations and groundwater composition to approach a steady state.

Aquifer sediments in the model are confined by and interleaved with fine-grained sediments that contain sedimentary organic matter. The organic matter decays gradually by microbial fermentation and anaerobic oxidation,

$$\underset{\text{organic matter}}{C_n \cdot m\,H_2O} \rightarrow \frac{n}{2}\,\underset{\text{acetate}}{CH_3COO^-} + \frac{n}{2}\,H^+ + (m-n)\,H_2O\,, \tag{37.9}$$

to simpler compounds, represented in the simulation by acetate.

At $t = 0$, acetate begins to diffuse from the fine-grained layers into the aquifer, where it can serve as the substrate for acetotrophic sulfate reduction,

$$CH_3COO^- + SO_4^{--} \rightarrow 2\,HCO_3^- + HS^-\,, \tag{37.10}$$

and acetoclastic methanogenesis,

$$CH_3COO^- + H_2O \rightarrow CH_4(aq) + HCO_3^-\,. \tag{37.11}$$

Acetate is added to the aquifer sediments at a rate of 4 μmol m^{-3} yr^{-1}, within the range of 2–5 μmol m^{-3} yr^{-1} observed by Park *et al.* (2006) in nonmarine sediments of a coastal plain aquifer.

The two functional groups of microbes in the aquifer, the sulfate-reducing bacteria and methanogens, are present initially in small amounts, just 10^{-6} mg kg^{-1}, but their populations can grow as they derive energy by metabolizing the acetate. The acetotrophic sulfate reducers proceed at a rate r (mol s^{-1}) given by Equation 19.16, the thermodynamically consistent dual Monod equation,

$$r = n_w\,k_+\,[X]\,\frac{m_{\text{Ac}}}{m_{\text{Ac}} + K'_{\text{D}}}\,\frac{m_{\text{SO}_4}}{m_{\text{SO}_4} + K'_{\text{A}}}\,F_{\text{T}}\,, \tag{37.12}$$

where m_{Ac} and m_{SO_4} are the molal concentrations of acetate and sulfate species, and K'_{D} and K'_{A} are the half-saturation constants (molal) for the electron donating and accepting half-reactions. The thermodynamic potential factor F_{T} is given by Equation 37.3, and Equation 37.7 describes growth and decay of the microbial population over the course of the simulation.

Following the calculations in Section 19.5, we take a rate constant k_+ for sulfate reduction of 10^{-9} mol mg^{-1} s^{-1}, a half-saturation constant K'_{D} for acetate of 70 μmolal, and a growth yield of 4300 mg mol^{-1} from a study of the kinetics of *Desulfobacter postgatei* by Ingvorsen *et al.* (1984). We set a value for K'_{A}, the half-saturation constant for sulfate, of 200 μmolal, as suggested by Ingvorsen *et al.* (1984) and Pallud and Van Cappellen (2006).

Taking the rate-limiting step in the electron transport chain to be trans-membrane proton translocation, which occurs about five times per sulfate consumed (Rabus *et al.*, 2006), the average stoichiometric number χ (entered into REACT as $\omega = 1/\chi$) for Reaction 37.10 is five. Sulfate reducers conserve about 45 kJ mol^{-1} of sulfate consumed (Qusheng Jin, unpublished data), so we set ΔG_{P} to this value and m to one.

For the methanogens, we use Equation 19.15, the thermodynamically consistent Monod equation,

$$r = n_w \, k_+ \, [X] \, \frac{m_{\mathrm{Ac}}}{m_{\mathrm{Ac}} + K'_{\mathrm{D}}} \, F_{\mathrm{T}} \,, \tag{37.13}$$

as the rate law. From an experimental study by Yang and Okos (1987) of the growth of *Methanosarcina sp.*, we take a rate constant k_+ of 2×10^{-9} mol mg^{-1} s^{-1}, a half-saturation constant K'_{D} of 20 mmolal, and a growth yield of 2000 mg mol^{-1}. (The notation K'_{D} here is arbitrary, since acetate serves in the metabolism as electron acceptor as well as donor.) The value assumed for the rate constant at 25 °C is one-half that observed at 35 °C, since reaction rate approximately doubles over this temperature interval (Huser *et al.*, 1982).

Experimental studies of *Methanosarcina* and current understanding of the organism's metabolic pathway allow us to estimate the parameters in the thermodynamic term (Qusheng Jin, personal communication). The methanogens conserve about 24 kJ (mol acetate)$^{-1}$, so we set ΔG_{P} to 48 kJ mol^{-1} and m to one-half. A double proton translocation occurs within the central metabolic pathway, furthermore, so, if we take these as the rate-limiting steps, the average stoichiometric number χ is two.

It is difficult to estimate the decay constant D from the results of laboratory experiments, since microbes in the natural environment are more likely to die from predation (e.g., Jurkevitch, 2007) than spontaneous decay. Instead, we figure a value from Equation 37.7, setting $\mathrm{d}[X]/\mathrm{d}t = 0$ to reflect the steady state. In this case, we see that the molal reaction rate, expressed per unit biomass,

$$\frac{r}{n_w \, [X]} = \frac{D}{Y} \,, \tag{37.14}$$

equals the ratio of the decay constant to growth yield. Substituting into this equation the rate law for sulfate-reducing bacteria (Eq. 37.12) and taking m_{Ac} to be $\ll K'_{\mathrm{D}}$ and $m_{\mathrm{SO_4}} \ll K'_{\mathrm{A}}$, the decay constant is given by

$$D = Y \, k_+ \, \frac{m_{\mathrm{Ac}}}{K'_{\mathrm{D}}} \, \frac{m_{\mathrm{SO_4}}}{K'_{\mathrm{A}}} \, F_{\mathrm{T}} \,. \tag{37.15}$$

If we choose as representative conditions 1 μmolal acetate, 10 μmolal sulfate, and a thermodynamic factor F_{T} of 0.4, and use the kinetic parameters already cited, D is about 10^{-9} s^{-1}, a value we carry for both functional groups of microbes.

To set up the simulation, we use the thermodynamic dataset from the calculation in Section 19.5, which was expanded to include mackinawite (FeS). As before, we suppress the iron sulfide minerals pyrite and troilite, and decouple acetate and methane from carbonate, and sulfide from sulfate. We set the aquifer to include a small amount of siderite, which serves as a sink for aqueous sulfide,

$$\underset{\textit{siderite}}{\mathrm{FeCO_3}} + \mathrm{HS^-} \rightarrow \underset{\textit{mackinawite}}{\mathrm{FeS}} + \mathrm{HCO_3^-} \,. \tag{37.16}$$

In this way, we avoid having the sulfide produced by the sulfate reducers accumulate in solution, inhibiting further reaction.

We begin in X1T by defining the initial system

```
read thermo+Mackinawite.tdat
suppress Pyrite, Troilite
decouple CH3COO-, CH4(aq), HS-

interval start =   0 m.y., fluid = recharge
interval end   = 0.1 m.y.
```

```
length = 200 km
Nx = 50
discharge = 10 m3/m2/yr
porosity = 30%
dispersivity = 1 m

scope initial
  Ca++    =   1.    mmolal
  HCO3-   =   2.    mmolal
  SO4--   =    .04  mmolal
  CH3COO- =    .001 umolal
  HS-     =    .001 umolal
  CH4(aq) =    .001 umolal
  swap Siderite for Fe++
  0.1 volume% Siderite
  pH = 7.5
  balance on HCO3-

scope recharge
  Ca++    =   1.    mmolal
  HCO3-   =   2.    mmolal
  SO4--   =    .04  mmolal
  CH3COO- =    .001 umolal
  HS-     =    .001 umolal
  CH4(aq) =    .001 umolal
  swap Siderite for Fe++
  pH = 7.5
  balance on HCO3-
```

We proceed to define the acetate source, set a forward-in-time solution (see Section 17.3.3), and prescribe the microbial reaction rates

(cont'd)

```
react 4 umol/m3yr CH3COO-
react 2 umol/m3yr Ca++

theta = 0
kinetic microbe-SRB \
  rxn = "CH3COO- + SO4-- -> 2*HCO3- + HS-", \
  biomass = 10^-6, growth_yield = 4300, decay_con = 10^-9, \
  ATP_energy = -45, ATP_number = 1, order1 = 1/5, \
  mpower(CH3COO-) = 1, mpowerD(CH3COO-) = 1, \
  mpower(SO4--)   = 1, mpowerA(SO4--)   = 1, \
  rate_con = 10^-9, KD = 70e-6, KA = 200e-6

kinetic microbe-Meth \
  rxn = "CH3COO- + H2O -> CH4(aq) + HCO3-", \
  biomass = 10^-6, growth_yield = 2000, decay_con = 10^-9, \
  ATP_energy = -48, ATP_number = 1/2, order1 = 1/2, \
  mpower(CH3COO-) = 1, mpowerD(CH3COO-) = 1, \
  rate_con = 2e-9, KD = 20e-3
```

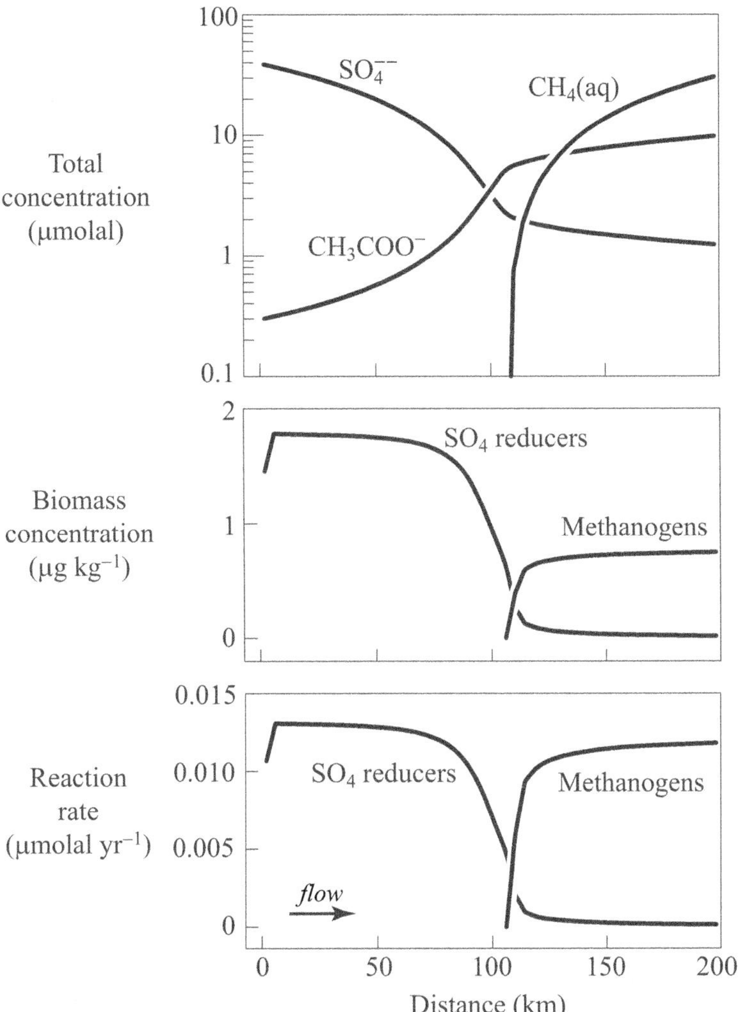

Figure 37.3 Steady-state distribution of microbial activity and groundwater composition in an aquifer hosting acetotrophic sulfate reduction and acetoclastic methanogenesis, obtained as the long-term solution to a reactive transport model.

```
dxplot .1 log
```

Typing `go` triggers the calculation.

Figures 37.3–37.4 show the results at the end of the simulation, after groundwater composition and microbial activity across the aquifer have approached steady state. Once the sulfate initially present is consumed or flushed from the aquifer, the only source of sulfate is in the recharging groundwater. With time in the simulation, sulfate-reducing bacteria grow into a community that consumes sulfate from the recharging groundwater and some of the acetate diffusing into the aquifer; the acetate and sulfate are consumed in equal molar proportions, according to Reaction 37.10. Biomass in the aquifer evolves at each point in the aquifer

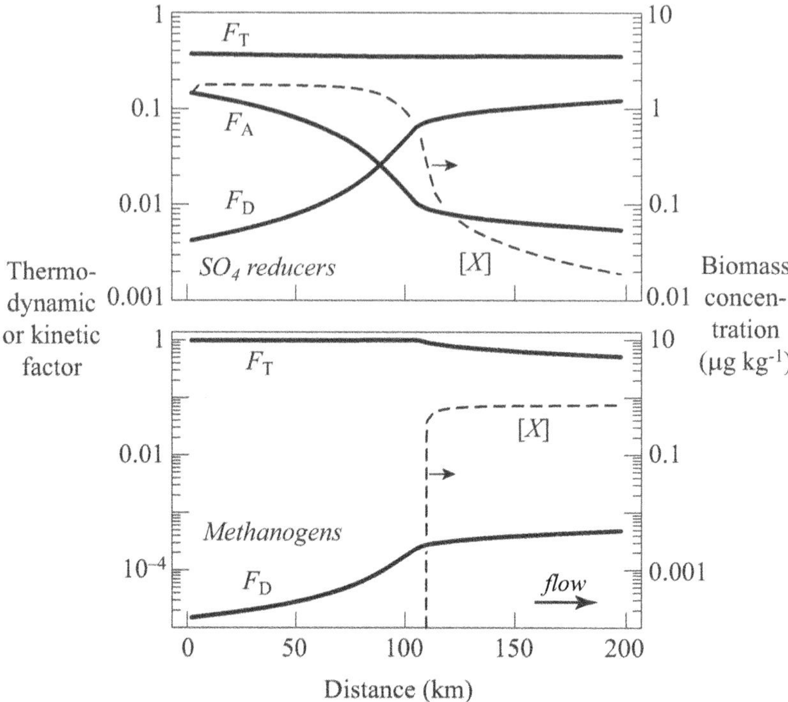

Figure 37.4 Factors controlling rates of microbial activity in the simulation depicted in Figure 37.3, for acetotrophic sulfate reduction (top) and acetoclastic methanogenesis (bottom). Factors include the thermodynamic potential factor F_T, kinetic factors $F_D = m_{Ac}/(m_{Ac} + K'_D)$ and $F_A = m_{SO_4}/(m_{SO_4} + K'_A)$, and biomass concentration $[X]$.

to the level needed for the concentrations of sulfate and acetate to satisfy Equation 37.15, the steady-state criterion.

Sulfate-reducing bacteria exclude methanogens completely from the upstream portion of the aquifer in an interesting way: they hold acetate concentration to a level at which acetoclastis proceeds at a rate insufficient to allow methanogens to grow as fast as they die. Specifically, as shown by the steady-state condition (Eq. 37.14), the criterion for maintaining a population of methanogens,

$$k_+ \frac{m_{Ac}}{K'_D} F_T = \frac{D}{Y}, \qquad (37.17)$$

is that reaction rate per unit biomass (the left side of the equation, from Equation 37.13) must match the ratio of the decay constant to growth yield. The sulfate reducers do not allow this and hence methanogens, even in the presence of a strong thermodynamic drive for acetoclastis (Fig. 37.4), cannot colonize this portion of the aquifer.

As groundwater flows along the aquifer, its sulfate content is gradually depleted by the sulfate reducers. The acetate added to the aquifer is not consumed completely by the microbes, so acetate concentration gradually rises, as required by Equation 37.15. About 100 km along the flow, acetate concentration rises to the point at which methanogens can maintain a population, according to Equation 37.17.

From this point, sulfate reducers are at a disadvantage. They must compete with methanogens for acetate, and do so in the presence of little sulfate, the paucity of which slows their metabolism (Fig. 37.4). Acetoclastis becomes the dominant metabolism downstream in the aquifer. Methanogens here grow to form a second microbiological zone, which they dominate. The zone is marked geochemically by a rise in acetate concentration and the accumulation in the flowing groundwater of dissolved methane.

Bethke *et al.* (2008) continued this line of reasoning by constructing models that allow a third functional group of microbes, the iron reducers, to populate the aquifer, and Park *et al.* (2009) used reactive transport modeling to illuminate the relationships between microbial activity and weathering in the Middendorf aquifer, in South Carolina, USA. Bethke *et al.* (2011) also approached the question of how the three functional groups interact in the subsurface, and control the chemistry of groundwater flowing there, from theoretical, experimental, and observational perspectives. The references provide considerable guidance on how quantitative models of microbial activity in the subsurface more elaborate than presented here might be constructed and interpreted.

Appendix A
Sources of Modeling Software

The following is a list, current at the time of publication, of sources of some of the most popular geochemical modeling software programs and packages. Some of the packages are available for download at no cost, whereas others may be licensed for a fee.

You should consult the software documentation or web page for details prior to seeking user assistance, keeping in mind that software authors do not necessarily provide this service themselves.

CHEAQS NEXT

origin: Wilko Verweij
internet: `www.cheaqs.eu`
reference: Verweij (2017), Verweij and Simonin (2020)

CHEMEQL

origin: Beat Müller
Eawag, Swiss Federal Institute of Aquatic Science and Technology
Kastanienbaum, Switzerland
internet: `www.eawag.ch/en/department/surf/projects/chemeql`
reference: Müller (2015)

CHESS, HYTEC

origin: Jan van der Lee
Centre de Géosciences, MINES ParisTech
Fontainebleau, France
reference: van der Lee *et al.* (2003), Sin *et al.* (2017)

CHIM-XPT

origin: Mark H. Reed
University of Oregon
Eugene, Oregon, USA
internet: `pages.uoregon.edu/palandri`
reference: Reed *et al.* (2016)

CRUNCHFLOW

origin: Carl Steefel
Lawrence Berkeley National Laboratory
Berkeley, California, USA
internet: `bitbucket.org/crunchflow/`
reference: Steefel *et al.* (2015)

EQ3/EQ6

origin: Thomas J. Wolery
Lawrence Livermore National Laboratory
Livermore, California, USA
internet: `www-gs.llnl.gov/-energy-homeland-security/geochemistry`
reference: Wolery (2003)

GEM-SELEKTOR, GEMS3K, and GEMSFITS

origin: Dmitrii Kulik and George Dan Miron
Paul Scherrer Institute
Villigen, Switzerland
internet: `gems.web.psi.ch`
reference: Wagner *et al.* (2012), Kulik *et al.* (2013), Miron *et al.* (2015)

GEOCHEM-EZ

origin: Jon E. Shaff
Cornell University
Ithaca, New York, USA
internet: `www.plantmineralnutrition.net/software/geochem_ez`
reference: Shaff *et al.* (2010)

The Geochemist's Workbench®

origin: University of Illinois
Urbana, Illinois, USA
internet: `www.GWB.com`
reference: Bethke *et al.* (2021)

JESS

origin: Peter May, Darren Rowland, Kevin Murray
Murdoch University
Perth, Western Australia, Australia
internet: `jess.murdoch.edu.au`

MEDUSA

origin: Ignasi Puigdomènech
KTH Royal Institute of Technology
Stockholm, Sweden
internet: `sites.google.com/site/chemdiagr`
reference: Puigdomènech *et al.* (2014)

MIN3P

origin: Ulrich Mayer
University of British Columbia
Vancouver, British Columbia, Canada
internet: `www.min3p.com`
reference: Mayer *et al.* (2002)

MINEQL+

origin: William Schecher
Environmental Research Software
Hallowell, Maine, USA
internet: `www.mineql.com`
references: Westall *et al.* (1976), Schecher and McAvoy (1994)

MINTEQA2

origin: US Environmental Protection Agency
National Exposure Research Laboratory
Athens, Georgia, USA
internet: `www.epa.gov/ceam/-minteqa2-equilibrium-speciation-model`
reference: Allison *et al.* (1991)

ORCHESTRA

origin: Hans Meeussen
internet: `orchestra.meeussen.nl`
reference: Meeussen (2003)

PFLOTRAN

origin: Peter Lichtner
internet: `www.pflotran.org`

PHREEQC, PHAST, and VS2DRT

origin: David Parkhurst
US Geological Survey
Lakewood, Colorado, USA
internet: `www.usgs.gov/software/phreeqc-version-3`
reference: Parkhurst and Appelo (2013), Parkhurst *et al.* (2010), Healy *et al.* (2018)

TOUGHREACT

origin: Lawrence Berkeley National Laboratory
Berkeley, California, USA
internet: `tough.lbl.gov/software/toughreact`
reference: Xu *et al.* (2014)

VISUAL MINTEQ

origin: Jon Petter Gustafsson
KTH Royal Institute of Technology
Stockholm, Sweden
internet: `vminteq.lwr.kth.se`

WATEQ4F

origin: US Geological Survey
Reston, Virginia, USA
internet: `water.usgs.gov/software/wateq4f.html`
reference: Ball and Nordstrom (1991)

WATCH

origin: Stefán Arnórsson, Sven Sigurdsson, Hördur Svavarsson
Iceland Geological Survey
Reykjavik, Iceland
internet: `en.isor.is/software`
reference: Arnórsson *et al.* (1982)

Appendix B

Evaluating the HMW Activity Model

The best way to fully understand the calculation procedure for the Harvie–Møller–Weare activity model (Harvie *et al.*, 1984) is to carry through a simple example by hand. In this appendix, we follow the steps in the procedure outlined in Tables 8.1–8.2, using the model coefficients given in Tables 8.3–8.6. A spreadsheet showing the calculation procedure below is installed along with GWB 2022 and later.

We take as an example a 6 molal NaCl solution containing 0.01 molal $CaSO_4$. Since the only species considered for this chemical system are Na^+, Cl^-, Ca^{++}, and SO_4^{--}, we can immediately write down the species molalities m_i along with their charges z_i:

	m_i	z_i
Na^+	6	+1
Cl^-	6	−1
Ca^{++}	0.01	+2
SO_4^{--}	0.01	−2

The only task left to us is to calculate the activity coefficients.

GIVEN DATA. The following data are model parameters from Tables 8.3–8.6.

$\beta^{(0)}_{MX}$	Cl^-	Ca^{++}	SO_4^{--}
Na^+	0.0765	–	0.01958
Cl^-	×	0.3159	–
Ca^{++}	×	×	0.2

$\beta^{(1)}_{MX}$	Cl^-	Ca^{++}	SO_4^{--}
Na^+	0.2664	–	1.113
Cl^-	×	1.614	–
Ca^{++}	×	×	3.1973

$\beta^{(2)}_{MX}$	Cl^-	Ca^{++}	SO_4^{--}
Na^+	0	–	0
Cl^-	×	0	–
Ca^{++}	×	×	−54.24

C^{ϕ}_{MX}	Cl^-	Ca^{++}	SO_4^{--}
Na^+	0.00127	–	0.00497
Cl^-	×	−0.00034	–
Ca^{++}	×	×	0

$\alpha^{(1)}_{MX}$	Cl^-	Ca^{++}	SO_4^{--}
Na^+	2	–	2
Cl^-	×	2	–
Ca^{++}	×	×	1.4

$\alpha^{(2)}_{MX}$	Cl^-	Ca^{++}	SO_4^{--}
Na^+	0	–	0
Cl^-	×	0	–
Ca^{++}	×	×	12

θ_{ij}	Cl^-	Ca^{++}	SO_4^{--}
Na^+	–	0.07	–
Cl^-	×	–	0.02
Ca^{++}	×	×	–

$\psi_{aa'\mathrm{Na}^+}$	Cl^-	Ca^{++}	SO_4^{--}
Na^+	–	–	–
Cl^-	×	–	0.0014
Ca^{++}	×	×	–

$\psi_{aa'\mathrm{Ca}^{++}}$	Cl^-	Ca^{++}	SO_4^{--}
Na^+	–	–	–
Cl^-	×	–	−0.018
Ca^{++}	×	×	–

$\psi_{cc'\mathrm{Cl}^-}$	Cl^-	Ca^{++}	SO_4^{--}
Na^+	–	−0.007	–
Cl^-	×	–	–
Ca^{++}	×	×	–

$\psi_{cc'\mathrm{SO}_4^{--}}$	Cl^-	Ca^{++}	SO_4^{--}
Na^+	–	−0.055	–
Cl^-	×	–	–
Ca^{++}	×	×	–

There are no neutral species; hence, no λ_{ni}.

STEP 1. The solution ionic strength I and total molal charge Z are

$$I = \tfrac{1}{2}\,[1 \times 6 \;+\; 1 \times 6 \;+\; 4 \times 0.01 \;+\; 4 \times 0.01] = 6.04 \text{ molal}$$

$$Z = 1 \times 6 \;+\; 1 \times 6 \;+\; 2 \times 0.01 \;+\; 2 \times 0.01 = 12.04 \text{ molal} \;.$$

STEP 2. Using the program in Table B.1, we calculate values for ${}^{E}\theta_{ij}(I)$ and ${}^{E}\theta'_{ij}(I)$:

${}^{E}\theta_{ij}(I)$	Cl^-	Ca^{++}	SO_4^{--}
Na^+	–	−0.05933	–
Cl^-	×	–	−0.05933
Ca^{++}	×	×	–

${}^{E}\theta'_{ij}(I)$	Cl^-	Ca^{++}	SO_4^{--}
Na^+	–	0.004861	–
Cl^-	×	–	0.004861
Ca^{++}	×	×	–

STEP 3. Values for functions $g(x)$ and $g'(x)$, taking $x = \alpha_{MX}^{(1)}\sqrt{I}$ and $\alpha_{MX}^{(2)}\sqrt{I}$, are

$g(\alpha_{MX}^{(1)}\sqrt{I})$	Cl^-	Ca^{++}	SO_4^{--}
Na^+	0.07919	–	0.07919
Cl^-	×	0.07919	–
Ca^{++}	×	×	0.1449

$g'(\alpha_{MX}^{(1)}\sqrt{I})$	Cl^-	Ca^{++}	SO_4^{--}
Na^+	−0.07186	–	−0.07186
Cl^-	×	−0.07186	–
Ca^{++}	×	×	−0.1129

Coefficent $\alpha_{MX}^{(2)}$ is set for Ca^{++}–SO_4^{--} only, yielding $g(\alpha_{MX}^{(2)}\sqrt{I})$ of 0.002299 for the pairing; the corresponding $g'(\alpha_{MX}^{(2)}\sqrt{I})$ is −0.002299.

Table B.1 *Program (ANSI C) for calculating* ${}^E\theta_{ij}(I)$ *and* ${}^E\theta'_{ij}(I)$ *at 25 °C*

```
#include <stdio.h>
#include <stdlib.h>
#include <math.h>

void calc_lambdas(const double is, double elambda[], double elambda1[]);
void calc_thetas(const int z1, const int z2, double elambda[], double elambda1[], double& etheta, double& etheta_prime);

int main() {
   double elambda[17], elambda1[17], etheta, etheta_prime, is;
   int z1, z2;

   while (printf("Enter I, z1, z2: ") && scanf("%lf %i %i", &is, &z1, &z2) == 3) {
      if (abs(z1) <= 4 && abs(z2) <= 4 && is > 0) {
         calc_lambdas(is, elambda, elambda1);
         calc_thetas(z1, z2, elambda, elambda1, etheta, etheta_prime);
         printf("E-theta(I) = %lf, E-theta'(I) = %f\n\n", etheta, etheta_prime);
      }
      else
         printf("Input data out of range\n");
   }
}

void calc_lambdas(const double is, double elambda[], double elambda1[]) {

   // Coefficients c1-c4 are used to approximate the integral function "J";
   //  aphi is the Debye-Huckel constant at 25 C

   const double c1 = 4.581, c2 = 0.7237, c3 = 0.0120, c4 = 0.528;
   const double aphi = 0.392;   /* Value at 25 C */

   // Calculate E-lambda terms for charge combinations of like sign,
   // using method of Pitzer (1975).

   for (int i = 1; i <= 4; i++) {
      for (int j = i; j <= 4; j++) {
         const int ij = i * j;
         const double zprod = (double)ij;
         const double x = 6.0 * zprod * aphi * sqrt(is);                        /* eqn 23 */
         const double jfunc = x / (4.0 + c1 * pow(x, -c2) * exp(-c3 * pow(x, c4)));     /* eqn 47 */

         const double t = c3 * c4 * pow(x, c4);
         const double dj = c1 * pow(x, (-c2 - 1.0)) * (c2 + t) * exp(-c3 * pow(x, c4));
         const double jprime = (jfunc / x) * (1.0 + jfunc * dj);

         elambda[ij] = zprod * jfunc / (4.0 * is);                              /* eqn 14 */
         elambda1[ij] = (3.0 * zprod * zprod * aphi * jprime / (4.0 * sqrt(is)) - elambda[ij]) / is;
      }
   }
}

void calc_thetas(const int z1, const int z2, double elambda[], double elambda1[],
                 double& etheta, double& etheta_prime) {

   // Calculate E-theta(I) and E-theta'(I) using method of Pitzer (1987)

   if (z1 * z2 < 0) {
      etheta = 0.0;
      etheta_prime = 0.0;
   }
   else {
      const int i = abs(z1);
      const int j = abs(z2);
      const double f1 = (double)i / (2.0 * (double)j);
      const double f2 = (double)j / (2.0 * (double)i);                          /* eqn A14 */
      etheta = elambda[i * j] - f1 * elambda[j * j] - f2 * elambda[i * i];
      etheta_prime = elambda1[i * j] - f1 * elambda1[j * j] - f2 * elambda1[i * i];
   }
}
```

STEP 4. The second virial coefficients for cation–anion pairs are

B_{MX}	Cl^-	Ca^{++}	SO_4^{--}
Na^+	0.09760	–	0.1077
Cl^-	×	0.4437	–
Ca^{++}	×	×	0.5386

B'_{MX}	Cl^-	Ca^{++}	SO_4^{--}
Na^+	−0.003169	–	−0.01324
Cl^-	×	−0.01920	–
Ca^{++}	×	×	−0.03909

B^{ϕ}_{MX}	Cl^-	Ca^{++}	SO_4^{--}
Na^+	0.07845	–	0.02774
Cl^-	×	0.3277	–
Ca^{++}	×	×	0.3024

STEP 5. The third virial coefficients for cation–anion pairs are

C_{MX}	Cl^-	Ca^{++}	SO_4^{--}
Na^+	0.0006350	–	0.001757
Cl^-	×	−0.0001202	–
Ca^{++}	×	×	0

STEP 6. The second virial coefficients for cation–cation and anion–anion pairs are

Φ_{ij}	Cl^-	Ca^{++}	SO_4^{--}
Na^+	–	0.01067	–
Cl^-	×	–	−0.03933
Ca^{++}	×	×	–

Φ'_{ij}	Cl^-	Ca^{++}	SO_4^{--}
Na^+	–	0.004861	–
Cl^-	×	–	0.004861
Ca^{++}	×	×	–

Φ^{ϕ}_{ij}	Cl^-	Ca^{++}	SO_4^{--}
Na^+	–	0.04003	–
Cl^-	×	–	−0.009970
Ca^{++}	×	×	–

STEP 7. From the above results, the value of F is

$$F = -1.2568 \ .$$

STEP 8. The ion activity coefficients are calculated,

	Na^+	Ca^{++}	Cl^-	SO_4^{--}
$z_i^2 F$	−1.2568	−5.0271	−1.2568	−5.0271
First sum	1.2194	5.3266	1.2259	1.4303
Second sum	−0.0002	−0.1273	−0.0007	−0.4226
Third sum	0.0001	−0.0011	−0.0004	−0.0033
Fourth sum	0.0230	0.0459	0.0230	0.0459
Fifth sum	0.0	0.0	0.0	0.0
$\ln \gamma_i$	−0.0145	0.2172	−0.0090	−3.9768
γ_i	0.9856	1.2425	0.9910	0.0187

STEP 9. The quantity $\sum_i m_i \ (\phi - 1)$ is calculated,

First term	−1.4735	
First sum	3.1211	
Second sum	−0.0002	
Third sum	−0.0001	
Fourth sum	0.0	
Fifth sum	0.0	
	1.6484	× 2 = 3.2968

STEP 10. The activity of water is given,

$\sum_i m_i$	12.02
$\left(\sum_i m_i\right) \phi$	15.3168
$\ln a_w$	−0.2759
a_w	0.7589

We can use program SPECE8 to quickly verify our calculations:

```
read thermo_hmw.tdat
Na+   = 6 molal
Cl-   = 6 molal
Ca++  = .01 molal
SO4-- = .01 molal
go
```

The values for γ_i and a_w in the program output can be compared to the results obtained in Steps 8 and 10.

Appendix C
Minerals in the LLNL Database

Mineral	Chemical formula	General type
Acanthite	Ag_2S	sulfide
Akermanite	$Ca_2MgSi_2O_7$	
Alabandite	MnS	sulfide
Albite	$NaAlSi_3O_8$	feldspar
Albite high	$NaAlSi_3O_8$	feldspar
Albite low	$NaAlSi_3O_8$	feldspar
Alstonite	$BaCa(CO_3)_2$	carbonate
Alunite	$KAl_3(OH)_6(SO_4)_2$	sulfate
Amesite-14A	$Mg_4Al_4Si_2O_{10}(OH)_8$	serpentine
Amrph^silica	SiO_2	silica
Analc-dehydr	$NaAlSi_2O_6$	zeolite
Analcime	$NaAlSi_2O_6 \cdot H_2O$	zeolite
Andalusite	Al_2SiO_5	
Andradite	$Ca_3Fe_2(SiO_4)_3$	garnet
Anglesite	$PbSO_4$	sulfate
Anhydrite	$CaSO_4$	sulfate
Annite	$KFe_3AlSi_3O_{10}(OH)_2$	mica
Anorthite	$CaAl_2Si_2O_8$	feldspar
Antarcticite	$CaCl_2 \cdot 6H_2O$	halide
Anthophyllite	$Mg_7Si_8O_{22}(OH)_2$	amphibole
Antigorite	$Mg_{24}Si_{17}O_{42.5}(OH)_{31}$	serpentine
Aragonite	$CaCO_3$	carbonate
Arcanite	K_2SO_4	sulfate
Arsenolite	As_2O_3	oxide
Arsenopyrite	$AsFeS$	sulfide
Artinite	$Mg_2CO_3(OH)_2 \cdot 3H_2O$	carbonate
Azurite	$Cu_3(CO_3)_2(OH)_2$	carbonate
Barite	$BaSO_4$	sulfate
Barytocalcite	$BaCa(CO_3)_2$	carbonate

(continues)

Mineral	Chemical formula	General type
Bassanite	$CaSO_4 \cdot .5H_2O$	sulfate
Bassetite	$Fe(UO_2)_2(PO_4)_2$	phosphate
Beidellit-Ca	$Ca_{.165}Al_{2.33}Si_{3.67}O_{10}(OH)_2$	smectite
Beidellit-H	$H_{.33}Al_{2.33}Si_{3.67}O_{10}(OH)_2$	smectite
Beidellit-K	$K_{.33}Al_{2.33}Si_{3.67}O_{10}(OH)_2$	smectite
Beidellit-Mg	$Mg_{.165}Al_{2.33}Si_{3.67}O_{10}(OH)_2$	smectite
Beidellit-Na	$Na_{.33}Al_{2.33}Si_{3.67}O_{10}(OH)_2$	smectite
Berlinite	$AlPO_4$	phosphate
Bieberite	$CoSO_4 \cdot 7H_2O$	sulfate
Birnessite	$Mn_8O_{19}H_{10}$	
Bischofite	$MgCl_2 \cdot 6H_2O$	chloride
Bixbyite	Mn_2O_3	oxide
Bloedite	$Na_2Mg(SO_4)_2 \cdot 4H_2O$	sulfate
Boehmite	$AlOOH$	hydroxide
Boltwood-Na	$Na_{.7}K_{.3}H_3OUO_2SiO_4 \cdot H_2O$	
Boltwoodite	$K(H_3O)UO_2(SiO_4)$	
Borax	$Na_2B_4O_5(OH)_4 \cdot 8H_2O$	borate
Boric acid	$B(OH)_3(c)$	
Bornite	Cu_5FeS_4	sulfide
Brezinaite	Cr_3S_4	sulfide
Brucite	$Mg(OH)_2$	hydroxide
Burkeite	$Na_6CO_3(SO_4)_2$	sulfide
Ca-Al Pyroxene	$CaAl_2SiO_6$	pyroxene
Calcite	$CaCO_3$	carbonate
Carnallite	$KMgCl_3 \cdot 6H_2O$	halide
Carnotite	$K_2(UO_2)_2(VO_4)_2$	vanadate
Cattierite	CoS_2	sulfide
Celestite	$SrSO_4$	sulfate
Cerussite	$PbCO_3$	carbonate
Chalcedony	SiO_2	silica
Chalcocite	Cu_2S	sulfide
Chalcopyrite	$CuFeS_2$	sulfide
Chamosite-7A	$Fe_2Al_2SiO_5(OH)_4$	7 Å clay
Chloromagnesite	$MgCl_2$	halide
Chloropyromorphite	$Pb_5(PO_4)_3Cl$	phosphate
Chrysotile	$Mg_3Si_2O_5(OH)_4$	serpentine
Cinnabar	HgS	sulfide
Claudetite	As_2O_3	oxide
Clinochl-14A	$Mg_5Al_2Si_3O_{10}(OH)_8$	chlorite
Clinochl-7A	$Mg_5Al_2Si_3O_{10}(OH)_8$	
Clinoptil-Ca	$CaAl_2Si_{10}O_{24} \cdot 8H_2O$	zeolite
Clinoptil-K	$K_2Al_2Si_{10}O_{24} \cdot 8H_2O$	zeolite

(continues)

Mineral	Chemical formula	General type
Clinoptil-Mg	$MgAl_2Si_{10}O_{24} \cdot 8H_2O$	zeolite
Clinoptil-Na	$Na_2Al_2Si_{10}O_{24} \cdot 8H_2O$	zeolite
Clinozoisite	$Ca_2Al_3Si_3O_{12}(OH)$	epidote
Coffinite	$USiO_4$	epidote
Colemanite	$Ca_2B_6O_8(OH)_6 \cdot 2H_2O$	borate
Copper	Cu	native element
Cordier^anhy	$Mg_2Al_4Si_5O_{18}$	
Cordier^hydr	$Mg_2Al_4Si_5O_{18} \cdot H_2O$	
Corundum	Al_2O_3	oxide
Covellite	CuS	sulfide
Cristobalite	SiO_2	silica
Cronstedt-7A	$Fe_4SiO_5(OH)_4$	serpentine
Cuprite	Cu_2O	oxide
Daphnite-14A	$Fe_5Al_2Si_3O_{10}(OH)_8$	chlorite
Daphnite-7A	$Fe_5Al_2Si_3O_{10}(OH)_8$	
Dawsonite	$NaAlCO_3(OH)_2$	carbonate
Diaspore	$AlHO_2$	hydroxide
Diopside	$CaMgSi_2O_6$	pyroxene
Dolomite	$CaMg(CO_3)_2$	carbonate
Dolomite-dis	$CaMg(CO_3)_2$	carbonate
Dolomite-ord	$CaMg(CO_3)_2$	carbonate
Enstatite	$MgSiO_3$	pyroxene
Epidote	$Ca_2FeAl_2Si_3O_{12}OH$	epidote
Epidote-ord	$Ca_2FeAl_2Si_3O_{12}OH$	epidote
Epsomite	$MgSO_4 \cdot 7H_2O$	sulfate
Eu	Eu	native element
Eucryptite	$LiAlSiO_4$	
Fayalite	Fe_2SiO_4	olivine
Ferrite-Ca	$CaFe_2O_4(c)$	
Ferrite-Cu	$CuFe_2O_4(c)$	
Ferrite-Mg	$MgFe_2O_4(c)$	
Ferrite-Zn	$ZnFe_2O_4$	
Ferrosilite	$FeSiO_3$	pyroxene
Fluorapatite	$Ca_5(PO_4)_3F$	phosphate
Fluorite	CaF_2	fluoride
Forsterite	Mg_2SiO_4	olivine
Galena	PbS	sulfide
Gaylussite	$CaNa_2(CO_3)_2 \cdot 5H_2O$	carbonate
Gehlenite	$Ca_2Al_2SiO_7$	
Gibbsite	$Al(OH)_3$	hydroxide
Goethite	FeOOH	hydroxide
Gold	Au	native element

(continues)

Mineral	Chemical formula	General type
Graphite	C	native element
Greenalite	$Fe_3Si_2O_5(OH)_4$	serpentine
Grossular	$Ca_3Al_2Si_3O_{12}$	garnet
Gummite	UO_3	oxide
Gypsum	$CaSO_4 \cdot 2H_2O$	sulfate
Haiweeite	$Ca(UO_2)_2(Si_2O_5)_3 \cdot 5H_2O$	
Halite	$NaCl$	halide
Hausmannite	Mn_3O_4	oxide
Hedenbergite	$CaFe(SiO_3)_2$	pyroxene
Hematite	Fe_2O_3	oxide
Hercynite	$FeAl_2O_4$	oxide
Heulandite	$CaAl_2Si_7O_{18} \cdot 6H_2O$	zeolite
Hexahydrite	$MgSO_4 \cdot 6H_2O$	sulfate
Hinsdalite	$PbAl_3(PO_4)(SO_4)(OH)_6$	phosphate
Huntite	$CaMg_3(CO_3)_4$	carbonate
Hydroboracite	$MgCa(B_6O_{11}) \cdot 6H_2O$	borate
Hydromagnesite	$Mg_5(CO_3)_4(OH)_2 \cdot 4H_2O$	carbonate
Hydrophilite	$CaCl_2$	halide
Hydroxyapatite	$Ca_5(PO_4)_3OH$	phosphate
Hydroxypyromorphite	$Pb_5(PO_4)_3OH$	phosphate
Illite	$K_{.6}Mg_{.25}Al_{2.3}Si_{3.5}O_{10}(OH)_2$	10 Å clay
Jadeite	$NaAl(SiO_3)_2$	pyroxene
Jarosite-K	$KFe_3(SO_4)_2(OH)_6$	sulfate
Jarosite-Na	$NaFe_3(SO_4)_2(OH)_6$	sulfate
K-feldspar	$KAlSi_3O_8$	feldspar
Kainite	$KMgClSO_4 \cdot 3H_2O$	sulfate
Kalicinite	$KHCO_3$	carbonate
Kalsilite	$KAlSiO_4$	feldspathoid
Kaolinite	$Al_2Si_2O_5(OH)_4$	7 Å clay
Kasolite	$PbUO_2SiO_4 \cdot H_2O$	
Kieserite	$MgSO_4 \cdot H_2O$	sulfate
Kyanite	Al_2SiO_5	
Larnite	Ca_2SiO_4	olivine
Laumontite	$CaAl_2Si_4O_{12} \cdot 4H_2O$	zeolite
Lawrencite	$FeCl_2$	halide
Lawsonite	$CaAl_2Si_2O_7(OH)_2 \cdot H_2O$	epidote
Leonhardtite	$MgSO_4 \cdot 4H_2O$	sulfate
Lime	CaO	oxide
Linnaeite	Co_3S_4	sulfide
Magnesite	$MgCO_3$	carbonate
Magnetite	Fe_3O_4	oxide
Malachite	$Cu_2CO_3(OH)_2$	carbonate

(continues)

Mineral	Chemical formula	General type
Manganite	$MnOOH$	hydroxide
Manganosite	MnO	oxide
Margarite	$CaAl_4Si_2O_{10}(OH)_2$	mica
Maximum Microcline	$KAlSi_3O_8$	feldspar
Melanterite	$FeSO_4 \cdot 7H_2O$	sulfate
Mercallite	$KHSO_4$	sulfate
Merwinite	$Ca_3Mg(SiO_4)_2$	olivine
Metacinnabar	HgS	sulfide
Minnesotaite	$Fe_3Si_4O_{10}(OH)_2$	mica
Mirabilite	$Na_2SO_4 \cdot 10H_2O$	sulfate
Misenite	$K_8H_6(SO_4)_7$	sulfate
Modderite	$CoAs$	arsenide
Molysite	$FeCl_3$	halide
Monohydrocalcite	$CaCO_3 \cdot H_2O$	carbonate
Monticellite	$CaMgSiO_4$	olivine
Mordenite-K	$KAlSi_5O_{12} \cdot 3H_2O$	zeolite
Mordenite-Na	$NaAlSi_5O_{12} \cdot 3H_2O$	zeolite
Muscovite	$KAl_3Si_3O_{10}(OH)_2$	mica
Nepheline	$NaAlSiO_4$	feldspathoid
Nesquehonite	$Mg(HCO_3)(OH) \cdot 2H_2O$	carbonate
Ningyoite	$CaU(PO_4)_2 \cdot 2H_2O$	phosphate
Nontronit-Ca	$Ca_{.165}Fe_2Al_{.33}Si_{3.67}O_{10}(OH)_2$	smectite
Nontronit-K	$K_{.33}Fe_2Al_{.33}Si_{3.67}O_{10}(OH)_2$	smectite
Nontronit-Mg	$Mg_{.165}Fe_2Al_{.33}Si_{3.67}O_{10}(OH)_2$	smectite
Nontronit-Na	$Na_{.33}Fe_2Al_{.33}Si_{3.67}O_{10}(OH)_2$	smectite
Orpiment	As_2S_3	sulfide
Paragonite	$NaAl_3Si_3O_{10}(OH)_2$	mica
Pargasite	$NaCa_2Al_3Mg_4Si_6O_{22}(OH)_2$	amphibole
Pentahydrite	$MgSO_4 \cdot 5H_2O$	sulfate
Petalite	$Li_2Al_2Si_8O_{20}$	feldspathoid
Phengite	$KAlMgSi_4O_{10}(OH)_2$	mica
Phlogopite	$KAlMg_3Si_3O_{10}(OH)_2$	mica
Pirssonite	$Na_2Ca(CO_3)_2 \cdot 2H_2O$	carbonate
Plumbogummite	$PbAl_3(PO_4)_2(OH)_5 \cdot H_2O$	phosphate
Portlandite	$Ca(OH)_2$	hydroxide
Prehnite	$Ca_2Al_2Si_3O_{10}(OH)_2$	mica
Przhevalskite	$Pb(UO_2)_2(PO_4)_2$	phosphate
Pseudowollastonite	$CaSiO_3$	
Pyrite	FeS_2	sulfide
Pyrolusite	MnO_2	oxide
Pyrophyllite	$Al_2Si_4O_{10}(OH)_2$	mica
Pyrrhotite	$Fe_{.875}S$	sulfide

(continues)

Mineral	Chemical formula	General type
Quartz	SiO_2	silica
Quicksilver	Hg	native element
Rankinite	$Ca_3Si_2O_7$	
Realgar	AsS	sulfide
Rhodochrosite	$MnCO_3$	carbonate
Rhodonite	$MnSiO_3$	pyroxene
Ripidolit-14A	$Fe_2Mg_3Al_2Si_3O_{10}(OH)_8$	chlorite
Ripidolit-7A	$Fe_2Mg_3Al_2Si_3O_{10}(OH)_8$	
Rutherfordine	UO_2CO_3	carbonate
Safflorite	$CoAs_2$	arsenide
Saleeite	$Mg(UO_2)_2(PO_4)_2$	phosphate
Sanidine high	$KAlSi_3O_8$	feldspar
Saponite-Ca	$Ca_{.165}Mg_3Al_{.33}Si_{3.67}O_{10}(OH)_2$	smectite
Saponite-H	$H_{.33}Mg_3Al_{.33}Si_{3.67}O_{10}(OH)_2$	smectite
Saponite-K	$K_{.33}Mg_3Al_{.33}Si_{3.67}O_{10}(OH)_2$	smectite
Saponite-Mg	$Mg_{3.165}Al_{.33}Si_{3.67}O_{10}(OH)_2$	smectite
Saponite-Na	$Na_{.33}Mg_3Al_{.33}Si_{3.67}O_{10}(OH)_2$	smectite
Scacchite	$MnCl_2$	halide
Schoepite	$UO_2(OH)_2 \cdot H_2O$	hydroxide
Scorodite	$FeAsO_4 \cdot 2H_2O$	arsenide
Sepiolite	$Mg_4Si_6O_{15}(OH)_2 \cdot 6H_2O$	
Siderite	$FeCO_3$	carbonate
Sillimanite	Al_2SiO_5	
Silver	Ag	native element
Sklodowskite	$Mg(UO_2)_2(SiO_4)_2O_6H_{14}$	
Smectite-Reykjanes	$Na_{.33}Ca_{.66}K_{.03}Mg_{1.29}Fe_{.68}\cdot$ $Mn_{.01}Al_{1.11}Si_{3.167}O_{10}(OH)_2$	smectite
Smectite-high-Fe-Mg	$Na_{.1}Ca_{.025}K_{.2}Mg_{1.15}Fe_{.7}\cdot$ $Al_{1.25}Si_{3.5}O_{10}(OH)_2$	smectite
Smectite-low-Fe-Mg	$Na_{.15}Ca_{.02}K_{.2}Mg_{.9}Fe_{.45}\cdot$ $Al_{1.25}Si_{3.75}O_{10}(OH)_2$	smectite
Smithsonite	$ZnCO_3$	carbonate
Soddyite	$(UO_2)_2(SiO_4) \cdot 2H_2O$	
Sphalerite	ZnS	sulfide
Spinel	Al_2MgO_4	oxide
Spodumene-a	$LiAlSi_2O_6$	pyroxene
Strengite	$FePO_4 \cdot 2H_2O$	phosphate
Strontianite	$SrCO_3$	carbonate
Sulfur-Rhmb	S	native element
Sylvite	KCl	halide
Tachyhydrite	$Mg_2CaCl_6 \cdot 12H_2O$	halide
Talc	$Mg_3Si_4O_{10}(OH)_2$	mica

(continues)

Mineral	Chemical formula	General type
Tenorite	CuO	oxide
Tephroite	Mn_2SiO_4	olivine
Thenardite	Na_2SO_4	sulfate
Thorianite	ThO_2	oxide
Todorokite	$Mn_7O_{12} \cdot 3H_2O$	oxide
Torbernite	$Cu(UO_2)_2(PO_4)_2$	phosphate
Tremolite	$Ca_2Mg_5Si_8O_{22}(OH)_2$	amphibole
Tridymite	SiO_2	silica
Troilite	FeS	sulfide
Tsumebite	$Pb_2Cu(PO_4)(OH)_3 \cdot 3H_2O$	phosphate
Tyuyamunite	$Ca(UO_2)_2(VO_4)_2$	vanadate
Uraninite	UO_2	oxide
Uranocircite	$Ba(UO_2)_2(PO_4)_2$	phosphate
Uranophane	$Ca(H_2O)_2(UO_2)_2(SiO_2)_2(OH)_6$	hydroxide
Vivianite	$Fe_3(PO_4)_2 \cdot 8H_2O$	phosphate
Wairakite	$CaAl_2Si_4O_{10}(OH)_4$	zeolite
Weeksite	$K_2(UO_2)_2(Si_2O_5)_3 \cdot 4H_2O$	
Whitlockite	$Ca_3(PO_4)_2$	phosphate
Witherite	$BaCO_3$	carbonate
Wollastonite	$CaSiO_3$	
Wurtzite	ZnS	sulfide
Wustite	$Fe_{.947}O$	oxide
Zoisite	$Ca_2Al_3Si_3O_{12}(OH)$	epidote

Appendix D
Nonlinear Rate Laws

As noted in Chapter 17, transition state theory does not require that kinetic rate laws take a linear form, although most kinetic studies have assumed that they do. The rate law for reaction of a mineral $A_{\vec{k}}$, for example, can be expressed in the general nonlinear form

$$r_{\vec{k}} = \text{sgn}\left(1 - \frac{Q_{\vec{k}}}{K_{\vec{k}}}\right) (A_S\, k_+)_{\vec{k}} \prod^{\vec{j}} (m_{\vec{j}})^{p_{\vec{j}\vec{k}}} \left| 1 - \left(\frac{Q_{\vec{k}}}{K_{\vec{k}}}\right)^{\omega} \right|^{\Omega} , \tag{D.1}$$

where ω and Ω are arbitrary exponents that are determined empirically (e.g., Steefel and Lasaga, 1994); nonlinear rate laws for redox reaction, complexation, and so on assume parallel forms.

In this equation, sgn is a function that borrows the sign of its argument; it equals positive one when the fluid is undersaturated and negative one when it is supersaturated. This equation resembles the linear form of the rate law (Eq. 17.2) except for the presence of the exponents ω and Ω. When the values of ω and Ω are set to one, the rate law reduces to its linear form.

To incorporate nonlinear rate laws into the solution procedure for tracing kinetic reaction paths (Section 17.3), we need to find the partial derivative of the reaction rate $r_{\vec{k}}$ with respect to the molalities m_i of the basis species A_i. The derivatives are given by

$$\begin{aligned} \frac{\partial r_{\vec{k}}}{\partial m_i} = \text{sgn}\left(1 - \frac{Q_{\vec{k}}}{K_{\vec{k}}}\right) \frac{(A_S\, k_+)_{\vec{k}}}{m_i} \prod^{\vec{j}} (m_{\vec{j}})^{p_{\vec{j}\vec{k}}} \times \\ \left\{ - \text{sgn}\left(1 - \frac{Q_{\vec{k}}}{K_{\vec{k}}}\right) \omega\, \Omega\, \nu_{i\vec{k}} \left(\frac{Q_{\vec{k}}}{K_{\vec{k}}}\right)^{\omega} \left| 1 - \left(\frac{Q_{\vec{k}}}{K_{\vec{k}}}\right)^{\omega} \right|^{\Omega-1} + \right. \\ \left. \sum_{\vec{j}} \nu_{i\vec{j}}\, p_{\vec{j}\vec{k}} \left| 1 - \left(\frac{Q_{\vec{k}}}{K_{\vec{k}}}\right)^{\omega} \right|^{\Omega} \right\} . \end{aligned} \tag{D.2}$$

This formula replaces Equation 17.17 in the calculation procedure.

To illustrate the effects of ω and Ω on reaction rates, we consider the reaction of quartz with deionized water, from Chapter 17. As before, we begin in REACT

```
time begin = 0 days, end = 5 days
T = 100
pH = 7
Cl-      = 10 umolal
Na+      = 10 umolal
```

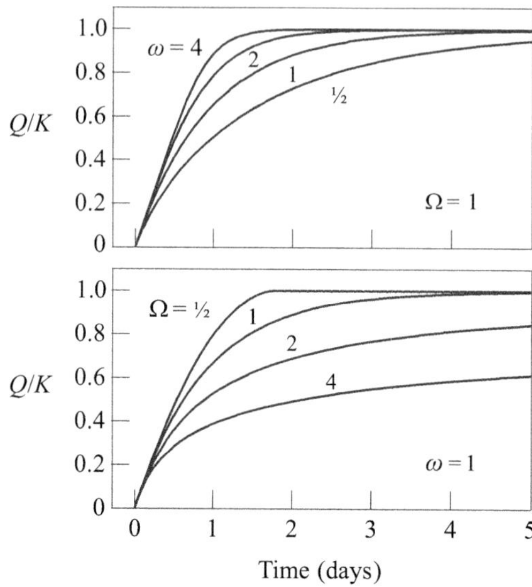

Figure D.1 Variation of quartz saturation with time as quartz sand reacts at 100 °C with deionized water, calculated according to nonlinear forms of a kinetic rate law using various values of ω and Ω.

```
SiO2(aq) =  1 umolal

react 5000 g Quartz
kinetic Quartz rate_con = 2.e-15  surface = 1000
```

and then set ω and Ω (keywords `order1` and `order2`, respectively) to differing values. For example,

```
(cont'd)
kinetic Quartz order1 = 1/2, order2 = 1
go
```

Figure D.1 shows the calculation results.

References

Aagaard, P. and H. C. Helgeson, 1982, Thermodynamic and kinetic constraints on reaction rates among minerals and aqueous solutions, I. Theoretical considerations. *American Journal of Science* **282**, 237–285.

Adamson, A. W. and A. P. Gast, 1997, *Physical Chemistry of Surfaces*, 6th ed. Wiley, New York.

Alemi, M. H., D. A. Goldhamer and D. R. Nelson, 1991, Modeling selenium transport in steady-state, unsaturated soil columns. *Journal of Environmental Quality* **20**, 89–95.

Allison, J. D., D. S. Brown and K. J. Novo-Gradac, 1991, MINTEQA2/ PRODEFA2, a geochemical assessment model for environmental systems, version 3.0 user's manual. US Environmental Protection Agency Report EPA/600/3–91/021.

Alvarez, P. J. J., P. J. Anid and T. M. Vogel, 1991, Kinetics of aerobic biodegradation of benzene and toluene in sandy aquifer material. *Biodegradation* **2**, 43–51.

Anderson, G. M., 2017, *Thermodynamics of Natural Systems*, 3rd ed. Cambridge University Press.

Anderson, G. M. and D. A. Crerar, 1993, *Thermodynamics in Geochemistry, The Equilibrium Model*. Oxford University Press.

Anderson, G. M. and G. Garven, 1987, Sulfate-sulfide-carbonate associations in Mississippi Valley-type lead-zinc deposits. *Economic Geology* **82**, 482–488.

Aplin, A. C. and E. A. Warren, 1994, Oxygen isotopic indications of the mechanisms of silica transport and quartz cementation in deeply buried sandstones. *Geology* **22**, 847–850.

Appelo, C. A. J., D. L. Parkhurst and V. E. A. Post, 2014, Equations for calculating hydrogeochemical reactions of minerals and gases such as CO_2 at high pressures and temperatures. *Geochimica et Cosmochimica Acta* **125**, 49–67.

Appelo, C. A. J. and D. Postma, 1999, A consistent model for surface complexation on birnessite (δ-MnO_2) and its application to a column experiment. *Geochimica et Cosmochimica Acta* **63**, 3039–3048.

Appelo, C. A. J. and D. Postma, 2005, *Geochemistry, Groundwater, and Pollution*, 2nd ed. Balkema, Rotterdam.

Arnórsson, S., E. Gunnlaugsson and H. Svavarsson, 1983, The chemistry of geothermal waters in Iceland, II. Mineral equilibria and independent variables controlling water compositions. *Geochimica et Cosmochimica Acta* **47**, 547–566.

Arnórsson, S., S. Sigurdsson and H. Svavarsson, 1982, The chemistry of geothermal waters in Iceland. I. Calculation of aqueous speciation from 0° to 370°C. *Geochimica et Cosmochimica Acta* **46**, 1513–1532.

Atkinson, K., W. Han and D. Steward, 2009, *Numerical Solution of Ordinary Differential Equations*. Wiley, New York.

Baccar, M. B. and B. Fritz, 1993, Geochemical modelling of sandstone diagenesis and its consequences on the evolution of porosity. *Applied Geochemistry* **8**, 285–295.

Baes, C. F., Jr. and R. E. Mesmer, 1976, *The Hydrolysis of Cations*. Wiley, New York.

Bahr, J. M. and J. Rubin, 1987, Direct comparison of kinetic and local equilibrium formulations for solute transport affected by surface reactions. *Water Resources Research* **23**, 438–452.

Ball, J. W, E. A. Jenne and D. K. Nordstrom, 1979, WATEQ2—a computerized chemical model for trace and major element speciation and mineral equilibria of natural waters. In E. A. Jenne (ed.), *Chemical Modeling in Aqueous Systems*, American Chemical Society, Washington DC, pp. 815–835.

Ball, J. W. and D. K. Nordstrom, 1991, User's manual for WATEQ4F, with revised thermodynamic data base and test cases for calculating speciation of major, trace, and redox elements in natural waters. US Geological Survey Open File Report 91–183.

Banfield, J. F. and K. H. Nealson (eds.), 1997, *Geomicrobiology: Interactions Between Microbes and Minerals*. Reviews in Mineralogy 35, Mineralogical Society of America, Washington DC.

Barton, P. B., Jr., P. M. Bethke and P. Toulmin, 3rd, 1963, Equilibrium in ore deposits. *Mineralogical Society of America Special Paper* **1**, 171–185.

Bear, J., 1972, *Dynamics of Fluids in Porous Media*. Elsevier, Amsterdam.

Bear, J., 1979, *Hydraulics of Groundwater*. McGraw-Hill, New York.

Bekins B. A., E. M. Godsy and E. Warren, 1999, Distribution of microbial physiologic types in an aquifer contaminated by crude oil. *Microbial Ecology* **37**, 263–275.

Bekins, B. A., E. Warren and E. M. Godsy, 1998, A comparison of zero-order, first-order, and Monod biotransformation models. *Ground Water* **36**, 261–268.

Belitz, K. and J. D. Bredehoeft, 1988, Hydrodynamics of Denver basin, explanation of subnormal fluid pressures. *American Association of Petroleum Geologists Bulletin* **72**, 1334–1359.

Benjamin, M. A., 2002, Modeling the mass-action expression for bidentate adsorption. *Environmental Science and Technology* **36**, 307–313.

Berger, T., F. Mathurin, J. P. Gustafsson, P. Peltola and M. E. Åström, 2015, The impact of fluoride on Al abundance and speciation in boreal streams. *Chemical Geology* **409**, 118–124.

Berner, R. A., 1980, *Early Diagenesis, A Theoretical Approach*. Princeton University Press, Princeton, New Jersey.

Bethke, C. M., 1992, The question of uniqueness in geochemical modeling. *Geochimica et Cosmochimica Acta* **56**, 4315–4320.

Bethke, C. M., 1997, Modelling transport in reacting geochemical systems. *Comptes Rendus de l'Académie des Sciences* **324**, 513–528.

Bethke, C. M. and P. V. Brady, 2000, How the K_d approach undermines groundwater cleanup. *Ground Water* **38**, 435–443.

Bethke, C. M., D. Ding, Q. Jin and R. A. Sanford, 2008, Origin of microbiological zoning in groundwater flows. *Geology* **36**, 739–742.

Bethke, C. M., B. Farrell and M. Sharifi, 2021, The Geochemist's Workbench® Release 16 (five volumes). Aqueous Solutions LLC, Champaign, Illinois.

Bethke, C. M., W. J. Harrison, C. Upson and S. P. Altaner, 1988, Supercomputer analysis of sedimentary basins. *Science* **239**, 261–267.

Bethke, C. M., M.-K. Lee and R. F. Wendlandt, 1992, Mass transport and chemical reaction in sedimentary basins, natural and artificial diagenesis. In M. Quintard and M. S. Todorovic (eds.), *Heat and Mass Transfer in Porous Media*. Elsevier, Amsterdam, pp. 421–434.

Bethke, C. M. and S. Marshak, 1990, Brine migrations across North America—the plate tectonics of groundwater. *Annual Review Earth and Planetary Sciences* **18**, 287–315.

Bethke, C. M., R. A. Sanford, M. F. Kirk, Q. Jin and T. M. Flynn, 2011, The thermodynamic ladder in geomicrobiology. *American Journal of Science* **311**, 183–210.

Bibby, R., 1981, Mass transport of solutes in dual-porosity media. *Water Resources Research* **17**, 1075–1081.

Bird, R. B., W. E. Stewart and E. N. Lightfoot, 2006, *Transport Phenomena*, revised 2nd ed. Wiley, New York.

Bischoff, J. L., J. A. Fitzpatrick and R. J. Rosenbauer, 1993, The solubility and stabilization of ikaite ($CaCO_3 \cdot 6H_2O$) from 0 °C to 25 °C, environmental and paleoclimatic implications for thinalite tufa. *Journal of Geology* **101**, 21–33.

Bischoff, J. L., D. B. Herbst and R. J. Rosenbauer, 1991, Gaylussite formation at Mono Lake, California. *Geochimica et Cosmochimica Acta* **55**, 1743–1747.

Bjorlykke, K. and P. K. Egeberg, 1993, Quartz cementation in sedimentary basins. *American Association of Petroleum Geologists Bulletin* **77**, 1538–1548.

Blesa, M. A., N. M. Figliolia, A. J. G. Maroto and A. E. Regazzoni, 1984, The influence of temperature on the interface magnetite–aqueous electrolyte solution. *Journal of Colloid and Interface Science* **101**, 410–418.

Block, J. and O. B. Waters, Jr., 1968, The $CaSO_4$–Na_2SO_4–NaCl–H_2O system at 25 °C to 100 °C. *Journal of Chemical and Engineering Data* **13**, 336–344.

Blowes, D. W., C. J. Ptacek, J. L. Jambor and C. G. Weisener, 2005, The geochemistry of acid mine drainage. In B. S. Lollar (ed.), *Environmental Geochemistry*, Elsevier, Amsterdam, pp. 149–204.

Blum, A. E. and L. L. Stillings, 1995, Feldspar dissolution kinetics. *Reviews in Mineralogy* **31**, 291–351.

Blum, J. S., A. B. Bindi, J. Buzelli, J. F. Stolz and R. S. Oremland, 1998, *Bacillus arsenicoselenatis*, sp. nov., and *Bacillus selenitireducens*, sp. nov.: two haloalkaliphiles from Mono Lake, California that respire oxyanions. *Archives Microbiology* **171**, 19–30.

Bogard, M. J. and P. A. del Giorgio, 2016, The role of metabolism in modulating CO_2 fluxes in boreal lakes. *Global Biogeochemical Cycles* **30**, 1509–1525.

Bolt, G. H. and W. H. Van Riemsdijk, 1982, Ion adsorption on inorganic variable charge constituents. In G. H. Bolt (ed.), *Soil Chemistry, B. Physico-Chemical Models*, 2nd ed. Elsevier, Amsterdam, pp. 459–504.

Boudart, M, 1976, Consistency between kinetics and thermodynamics. *Journal of Physical Chemistry* **80**, 2869–2870.

Boulding, J. R., 1990, Assessing the geochemical fate of deep-well-injected hazardous waste. US Environmental Protection Agency Report EPA/625/6–89/025a.

Bourcier, W. L., 1985, Improvements to the solid solution modeling capabilities of the EQ3/6 geochemical code. Lawrence Livermore National Laboratory Report UCID-20587.

Bowers, T. S. and H. P. Taylor, Jr., 1985, An integrated chemical and stable isotope model of the origin of midocean ridge hot spring systems. *Journal of Geophysical Research* **90**, 12583–12606.

Bowers, T. S., K. L. Von Damm and J. H. Edmond, 1985, Chemical evolution of mid-ocean ridge hot springs. *Geochimica et Cosmochimica Acta* **49**, 2239–2252.

Boynton, F. P., 1960, Chemical equilibrium in multicomponent polyphase systems. *Journal of Chemical Physics* **32**, 1880–1881.

Brady, J. B., 1975, Chemical components and diffusion. *American Journal of Science* **275**, 1073–1088.

Brady, P. V. and C. M. Bethke, 2000, Beyond the K_d approach. *Ground Water* **38**, 321–322.

Brady, P. V., M. V. Brady and D. J. Borns, 1998, *Natural Attenuation, CERCLA, RBCA's, and the Future of Environmental Remediation*. CRC Press, Boca Raton, Florida.

Brady, P. V. and J. V. Walther, 1989, Controls on silicate dissolution rates in neutral and basic pH solutions at 25 °C. *Geochimica et Cosmochimica Acta* **53**, 2823–2830.

Brantley, S. L., 1992, Kinetics of dissolution and precipitation—experimental and field results. In Y. K. Kharaka and A. S. Maest (eds.), *Water–Rock Interaction*. Balkema, Rotterdam, pp. 3–6.

Brantley, S. L., D. A. Crerar, N. E. Møller and J. H. Weare, 1984, Geochemistry of a modern marine evaporite, Bocana de Virrilá, Peru. *Journal of Sedimentary Petrology* **54**, 447–462.

Brantley, S. L., J. D. Kubicki and A. F. White (eds.), 2008, *Kinetics of Water-Rock Interaction*. Springer-Verlag, New York.

Brendler, V., A. Vahle, T. Arnold, G. Bernhard and T. Fanghänel, 2003, RES^{3}T-Rossendorf expert system for surface and sorption thermodynamics. *Journal of Contaminant Hydrology* **61**, 281–291.

Brewer, P. G., 1975, Minor elements in sea water. In J. P. Riley and G. Skirrow (eds.), *Chemical Oceanography*. Academic Press, New York.

Brezonik, P. L., 1994, *Chemical Kinetics and Process Dynamics in Aquatic Systems*. Lewis Publishers, Boca Raton, Florida.

Brinkley, S. R., Jr., 1947, Calculation of the equilibrium composition of systems of many components. *Journal of Chemical Physics* **15**, 107–110.

Brinkley, S. R., Jr., 1960, Discussion of "A brief survey of past and curent methods of solution for equilibrium composition" by H. E. Brandmaier and J. J. Harnett. In G. S. Bahn and E. E. Zukoski (eds.), *Kinetics, Equilibria and Performance of High Temperature Systems*. Butterworths, Washington DC, p. 73.

Brower, R. D., A. P. Visocky, I. G. Krapac *et al.*, 1989, *Evaluation of Underground Injection of Industrial Waste in Illinois*. Illinois Scientific Surveys Joint Report 2.

Brown, T. H. and B. J. Skinner, 1974, Theoretical prediction of equilibrium phase assemblages in multicomponent systems. *American Journal of Science* **274**, 961–986.

Bunge, A. L. and C. J. Radke, 1982, Migration of alkaline pulses in reservoir sands. *Society of Petroleum Engineers Journal* **22**, 998–1012.

Butler, G. P., 1969, Modern evaporite deposition and geochemistry of coexisting brines, the sabkha, Trucial Coast, Arabian Gulf. *Journal of Sedimentary Petrology* **39**, 70–89.

Carnahan, B., H. A. Luther and J. O. Wilkes, 1969, *Applied Numerical Methods*. Wiley, New York.

Carpenter, A. B., 1980, The chemistry of dolomite formation I: the stability of dolomite. *Society of Economic Paleontologists and Mineralogists Special Publication* **28**, 111–121.

Carroll, S. A. and J. V. Walther, 1990, Kaolinite dissolution at 25 °C, 60 °C, and 80 °C. *American Journal of Science* **290**, 797–810.

Casey, W. H. and J. R. Rustad, 2007, Reaction dynamics, molecular clusters, and aqueous geochemistry. *Annual Review of Earth and Planetary Science* **35**, 21–46.

Casey, W. H. and T. W. Swaddle, 2003, Why small? The use of small inorganic clusters to understand mineral surface and dissolution reactions in geochemistry. *Reviews of Geophysics* **41**, 4-1–4-20.

Cederberg, G. A., R. L. Street and J. O. Leckie, 1985, A groundwater mass transport and equilibrium chemistry model for multicomponent systems. *Water Resources Research* **21**, 1095–1104.

Chapelle, F. H., 2001, *Ground-Water Microbiology and Geochemistry*, 2nd ed. Wiley, New York.

Chapelle, F. H. and D. R. Lovley, 1992, Competitive exclusion of sulfate reduction by Fe(III)-reducing bacteria: a mechanism for producing discrete zones of high-iron ground water. *Ground Water* **30**, 29–36.

Chapman, B. M., D. R. Jones and R. F. Jung, 1983, Processes controlling metal ion attenuation in acid mine drainage streams. *Geochimica et Cosmochimica Acta* **47**, 1957–1973.

Cheng, H. P. and G. T. Yeh, 1998, Development of a three-dimensional model of subsurface flow, heat transfer, and reactive chemical transport: 3DHYDROGEOCHEM. *Journal of Contaminant Hydrology* **34**, 47–83.

Chorover, J. and M. L. Brusseau, 2008, Kinetics of sorption–desorption. In S. L. Brantley, J. D. Kubicky and A. F. White (eds.), *Kinetics of Water-Rock Interaction*. Springer, New York, pp. 109–149.

Cicconi, M. R., R. Moretti and D. R. Neuville, 2020, Earth's electrodes. *Elements* **16**, 157–160.

Clayton, J. L. and P. J. Swetland, 1980, Petroleum generation and migration in Denver basin. *American Association of Petroleum Geologists Bulletin* **64**, 1613–1633.

Cole, D. R. and H. Ohmoto, 1986, Kinetics of isotopic exchange at elevated temperatures and pressures. *Reviews in Mineralogy* **16**, 41–90.

Coudrain-Ribstein, A. and P. Jamet, 1989, Choix des composantes et spéciation d'une solution. *Comptes Rendus de l'Académie des Sciences* **309-II**, 239–244.

Cox, B. G., 1994, *Modern Liquid Phase Kinetics*. Oxford University Press.

Crank, J., 1975, *The Mathematics of Diffusion*, 2nd ed. Oxford University Press.

Crerar, D. A., 1975, A method for computing multicomponent chemical equilibria based on equilibrium constants. *Geochimica et Cosmochimica Acta* **39**, 1375–1384.

Criaud, A., C. Fouillac and B. Marty, 1989, Low enthalpy geothermal fluids from the Paris basin. 2. Oxidation-reduction state and consequences for the prediction of corrosion and sulfide scaling. *Geothermics* **18**, 711–727.

Davies, C. W., 1962, *Ion Association*. Butterworths, Washington DC.

Davis, A., R. L. Olsen and D. R. Walker, 1991, Distribution of metals between water and entrained sediment in streams impacted by acid mine drainage, Clear Creek, Colorado, U.S.A. *Applied Geochemistry* **6**, 333–348.

Davis, J. A., R. O. James and J. O. Leckie, 1978, Surface ionization and complexation at the oxide/water interface. 1. Computation of electrical double layer properties in simple electrolytes. *Journal of Colloid and Interface Science* **63**, 480–499.

Davis, J. A. and D. B. Kent, 1990, Surface complexation modeling in aqueous geochemistry. *Reviews in Mineralogy* **23**, 177–260.

Davis, J. A. and J. O. Leckie, 1980, Surface ionization and complexation at the oxide/water interface. 3. Adsorption of anions. *Journal of Colloid and Interface Science* **74**, 32–43.

Degens, E. T. and D. A. Ross (eds.), 1969, *Hot Brines and Recent Heavy Metal Deposits in the Red Sea*. Springer-Verlag, New York.

Delany, J. M. and S. R. Lundeen, 1989, The LLNL thermochemical database. Lawrence Livermore National Laboratory Report UCRL-21658.

Delany, J. M. and T. J. Wolery, 1984, Fixed-fugacity option for the EQ6 geochemical reaction path code. Lawrence Livermore National Laboratory Report UCRL-53598.

Deloule, E., 1982, The genesis of fluorspar hydrothermal deposits at Montroc and Le Burc, The Tarn, as deduced from fluid inclusion analysis. *Economic Geology* **77**, 1867–1874.

Denbigh, K., 1981, *The Principles of Chemical Equilibrium*, 4th ed. Cambridge University Press.

Domenico, P. A. and F. W. Schwartz, 1998, *Physical and Chemical Hydrogeology*, 2nd ed. Wiley, New York.

Dongarra, J. J., C. B. Moler, J. R. Bunch and G. W. Stewart, 1979, *Linpack Users' Guide*. Society for Industrial and Applied Mathematics, Philadelphia.

Dove, P. M. and D. A. Crerar, 1990, Kinetics of quartz dissolution in electrolyte solutions using a hydrothermal mixed flow reactor. *Geochimica et Cosmochimica Acta* **54**, 955–969.

Drever, J. I., 1988, *The Geochemistry of Natural Waters*, 2nd ed. Prentice-Hall, Englewood Cliffs, New Jersey.

Dria, M. A., R. S. Schedchter and L. W. Lake, 1988, An analysis of reservoir chemical treatments. *SPE Production Engineering* **3**, 52–62.

Driscoll, C. T. Jr., J. P. Baker, J. J. Bisogni, Jr. and C. L. Schofieldt, 1980, Effect of aluminium speciation on fish in dilute acidified waters. *Nature* **284**, 161–164.

Druhan, J. and C. Tournassat (eds.), 2019, *Reactive Transport in Natural and Engineered Systems*. Reviews in Mineralogy and Geochemistry 85, Mineralogical Society of America, Washington DC.

Drummond, S. E. and H. Ohmoto, 1985, Chemical evolution and mineral deposition in boiling hydrothermal systems. *Economic Geology* **80**, 126–147.

Dzombak, D. A. and F. M. M. Morel, 1987, Adsorption of inorganic pollutants in aquatic systems. *Journal of Hydraulic Engineering* **113**, 430–475.

Dzombak, D. A. and F. M. M. Morel, 1990, *Surface Complexation Modeling*. Wiley, New York.

Esters, L., S. Landwehr, G. Sutherland *et al.*, 2017, Parameterizing air-sea gas transfer velocity with dissipation. *Journal of Geophysical Research: Oceans* **122**, 3041–3056.

Eugster, H. P., C. E. Harvie and J. H. Weare, 1980, Mineral equilibria in the six-component seawater system, Na–K–Mg–Ca–SO_4–Cl–H_2O, at 25 °C. *Geochimica et Cosmochimica Acta* **44**, 1335–1347.

Eugster, H. P. and B. F. Jones, 1979, Behavior of major solutes during closed-basin brine evolution. *American Journal of Science* **279**, 609–631.

Faure, G., 1986, *Principles of Isotope Geology*, 2nd ed. Wiley, New York.

Felmy, A. R. and J. H. Weare, 1986, The prediction of borate mineral equilibria in natural waters: application to Searles Lake, California. *Geochimica et Cosmochimica Acta* **50**, 2771–2783.

Ficklin, W. H., G. S. Plumlee, K. S. Smith and J. B. McHugh, 1992, Geochemical classification of mine drainages and natural drainages in mineralized areas. In Y. K. Kharaka and A. S. Maest (eds.), *Water–Rock Interaction*. Balkema, Rotterdam, pp. 381–384.

Fouke, B. W., G. T. Bonheyo, E. Sanzenbacher and J. Frias-Lopez, 2003, Partitioning of bacterial communities between travertine depositional facies at Mammoth Hot Springs, Yellowstone National Park, USA. *Canadian Journal Earth Sciences* **40**, 1531–1548.

Fournier, R. O., 1977, Chemical geothermometers and mixing models for geothermal systems. *Geothermics* **5**, 41–50.

Fournier, R. O and R. W. Potter II, 1979, Magnesium correction to the Na-K-Ca chemical geothermometer. *Geochimica et Cosmochimica Acta* **43**, 1543–1550.

Fournier, R. O. and J. J. Rowe, 1966, Estimation of underground temperatures from the silica content of water from hot springs and wet-steam wells. *American Journal of Science* **264**, 685–697.

Fournier, R. O. and A. H. Truesdell, 1973, An empirical Na-K-Ca geothermometer for natural waters. *Geochimica et Cosmochimica Acta* **37**, 1255–1275.

Freeze, R. A. and J. A. Cherry, 1979, *Groundwater*. Prentice Hall, Englewood Cliffs, New Jersey.

Ganguly, J., 2020, *Thermodynamics in Earth and Planetary Sciences*, 2nd ed., Cambridge University Press.

Garrels, R. M. and F. T. Mackenzie, 1967, Origin of the chemical compositions of some springs and lakes. *Equilibrium Concepts in Natural Waters, Advances in Chemistry Series* **67**, American Chemical Society, Washington DC, pp. 222–242.

Garrels, R. M. and M. E. Thompson, 1962, A chemical model for sea water at 25 °C and one atmosphere total pressure. *American Journal of Science* **260**, 57–66.

Garven, G. and R. A. Freeze, 1984, Theoretical analysis of the role of groundwater flow in the genesis of stratabound ore deposits: 2, Quantitative results. *American Journal of Science* **284**, 1125–1174.

Gelhar, L. W., 1986, Stochastic subsurface hydrology from theory to applications. *Water Resources Research* **22**, 135-S–145-S.

Gerke, H. H. and M. T. van Genuchten, 1993, A dual-porosity model for simulating the preferential movement of water and solutes in structured porous media. *Water Resources Research* **29**, 305–319.

Giffaut, E., M. Grivé, Ph. Blanc *et al.*, 2014, Andra thermodynamic database for performance assessment: ThermoChimie. *Applied Geochemistry* **49**, 225–236.

Giggenbach, W. F., 1988, Geothermal solute equilibria, derivation of Na-K-Mg-Ca geoindicators. *Geochimica et Cosmochimica Acta* **52**, 2749–2765.

Glynn, P. D., E. J. Reardon, L. N. Plummer and E. Busenberg, 1990, Reaction paths and equilibrium endpoints in solid-solution aqueous-solution systems. *Geochimica et Cosmochimica Acta* **54**, 267–282.

Greenberg, J. P. and N. Møller, 1989, The prediction of mineral solubilities in natural waters, a chemical equilibrium model for the Na–K–Ca–Cl–SO_4–H_2O system to high concentration from 0 to 250 °C. *Geochimica et Cosmochimica Acta* **53**, 2503–2518.

Greenwood, H. J., 1975, Thermodynamically valid projections of extensive phase relationships. *American Mineralogist* **60**, 1–8.

Grenthe, I., A. V. Plyasunov and K. Spahiu, 1997, Estimations of medium effects on thermodynamic data. In I. Grenthe and I. Puigdomènech (eds.), *Modelling in Aquatic Chemistry*, NEA OECD Publications, Paris, 325–426.

Guggenheim, E. A., 1967, *Thermodynamics, an Advanced Treatment for Chemists and Physicists*, 5th ed. North-Holland, Amsterdam.

Gunnarsson, M., Z. Abbas, E. Ahlberg, S. Gobom and S. Nordholm, 2002, Corrected Debye–Hückel analysis of surface complexation, II. A theory of surface charging. *Journal of Colloid and Interface Science* **249**, 52–61.

Gupta, S. S. and K. G. Bhattacharyya, 2011, Kinetics of adsorption of metal ions on inorganic materials: a review. *Advances in Colloid and Interface Science* **162**, 39–58.

Haas, J. L., Jr. and J. R. Fisher, 1976, Simultaneous evaluation and correlation of thermodynamic data. *American Journal of Science* **276**, 525–545.

Hardie, L. A., 1987, Dolomitization, a critical view of some current views. *Journal of Sedimentary Petrology* **57**, 166–183.

Hardie, L. A., 1991, On the significance of evaporites. *Annual Review Earth and Planetary Sciences* **19**, 131–168.

Hardie, L. A. and H. P. Eugster, 1970, The evolution of closed-basin brines. *Mineralogical Society of America Special Paper* **3**, 273–290.

Harrison, W. J., 1990, Modeling fluid/rock interactions in sedimentary basins. In T. A. Cross (ed.), *Quantitative Dynamic Stratigraphy*. Prentice Hall, Englewood Cliffs, New Jersey, pp. 195–231.

Harvie, C. E., J. P. Greenberg and J. H. Weare, 1987, A chemical equilibrium algorithm for highly non-ideal multiphase systems: free energy minimization. *Geochimica et Cosmochimica Acta* **51**, 1045–1057.

Harvie, C. E., N. Møller and J. H. Weare, 1984, The prediction of mineral solubilities in natural waters: the Na–K–Mg–Ca–H–Cl–SO_4–OH–HCO_3-CO_3–CO_2–H_2O system to high ionic strengths at 25 °C. *Geochimica et Cosmochimica Acta* **48**, 723–751.

Harvie, C. E. and J. H. Weare, 1980, The prediction of mineral solubilities in natural waters: the Na–K–Mg–Ca–Cl–SO_4–H_2O system from zero to high concentration at 25 °C. *Geochimica et Cosmochimica Acta* **44**, 981–997.

Harvie, C. E., J. H. Weare, L. A. Hardie and H. P. Eugster, 1980, Evaporation of seawater: calculated mineral sequences. *Science* **208**, 498–500.

Hay, R. L., 1963, Stratigraphy and zeolitic diagenesis of the John Day formation of Oregon. *University of California Publications in Geological Sciences*, Berkeley, California, pp. 199–261.

Hay, R. L., 1966, Zeolites and zeolitic reactions in sedimentary rocks. *Geological Society of America Special Paper* **85**.

Hayes, J. B., 1979, Sandstone diagenesis—the hole truth. *Society of Economic Paleontologists and Mineralogists Special Publication* **26**, 127–139.

Hayes, K. F. and J. O. Leckie, 1987, Mechanism of lead ion adsorption at the goethite–water interface. *ACS Symposium Series* **323**, 114–141.

Hayes, K. F., C. Papelis and J. O. Leckie, 1988, Modeling ionic strength effects on anion adsorption at hydrous oxide/solution interfaces. *Journal of Colloid and Interface Science* **125**, 717–726.

Hayes, K. F., G. Redden, E. Wendell and J. O. Leckie, 1991, Surface complexation models: an evaluation of model parameter estimation using FITEQL and oxide mineral titration data. *Journal of Colloid and Interface Science* **142**, 448–469.

He, S., J. E. Oddo and M. B. Tomson, 1994, The inhibition of gypsum and barite nucleation in NaCl brines at temperatures from 25 to 90 °C. *Applied Geochemistry* **9**, 561–567.

Healy, R. W., S. S. Haile, D. L. Parkhurst and S. R. Charlton, 2018, VS2DRTI simulating heat and reactive solute transport in variably saturated porous media. *Groundwater* **56**, 810–815.

Helgeson, H. C., 1968, Evaluation of irreversible reactions in geochemical processes involving minerals and aqueous solutions, I. Thermodynamic relations. *Geochimica et Cosmochimica Acta* **32**, 853–877.

Helgeson, H. C., 1969, Thermodynamics of hydrothermal systems at elevated temperatures and pressures. *American Journal of Science* **267**, 729–804.

Helgeson, H. C., 1970, A chemical and thermodynamic model of ore deposition in hydrothermal systems. *Mineralogical Society of America Special Paper* **3**, 155–186.

Helgeson, H. C., T. H. Brown, A. Nigrini and T. A. Jones, 1970, Calculation of mass transfer in geochemical processes involving aqueous solutions. *Geochimica et Cosmochimica Acta* **34**, 569–592.

Helgeson, H. C., J. M. Delany, H. W. Nesbitt and D. K. Bird, 1978, Summary and critique of the thermodynamic properties of rock-forming minerals. *American Journal of Science* **278-A**, 1–229.

Helgeson, H. C., R. M. Garrels and F. T. Mackenzie, 1969, Evaluation of irreversible reactions in geochemical processes involving minerals and aqueous solutions, II. Applications. *Geochimica et Cosmochimica Acta* **33**, 455–481.

Helgeson, H. C. and D. H. Kirkham, 1974, Theoretical prediction of the thermodynamic behavior of aqueous electrolytes at high pressures and temperatures, II. Debye–Hückel parameters for activity coefficients and relative partial molal properties. *American Journal of Science* **274**, 1199–1261.

Helgeson, H. C., D. H. Kirkham and G. C. Flowers, 1981, Theoretical prediction of the thermodynamic behavior of aqueous electrolytes at high temperatures and pressures, IV. Calculation of activity coefficients, osmotic coefficients, and apparent molal and standard and relative partial molal properties to 600 °C and 5 kB. *American Journal of Science* **281**, 1249–1516.

Hem, J. D., 1985, Study and interpretation of the chemical characteristics of natural water. *US Geological Survey Water-Supply Paper* **2254**.

Hemley, J. J., G. L. Cygan and W. M. d'Angelo, 1986, Effect of pressure on ore mineral solubilities under hydrothermal conditions. *Geology* **14**, 377–379.

Henley, R. W., 1984, Chemical structure of geothermal systems. *Reviews in Economic Geology* **1**, 9–28.

Hiemstra, T. and Van Riemsdijk, W. H., 1996, A surface structural approach to ion adsorption: the charge distribution (CD) model. *Journal of Colloid and Interface Science* **179**, 488–508.

Hill, C. G., Jr., 1977, *An Introduction to Chemical Engineering Kinetics and Reactor Design*. Wiley, New York.

Hoffmann, R., 1991, Hot brines in the Red Sea. *American Scientist* **79**, 298–299.

Holland, H. D., 1978, *The Chemistry of the Atmosphere and Oceans*. Wiley, New York.

Hostettler, J. D., 1984, Electrode electrons, aqueous electrons, and redox potentials in natural waters. *American Journal of Science* **284**, 734–759.

Hubbert, M. K., 1940, The theory of ground-water motion. *Journal of Geology* **48**, 785–944.

Huber, J. A., D. A. Butterfield and J. A. Baross, 2003, Bacterial diversity in a subseafloor habitat following a deep-sea volcanic eruption. *FEMS Microbiology Ecology* **43**, 393–409.

Hubert, J. F., 1960, Petrology of the Fountain and Lyons formations, Front Range, Colorado. *Colorado School of Mines Quarterly* **55**.

Hunt, J. M., 1990, Generation and migration of petroleum from abnormally pressured fluid compartments. *American Association of Petroleum Geologists Bulletin* **74**, 1–12.

Huser, B. A., K. Wuhrmann and A. J. B. Zehnder, 1982, *Methanothrix soehngenii* gen. nov. sp. nov., a new acetotrophic non-hydrogen-oxidizing methane bacterium. *Archives of Microbiology* **132**, 1–9.

Hutcheon, I., 1984, A review of artificial diagenesis during thermally enhanced recovery. In D. A. MacDonald and R. C. Surdam (eds.), *Clastic Diagenesis*. American Association of Petroleum Geologists, Tulsa, Oklahoma, pp. 413–429.

HZDR, 2021, *RES^{3}T-Rossendorf Expert System for Surface and Sorption Thermodynamics*. Helmholtz-Zentrum Dresden-Rossendorf, `www.hzdr.de/res3t`.

Ingvorsen, K., A. J. B. Zehnder and B. B. Jørgensen, 1984, Kinetics of sulfate and acetate uptake by *Desulfobacter postgatei*. *Applied and Environmental Microbiology* **47**, 403–408.

Interscience Publishers, 1954, *The Collected Papers of P. J. W. Debye*. Interscience Publishers, Inc., New York.

ISO, 1985, *International Standard ISO–7888: Water quality–Determination of electrical conductivity*. International Organization for Standardization, Geneva, Switzerland.

Janecky, D. R. and W. E. Seyfried, Jr., 1984, Formation of massive sulfide deposits on oceanic ridge crests: incremental reaction models for mixing between hydrothermal solutions and seawater. *Geochimica et Cosmochimica Acta* **48**, 2723–2738.

Janecky, D. R. and W. C. Shanks, III, 1988, Computational modeling of chemical and sulfur isotopic reaction processes in seafloor hydrothermal systems, chimneys, massive sulfides, and subjacent alteration zones. *Canadian Mineralogist* **26**, 805–825.

Jankowski, J. and G. Jacobson, 1989, Hydrochemical evolution of regional groundwaters to playa brines in central Australia. *Journal of Hydrology* **108**, 123–173.

Jarraya, F. and M. El Mansar, 1987, *Modelisation Simplifié du Gisement de Saumare á Sebkhat el Melah á Zarzis*. Projet de fin d'etudes, Ecole Nationale d'Ingénieurs de Tunis, Tunis, Tunisia.

Javandel, I., C. Doughty and C. F. Tsang, 1984, *Groundwater Transport: Handbook of Mathematical Models*. American Geophysical Union, Washington DC.

Jenne, E. A. (ed.), 1998, *Adsorption of Metals by Geomedia*. Academic Press, New York.

Jennings, H. Y., Jr., C. E. Johnson, Jr. and C. D. McAuliffe, 1974, A caustic waterflooding process for heavy oils. *Journal of Petroleum Technology* **26**, 1344–1352.

Jin, Q., 2007, Control of hydrogen partial pressures on the rates of syntrophic microbial metabolisms: a kinetic model for butyrate fermentation. *Geobiology* **5**, 35–48.

Jin, Q. and C. M. Bethke, 2002, Kinetics of electron transfer through the respiratory chain. *Biophysical Journal* **83**, 1797–1808.

Jin, Q. and C. M. Bethke, 2003, A new rate law describing microbial respiration. *Applied and Environmental Microbiology* **69**, 2340–2348.

Jin, Q. and C. M. Bethke, 2005, Predicting the rate of microbial respiration in geochemical environments. *Geochimica et Cosmochimica Acta* **69**, 1133–1143.

Jin, Q. and C. M. Bethke, 2007, The thermodynamics and kinetics of microbial metabolism. *American Journal of Science* **307**, 643–677.

Jin, Q. and C. M. Bethke, 2009, Cellular energy conservation and the rate of microbial sulfate reduction. *Geology* **37**, 1027–1030.

Johnson, C. A., 1986, The regulation of trace element concentrations in river and estuarine waters with acid mine drainage, the adsorption of Cu and Zn on amorphous Fe oxyhydroxides. *Geochimica et Cosmochimica Acta* **50**, 2433–2438.

Johnson, J. W., E. H. Oelkers and H. C. Helgeson, 1991, SUPCRT92: a software package for calculating the standard molal thermodynamic properties of minerals, gases, aqueous species, and reactions from 1 to 5000 bars and 0° to 1000 °C. Earth Sciences Department, Lawrence Livermore Laboratory.

Jurkevitch, E., 2007, Predatory behaviors in bacteria—diversity and transitions. *Microbe* **2**, 67–73.

Karamalidis, A. K. and D. A. Dzombak, 2010, *Surface Complexation Modeling: Gibbsite*. Wiley, New York.

Karpov, I. K. and L. A. Kaz'min, 1972, Calculation of geochemical equilibria in heterogeneous multicomponent systems. *Geochemistry International* **9**, 252–262.

Karpov, I. K., L. A. Kaz'min and S. A. Kashik, 1973, Optimal programming for computer calculation of irreversible evolution in geochemical systems. *Geochemistry International* **10**, 464–470.

Kashefi, K. and D. R. Lovley, 2003, Extending the upper temperature limit for life. *Science* **301**, 934.

Kastner, M., 1984, Control of dolomite formation. *Nature* **311**, 410–411.

Keenan, J. H., F. G. Keyes, P. G. Hill and J. G. Moore, 1969, *Steam Tables, Thermodynamic Properties of Water Including Vapor, Liquid, and Solid Phases*. Wiley, New York.

Kelley, D. S., J. A. Baross and J. R. Delaney, 2002, Volcanoes, fluids, and life at mid-ocean ridge spreading centers. *Annual Review of Earth and Planetary Sciences* **30**, 385–491.

Kennedy, V. C., G. W. Zellweger and B. F. Jones, 1974, Filter pore-size effects on the analysis of Al, Fe, Mn, and Ti in water. *Water Resources Research* **10**, 785–790.

Kharaka, Y. K. and I. Barnes, 1973, SOLMNEQ: solution-mineral equilibrium computations. US Geological Survey Computer Contributions Report PB-215-899.

Kharaka, Y. K., W. D. Gunter, P. K. Aggarwal, E. H. Perkins and J. D. DeBraal, 1988, SOLMINEQ.88, a computer program for geochemical modeling of water–rock interactions. US Geological Survey Water Resources Investigation Report 88–4227.

Kim, C., Q. Zhou, B. Deng, E. C. Thornton and H. Xu, 2001, Chromium(VI) reduction by hydrogen sulfide in aqueous media: stoichiometry and kinetics. *Environmental Science and Technology* **35**, 2219–2225.

Kim, S.-B., I. Hwang, D.-J. Kim, S. Lee and W. A. Jury, 2003, Effect of sorption on benzene biodegradation in sandy soil. *Environmental Toxicology and Chemistry* **22**, 2306–2311.

King, T. V. V. (ed.), 1995, Environmental considerations of active and abandoned mine lands: lessons from Summitville, Colorado. *US Geological Survey Bulletin* **2220**.

Kirk, M. F., T. R. Holm, J. Park *et al.*, 2004, Bacterial sulfate reduction limits natural arsenic contamination of groundwater. *Geology* **32**, 953–956.

Knapp, R. B., 1989, Spatial and temporal scales of local equilibrium in dynamic fluid-rock systems. *Geochimica et Cosmochimica Acta* **53**, 1955–1964.

Knauss, K. G. and T. J. Wolery, 1986, Dependence of albite dissolution kinetics on pH and time at 25 °C and 70 °C. *Geochimica et Cosmochimica Acta* **50**, 2481–2497.

Knauss, K. G. and T. J. Wolery, 1988, The dissolution kinetics of quartz as a function of pH and time at 70 °C. *Geochimica et Cosmochimica Acta* **52**, 43–53.

Konhauser, K., 2007, *Introduction to Geomicrobiology*. Blackwell, Malden, Massachusetts.

Kulik, D. A., T. Wagner, S. V. Dmytrieva *et al.*, 2013, GEM-SELEKTOR geochemical modeling package: numerical kernel GEMS3K for coupled simulation codes. *Computational Geosciences* **17**, 1–24.

Lafon, G. M., G. A. Otten and A. M. Bishop, 1992, Experimental determination of the calcite-dolomite equilibrium below 200 °C; revised stabilities for dolomite and magnesite support near-equilibrium dolomitization models. *Geological Society of America Abstracts with Programs* **24**, A210-A211.

Land, L. S. and G. L. Macpherson, 1992, Geothermometry from brine analyses: lessons from the Gulf Coast, U.S.A. *Applied Geochemistry* **7**, 333–340.

Lasaga, A. C., 1981a, Rate laws of chemical reactions. In A. C. Lasaga and R. J. Kirkpatrick (eds.), *Kinetics of Geochemical Processes*. Mineralogical Society of America, Washington DC, pp. 1–68.

Lasaga, A. C., 1981b, Transition state theory. In A. C. Lasaga and R. J. Kirkpatrick (eds.), *Kinetics of Geochemical Processes*. Mineralogical Society of America, Washington DC, pp. 135–169.

Lasaga, A. C., 1984, Chemical kinetics of water–rock interactions. *Journal of Geophysical Research* **89**, 4009–4025.

Lasaga, A. C., 1998, *Kinetic Theory in the Earth Sciences*. Princeton University Press, Princeton, New Jersey.

Lasaga, A. C. and D. M. Rye, 1993, Fluid flow and chemical reaction kinetics in metamorphic systems. *American Journal of Science* **293**, 361–404.

Lasaga, A. C., J. M. Soler, J. Ganor, T. E. Burch and K. L. Nagy, 1994, Chemical weathering rate laws and global geochemical cycles. *Geochimica et Cosmochimica Acta* **58**, 2361–2386.

Leach, D. L., G. S. Plumlee, A. H. Hofstra *et al.*, 1991, Origin of late dolomite cement by CO_2-saturated deep basin brines: evidence from the Ozark region, central United States. *Geology* **19**, 348–351.

Leamnson, R. N., J. Thomas, Jr. and H. P. Ehrlinger, III, 1969, A study of the surface areas of particulate microcrystalline silica and silica sand. *Illinois State Geological Survey Circular* **444**.

Lee, M.-K. and C. M. Bethke, 1994, Groundwater flow, late cementation, and petroleum accumulation in the Permian Lyons sandstone, Denver basin. *American Association of Petroleum Geologists Bulletin* **78**, 217–237.

Lee, M.-K. and C. M. Bethke, 1996, A model of isotope fractionation in reacting geochemical systems. *American Journal of Science* **296**, 965–988.

Levandowski, D. W., M. E. Kaley, S. R. Silverman and R. G. Smalley, 1973, Cementation in Lyons sandstone and its role in oil accumulation, Denver basin, Colorado. *American Association of Petroleum Geologists Bulletin* **57**, 2217–2244.

Levenspeil, O., 1972, *Chemical Reaction Engineering*, 2nd ed. Wiley, New York.

Lichtner, P. C., 1985, Continuum model for simultaneous chemical reactions and mass transport in hydrothermal systems. *Geochimica et Cosmochimica Acta* **49**, 779–800.

Lichtner, P. C., 1988, The quasi-stationary state approximation to coupled mass transport and fluid–rock interaction in a porous medium. *Geochimica et Cosmochimica Acta* **52**, 143–165.

Lichtner, P. C., 1996, Continuum formulation of multicomponent–multiphase reactive transport. *Reviews in Mineralogy* **34**, 1–81.

Lichtner, P. C., E. H. Oelkers and H. C. Helgeson, 1986, Interdiffusion with multiple precipitation/dissolution reactions: transient model and the steady-state limit. *Geochimica et Cosmochimica Acta* **50**, 1951–1966.

Lichtner, P. C., C. I. Steefel and E. H. Oelkers (eds.), 1996, *Reactive Transport in Porous Media.* Reviews in Mineralogy 34, Mineralogical Society of America, Washington DC.

Lico, M. S., Y. K. Kharaka, W. W. Carothers and V. A. Wright, 1982, Methods for collection and analysis of geopressured geothermal and oil field waters. *US Geological Survey Water Supply Paper* **2194**.

Liger, E., L. Charlet and P. Van Cappellen, 1999, Surface catalysis of uranium(VI) reduction by iron(II). *Geochimica et Cosmochimica Acta* **63**, 2939–2955.

Limousin, G., J.-P. Gaudet, L. Charlet, S. Szenknect, V. Barthès and M. Krimissa, 2007, Sorption isotherms: a review on physical bases, modeling and measurement. *Applied Geochemistry* **22**, 249–275.

Lind, C. J. and J. D. Hem, 1993, Manganese minerals and associated fine particulates in the streambed of Pinal Creek, Arizona, U.S.A.: a mining-related acid drainage problem. *Applied Geochemistry* **8**, 67–80.

Lindberg, R. D. and D. D. Runnells, 1984, Groundwater redox reactions: an analysis of equilibrium state applied to Eh measurements and geochemical modeling. *Science* **225**, 925–927.

Liss, P. S., 1983, Gas transfer: experiments and geochemical implications. In P. S. Liss and W. G. N. Slinn (eds.), *Air-Sea Exchange of Gases and Particles.* NATO Science Series C-108, Springer Netherlands, Heidelberg, pp. 241–298.

Liu, C. W. and T. N. Narasimhan, 1989a, Redox-controlled multiple-species reactive chemical transport, 1. Model development. *Water Resources Research* **25**, 869–882.

Liu, C. W. and T. N. Narasimhan, 1989b, Redox-controlled multiple-species reactive chemical transport, 2. Verification and application. *Water Resources Research* **25**, 883–910.

Lützenkirchen, J. (ed.), 2006, *Surface Complexation Modelling*, Academic Press, San Diego, California.

Madigan, M. and J. Martinko, 2017, *Brock Biology of Microorganisms*, 15th ed. Pearson, New York.

Maher, K. and K. U. Mayer (eds.), 2019, Reactive transport modeling. *Elements* **15**, Mineralogical Society of America, Washington DC.

Malmberg, C. G. and A. A. Maryott, 1956, Dielectric constant of water from 0° to 100°C. *Journal of Research of the National Bureau of Standards* **56**, 1–8.

Malmstrom, M. E., G. Destouni and P. Martinet, 2004, Modeling expected solute concentration in randomly heterogeneous flow systems with multicomponent reactions. *Environmental Science and Technology* **38**, 2673–2679.

March, R., F. Doster and S. Geiger, 2018, Assessment of CO_2 storage potential in naturally fractured reservoirs with dual-porosity models. *Water Resources Research* **54**, 1650–1668.

Marshall, W. L. and R. Slusher, 1966, Thermodynamics of calcium sulfate dihydrate in aqueous sodium chloride solutions, 0–110°. *Journal of Physical Chemistry* **70**, 4015–4027.

Martin, C. A., 1965, Denver basin. *American Association of Petroleum Geologists Bulletin* **49**, 1908–1925.

Mathur, S. S. and D. A. Dzombak, 2006, Surface complexation modeling: goethite. In J. Lützenkirchen (ed.), *Surface Complexation Modelling*, Academic Press, San Diego, California, pp. 443–468.

Mattes, B. W. and E. W. Mountjoy, 1980, Burial dolomitization of the Upper Devonian Miette Buildup, Jasper National Park, Alberta. In D. H. Zenger, J. B. Dunham and R. L. Effington (eds.), *Concepts and Models of Dolomitization. SEPM Special Publication* **28**, 259–297.

May, H., 1992, The hydrolysis of aluminum, conflicting models and the interpretation of aluminum geochemistry. In Y. K. Kharaka and A. S. Maest (eds.), *Water–Rock Interaction*. Balkema, Rotterdam, pp. 13–21.

Mayer, K. U., E. O. Frind and D. W. Blowes, 2002, Multicomponent reactive transport modeling in variably saturated porous media using a generalized formulation for kinetically controlled reactions. *Water Resources Research* **38**, 13-1–13-21.

Mayo, A. L., P. J. Nielsen, M. Loucks and W. H. Brimhall, 1992, The use of solute and isotopic chemistry to identify flow patterns and factors which limit acid mine drainage in the Wasatch Range, Utah. *Ground Water* **30**, 243–249.

McCleskey, R. B., D. K. Nordstrom, J. N. Ryan and J. W. Ball, 2012, A new method of calculating electrical conductivity with applications to natural waters. *Geochimica et Cosmochimica Acta* **77**, 369–382.

McCollom, T. M. and E. L. Shock, 1997, Geochemical constraints on chemolithoautotrophic metabolism by microorganisms in seafloor hydrothermal systems. *Geochimica et Cosmochimica Acta* **61**, 4375–4391.

McConaghy, J. A., G. H. Chase, A. J. Boettcher and T. J. Major, 1964, Hydrogeologic data of the Denver basin, Colorado. *Colorado Groundwater Basic Data Report* **15**.

McCoy, A. W., III, 1953, Tectonic history of Denver basin. *American Association of Petroleum Geologists Bulletin* **37**, 1873–1893.

McDuff, R. E. and F. M. M. Morel, 1980, The geochemical control of seawater (Sillen revisited). *Environmental Science and Technology* **14**, 1182–1186.

Meeussen, J. C. L., 2003, ORCHESTRA: an object-oriented framework for implementing chemical equilibrium models. *Environmental Science and Technology* **37**, 1175–1182.

Mehnert, E., C. R. Gendron and R. D. Brower, 1990, Investigation of the hydraulic effects of deep-well injection of industrial wastes. *Illinois State Geological Survey Environmental Geology* **135**, 100 p.

Merino, E., D. Nahon and Y. Wang, 1993, Kinetics and mass transfer of pseudomorphic replacement, application to replacement of parent minerals and kaolinite by Al, Fe, and Mn oxides during weathering. *American Journal of Science* **293**, 135–155.

Meyers, W. J. and K. C. Lohmann, 1985, Isotope geochemistry of regional extensive calcite cement zones and marine components in Mississippian limestones, New Mexico. In N. Schneidermann and P. M. Harris (eds.), *Carbonate Cements. SEPM Special Publication* **36**, 223–239.

Michard, G., C. Fouillac, D. Grimaud and J. Denis, 1981, Une méthode globale d'estimation des températures des réservoirs alimentant les sources thermales, exemple du Massif Centrale Français. *Geochimica et Cosmochimica Acta* **45**, 1199–1207.

Michard, G. and E. Roekens, 1983, Modelling of the chemical composition of alkaline hot waters. *Geothermics* **12**, 161–169.

Miron, G. D., D. A. Kulik, S. V. Dmytrieva and T. Wagner, 2015, GEMSFITS: code package for optimization of geochemical model parameters and inverse modeling. *Applied Geochemistry* **55**, 28–45.

Møller, N., 1988, The prediction of mineral solubilities in natural waters: a chemical equilibrium model for the Na–Ca–Cl–SO_4–H_2O system, to high temperature and concentration. *Geochimica et Cosmochimica Acta* **52**, 821–837.

Morel, F. M. M., 1983, *Principles of Aquatic Chemistry*. Wiley, New York.

Morel, F. and J. Morgan, 1972, A numerical method for computing equilibria in aqueous chemical systems. *Environmental Science and Technology* **6**, 58–67.

Morgan, J. J., 1967, Chemical equilibria and kinetic properties of manganese in natural waters. In S. D. Faust and J. V. Hunter (eds.), *Principles and Applications of Water Chemistry*. Wiley, New York.

Morse, J. W. and W. H. Casey, 1988, Ostwald processes and mineral paragenesis in sediments. *American Journal of Science* **288**, 537–560.

Moses, C. O., D. K. Nordstrom, J. S. Herman and A. L. Mills, 1987, Aqueous pyrite oxidation by dissolved oxygen and by ferric iron. *Geochimica et Cosmochimica Acta* **51**, 1561–1571.

Mottl, M. J. and T. F. McConachy, 1990, Chemical processes in buoyant hydrothermal plumes on the East Pacific Rise near 21 ° N. *Geochimica et Cosmochimica Acta* **54**, 1911–1927.

Müller, B., 2015, CHEMEQL V3.2, A program to calculate chemical speciation equilibria, titrations, dissolution, precipitation, adsorption, kinetics, pX-pY diagrams, solubility diagrams. Limnological Research Center EAWAG, Kastanienbaum, Switzerland.

Nagy, K. L., 1995, Dissolution and precipitation kinetics of sheet silicates. *Reviews in Mineralogy* **31**, 173–233.

Nagy, K. L., A. E. Blum and A. C. Lasaga, 1991, Dissolution and precipitation kinetics of kaolinite at 80 °C and pH 3: the dependence on solution saturation state. *American Journal of Science* **291**, 649–686.

Nagy, K. L. and A. C. Lasaga, 1992, Dissolution and precipitation kinetics of gibbsite at 80 °C and pH 3, the dependence on solution saturation state. *Geochimica et Cosmochimica Acta* **56**, 3093–3111.

Neuman, S. P., 1990, Universal scaling of hydraulic conductivities and dispersivities in geologic media. *Water Resources Research* **26**, 1749–1758

Newman, J. and K. E. Thomas-Alyea, 2004, *Electrochemical Systems*, 3rd ed. Wiley, Hoboken, New Jersey.

Nordeng, S. H. and D. F. Sibley, 1994, Dolomite stoichiometry and Ostwald's step rule. *Geochimica et Cosmochimica Acta* **58**, 191–196.

Nordstrom, D. K., 1982, Aqueous pyrite oxidation and the consequent formation of secondary iron minerals. In *Acid Sulfate Weathering. Soil Science Society of America Special Publication* **10**, 37–56.

Nordstrom, D. K., E. A. Jenne and J. W. Ball, 1979, Redox equilibria of iron in acid mine waters. In E. A. Jenne (ed.), *Chemical Modeling in Aqueous Systems*, American Chemical Society, Washington DC, pp. 51–79.

Nordstrom, D. K., R. H. McNutt, I. Puigdoménech, J. A. T. Smellie and M. Wolf, 1992, Ground water chemistry and geochemical modeling of water–rock interactions at the Osamu Utsumi mine and the Morro do Ferro analogue study sites, Poços de Caldas, Minas Gerais, Brazil. *Journal of Geochemical Exploration* **45**, 249–287.

Nordstrom, D. K. and J. L. Munoz, 1994, *Geochemical Thermodynamics*, 2nd ed. Blackwell, Boston.

O'Connell, J., and J. Haile, 2005, *Thermodynamics: Fundamentals for Applications*. Cambridge University Press.

Okereke, A. and S. E. Stevens, Jr., 1991, Kinetics of iron oxidation by *Thiobacillus ferrooxidans*. *Applied and Environmental Microbiology* **57**, 1052–1056.

O'Neil, J. R., 1987, Preservation of H, C, and O isotopic ratios in the low temperature environment. In T. K. Kyser (ed.), *Stable Isotope Geochemistry of Low Temperature Processes. Mineralogical Society of Canada Short Course* **13**, 85–128.

Oreskes, N., K. Shrader-Frechette and K. Belitz, 1994, Verification, validation, and confirmation of numerical models in the Earth sciences. *Science* **263**, 641–646.

Ortoleva, P. J., E. Merino, C. Moore and J. Chadam, 1987, Geochemical self-organization, I: Reaction-transport feedbacks and modeling approach. *American Journal of Science* **287**, 979–1007.

Pace, M. L. and Y. T. Prairie, 2005, Respiration in lakes. In P. A. del Giorgio and J. le B. Williams (eds.), *Respiration in Aquatic Ecosystems*, Oxford University Press, pp. 103–121.

Pačes, T., 1975, A systematic deviation from Na–K–Ca geothermometer below 75 °C and above 10^{-4} atm P_{CO_2}. *Geochimica et Cosmochimica Acta* **39**, 541–544.

Pačes, T., 1983, Rate constants of dissolution derived from the measurements of mass balance in hydrological catchments. *Geochimica et Cosmochimica Acta* **47**, 1855–1863.

Pallud, C. and P. Van Cappellen, 2006, Kinetics of microbial sulfate reduction in estuarine sediments. *Geochimica et Cosmochimica Acta* **70**, 1148–1162.

Panikov, N. S., 1995, *Microbial Growth Kinetics*. Chapman and Hall, London.

Park, J., R. A. Sanford and C. M. Bethke, 2006, Geochemical and microbiological zonation of the Middendorf aquifer, South Carolina. *Chemical Geology* **230**, 88–104.

Park, J., R. A. Sanford and C. M. Bethke, 2009, Microbial activity and chemical weathering in the Middendorf Aquifer, South Carolina. *Chemical Geology* **258**, 232–241.

Parkhurst, D. L., 1995, User's guide to PHREEQC, a computer model for speciation, reaction-path, advective-transport and inverse geochemical calculations. US Geological Survey Water-Resources Investigations Report 95–4227.

Parkhurst, D. L. and C. A. J. Apello, 2013, Description of input and examples for PHREEQC version 3–A computer program for speciation, batch-reaction, one-dimensional transport, and inverse geochemical calculations. *US Geological Surve Techniques and Methods* **6–A43**.

Parkhurst, D. L., K. L. Kipp and S. R. Charlton, 2010, PHAST version 2. A program for simulating ground-water flow, solute transport, and multicomponent geochemical reactions. *US Geological Survey Techniques and Methods* **6–A35**.

Parkhurst, D. L., D. C. Thorstenson and L. N. Plummer, 1980, PHREEQE—a computer program for geochemical calculations. US Geological Survey Water-Resources Investigations Report 80–96.

Pawlowicz, R., 2008, Calculating the conductivity of natural waters. *Limnology and Oceanography: Methods* **6**, 489–501.

Peaceman, D. W., 1977, *Fundamentals of Numerical Reservoir Simulation*. Elsevier, Amsterdam.

Peng, D.-Y. and D. B. Robinson, 1976, A new two-constant equation of state. *Industrial and Engineering Chemistry Fundamentals* **15**, 59–64.

Perkins, E. H. and T. H. Brown, 1982, Program PATH, calculation of isothermal and isobaric mass transfer. University of British Columbia, unpublished manuscript.

Perthuisot, J. P., 1980, Sebkha el Melah near Zarzis: a recent paralic salt basin (Tunisia). In G. Busson (ed.), *Evaporite Deposits, Illustration and Interpretation of Some Environmental Sequences*, Editions Technip, Paris, pp. 11–17, 92–95.

Phillips, O. M., 1991, *Flow and Reaction in Permeable Rocks*. Cambridge University Press.

Phillips, S. L., A. Igbene, J. A. Fair and H. Ozbek, 1981, A technical databook for geothermal energy utilization. Lawrence Berkeley Laboratory Report LBL-12810.

Pitzer, K. S., 1975, Thermodynamics of electrolytes, V. Effects of higher order electrostatic terms. *Journal of Solution Chemistry* **4**, 249–265.

Pitzer, K. S., 1979, Theory: ion interaction approach. In R. M. Pytkowitz (ed.), *Activity Coefficients in Electrolyte Solutions*, vol. 1. CRC Press, Boca Raton, Florida, pp. 157–208.

Pitzer, K. S., 1987, A thermodynamic model for aqueous solutions of liquid-like density. *Reviews in Mineralogy* **17**, 97–142.

Pitzer, K. S., 1991, Ion interaction approach: theory and data correlation. In K. S. Pitzer (ed.), *Activity Coefficients in Electrolyte Solutions*, 2nd ed., CRC Press, Boca Raton, Florida, pp. 75–154.

Pitzer, K. S. and L. Brewer, 1961, Revised edition of *Thermodynamics*, by G. N. Lewis and M. Randall, 2nd ed., McGraw-Hill, New York.

Plankey, B. J., H. H. Patterson and C. S. Cronan, 1986, Kinetics of aluminum fluoride complexation in acidic waters. *Environmental Science and Technology* **20**, 160–165.

Plumlee, G., 1994a, USGS assesses the impact of Summitville. *USGS Office of Mineral Resources Newsletter* **5(2)**, 1–2.

Plumlee, G., 1994b, Environmental geology models of mineral deposits. *Society of Economic Geologists Newsletter* **16**, 5–6.

Plumlee, G. S., M. B. Goldhaber and E. L. Rowan, 1995, The potential role of magmatic gases in the genesis of Illinois-Kentucky fluorspar deposits, implications from chemical reaction path modeling. *Economic Geology* **90**, 999–1011.

Plumlee, G. S., K. S. Smith, W. H. Ficklin and P. H. Briggs, 1992, Geological and geochemical controls on the composition of mine drainages and natural drainages in mineralized areas. In Y. K. Kharaka and A. S. Maest (eds.), *Water–Rock Interaction*. Balkema, Rotterdam, pp. 419–422.

Plummer, L. N., 1992, Geochemical modeling of water–rock interaction: past, present, future. In Y. K. Kharaka and A. S. Maest (eds.), *Water–Rock Interaction*. Balkema, Rotterdam, pp. 23–33.

Plummer, L. N., D. L. Parkhurst, G. W. Fleming and S. A. Dunkle, 1988, PHRQPITZ, a computer program incorporating Pitzer's equations for calculation of geochemical reactions in brines. US Geological Survey Water-Resources Investigations Report 88–4153.

Plyasunov, A. B. and E. S. Popova, 2013, Temperature dependence of the parameter of the SIT model for activity coefficients of 1:1 electrolytes. *Journal of Solution Chemistry* **42**, 1320–1335.

Poling, B. E., J. M. Prausnitz and J. P. O'Connell, 2001, *The Properties of Gases and Liquids*, 5th ed. McGraw-Hill, New York.

Prigogine, I. and R. Defay, 1954, *Chemical Thermodynamics*, D. H. Everett (trans.). Longmans, London.

Puigdomènech, I., E. Colàs, M. Grivé, I. Campos and D. García, 2014, A tool to draw chemical equilibrium diagrams using SIT: applications to geochemical systems and radionuclide solubility. In L. Duro, J. Giménez, I. Casas and J. de Pablo (eds.), *Scientific Basis for Nuclear Waste Management XXXVII*, Materials Research Society, Warrendale, Pennsylvania, 111–116.

Rabus R., T. A. Hansen and F. Widdel, 2006, Dissimilatory sulfate- and sulfur-reducing prokaryotes. In *The Prokaryotes: An Evolving Electronic Resource for the Microbiological Community*, http://www.springerlink.-com/content/n1084686101028pj/fulltext.pdf, Springer, New York.

Reardon, E. J., 1981, K_d's—can they be used to describe reversible ion sorption reactions in contaminant migration? *Ground Water* **19**, 279–286.

Reed, M. and N. Spycher, 1984, Calculation of pH and mineral equilibria in hydrothermal waters with application to geothermometry and studies of boiling and dilution. *Geochimica et Cosmochimica Acta* **48**, 1479–1492.

Reed, M. H., 1977, Calculations of hydrothermal metasomatism and ore deposition in submarine volcanic rocks with special reference to the West Shasta district, California. Ph.D. dissertation, University of California, Berkeley.

Reed, M. H., 1982, Calculation of multicomponent chemical equilibria and reaction processes in systems involving minerals, gases and an aqueous phase. *Geochimica et Cosmochimica Acta* **46**, 513–528.

Reed, M. H., N. F. Spycher and J. Palandri, 2016, Users guide for CHIM-XPT: a program for computing reaction processes in aqueous-mineral-gas systems and MINTAB guide. Version 2.50. `pages.uoregon.edu/palandri/data/chim-xpt%20guide%20V.2.50.pdf`.

Rice, E. W., R. B. Baird and A. D. Eaton (eds.), 2017, *Standard Methods for the Examination of Water and Wastewater*, 23rd ed. American Public Health Association, Washington DC.

Richtmyer, R. D., 1957, *Difference Methods for Initial-Value Problems.* Wiley-Interscience, New York.

Rimstidt, J. D., 2014, *Geochemical Rate Models*, Cambridge University Press.

Rimstidt, J. D. and H. L. Barnes, 1980, The kinetics of silica-water reactions. *Geochimica et Cosmochimica Acta* **44**, 1683–1700.

Robie, R. A., B. S. Hemingway and J. R. Fisher, 1979, Thermodynamic properties of minerals and related substances at 298.15 K and 1 bar (10^5 pascals) pressure and at higher temperatures. *US Geological Survey Bulletin* **1452** (corrected edition).

Robinson, R. A. and R. H. Stokes, 1968, *Electrolyte Solutions.* Butterworths, London.

Roden, E. E. and R. G. Wetzel, 2003, Competition between Fe(III)-reducing and methanogenic bacteria for acetate in iron-rich freshwater sediments. *Microbial Ecology* **45**, 252–258.

Rosing, M. T., 1993, The buffering capacity of open heterogeneous systems. *Geochimica et Cosmochimica Acta* **57**, 2223–2226.

Rowan, E., 1991, Un modéle géochimique, thérmique-hydrogéologique et tectonique pour la genése des gisements filoniens de fluorite de l'Albigeois, sud-ouest du Massif Central, France. Mémoire de DEA, Université Pierre et Marie Curie (Paris IV), Paris.

Rowland, D., E. Königsberger, G. Hefner and P. M. May, 2015, Aqueous electrolyte solution modelling: some limitations of the Pitzer equations. *Applied Geochemistry* **55**, 170–183.

Rubin, J., 1983, Transport of reacting solutes in porous media, relationship between mathematical nature of problem formulation and chemical nature of reactions. *Water Resources Research* **19**, 1231–1252.

Runnells, D. D., 1969, Diagenesis, chemical sediments, and the mixing of natural waters. *Journal of Sedimentary Petrology* **39**, 1188–1201.

Runnells, D. D. and R. D. Lindberg, 1990, Selenium in aqueous solutions: the impossibility of obtaining a meaningful Eh using a platinum electrode, with implications for modeling of natural waters. *Geology* **18**, 212–215.

Sahai, N. and D. A. Sverjensky, 1997, Evaluation of internally consistent parameters for the triple-layer model by systematic analysis of oxide surface titration data. *Geochimica et Cosmochimica Acta* **61**, 2801–2826.

Santore, R. C., A. C. Ryan, F. Kroglund *et al.*, 2018, Development and application of a biotic ligand model for predicting the chronic toxicity of dissolved and precipitated aluminum to aquatic organisms. *Environmental Toxicology and Chemistry* **37**, 70–79.

Schecher, W. D. and D. C. McAvoy, 1994, MINEQL+, *A Chemical Equilibrium Program for Personal Computers, User's Manual*, version 3.0. Environmental Research Software, Inc., Hallowell, Maine.

Schrenk, M. O., D. S. Kelley, J. R. Delaney and J. A. Baross, 2003, Incidence and diversity of microorganisms within the walls of an active deep-sea sulfide chimney. *Applied and Environmental Microbiology* **69**, 3580–3592.

Shaff, J. E., B. A. Schultz, E. J. Craft, R. T. Clark and L. V. Kochian, 2010, GEOCHEM-EZ: a chemical speciation program with greater power and flexibility. *Plant and Soil* **330**, 207–214.

Shanks, W. C., III, and J. L. Bischoff, 1977, Ore transport and deposition in the Red Sea geothermal system, a geochemical model. *Geochimica et Cosmochimica Acta* **41**, 1507–1519.

Sherlock, E. J., R. W. Lawrence and R. Poulin, 1995, On the neutralization of acid rock drainage by carbonate and silicate minerals. *Environmental Geology* **25**, 43–54.

Shock, E. L., 1988, Organic acid metastability in sedimentary basins. *Geology* **16**, 886–890 (correction, **17**, 572–573).

Sibley, D. F. and H. Blatt, 1976, Intergranular pressure solution and cementation of the Tuscarora orthoquartzite. *Journal of Sedimentary Petrology* **46**, 881–896.

Sillén, L. G., 1967, The ocean as a chemical system. *Science* **156**, 1189–1197.

Simoes, M. C., K. J. Hughes, D. B. Ingham, L. Ma and M. Pourkashanian, 2017, Temperature dependence of the parameters in the Pitzer equations. *Journal of Chemical and Engineering Data* **62**, 2000–2013.

Sin, I., V. Lagneau and J. Corvisier, 2017, Integrating a compressible multicomponent two-phase flow into an existing reactive transport simulator. *Advances in Water Resources* **100**, 62–77.

Skirrow, G., 1965, The dissolved gases—carbon dioxide. In J. P. Riley and G. Skirrow (eds.), *Chemical Oceanography*, Academic Press, London, pp. 227–322.

Smith, G. D., 1986, *Numerical Solution of Partial Differential Equations: Finite Difference Methods*, 3rd ed. Oxford University Press.

Smith, K. S., W. H. Ficklin, G. S. Plumlee and A. L. Meier, 1992, Metal and arsenic partitioning between water and suspended sediment at mine-drainage sites in diverse geologic settings. In Y. K. Kharaka and A. S. Maest (eds.), *Water–Rock Interaction*. Balkema, Rotterdam, pp. 443–447.

Smith, W. R. and R. W. Missen, 1982, *Chemical Reaction Equilibrium Analysis: Theory and Algorithms*. Wiley, New York.

Snoeyink, V. L. and D. Jenkins, 1980, *Water Chemistry*. Wiley, New York.

Sorbie, K. S., M. Yuan and M. M. Jordan, 1994, Application of a scale inhibitor squeeze model to improve field squeeze treatment design (SPE paper 28885). *European Petroleum Conference Proceedings Volume* (vol. 2 of 2), Society of Petroleum Engineers, Richardson, Texas, pp. 179–191.

Søreide, I. and C. H. Whitson, 1992, Peng–Robinson predictions for hydrocarbons, CO_2, N_2, and H_2S with pure water and NaCl brine. *Fluid Phase Equilibria* **77**, 217–240.

Sørensen, S. P. L., 1909, Enzymstudier, II., Om Maalingen og Betydningen af Brintionkoncentration ved enzymatiske Processer. *Meddelelser fra Carlsberg Laboratoriet* **8**, 1–153.

Sposito, G., 2016, *The Chemistry of Soils*, 3rd ed. Oxford University Press.

Spycher, N. F. and M. H. Reed, 1988, Fugacity coefficients of H_2, CO_2, CH_4, H_2O and of H_2O–CO_2–CH_4 mixtures, a virial equation treatment for moderate pressures and temperatures applicable to calculations of hydrothermal boiling. *Geochimica et Cosmochimica Acta* **52**, 739–749.

Steefel, C. I., C. A. J. Appelo, B. Arora *et al.*, 2015, Reactive transport codes for subsurface environmental simulation. *Computational Geosciences* **19**, 445–478.

Steefel, C. I., D. J. DePaolo and P. C. Lichtner, 2005, Reactive transport modeling: an essential tool and a new research approach for the Earth sciences. *Earth and Planetary Science Letters* **240**, 539–558.

Steefel, C. I. and A. C. Lasaga, 1992, Putting transport into water–rock interaction models. *Geology* **20**, 680–684.

Steefel, C. I. and A. C. Lasaga, 1994, A coupled model for transport of multiple chemical species and kinetic precipitation/dissolution reactions with application to reactive flow in single phase hydrothermal systems. *American Journal of Science* **294**, 529–592.

Steefel, C. I. and K. T. B. MacQuarrie, 1996, Approaches to modeling of reactive transport in porous media. *Reviews in Mineralogy* **34**, 85–129.

Steefel, C. I. and P. Van Cappellen, 1990, A new kinetic approach to modeling water–rock interaction, the role of nucleation, precursors, and Ostwald ripening. *Geochimica et Cosmochimica Acta* **54**, 2657–2677.

Steefel, C. I. and S. B. Yabusaki, 1996, OS3D/GIMRT, Software for multicomponent-multidimensional reactive transport: user's manual and programmer's guide. Report PNL-11166, Pacific Northwest National Laboratory, Richland, Washington.

Stumm, W., 1992, *Chemistry of the Solid-Water Interface*. Wiley, New York.

Stumm, W. and J. J. Morgan, 1996, *Aquatic Chemistry, Chemical Equilibria and Rates in Natural Waters*, 3rd ed. Wiley, New York.

Stumm, W. and R. Wollast, 1990, Coordination chemistry of weathering, kinetics of the surface-controlled dissolution of oxide minerals. *Reviews of Geophysics* **28**, 53–69.

Sudicky, E. A. and E. O. Frind, 1982, Contaminant transport in fractured porous media: analytical solutions for a system of parallel fractures. *Water Resources Research* **18**, 1634–1642.

Sung, W. and J. J. Morgan, 1981, Oxidative removal of Mn(II) from solution catalysed by the γ-FeOOH (lepidocrocite) surface. *Geochimica et Cosmochimica Acta* **45**, 2377–2383.

Surdam, R. C. and J. R. Boles, 1979, Diagenesis of volcanic sandstones. *Society of Economic Paleontologists and Mineralogists Special Publication* **26**, 227–242.

Sverjensky, D. A., 1984, Oil field brines as ore-forming solutions. *Economic Geology* **79**, 23–37.

Sverjensky, D. A., 1987, The role of migrating oil field brines in the formation of sediment-hosted Cu-rich deposits. *Economic Geology* **82**, 1130–1141.

Sverjensky, D. A., 1993, Physical surface-complexation models for sorption at the mineral–water interface. *Nature* **364**, 776–780.

Sverjensky, D. A., 2003, Standard states for the activities of mineral surface sites and species. *Geochimica et Cosmochimica Acta* **67**, 17–28.

Sverjensky, D. A., 2006, Prediction of the speciation of alkaline earths adsorbed on mineral surfaces in salt solutions. *Geochimica et Cosmochimica Acta* **70**, 2427–2453.

Sydansk, R. D., 1982, Elevated-temperature caustic/sandstone interaction, implications for improving oil recovery. *Society of Petroleum Engineers Journal* **22**, 453–462.

Tan, K. L. and B. H. Hameed, 2017, Insight into the adsorption kinetics models for the removal of contaminants from aqueous solutions. *Journal of the Taiwan Institute of Chemical Engineers* **74**, 25–48.

Tang, D. H., E. A. Sudicky and E. O. Frind, 1981, Contaminant transport in fractured porous media: analytical solution for a single fracture. *Water Resources Research* **17**, 555–564.

Taylor, B. E., M. C. Wheeler and D. K. Nordstrom, 1984, Isotope composition of sulphate in acid mine drainage as measure of bacterial oxidation. *Nature* **308**, 538–541.

Thauer, R. K., K. Jungerman and K. Decker, 1977, Energy conservation in chemotrophic anaerobic bacteria. *Bacteriological Reviews* **41**, 100–180.

Thompson, J. B., Jr., 1959, Local equilibrium in metasomatic processes. In P. H. Abelson (ed.), *Researches in Geochemistry*, Wiley, New York, pp. 427–457.

Thompson, J. B., Jr., 1970, Geochemical reaction and open systems. *Geochimica et Cosmochimica Acta* **34**, 529–551.

Thompson, J. B., Jr., 1982, Composition space: an algebraic and geometric approach. *Reviews in Mineralogy* **10**, 1–31.

Thompson, M. E., 1992, The history of the development of the chemical model for seawater. *Geochimica et Cosmochimica Acta* **56**, 2985–2987.

Thorstenson, D. C., 1984, The concept of electron activity and its relation to redox potentials in aqueous geochemical systems. US Geological Survey Open File Report 84–072, 45 p.

Thorstenson, D. C., D. W. Fisher and M. G. Croft, 1979, The geochemistry of the Fox Hills-Basal Hell Creek aquifer in southwestern North Dakota and northwestern South Dakota. *Water Resources Research* **15**, 1479–1498.

Todd, A. C. and M. Yuan, 1990, Barium and strontium sulphate solid-solution formation in relation to North Sea scaling problems. *SPE Production Engineering* **5**, 279–285.

Todd, A. C. and M. Yuan, 1992, Barium and strontium sulphate solid-solution scale formation at elevated temperatures. *SPE Production Engineering* **7**, 85–92.

Truesdell, A. H. and B. F. Jones, 1974, WATEQ, a computer program for calculating chemical equilibria of natural waters. *US Geological Survey Journal of Research* **2**, 233–248.

Tsonopoulos, C., 1974, An empirical correlation of second virial coefficients. *AIChE Journal* **20**, 263–272.

Tsonopoulos, C. and J. L. Heidman, 1990, From the virial to the cubic equation of state. *Fluid Phase Equilibria* **57**, 261–276.

Turner, D. R. and S. A. Sassman, 1996, Approaches to sorption modeling for high-level waste performance assessment. *Journal of Contaminant Hydrology* **21**, 311–332.

Valocchi, A. J., 1985, Validity of the local equilibrium assumption for modeling sorbing solute transport through homogeneous soils. *Water Resources Research* **21**, 808–820.

Valocchi, A. J, R. L. Street and P. V. Roberts, 1981, Transport of ion-exchanging solutes in groundwater, chromatographic theory and field simulation. *Water Resources Research* **17**, 1517–1527.

van der Lee, J., L. De Windt, V. Lagneau and P. Goblet, 2003, Module oriented modeling of reactive transport with HYTEC. *Computers and Geosciences* **29**, 265–275.

Van Dover, C. L., 2000, *The Ecology of Deep-Sea Hydrothermal Vents*. Princeton University Press, Princeton, New Jersey.

Van Zeggeren, F. and S. H. Storey, 1970, *The Computation of Chemical Equilibria*. Cambridge University Press.

Verweij, W., 2017, Manual for CHEAQS NEXT, a program for calculating CHemical Equilibria in AQuatic Systems. `http://www.cheaqs.eu/manual.pdf`.

Verweij, W. and J.-P. Simonin, 2020, Implementing the mean spherical approximation model in the speciation code CHEAQS NEXT at high salt concentrations. *Journal of Solution Chemistry* **49**, 1319–1327.

Viani, B. E. and C. J. Bruton, 1992, Modeling ion exchange in clinoptilolite using the EQ3/6 geochemical modeling code. In Y. K Kharaka and A. S Maest (eds.), *Water–Rock Interaction*. Balkema, Rotterdam, pp. 73–77.

Von Damm, K. L., J. M. Edmond, B. Grant and C. I. Measures, 1985, Chemistry of submarine hydrothermal solutions at 21 °N, East Pacific Rise. *Geochimica et Cosmochimica Acta* **49**, 2197–2220.

Wagner, T., D. A. Kulik, F. F. Hingerl and S. V. Dmytrieva, 2012, GEM-SELEKTOR geochemical modeling package: TSOLMOD library and data interface for multicomponent phase models. *Canadian Mineralogist* **50**, 1173–1195.

Wang, Z. and D. E. Giammar, 2013, Mass action expressions for bidentate adsorption in surface complexation modeling: theory and practice. *Environmental Science and Technology* **47**, 3982–3996.

Warga, J., 1963, A convergent procedure for solving the thermo-chemical equilibrium problem. *Journal Society Industrial and Applied Mathematicians* **11**, 594–606.

Wat, R. M. S., K. S. Sorbie, A. C. Todd, P. Chen and P. Jiang, 1992, Kinetics of $BaSO_4$ crystal growth and effect in formation damage (SPE paper 23814). *Proceedings of the Society of Petroleum Engineers International Symposium on Formation Damage Control*, Lafayette, Louisiana, February 26–27, 1992, pp. 429–437.

Weare, J. H., 1987, Models of mineral solubility in concentrated brines with application to field observations. *Reviews in Mineralogy* **17**, 143–176.

Webb, J. R., I. R. Santos, D. T. Maher and K. Finlay, 2019, The importance of aquatic carbon fluxes in net ecosystem carbon budgets: a catchment-scale review. *Ecosystems* **22**, 508–527.

Wedepohl, K. H., C. W. Correns, D. M. Shaw, K. K. Turekian and J. Zemann (eds.), 1978, *Handbook of Geochemistry*, volumes II/1 and II/2, Springer-Verlag, Berlin.

Wehrli, B. and W. Stumm, 1988, Oxygenation of vanadyl(IV). Effect of coordinated surface hydroxyl groups and OH. *Langmuir* **4**, 753–758.

Wehrli, B. and W. Stumm, 1989, Vanadyl in natural waters: adsorption and hydrolysis promote oxygenation. *Geochimica et Cosmochimica Acta* **53**, 69–77.

Weng, L., W. H. Van Riemsdijk and T. Hiemstra, 2008, Cu^{2+} and Ca^{2+} adsorption to goethite in the presence of fulvic acids. *Geochimica et Cosmochimica Acta* **72**, 5857–5870.

Westall, J., 1980, Chemical equilibrium including adsorption on charged surfaces. In M. C. Kavanaugh and J. O. Leckie (eds.), *Advances in Chemistry Series* **189**, American Chemical Society, Washington DC, pp. 33–44.

Westall, J., 2002, Geochemical equilibrium and the interpretation of Eh. In Wilkin, R. T., R. D. Ludwig and

R. G. Ford (eds.), *Workshop on Monitoring Oxidation-Reduction Processes for Ground-water Restoration, Workshop Summary*, Report EPA/600/R-02/002, US Environmental Protection Agency, Ada, Oklahoma, 21–23.

Westall, J. C. and H. Hohl, 1980, A comparison of electrostatic models for the oxide/solution interface. *Advances in Colloid Interface Science* **12**, 265–294.

Westall, J. C., J. L. Zachary and F. F. M. Morel, 1976, MINEQL, a computer program for the calculation of chemical equilibrium composition of aqueous systems. Technical Note 18, R. M. Parsons Laboratory, Department of Civil and Environmental Engineering, Massachusetts Institute of Technology, Cambridge, Massachusetts.

White, A. F., 1995, Chemical weathering rates of silicate minerals in soils. *Reviews in Mineralogy* **31**, 407–461.

White, D., 2007, *The Physiology and Biochemistry of Prokaryotes*, 3rd ed. Oxford University Press.

White, D. E., 1970, Geochemistry applied to the discovery, evaluation, and exploitation of geothermal energy resources. *Geothermics* **2**, 58–80.

White, W. B., 1967, Numerical determination of chemical equilibrium and the partitioning of free energy. *Journal of Chemical Physics* **46**, 4171–4175.

White, W. B., S. M. Johnson and G. B. Dantzig, 1958, Chemical equilibrium in complex mixtures. *Journal of Chemical Physics* **28**, 751–755.

Williamson, M. A. and J. D. Rimstidt, 1994, The kinetics and electrochemical rate-determining step of aqueous pyrite oxidation. *Geochimica et Cosmochimica Acta* **58**, 5443–5454.

Wolery, T. J., 1978, Some chemical aspects of hydrothermal processes at mid-ocean ridges, a theoretical study, I., Basalt-sea water reaction and chemical cycling between the oceanic crust and the oceans, II. Calculation of chemical equilibrium between aqueous solutions and minerals. Ph.D. dissertation, Northwestern University, Evanston, Illinois.

Wolery, T. J., 1979, Calculation of chemical equilibrium between aqueous solution and minerals: the EQ3/6 software package. Lawrence Livermore National Laboratory Report UCRL-52658.

Wolery, T. J., 1983, EQ3NR, a computer program for geochemical aqueous speciation-solubility calculations: user's guide and documentation. Lawrence Livermore National Laboratory Report UCRL-53414.

Wolery, T. J., 1992a, EQ3/EQ6, a software package for geochemical modeling of aqueous systems, package overview and installation guide (version 7.0). Lawrence Livermore National Laboratory Report UCRL-MA-110662 (1).

Wolery, T. J., 1992b, EQ3NR, a computer program for geochemical aqueous speciation-solubility calculations: theoretical manual, user's guide, and related documentation (version 7.0). Lawrence Livermore National Laboratory Report UCRL-MA-110662 (3).

Wolery, T. J., 1995, Letter report: progress in developing EQ3/6 for modeling boiling processes. Lawrence Liverment National Laboratory Report UCRL-ID-130356.

Wolery, T. J., 2003, EQ3/6 version 8.0 software users manual. Document 10813-UM-8.0-00, U.S. Department of Energy, Office of Civilian Radioactive Waste Management, Las Vegas, Nevada.

Wolery, T. J. and L. J. Walters, Jr., 1975, Calculation of equilibrium distributions of chemical species in aqueous solutions by means of monotone sequences. *Mathematical Geology* **7**, 99–115.

Xu, T., E. Sonnenthal, N. Spycher and L. Liange, 2014, TOUGHREACT V3.0 reference manual: a parallel simulation program for non-isothermal multiphase geochemical reactive transport. Report LBNL-DRAFT, Lawrence Berkeley National Laboratory, Berkeley, California.

Yabusaki, S. B., C. I. Steefel and B. D. Wood, 1998, Multidimensional, multicomponent subsurface reactive transport in non-uniform velocity fields: code verification using an advective reactive streamtube approach. *Journal of Contaminant Hydrology* **30**, 299–331.

Yang, S. T. and M. R. Okos, 1987, Kinetic study and mathematical modeling of methanogenesis of acetate using pure cultures of methanogens. *Biotechnology and Bioengineering* **30**, 661–667.

Yeh, G. T. and V. S. Tripathi, 1989, A critical evaluation of recent developments in hydrogeochemical transport models of reactive multi-chemical components. *Water Resources Research* **25**, 93–108.

Yoneda, H., 1958, Stability of cobalt (III) and chromium (III) ammine complexes in a strongly alkaline solution. *Bulletin of the Chemical Society of Japan* **31**, 209–213.

Yuan, D. and A. C. Todd, 1991, Prediction of sulphate scaling tendency in oilfield operations. *SPE Petroleum Engineering* **6**, 63–72.

Yuan, M., A. C. Todd and K. S. Sorbie, 1994, Sulphate scale precipitation arising from seawater injection, a prediction study. *Marine and Petroleum Geology* **11**, 24–30.

Zabala (de), E. F., J. M. Vislocky, E. Rubin and C. J. Radke, 1982, A chemical theory for linear alkaline flooding. *Society of Petroleum Engineers Journal* **22**, 245–258.

Zehnder, A. J. B., B. A. Huser, T. D. Brock and K. Wuhrmann, 1980, Characterization of an acetate decarboxylating, non-hydrogen-utilizing methane bacterium. *Archives of Microbiology* **124**, 1–11.

Zeleznik, F. J. and S. Gordon, 1960, An analytical investigation of three general methods of calculating chemical equilibrium compositions. NASA Technical Note D-473, Washington DC.

Zeleznik, F. J. and S. Gordon, 1968, Calculation of complex chemical equilibria. *Industrial and Engineering Chemistry* **60**, 27–57.

Zhang, W. and E. J. Bouwer, 1997, Biodegradation of benzene, toluene and naphthalene in soil-water slurry microcosms. *Biodegradation* **8**, 167–175.

Zhang. Y., 2008, *Geochemical Kinetics*. Princeton University Press, Princeton, New Jersey.

Zhao, H., H. Liu and J. Qu, 2009, Effect of pH on the aluminum salts hydrolysis during coagulation process: formation and decomposition of polymeric aluminum species. *Journal of Colloid and Interface Science* **330**, 105–112.

Zheng, C. and G. D. Bennett, 2002, *Applied Contaminant Transport Modeling*, 2nd ed. Wiley, New York.

Zhu, C. and G. Anderson, 2002, *Environmental Applications of Geochemical Modeling*. Cambridge University Press.

Zierenberg, R. A., M. W. W. Adams and A. J. Arp, 2000, Life in extreme environments: hydrothermal vents. *Proceedings National Academy of Sciences* **97**, 12 961–12 962.

Index